THERMES 2002

THERMAL CHALLENGES IN NEXT GENERATION ELECTRONIC SYSTEMS

PROCEEDINGS OF THE INTERNATIONAL CONFERENCE THERMES 2002
SANTA FE, NEW MEXICO, USA, 13 – 16 JANUARY 2002

Thermal Challenges in Next Generation Electronic Systems

Edited by

Yogendra K. Joshi
Georgia Institute of Technology, School of Mechanical Engineering, Atlanta, Georgia, USA

Suresh V. Garimella
Purdue University, School of Mechanical Engineering, West Lafayette, Indiana, USA

Millpress

MILLPRESS ROTTERDAM NETHERLANDS 2002

© Published for the United Engineering Foundation, Inc., New York, NY, USA
www.engfnd.org

Cover design: Millpress

The graphic represents the wick pattern of a multi-source, edge-cooled micro-heat pipe. The wick is anisotropic and creates cooling zones for each heat source eliminating source shadowing. A photolithographic mask is then created from the design to control the electroplating of the wick. The concept was developed at Sandia National Laboratories in Albuquerque, New Mexico. (see page 273)

The texts of the various papers have been submitted as electronic files by the individual authors and have been processed by the publisher under supervision of the editors. While the extended abstracts have been reviewed by the editors, the complete papers are unreviewed.

Published by Millpress Science Publishers, P.O. Box 84118, 3009 CC Rotterdam, Netherlands
Tel.: +31 10 421 26 97; Fax: +31 10 209 45 27; www.millpress.com

ISBN 90 77017 03 8

Printed in the Netherlands

Thermal Challenges in Next Generation Electronic Systems, Joshi & Garimella (eds)
© 2002 Millpress, Rotterdam, ISBN 90-77017-03-8

Table of contents

Compact models and system-level challenges

Advanced design methodologies

Analysis and testing methods

Miscellaneous

Thermal Challenges in Next Generation Electronic Systems, Joshi & Garimella (eds)
© 2002 Millpress, Rotterdam, ISBN 90-77017-03-8

Preface

As electronic products become faster and incorporate greater functionality, they are also shrinking in size and weight, with continuing pressures for cost reduction. Thermal issues are key in electronic product development at all levels of the electronic product hierarchy, from the chip to the ultimate system. Shrinking system sizes are resulting in increasing volumetric heat generation rates and surface heat fluxes in many products. This has resulted in a significant interest in ultra-compact high heat removal thermal management devices. Additional thermal management challenges arise as more and more electronic systems are employed in harsh environments, subject to large variations in ambient temperatures and external thermal loads. A synergistic activity is thermal characterization through computations. High-fidelity modeling schemes are being sought, as development cycles shrink to a period of months for many portable products.

The THERMES 2002 workshop focuses on the recent advances in thermal management and characterization schemes, with broad U.S. and international participation. Keynote and invited talks in selected emerging areas are complemented with contributed papers, as well as panel discussions focusing on technology and market trends and identification of research challenges. The three-day workshop is organized in a serial format. A keynote lecture on each day is complemented by an invited talk leading each session of contributed papers. A number of panel discussions are included to elicit audience participation on research topics of current interest. The workshop program consists of 3 keynote talks, 9 invited talks, and 35 contributed talks.

THERMES 2002 would not have been possible without the valuable support from a number of individuals and organizations. Professor F. Schmidt (Pennsylvania State University and United Engineering Foundation) was the catalyst in the initiation of a proposal to the United Engineering Foundation (UEF) to organize THERMES 2002. Additional support was provided by the US National Science Foundation (Dr. S. Thynell, Program Manager). We are thankful to the organizing committee members for their valuable input at various stages of the conference planning. In particular, Dr. R. Mahajan, Mr. C. Patel, and Prof. B. Sammakia were instrumental in arranging the keynote talks, and Prof. K.C. Toh in coordinating the panel sessions. The United Engineering Foundation took care of the logistical arrangements at Santa Fe, as well as program mailings and registration. We are particularly appreciative of the excellent service and enthusiastic assistance provided by the publisher of these proceedings, Mr. Jack Lewis of Millpress, Netherlands. The success of an event such as this is ultimately dependent on the enthusiasm and involvement of the participants. We very much hope that the important workshop theme and a schedule that allows for liberal discussion times, coupled with the exciting location of Santa Fe, will produce an event of lasting significance.

Yogendra Joshi   Suresh V. Garimella

Thermal Challenges in Next Generation Electronic Systems, Joshi & Garimella (eds)
© 2002 Millpress, Rotterdam, ISBN 90-77017-03-8

Panel sessions

The UEF Conferences are specifically designed to enhance interaction among participants and prompt dialogue on recent developments in a particular field of engineering. The organizers of THERMES 2002 have therefore put together five panels on wide-ranging but topical issues to encourage participants to actively generate ideas and disseminate information on their ongoing investigations. The panels will be moderated by key investigators drawn from academia, research institutes and industrial research centers in the US, Europe and Asia.

C. Patel (HP Laboratories) and A. Ortega (U. of Arizona) lead a discussion on "*The Hardware Needs of Future High Performance & Internet Computing Workloads and the Implications on Cooling and Physical Design*." The changing computer workloads, the advent of internet computing, miniaturization of silicon technologies, growth in performance and power dissipation has had a profound effect on computing hardware needs. The panelists, including R. Friedrich (HP Labs) and Ravi Mahajan (Intel), present their views on high performance computing of the future and its implications on cooling and physical design.

J. Lohan (Galway-Mayo Inst. of Tech., Ireland) and M. Baelmans (KU Leuven, Belgium) head a panel, which also comprises P. Rodgers (a consultant, Ireland) and G. Kromann (Motorola), to look into "*Component through System Level Numerical Modeling: Requirements, Reality and Future?* " In spite of increased computing power, are thermal designers confident that they can achieve effective cooling solutions for their electronic system through numerically modeling, thus reducing or avoiding time consuming and expensive experimentation? The panel (i) reviews recent advances in numerical modeling techniques and contrasts these against the designers' expectations from the competing perspectives of prediction accuracy, product design time, ease of use and cost; (ii) highlights the need for recognized benchmarks at the component, PCB and system levels; and (iii) outlines future possibilities for hardware, software and numerical tools.

Ever since the pioneering work of Tuckerman and Pease, there has been the expectation that microchannel heat exchangers will be the breakthrough that will enable thermal management engineers to keep up with the frenetic pace of advances in electronic devices. A panel led by V. Carey (UC Berkeley) and K.C. Toh (NTU Singapore) explores "*Microscale Cooling Systems – Its Successes, Barriers and Integration with Electronic Devices*." How well the technology has lived up to its promise, what the technological and experimental successes and barriers are, and especially the degree to which this technology is ready for integration into the next generation of electronic packages, are some of the issues explored.

Manufacturing plays a key role in our lives as evidenced by the pervasiveness of the products around us. Making these products involve many processes, from metal drawing to semiconductor chip fabrication, where the transport phenomena are complex and need to be understood. A panel led by R. Mahajan (U. of Colorado) and B. Sammakia (SUNY Binghamton) explores the underlying "*Thermal Issues in Manufacturing*" and examines the range and scope of the phenomena that arise in the manufacture of products covering extremes of length scales, temperatures and materials.

The conference would not have been complete without a consideration of "*Thermal Management Roadmaps*". The challenges posed by high chip heat fluxes and ever more stringent performance and reliability constraints continue to make thermal management a key enabling technology in the development of microelectronic systems for the 21st century. A. Bar-Cohen (U. of Maryland) and Ravi Mahajan (Intel) lead a panel which focuses on the thermal packaging efforts that must be performed in the context of the salient trends and parameters that characterize the IC technology and the electronic products industry.

KEYNOTE AND INVITED LECTURES

Thermal Challenges in Next Generation Electronic Systems, Joshi & Garimella (eds)
© 2002 Millpress, Rotterdam, ISBN 90-77017-03-8

Towards planetary scale computing
Technical challenges for next generation Internet computing

Rich Friedrich & Chandrakant Patel
Hewlett-Packard Laboratories, Palo Alto, CA, USA

EXTENDED ABSTRACT: In the not-too-distant future, billions of people, places and things could all be connected to each other and to useful services through the Internet. In this world scalable, cost-effective information technology capabilities will need to be provisioned as a service, delivered as a service, metered and managed as a service, and purchased as a service. HP refers to this world as service-centric computing. Consequently, processing and storage will be accessible via a utility where customers pay for what they need when they need it. This processing and storage utility will become as ubiquitous as electrical and water utilities are today.

Within HP Labs, we are exploring the penultimate manifestation of this service-centric computing strategy -- *planetary scale computing* -- needed to power the service-centric computing world. In this model, large-scale data centers are networked around the world allocating processing power and data storage when and where it is needed as it is needed. Customers will pay for services based on the amount they use, much like a public utility.

The technical requirements for planetary scale computing will stretch the IT infrastructure beyond its elastic limit. Current and emerging services will consume 10-100 times more processing, communications and storage as they support a range of applications that include rich media, bioinformatics, and pervasive sensors. Scalability, density and complexity hurdles must be overcome. Technology and process innovation are required to deploy and operate this infrastructure to flexibly and economically support 21st century applications. New computing and storage architectures are required. As the scale of the numbers of servers in a data center grows and as the servers become more densely packed in racks, it is natural to think of *the data center as the computer*.

HP Labs is working on a number of approaches to these technical challenges. One is the *programmable data center* that provides automated "infrastructure on demand" with little or no operator intervention. For example, retail Web sites typically need much more processing and storage during the year-end holidays than they need the rest of the year, but the Internal Revenue Service needs more computing resource in April. These organizations might benefit from securing additional computing and storage when they need it, rather than to acquire it and have most of it remain idle the rest of the time. Furthermore, data center physical resources such as power and cooling must also be managed more efficiently than today. Key research topics include next generation large-scale computing, communications and storage architectures.

In a programmable data center, the infrastructure is physically wired once, but can be rewired programmatically, to meet the changing needs of customers and services. This includes linking these large data centers so that resources can be optimized for a region, a nation or around the world. When it's "off-hours" in one area, data centers there could be used more fully by providing services for users in other parts of the world, and vice-versa.

Managing this flexible and robust infrastructure is also critical. Researchers are defining the "data center operating system" that will automatically control the infrastructure. Introducing a virtualization layer between services and physical resources is the key here since it can minimize complexity and enable automation. The data center OS will consist of two parts: a service control component that brokers service demand with resource capacity, and a resource control component that will intelligently provision physical resources, including power and cooling, via the virtualization layer. These components will work together to describe, install, configure, deploy, monitor and assure services on a global scale.

Additional challenges face the designers of the next generation Internet computers. In particular we will examine the thermal and mechanical issues of future integrated circuits, densely packaged racks of servers, and large scale data centers that contain as many as 50,000 servers in a 100,000 square foot building. The thermo-mechanical challenges stem from high power density at chip level and the need to package the systems in a dense configuration. As an example, today's data center, with 1000 racks, over 30,000 square feet, requires 10 MW of power to power the computing infrastructure. A 100,000 square foot planetary data scale data center of tomorrow will require 50 MW of power for the computing infrastructure. The cooling for this 50 MW planetary scale data center will consume an additional 25 MW of power. Such a data center, with five thousand 10kW racks, would cost ~$45 million per year (@ $100/MWh) just to power the servers and $22 million per year to power the cooling for the data center. Of course the energy expense would be much larger in an energy crisis, like one faced by California in the summer of 2001.

The role of thermal and mechanical architecture is critical to ensure efficient power management, cooling and physical data center design. These trends motivate the following thermal mechanical research questions:

- What innovative thermo-mechanical designs are required for the next generation of chips, racks and data centers? How should thermo-mechanical design influence large-scale Internet server and storage architectures?
- What are the total energy costs over the lifetime of a product? This should include the original manufacture of thermal mechanical apparatus, on-going operational and final disposal costs.
- What thermo-mechanical principles can be employed to improve the operation and control of large-scale data centers?
- What techniques will allow us to use and re-use energy more efficiently?
- What is the ideal data center design from a thermo-mechanical perspective?

In addition to thermo-mechanical design, one must consider the power delivery means to the data center. With the proliferation of data centers all over the world, the high power demands will require us to devise a distributed energy source scheme to meet the power needs. Energy delivered from conventional sources to urban areas did not account for a world of pervasive computing. The use of energy from conventional sources will have great impact on conventional users of power. Therefore, use of energy from a multiplicity of alternate sources, and indeed re-use of energy from waste heat generated from data centers, is an important research question.

In summary, the world is moving towards a service-centric view of computing and storage. Innovations in the IT infrastructure are required to increase scalability, improve density, minimize complexity and minimize cost. Furthermore, new approaches to power management and cooling optimization are required to meet the demands of these 21st century computing centers.

Thermal Challenges in Next Generation Electronic Systems, Joshi & Garimella (eds)
© 2002 Millpress, Rotterdam, ISBN 90-77017-03-8

Design and optimization of air-cooled heat sinks for sustainable development

Avram Bar-Cohen
TherMNS Laboratory, Department of Mechanical Engineering, University of Maryland, College Park, MD, USA

Madhusudan Iyengar
Laboratory for the Thermal Management of Micro-Nano Systems, Dept. of Mechanical Engineering, University of Minnesota, MN, USA

Keywords: heat sinks, plate-fin, optimization, least-material, least-energy, sustainable development

ABSTRACT: The development of heat sinks for microelectronic applications, which are compatible with sustainable development, involves the achievement of a subtle balance between a superior thermal design, minimum material consumption, and minimum pumping power. Due to the rapid proliferation of electronic systems, substantial material streams and energy consumption rates are associated with the cooling of computers, as well as other categories of electronic equipment. This presentation explores the potential for the least-energy optimization of natural- and forced-convection cooled rectangular plate heat sinks. The results are evaluated in terms of a heat sink Coefficient of Performance, relating the cooling capability to the energy invested. Guidelines for "sustainable" heat sink designs are suggested.

1 INTRODUCTION

1.1 *Energy consumption rates*

As a result of the explosion in information technology, some 600 million computers [Computer Industry Almanac (1999)] are already in use world-wide, with personal computers constituting approximately half of the total, and expected to grow to some 400 million by the end of 2001 [Bar-Cohen et al. (2000)]. High performance, microprocessor-driven desk top computers typically dissipate nearly 50W, using an aluminum heat sink and a small, 1.5W fan to achieve junction-to-air resistances in the range of 1.5K/W. The substantial material stream and energy consumption rate associated with the cooling of these desktop computers, as well as other categories of computers and electronic equipment, threatens to dramatically increase the depletion rate of key resources and lends urgency to the “resource-constrained" optimization of these air-cooled heat sinks. The effort discussed herein deals with the development of a design and optimization methodology for "sustainable" forced and natural-convection cooled, plate fin, heat sinks. This methodology seeks to maximize the thermal energy that can be extracted from a specified space, while minimizing the material and energy consumed in the fabrication and operation of the specified heat sink.

Data provided by the LCA Committee of the Japanese aluminum industry (1999) suggests that approximately 85 kWh/kg are required to form, assemble, and transport aluminum heat sinks, including some allowance for re-cycled material. Combining this “invested” work with the energy consumed in the operation of the fans, the total energy required for the thermal management of a typical microprocessor can be expressed as:

$$E_T = 85\ M + (PP)(t_l) \tag{1}$$

where, E_T [kWh] is the energy used for cooling, M [kg] is the heat sink mass, PP [W] is the fan pumping power, and t_l [hrs] is the operating lifetime of the fan.

An assumed 100g heat sink, used to cool a 50W component with 0.5W of fan-provided pumping power, operating 2000 hours per year for 3 years, would consume a total of 11.5kW-hr (or some 3.8kW-hr/year), exclusive of inefficiencies associated with the operation of the fan. Thus, the amount of operating energy used in the cooling of the approximately 400 million microprocessor-driven personal computers units operating in 2001 would reach some 400,000MW-hours and the invested work some 1.1million MW-hours for a total of 1.5 Terra (1x1012) Watt-hours per year. Moreover, approximately 10 million kg of aluminum will be used in 2001 for the fabrication of heat sinks.

1.2 *Thermal performance metrics*

To determine the effectiveness of various heat sink designs, it is necessary to establish thermal performance metrics, which take into consideration the overall heat dissipation, the fin material consumed, and the heat sink volume, as well as the energy required to fabricate and operate the heat sink [Iyengar and Bar-Cohen (1998)]. In the present study the heat sink base has been assumed isothermal, as may occur with a small base plate, or with the use of a heat pipe or vapor chamber base plate as shown by Zhao and Avedisian (1997), and Garner and Toth (1997). This study, thus, focuses on the cooling capability of the fins. In a high-performance aluminum heat sink, the base thickness is typically in the range of 5mm-10mm and may represent between 25% and 50% of the total mass of the heat sink [Ben-Achour and Bar-Cohen (1998)]. This mass is not included in the calculations that follow.

The thermal resistance, R_{hs}, is the most commonly used thermal characteristic of heat sinks, i.e.

$$R_{hs} = \theta_b / q \quad (K/W) \tag{2}$$

where θ_b is the excess temperature of the heat sink base (K), and q is the total heat sink heat dissipation (W). It is to be expected that the R_{hs} will be highly dependent on the flow rate, decreasing asymptotically towards the conduction limit, as the flow rate increases. While system designers may find R_{hs} the most useful of the heat sink metrics, it usually masks the effect of heat sink area, as well as volume and material choice, on thermal performance.

The base area effect can be captured using an array heat transfer coefficient, h_a. This is, perhaps, the single most useful thermal metric and can be expressed as,

$$h_a = q/LW\theta_b \quad (W/m^2\text{-}K) \tag{3}$$

where L and W are the length and width of the heat sink base, respectively.

The thermal efficacy of fin material utilization can be characterized by the mass specific heat transfer coefficient, which is given by,

$$h_m = q/\rho_{fin} V_{fin} \theta_b \quad (W/kg\text{-}K) \tag{4}$$

where ρ_{fin} and V_{fin} are the density and volume of fin material, respectively.

The "space claim" heat transfer coefficient, h_{sc}, which represents the volumetric efficiency of the heat sink, may be calculated using,

$$h_{sc} = q/LW\theta_b \quad (W/m^3\text{-}K) \tag{5}$$

where H is the fin height.

By analogy to refrigeration systems, it is helpful to define a Coefficient-Of-Performance (COP) for the heat sink, relating the cooling rate, q, to the fan pumping power, PP, i.e.,

$$COP = q / PP \tag{6}$$

where PP is the pumping power, (not including the fan inefficiency) equal to,

$$PP = V_{air}\,\Delta P \tag{7}$$

where V_{air} is the volumetric airflow rate and ΔP is the pressure drop across the heat sink.

Extending the COP concept to address issues of sustainability, it is possible to define a COP_T, relating the cooling capability to the total energy invested in the operation (as above) as well as the formation and fabrication of the heat sink., i.e.

$$COP_T = q\,/\,(E_T/\,t_l) \tag{8}$$

where $E_T = [E_M + E_{PP}]$, and where ($E_M = 85M$) and $E_{PP} = (PP\ t_l)$, are the energy investments in the fin mass and fluid pumping power, respectively. The allocation of the invested energy between heat sink formation/fabrication and pumping work can be expressed in terms of an allocation factor, ξ_{PP}, defined as the ratio of pumping work to the total invested energy, E_{PP}/E_T. Thus, in a natural convection array, ξ_{PP}, is identically zero, while forced convection cooling of an unfinned surface would by definition, have a ξ_{PP} of unity. Consequently, the least-energy optimization of passively cooled heat sinks ($\xi_{PP} = 0$), must focus on the least-material design pioneered by A. D. Kraus [Kraus et al. (2001)] and described in previous publications [Iyengar and Bar-Cohen (1998) and Bar-Cohen et al. (2001)], while for forced convection of unfinned surfaces ($\xi_{PP} = 1$), attention must be devoted to minimizing pumping power and entropy generation as demonstrated by Bejan (1995). More generally however, the allocation of available energy between pumping work and formation/fabrication work, must be considered a design variable.

2 HEAT SINK THERMOFLUID FUNDAMENTALS

2.1 *Least-material optimization*

The desire to minimize the cost, weight, and volume of commercial heat sinks can often be addressed by the design of volumetrically efficient, least-material fins. The use of these optimal individual fins to compose multiple fin arrays has been described in Bar-Cohen et al. (2001) and Iyengar and Bar-Cohen (1998) for natural convection and in Iyengar and Bar-Cohen (2001) for forced convection cooling applications. In such fins the ratio of fin height to thickness has been selected to minimize the mass required to dissipate the specified heat flow. Following the pioneering work of Kern and Kraus (1972), the least-material approach provides a functional dependence between the fin thickness, t and the fin height H, for individual fins, that is expressed as,

$$t = 0.993 h_{fin} H^2 / k_{fin} \tag{9}$$

where k_{fin} is the fin thermal conductivity and h_{fin} is the average fin heat transfer coefficient.

For such plate fins, the fin efficiency, subject to the usual assumptions [Kern and Kraus (1972)], can be found via the familiar hyperbolic relation, with the fin height, H, used as the length scale, and given by,

$$\eta = [\tanh(mH)]/[mH] \tag{10}$$

where m is the fin parameter and is defined as,

$$m = [2h_{fin}\,/\,k_{fin}\,t]^{1/2} \tag{11}$$

Kern and Kraus (1972) showed that the least-material plate fin is characterized by a unique value of the mH product, equaling 1.4192. Since the fin efficiency, of such a fin, is solely dependent on the mH product, the efficiency of the least-material plate fin is a fixed value, equal to 0.627. It is this value of fin efficiency that is used in the present study to determine the preferred fin aspect ratio. Eq. (9) reveals that the fin aspect ratio, H/t, is affected by the variation in the heat transfer coefficient, attaining smaller values as h_{fin} increases.

2.2 *Forced convection in plate fin heat sinks*

The analytical methodology developed by Holahan et al. (1996) for calculating the thermal performance and pressure drop in fully shrouded, laminar, parallel plate heat sinks has been utilized to characterize the thermofluid performance of the side-inlet-side-exit (SISE) heat sink configurations depicted in Figure 1.

The pressure drop across the heat sink is determined with correlations for laminar duct flow, as,

$$\Delta P = \Delta P_{fd} + \frac{1}{2}\rho v_{in}^2 K(\infty) + \frac{1}{2}\rho v_{in}^2 K_c + \frac{1}{2}\rho v_{out}^2 K_e \tag{12}$$

where ΔP_{fd} is the frictional loss (Kakac et al., 1987); K_c and K_e are contraction and expansion coefficients, respectively; a term $K(\infty)$ is for hydrodynamic developing flat duct flow due (Kakac, et al. 1987), ρ is fluid density, and v_{in} and v_{out} are inlet and outlet velocities, respectively.

In this semi-analytical approach, the local heat transfer coefficient along segments of the fin surface is obtained from the correlation for developing thermal and hydrodynamic laminar flow in parallel plate channels, with uniform wall temperature, provided in Kakac et al. (1987) and given by,

$$h_{fin,local} = \frac{k_{air}}{2s}[7.55 + (0.024X^{-1.14})\frac{0.0179Pr^{0.17}X^{-0.64} - 0.14}{(1 + 0.0358Pr^{0.17}X^{-0.64})^2}] \tag{13}$$

In eq. (13), X is the dimensionless axial distance, and is given by,

$$X = \frac{[x\nu]}{4s^2 U_m Pr} \tag{14}$$

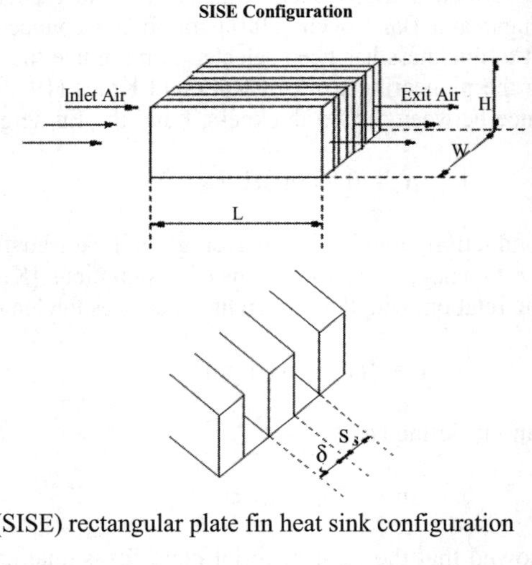

Figure 1, Side-Inlet-Side-Exit (SISE) rectangular plate fin heat sink configuration

The heat transfer rate from a single segment is then found using the superposition of a kernel function determined from the method of images. A simple heat balance on the fluid flowing in each inter-fin channel can then be used, in an iterative fashion, to determine the local air temperature. The heat dissipation from the heat sink array, q, is then found by the summation of patch heat transfer from all the segments. A careful analysis of the resulting heat dissipation rates, throughout the parametric space of interest, can then be used to guide the designer to the most thermally advantageous combinations of air flow characteristics and fin geometries. Results obtained with this approach, was shown to give good agreement with experimental results [Holahan et al. (1996)].

2.3 *Candidate heat sink configurations*

In the design of a forced convection heat sink, care must be taken to balance the pressure drop of the heat sink with the pressure head available from the selected fan, at the prescribed flow rate. This balance is represented most conveniently on a two-dimensional map, relating the pressure and the volumetric flow rate. In the coordinates of such a map, it is possible to represent both the "fan curve" - shown as several convex lines in Fig. 2 - and the hydraulic resistance of the heat sink - represented by several concave curves in Fig. 2. The fan curve farthest from the origin of this graph represents the highest pumping power choice, while the resistance curve closest to the horizontal axis represents the heat sink most easily traversed by the airflow. It may thus be seen that, for a given heat sink, as the fan pumping power increases operation shifts to higher and higher flow rates. Alternatively, for a given fan, progressively higher flow resistance heat sinks will operate at lower and lower volumetric flow rates. The points of intersection between these two sets of curves mark the possible operating points of the heat sink - fan systems considered, with each point representing a fan-heat sink combinations, with a particular thermal capability and fin mass. In pursuing the design of a "sustainable" heat sink, it is important to identify the combination that yields the maximum heat transfer within a specific design domain, subject to imposed design and operating constraints. It must also be recognized that the required operation parameters and array geometries may not always be achievable using existing technology.

For purposes of this study, eqs. (3-5) and eq. (12, 13), were used to perform a parametric search of different fin geometries was carried out for an allowable pressure drop between 20 and 80 Pa and for the volumetric air flow rate between 0.01 and 0.04 m^3/s. The highest heat transfer rates and the rates associated with the most efficient utilization of the heat sink material, for each combination of air flow rate and pressure drop, are shown in Figs. 4-5 and Table 1. Fig. 2 graphically describes the space investigated, also showing performance curves for some ideal fans and heat sinks.

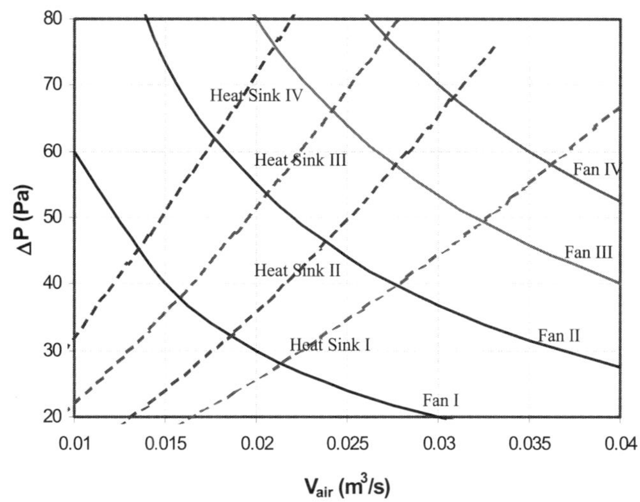

Figure 2, Parametric optimization design space

In succeeding sections of this paper an advanced heat sink application, occupying a volume of 500cm^3 [0.1m×0.1m×0.05m], with an excess base temperature of 25 K, is used to illustrate the design methodology. Information gathered from several prominent heat sink manufacturers, and discussed in greater detail in Iyengar and Bar-Cohen (2001), can serve to establish the geometric constraints, which are encountered in today's heat sinks. As described in the previous sections, volumetric flow rates and pressure drops across the heat sinks were varied from 0.01 to 0.04 m^3/s and 20 to 80 Pa, respectively, yielding pumping powers ranging from 0.2 W to 3.2 W. The COP and COP_T (relating to the total energy invested in the heat sink) are examined for the maximum heat sink design, least material design, and for typical manufacturing processes.

3 FORCED CONVECTION PLATE-FIN HEAT SINKS

3.1 *Maximum heat transfer design*

For a given operating condition, described by ΔP and V_{air}, designs with very small and very large numbers of fins, N, yield low values of h_a; the former due to limited surface area, A_{fin}, and the latter caused by the highly inefficient thin fins. An intermediate value can thus be found which maximizes the N x [$\eta h_{fin,} A_{fin}$] product and thereby the h_a value. This behavior can be seen clearly in Figure 3, where h_a, calculated using eq.(3) for a specified operating condition and heat sink volume, is shown as a function of the number of fins. For each value of N, there is a corresponding fin-to-fin spacing, s, and fin thickness, t , which meets the pressure drop requirement at the specified flow rate. In Figure 3, it may be seen that h_a, for ΔP = 20 Pa and V_{air} = 0.01 m^3/s, reaches a maximum value of nearly 743 W/m^2-K at an intermediate number of 40 fins. With an excess base temperature of 25 K, this heat sink is thus capable of dissipating nearly 186 W. This configuration requires a fin thickness of 0.57 mm and an inter-fin spacing of 1.93 mm, for a total heat sink mass of 0.308 kg. From Figure 3, maximum heat transfer design for the ΔP = 80 Pa and V_{air} = 0.04 m^3/s which weighs 0.198 kg, is seen to dissipate 223 W for an excess base temperature of 25 K, fin thickness of 0.41 mm and fin spacing of 2.37 mm.

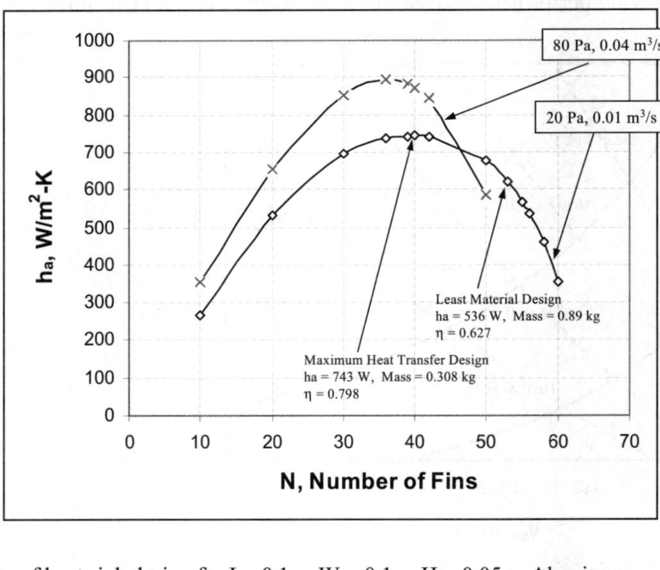

Figure 3, Optimization of heat sink design for L= 0.1m, W = 0.1m. H = 0.05m, Aluminum

Following this procedure, it is possible to identify the configuration yielding the maximum heat transfer capability for each operating condition, for a specific heat sink space claim, fin material, and base-to-ambient excess temperature difference. Figure 4, shows the surface of maximum h_a, over the range of operating points yielding a maximum h_a value of 1242 W/m^2-K (311 W at 25 K excess base temperature) occurring at a ΔP of 80 Pa and a V_{air} of 0.02 m^3/s. The prescribed geometry has a mass of 0.365 kg, and requires 51 fins that are 0.53 mm thick and spaced at 1.43 mm. The corresponding value of the mass specific heat transfer coefficient, h_m, given by eq. (4) was determined to be 34 W/kg-K. The h_a for a fixed V_{air} is found to increase with increasing ΔP, although for most of the cases there was found to be an optimum V_{air} that maximized the h_a for a fixed ΔP. For $\Delta P = 20$ Pa and $V_{air} = 0.01$ m^3/s, there were no feasible geometries.

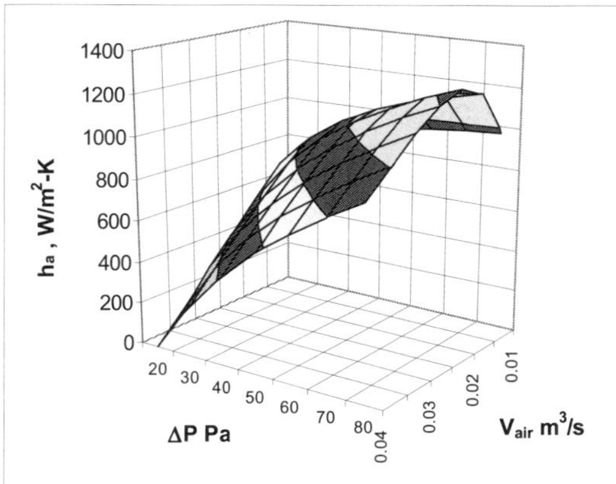

Figure 4, Surface for array heat transfer coefficient, h_a, maximum heat transfer, L = 0.1m, W = 0.1m, H = 0.05m, Aluminum

3.2 *Least-material optimization*

Returning to Figure 3 and applying the least-material fin efficiency criteria (= 0.627), the least-material design for ΔP of 20 Pa and V_{air} of 0.01 m^3/s, can be found to occur at 51 fins (fin thickness = 0.12 mm and fin spacing = 1.67 mm), yielding an h_a value of approximately 537 W/m^2-K. While this value is 28% less than the maximum achievable value at these flow conditions, at 89 grams it requires 71% less mass, indicating considerable mass savings for a moderate loss in thermal performance. This is reflected in the improvement in the mass specific heat transfer coefficient by a factor of 2.5, from 24.1 W/kg-K to 60.3 W/kg-K for this least-material design. However this design will demand significantly thinner fins, at 0.12 mm, as compared to the maximum heat transfer design using 0.57mm fins.

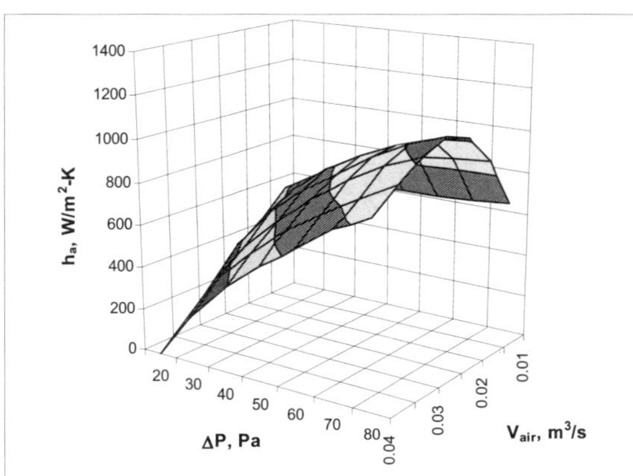

Figure 5, Surface for array heat transfer coefficient, least-material optimization, L = 0.1m, W = 0.1m, H = 0.05m, Aluminum

The least-material configurations appear to provide an effective heuristic for initiating the design of an energy-efficient "sustainable" heat sink. The h_a surface for the least-material designs is displayed in Figure 5, showing the maximum of 1071 W/m^2-K (268W, θ_b = 25 K) with a h_m value of 59.5 W/kg-K, for flow conditions of ΔP = 80 Pa and V_{air} = 0.025 m^3/s. The recommended array geometry weighs 0.18 kg, using 60 fins that are 0.22 mm thick and spaced 1.45 mm apart. Compared to the corresponding maximum heat transfer design, this represents a 75% improvement in h_m, while the heat transfer characterized by h_a diminishes by only 14%.

Such an analysis for other fin materials, specifically copper and magnesium, was found to result in similar trends to those seen in Figures 4 and 5. A summary of the results obtained is presented in Table 1. Looking closely at Table 1, it can seen that the Copper arrays show higher values for h_a, than the Magnesium and Aluminum heat sinks by a factor of 1.3 and 1.2, respectively, for the maximum heat transfer designs, and by approximately 1.1 for the least-material results. However, due to their considerably lower density, the Mg and Al designs are considerably better with respect to h_m, by a factor of 3.2 and 2.4, respectively for the maximum heat transfer cases, and by 2.0 and 1.7, respectively, for the least-material solutions.

Comparing the improvement in the h_m values in adopting the least-material philosophy for the different materials, the greatest increase of 150% was observed for the Cu arrays, while smaller yet significant increases of 75% and 56% can be seen for the Al and Mg arrays respectively. Dimensional requirements, for the flow conditions considered, suggest the range of fin thickness for the maximum heat transfer designs to be between 0.257 mm and 0.759 mm, and for the least-material designs to be between 0.0697 mm and 0.251 mm. Fin spacing requirements were always greater than 0.8 mm for the maximum heat transfer and least-material designs.

Table 1, Thermal performance optimization, forced convection arrays, L=W=0.1m, H=0.05m, θ_B = 25K

	Maximum Heat Transfer Design			*Least-Material Optimization*		
	Mg	Al	Cu	Mg	Al	Cu
ΔP, Pa	80	80	80	80	80	80
V_{air}, m^3/s	0.02	0.02	0.02	0.025	0.025	0.025
q_{pp}, W	1.6	1.6	1.6	2	2	2
ρ_{fin}, kg/m^3	1740	2700	8933	1740	2700	8933
k_{fin}, W/m-K	156	200	400	156	200	400
Geometry						
δ, mm	0.58	0.53	0.42	0.29	0.22	0.11
s, mm	1.46	1.43	1.37	1.49	1.45	1.36
N	49	51	56	56	60	68
Mass, kg	0.248	0.365	1.043	0.142	0.18	0.325
H/δ	86.2	94.3	119	172.4	227.3	454.6
H/s	34.3	35	36.5	33.6	34.5	36.8
Metrics						
q, W	287	310.5	372.2	256.2	267.8	290.3
h_a, W/m^2-K	1147.8	1242	1488.8	1024.8	1071	1161
h_m, W/kg-K	46.3	34	14.3	72.2	59.5	35.7
h_{sc}, W/m^3-K	22956	2484	29776	20496	21420	23220
COP	179.4	194.1	232.6	128.1	133.9	145.2

3.3 *Pumping power considerations*

To examine the pumping power influence on the performance of these air-cooled heat sinks, a parametric study was performed using the analytical model described in the previous section and focusing on the designs yielding maximum heat sink thermal transport. The three pumping powers chosen, 0.6 W, 1.6 W and 2.1 W, respectively, represent the low, medium and high ranges of the pumping powers addressed in this study. For a given pumping power, the best thermal performance attainable for each of the pressure drop and flow rate pairs was determined. The results, for each of the three pumping powers, expressed in terms of the array heat transfer coefficient, is depicted in Figure 6 (a, b, c). As might have been anticipated, the larger the pumping power the higher the maximum heat transfer rate attainable by a heat sink. However, at each pumping power the heat transfer rate is dependent on the flow rate, as well as the number of fins, and it is, thus, possible for the best configuration at a lower pumping power to exceed the performance of an ordinary configuration at a higher pumping power. This figure also reveals the existence of an optimum number of fins - which maximizes the heat transfer rate - for each combination of volumetric flow rate and pressure drop. Although at very low pumping powers, the very best heat transfer rates are associated with low volumetric flow rates and high pressure drops, as the pumping power increases, it is beneficial to operate the heat sink at progressively higher volumetric flow rates.

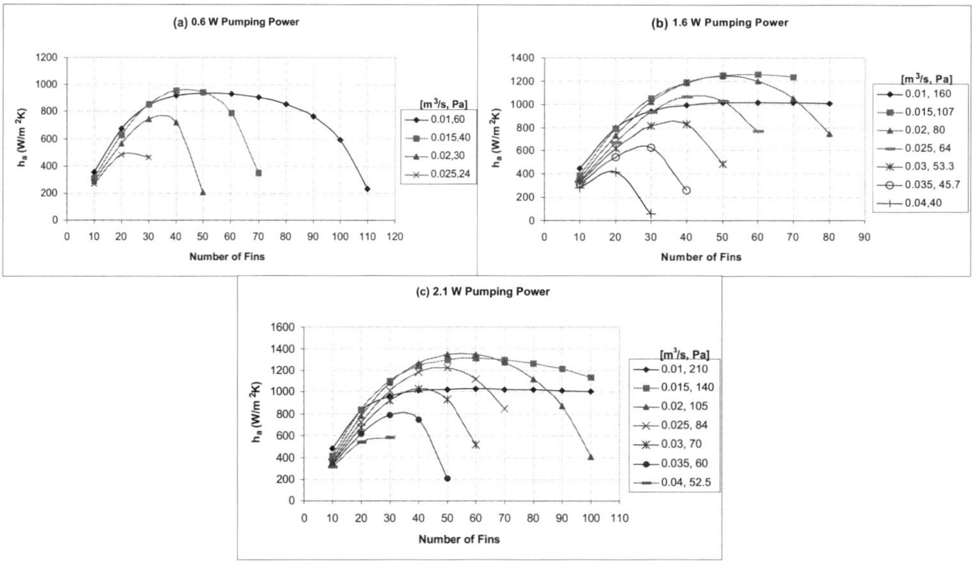

Figure 6, Array heat transfer coefficient for different pumping powers, (a) 0.6 W, (b) 1.6W, and (c) 2.1 W

Examining Figure 6(a) in more detail, it may be seen that across nearly the entire range of fin counts, the maximum heat transfer would be achieved by a flow rate of 0.01 m^3/s and pressure head of 60 Pa. However, in the range of 30-50 fins, the maximum heat transfer is achieved by the 0.015 m^3/s and 40 Pa condition, representing a "cross-over" in the h_a Vs N plane. Such cross-overs become more common as the pumping power increases, as clearly visible in Figures 6(b) and 6(c). Interestingly, at each pumping power, for the low volumetric flow rate and very high pressure drop, the heat transfer reaches a plateau, where thermal performance is independent - or nearly so - of the number of fins.

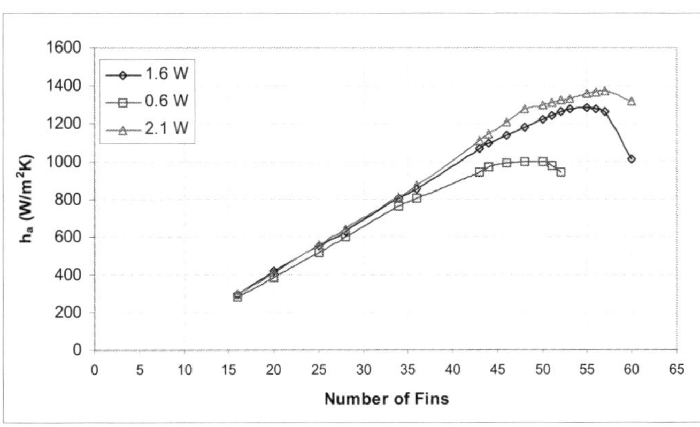

Figure 7, Optimum pumping power

The optimization of the pumping power is shown in Figure 7. The three pumping powers already discussed above were chosen for this analysis. Each point on the plot is defined by a specific operating condition and represents the maximum that can be achieved for the specified conditions. There is an optimum value of 57, 55 and 48 fins for the pumping powers of 2.1W, 1.6 W and 0.6 W, respectively, and the heat transfer improves by 22 % and 27 % across this range. Thus, at the upper-limit of heat sink cooling capability, there is a benefit to the application of greater pumping power. However, as can be clearly seen in Figure 7, up to some 35 fins, there is no benefit to higher pumping power; with the appropriate 0.6W pumping power combination of flow rate and pressure drop yielding essentially the identical heat transfer rate to that attained by 2.1W of pumping power.

3.4 *Coefficient-of-Performance (COP)*

Figure 8 shows the surface plot of the COP, defined in eq.(6) as the ratio of the heat transferred to the pumping power to the volumetric flow rates and pressure drops, for both the maximum heat transfer rate and the least material configurations, previously discussed. It should be emphasize that in the present study the pumping power doesn't include the fan efficiency.

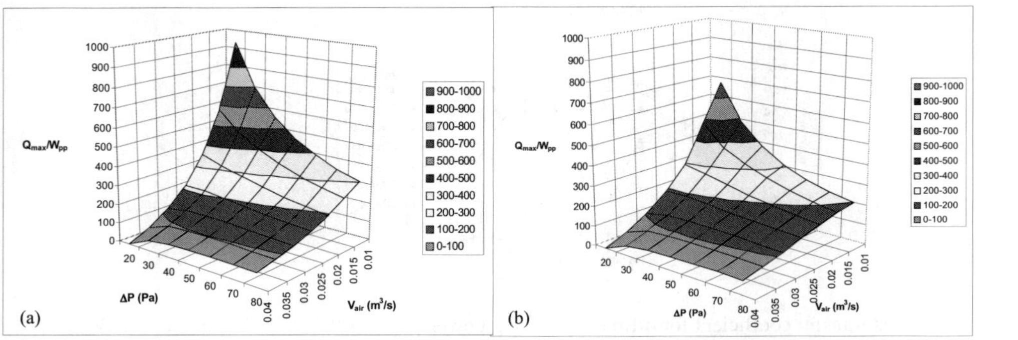

Figure 8, (a) COP for maximum heat transfer design, (b) COP for least material design

It may be noted that the general pattern is the same for the maximum and least-material heat sinks, with the peak COP occurring at the lowest values of pressure drop and flow rate and then decreasing rather steeply with increasing volumetric flow rate and the heat sink pressure drop. A COP comparison for these two design categories, using two extreme operating points ($0.01\ m^3/s$ and 20 Pa and $0.04\ m^3/s$ and 80 Pa), shows a substantial COP advantage for the maximum design at the lower flow rate and pressure drop with a COP of 930 Vs 670 and a far weaker advantage at the high end, COP of 69 Vs 66. Moreover, comparing pumping powers for the best heat transfer achievable, the least-material design requires 2 W (heat transfer 267 W) as compared to 1.6 W (heat transfer of 310 W) for the maximum heat transfer design. Therefore, it appears that the design of these high performance heat sinks provides an opportunity for a trade-off between heat sink mass and pumping power.

The pumping power performance coefficient, COP, for the copper and magnesium fins is shown in Table 1. As a general trend, the COP was again found to be lower for the least-material array as compared to the best maximum heat transfer designs. It is, thus, clear that the sizable improvements in material utilization obtained in the least-material designs, generates some additional burden in terms of energy utilization for operation.

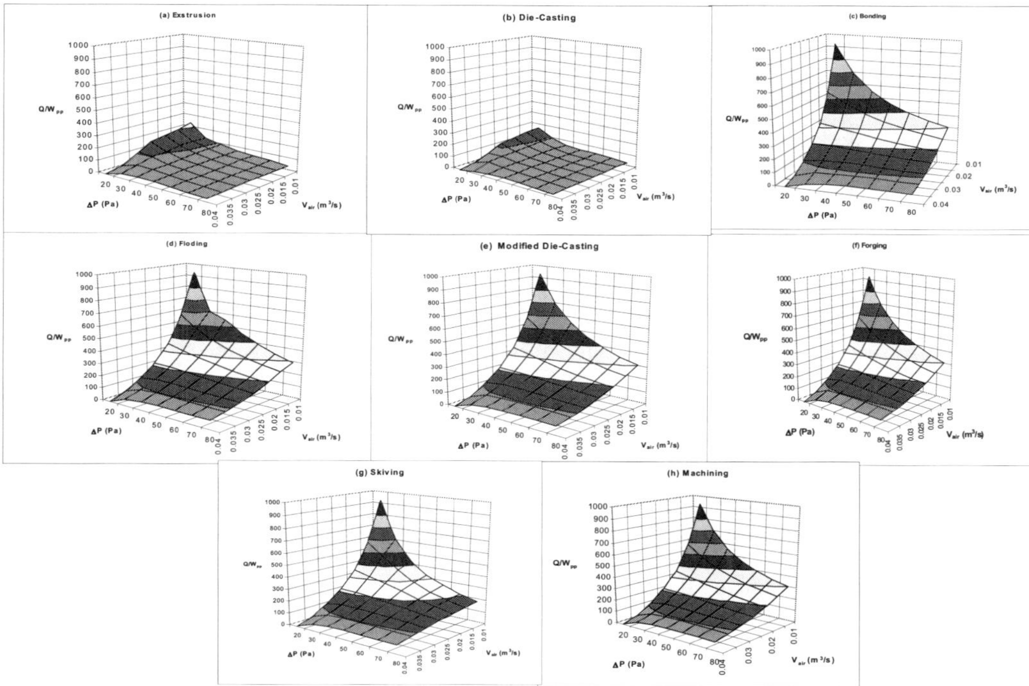

Figure 9, COP for design for manufacturability

4 TOTAL COEFFICIENT OF PERFORMANCE (COP_T)

The COP_T values in the range of volumetric flows and pressure drops investigated in this study are illustrated in Figure 10. The study herein considers the fins only. As may be seen in Figure 10(a) and 10(b), the maximum obtained COP_T is 94 for least material designs and approximately 52 for the maximum heat transfer designs, Figure 10 (a). The existence of optimal COP region it can be clearly seen for maximum heat transfer design, in Figure 10(b). Thus, comparing the COP_T of the least material design to the maximum heat transfer design, the additional mass used in the latter configurations predictably lowers the COP_T by some 42%. Interestingly, the COP_T for the maximum configurations display a "saddle" surface with a value of approximately 45 covering a broad segment of the parametric range.

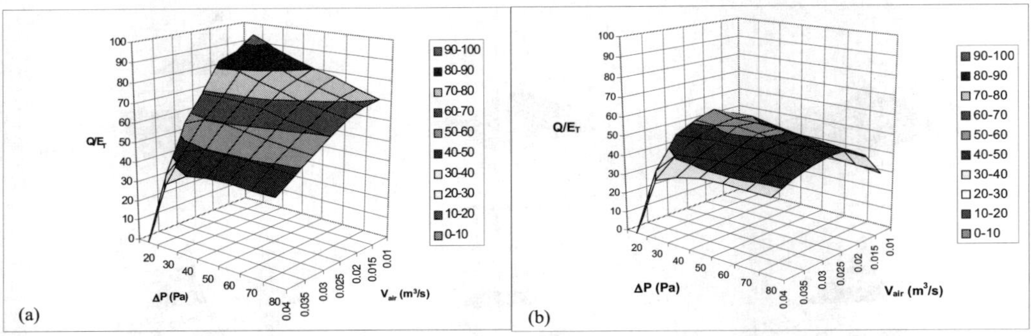

Figure 10: Total Coefficient of Performance for (a) least material design, and (b) maximum heat sink design

Figure 11 illustrates the coefficient of performance for two manufacturing processes, skiving and extrusion. It should be mentioned that COP_T for skiving was based on fabrication energy used for extrusion process. The maximum COP_T for skiving and extrusion is 52 and 23 respectively. It should be noticed that the maximum COP_T for skiving process and the maximum COP_T for maximum heat transfer design are the same.

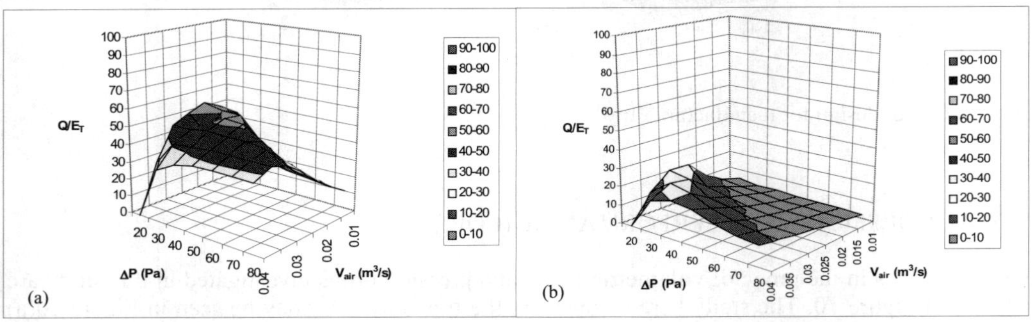

Figure 11,Total Coefficient of Performance for (a) skiving, and (b) extrusion

Comparing the energy needed to fabricate the heat sink to the energy consumed in operating the fan-heat sink combination, it can be concluded that the fabrication energy cost is higher then operating energy for volumetric flow rate from 0.01 m^3/s up to 0.03m^3/s for the range of pressure drops (20 Pa - 80 Pa) used in this study, as depicted in Figure 12. At 0.03 m^3/s these two costs become approximately the same. The operating energy cost becomes more significant than the fabrication energy cost for volumetric flow rate of 0.04 m^3/s. For the extrusion process, the fabrication energy cost exceeds the operating energy cost, for the entire operating domain.

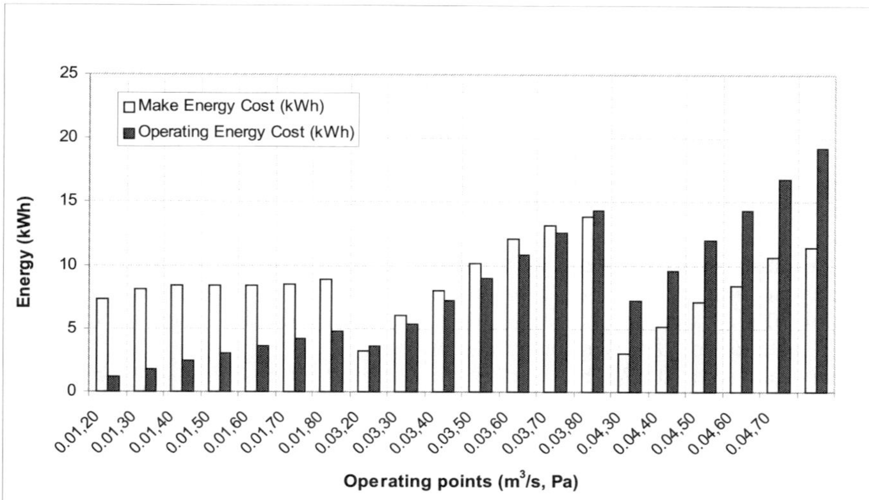

Figure 12, Make and operating energy cost comparison for least material design

5 NATURAL CONVECTION PLATE-FIN HEAT SINKS

Heat transfer from a fin array, as depicted in Figure 13(a) is the sum of heat dissipation from the fins and from the exposed, unfinned area on the array base, and can be calculated using the model developed by Iyengar and Bar-Cohen (1998) and Bar-Cohen et al. (2001), given as,

$$q_T = n_{fin}\theta_b(q_{fin} + h_{base}A_b) \tag{15}$$

Applying the Murray-Gardner assumptions [Kern and Kraus (1972)], and assuming an insulated fin tip, the heat dissipation capability of a single fin can be expressed as,

$$q_{fin} = Lk_{fin}t\theta_b m\,(\tanh mH) \tag{16}$$

where m is a fin parameter equal to $(2h_{fin}/k_{fin}t)^{1/2}$.

The prevailing heat transfer coefficients on fin surfaces, are intermediate between the isolated plate and fully developed limits, and can be found from correlations for natural convection in parallel, vertical plate channels presented by Bar-Cohen and Rohsenow (1984). When applied to non-isothermal plates, as encountered in rectangular plate-fin arrays, this composite Nusselt number correlation takes the form,

$$Nu_{fin} = h_{fin}s/k_f = [576/(\eta_{fin}El)^2 + 2.873/(\eta_{fin}El)^{1/2}]^{1/2} \tag{17}$$

where El is the Elenbaas number [Elenbaas, (1942)] and is equal to $El = (g\beta\theta_b Pr\, s^4)/L\nu$.

6 COMPARISON BETWEEN NATURAL- AND FORCED-CONVECTION HEAT SINKS

6.1 *Central trends*

Figure 13(c) displays the range of the COP_T for forced and natural convection cooling of the 10×10×5 cm heat sink, for an invested work of 10 kWh and an life of 6000 hours. Each line in Figure 13(c) represents a specific value of ξ_{PP}, and displays the variation of COP_T with the fin density. Each ξ_{PP} relates to a specific combination of fin mass, M, and pumping power, PP. A closer examination of the finned forced convection results, yields trends showing each forced convection curve to rise steeply with fin density towards a peak value, and then decreases gradually. In addition, the existence of an optimal ξ_{PP} value can also be seen, which would suggest the most favorable distribution of existing energy resources between heat sink manufacturing and operation, over a fixed product life cycle. For the design space explored in Figure 13(c), COP_T for an input of 10 kWh for 6000 h operation is maximized to yield a value of 92 for a ξ_{PP} of 0.042, representing 88g of fin material and 0.42W of pumping power.

A special case where the given 10 kWh has been entirely utilized for pumping air over the bare base surface, in the absence of any fin material, shows COP_T values comparable to that for finned natural convection. Finned forced convection arrays substantially outperform the natural convection and unfinned channels in terms of "cooling returns" on invested work, thus demonstrating the important benefits of so called 'energy intensive' forced air solutions. It must be noted, however, that under certain circumstances including remote locations, high noise, or expensive power; passive cooling will remain the technique of choice. The reader is reminded that in this study the fluid pumping power does not account for the efficiency of the fan, which when introduced may be expected to diminish the advantage shown for forced convection.

6.2 *Operating life*

It is instructive to explore the effect of "duty cycle" on the COP_T and the optimum design. Three different "duty cycles", namely sporadic, periodic, and continuous, have been chosen assuming the heat sink life cycle to last three years. The results for duty cycles are illustrated in Figure 14 for 1500h and 26208h. For finned forced convection utilizing 10 kWh of work, the maximum value of COP_T is seen to improve by a factor of 10, from 29 to 304, when the life cycle is increased by a factor of 17.5, from 1500 to 26208 hours. However, in the absence of the any pumping power usage, the natural convection configuration COP_T is seen to increase linearly from 2.1 to 36.9, for t_l values of 1500h and 26208h, but still fall well below the best forced convection COP_T value.

6.3 *Available energy for input cooling energy*

The impact of available energy on the COP_T was investigated for five different E_T values, of 10 [Fig. 13(c)], 15, 20, 25, and 30 kWh, respectively. A trend of reduced COP_T, with increasing E_T values, may be seen in Fig. 15(a), which shows the surface for maximum values for COP_T for each of the E_T and ξ_{PP} considered. The results suggest diminishing returns for energy investment in cooling from a heat sink volume. The value of the optimal resource allocation ratio ($\xi_{PP,\ opt}$) that maximizes COP_T is found to be weakly dependent on E_T, and appears to fall in the range of the conditions considered. On repeating the calculations presented in Figs. 13(c) and 14, for different input cooling energies of 20 and 30 kWh, respectively, the optimal values of ξ_{PP} were determined for a range of E_T and t_l, and are exhibited in Figure 15(b).

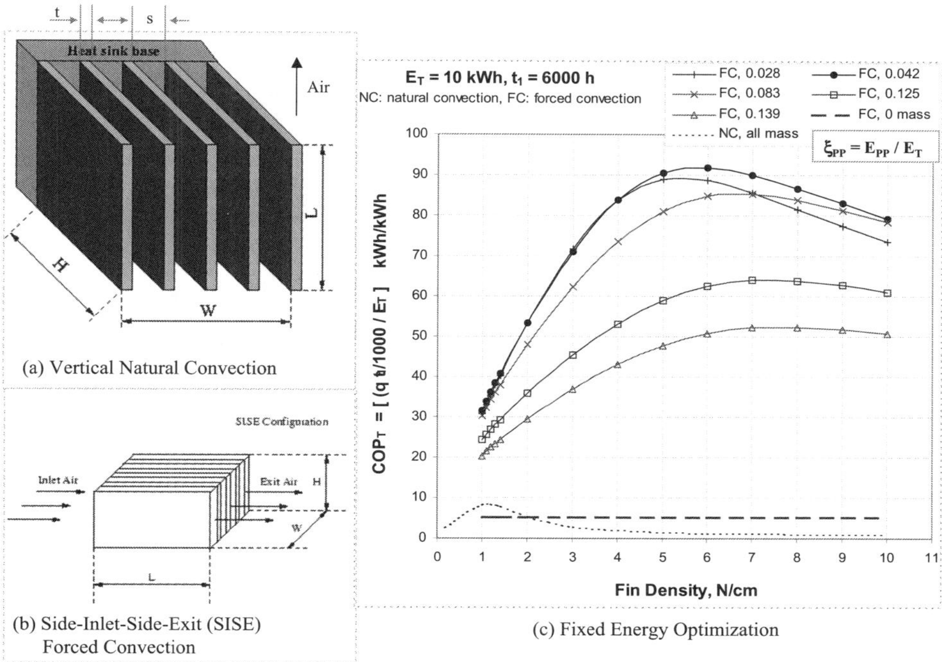

Figure 13: (a) Rectangular Plate-Fin Heat Sink Array Schematic for Vertical Natural Convection, (b) Side-Inlet-Side-Exit (SISE) Forced Convection, (c) COP_T for Rectangular Plate-Fin Arrays, [E_T = 10 kWh, Regular duty cycle, t_1 = 6000 h {3 years, 50 wk., 5 days/wk., 8 h/day}, L = 0.1m, W = 0.1m, H = .05m, θ_B = 25K, Aluminum]

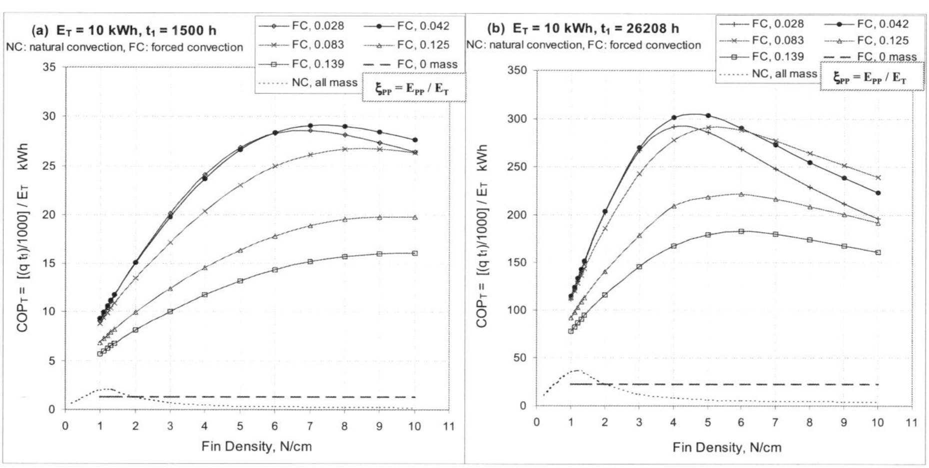

Figure 14: Effect of Duty Cycle on COP_T for Rectangular Plate-Fin Arrays, (a) High duty cycle, t_1 = 26208 h, (b) Low duty cycle, t_1 = 1500 h, [E_T = 10 kWh, 3 yr., L = 0.1m, W = 0.1m, H = 0.05m, θ_B = 25K, Aluminum]

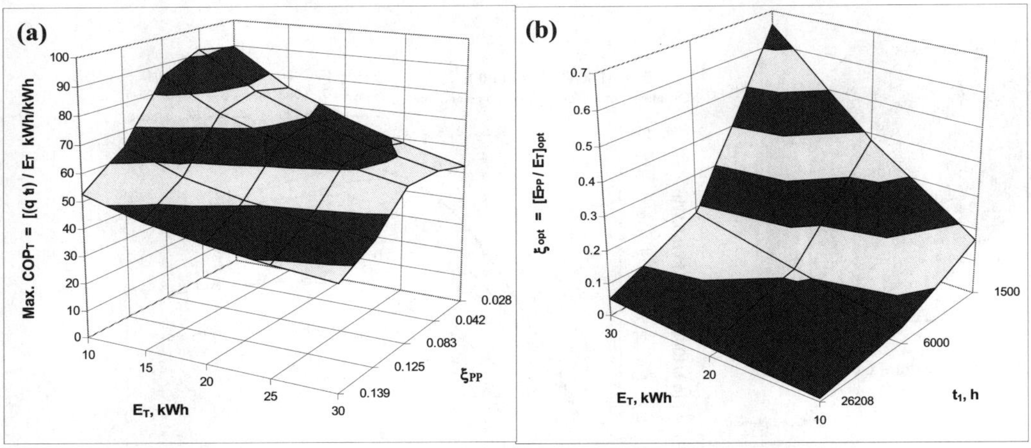

Fig. 15: Rectangular Plate-Fin Arrays, Forced Convection; (a) Effect of Total Available Energy and Allocation ratio on Maximum COP_T, [t_l = 6000 h], (b) Surface for Optimal ξ_{PP} with Total Input Cooling Energy, E_T, and Duty Cycle of Operation, t_l; [L = 0.1m, W = 0.1m, H = 0.05m, θ_B = 25K, Aluminum]

7 CONCLUSION

The foregoing has presented a methodology for determining the "least-energy" design of air-cooled heat sinks used in the thermal management of electronic components. It has been shown that the energy invested in the formation and fabrication of such heat sinks, including both maximum heat transfer and "least-material" designs, can far exceed the pumping power provided by the commonly used fans. Consequently, the design and optimization of forced convection heat sinks, which will promote sustainable development of this earth, must seek to minimize material consumption without leading to excessive pumping power requirements. A thermal Coefficient-of-Performance (COP_T), relating the cooling capability of the heat sink to the energy invested in both its fabrication/formation and operation, is used to determine the degree of sustainability of specific heat sink designs.

A detailed analysis of air-cooled heat sinks for the thermal management of high power electronic components, with an allowed "space claim" of 500cm^3 and in the operating range of 20 to 80 Pa pressure drop and 0.01 to 0.04 m^3/s of air flow rate, was performed to determine the thermal performance, the COP - based on pumping power alone, and the total-energy COP. The results reveal that to achieve the maximum thermal performance in an air-cooled heat sink, it is necessary to develop high fin-count heat sink designs with high aspect ratio fins, which will operate at high pressure drops and moderate air flow rates. High thermal performance "least-material" designs for these same applications will require substantially thinner - and , therefore, higher aspect ratio - fins, high pressure drops and high volumetric flow rates. Alternatively, if the COP_T is to be maximized, while maintaining high rates of cooling, the heat sinks must be designed with least-material fins and operated with fans providing moderate air flow rates and high-pressure heads. In the parametric range studied, it was found possible to generate high-performance designs which would yield COP_T values in the range of 60 to 80, or nearly twice that achievable with designs that ignore the energy invested in the heat sink materials. However, further improvements in heat sink manufacturing processes and available fans will be needed to achieve this level of performance.

This study demonstrates the usefulness of the COP_T metric to enable the least-energy optimization of plate-fin convective heat sinks, resulting in the maximization of the cooling capability that can be achieved in a specified volume, while minimizing the material and energy consumed in the fabrication and operation of the specified heat sink. The analysis show the forced convection arrays substantially outperform the natural convection designs, for a fixed available input cooling energy and life cycle of duration. For forced convection configurations, an optimal resource allocation ratio, was identified, that provides the most favorable distribution of existing energy resources between heat sink manufacturing and operation, over a fixed product life cycle.

ACKNOWLEDGEMENTS

The authors are indebted to A.D. Kraus for the inspiration derived from his pioneering development of "least-material" fin designs.

REFERENCES

Bar-Cohen, A., Iyengar, M., and Kraus, A. D., 2001, Design of Optimum Plate-Fin Natural Convective Heat Sinks, accepted for publication to the ASME Journal of Electronic Packaging.

Bar-Cohen, A., and Rohsenow, W., 1984, Thermally Optimum Spacing of Vertical, Natural Convection Cooled, Parallel Plates, Journal of Heat Transfer, Vol. 106, pp. 116-122.

Bar-Cohen, A., Prstic, S., Yazawa, K, Iyengar, M., 2000, Design and Optimization of Forced Convection Heat Sinks for Sustainable Development, Proceedings of the Euro Conference: New and Renewable Technologies for Sustainable Development, June 26-29, Madeira Island, Portugal.

Ben Achour, M. F., and Bar-Cohen, A., 1999, Heat Sink Optimization for Maximum Performance and Minimum Mass, Proceedings of the Interpack Conference, ASME Advances in Electronic Packaging, EEP-Vol.26-1, pp. 737-744.

Bejan, A., 1995, Entropy Generation Minimization, CRC Press, New York.

Computer Industry Almanac Inc., 1999, "Nearly 600 Million Computers-in-Use in Year 2000" Press release 11/03 1999

Elenbaas, W., Heat Dissipation of Parallel Plates by Free Convection, 1942, Physica, vol.9, No.1, pp. 665-671.

Garner, S. D., and Toth, J. E., 1997, Heat Pipes: A Practical and Cost Effective Method for Maximizing Heat Sink Effectiveness, Proceedings of the ASME Advances in Electronic Packaging, EEP-Vol. 19-2, pp. 1897-1902.

Holahan, M. F. , Kang S. S., Bar-Cohen, A., 1996, A Flowstream Based Analytical Model for Design of Parallel Plate Heat Sinks, ASME Proceedings of the 31st National Heat Transfer Conference, HTD-Vol. 329, Volume 7, pp 63-71.

Iyengar, M., and Bar-Cohen, A., 1998, Least-Material Optimization of Vertical Pin-Fin, Plate-Fin, and Triangular-Fin Heat Sinks in Natural Convective Heat Transfer, Proceedings of the InterSociety Conference on Thermomechanical Phenomena in Electronic Systems (ITHERM), May'1998, Seattle, USA, pp. 295-302.

Iyengar, M., and Bar-Cohen, A., 2001, Design for Manufacturability of SISE Plate-Fin Forced Convection Heat Sinks, Published in IEEE-CPMT, June.

Kakac, S., Shah, R. K., Aung, W., 1987, Handbook of Single-Phase Convective Heat Transfer, Second Edition, Wiley & Sons, New York, pp. 3-35, 3-42.

Kern, D. Q., and Kraus, A. D., 1972, Extended Surface Heat Transfer, McGraw-Hill, New York.

Kraus, A. D., Aziz, A., Welty, J., 2001, Extended Surface Heat Transfer, John Wiley & Sons, New York.

LCA committee report, 1999, Japan Aluminum Association

Zhao, Z. and Avedisian, C. T., 1997, Enhancing Forced Air Convection Heat Transfer from an Array of Parallel Plate Fins using a Heat Pipe, Int. J. Heat Mass Transfer, Vol. 40, No. 13, pp. 3135-3147.

Thermal Challenges in Next Generation Electronic Systems, Joshi & Garimella (eds)
© 2002 Millpress, Rotterdam, ISBN 90-77017-03-8

Temperature measurement of semiconductor devices

D.L. Blackburn
National Institute of Standards and Technology, Gaithersburg, MD, USA

Keywords: integrated circuits, semiconductor devices, temperature measurement, thermal resistance, transistors

EXTENDED ABSTRACT: The ability to measure accurately the temperature of operating microelectronic devices has become more important than it ever has been. It has always been true that extreme temperature places limits on the operating range of nearly all devices, but today, increasing power dissipation and power densities threaten to create temperatures that block continued progress according to 'Moore's Law.' Clearly, new and innovative methods for cooling chips and packages must be found along with new materials and circuit designs, and architectures for decreasing the power dissipation and operating temperature. Equally as clear, we must have accurate and well-understood methods for measuring the temperature of devices to aid in the development of these new techniques and materials.

The reasons for desiring to know the operating temperature of a semiconductor device can be divided into the following three broad categories:

- Predict reliability or operating life of device
- Measure material/device thermal properties insitu
- Confirm or determine the operating limits or thermal performance of a device

This paper will discuss the various methods that have been used for measuring the operating temperature of semiconductor devices. The methods include those that use the device itself to measure its own temperature (using temperature sensitive device electrical parameters, for instance), those that use a physically contacting 'thermometer' of some type (microprobes, for instance), and those that require no extra physical contact to the device (such as optical techniques).

The paper begins with a discussion of the various reasons for measuring the temperature of a device and how these reasons may decide the type of method that is used as well as the details of how the method is actually applied to the measurement. Next, there is a discussion of the temperature-sensitive materials properties and device characteristics that are important in the various measurement methods. The electrical and optical properties of semiconductors and metals are included, as are device characteristics such as junction voltage-current relationships, device gain, breakdown voltages, and general ac and dc output characteristics. A variety of temperature measurement techniques are discussed next, including their physical basis for temperature measurement, their relative accuracy and sensitivity as a thermometer, and their situational applicability.

Table 1 lists a few of the different methods, divided into optical, electrical, and contacting methods. The first three optical methods require an 'active' probe, in that an incident optical probe is required and the reflected or emitted light is measured. The fourth, infrared emission, is a 'passive' technique in that no incident radiation is required, but only thermally emitted radiation is measured. Most of the electrical parameters listed can be measured in a variety of ways, including dc, ac, pulsed, and switched. The condition under which the parameter is measured (dc, ac, etc.) has a very strong influence on the accuracy and situational applicability of the results.

Table 1. List of some temperature measurement methods for semiconductor devices

Optical	Micro Raman Spectroscopy Reflectance Photo Emission Infrared Emission
Electrical	Gain Junction Voltages Output Conductance Resistance Breakdown Voltage Leakage Currents
Contacting	Microprobe (thermistor, thermocouple)

Table 2: List of some advantages and disadvantages of generic measurement types

Technique	Advantages	Disadvantages
Optical	• Temperature map possible. • Generally non-contacting, minimally invasive. • Potentially very good spatial resolution.	• Chip surface must be visible. • Generally expensive and complex. • Device cannot be normally packaged.
Electrical	• No contact, non-invasive (can be done on a packaged device). • Generally inexpensive, simple.	• Generally a non-specific temperature average. • Electrical and other interferences.
Microprobes	• Temperature map possible. • Excellent spatial resolution.	• Contact to surface required. • Expensive and complex. • Device cannot be normally packaged.

Table 2 lists some of the advantages and disadvantages of the generic techniques. The major advantage of the Electrical techniques are that they can all be performed on packaged devices because the only contact required to the devices is to their electrical leads. The biggest disadvantage of these methods is that they all yield some usually non-specific average chip temperature. The major advantages of the optical and microprobe methods are that temperature maps can be produced and most have the potential for very good (micrometer or sub-micrometer) spatial resolution. The greatest disadvantage of these methods is that the surface of the device must be accessible and the device cannot be packaged in its usual, real application manner.

Reliable data from appropriate and reliable measurements form the foundation for all design and quality control efforts, from research to implementation. It is the goal of this paper to provide the scientists and engineers concerned with thermal management of semiconductor devices the information that is needed to select the appropriate temperature measurement methods for their application.

Thermal Challenges in Next Generation Electronic Systems, Joshi & Garimella (eds)
 ISBN 90-77017-03-8

Thermal issues in MEMS and microscale systems

Don L. DeVoe
Department of Mechanical Engineering and Institute for Systems Research, University of Maryland, College Park, MD, USA

Keywords: MEMS, microhotplate, microfluidics, thermomechanical actuation

ABSTRACT: A brief review of thermal issues in MEMS technology is presented, including an overview of traditional and emerging applications for microscale thermal systems. Issues relating to fundamental limitations and opportunities in this area are discussed. The use of thermal phenomena in three specific microsystems are presented in detail, namely microhotplate chemical sensors, microfluidic systems, and electrothermal micromotors, and future directions in microscale and nanoscale thermal systems are addressed.

1 INTRODUCTION

MicroElectroMechanical Systems (MEMS) technology generally involves coupled-field behavior, combining two or more energy domains such as mechanical, electrical, optical, or thermal. The thermal domain is arguably the most prevalent in MEMS devices, either by intentional design or undesirable coupling. Unlike thermal issues in electronic packaging relating to efficient heat removal, thermal challenges in MEMS devices encompass a wide range of issues, from the harnesing of thermal energy for actuation in micropumps and microvalves, to the ultimate sensor noise limits imposed by inherent thermo-mechanical coupling in microsensors.

One of the earliest and most successful applications of thermal MEMS is the use of silicon microfabrication for producing thermal inkjet print heads (Bassous et al., 1977; Peeters et al., 1997). MEMS technology has been a major enabler in this field since the mid-1970s, allowing significant reductions in cost while increasing the performance of these systems. The operating principles are based on a combination of thermodynamics and fluid dynamics using silicon micromachined components. In these devices, a vapor bubble is generated and expanded within a silicon chamber using an integrated microheater, forcing ink to be expelled from an exit nozzle. A similar technique has been applied to micromachined fuel injectors for automotive applications (Tseng et al., 1996). A novel design using multiple thermal elements for controlled, rapid droplet generation was also recently demonstrated (Tseng et al., 1998).

Bulk silicon etching has also been extensively applied to thermal isolation of passive and active on-chip components. An early application in this area is bulk-etched silicon beams for thermal isolation of diode detectors in a high-bandwidth frequency synthesizer produced by Hewlett-Packard (O'Neill, 1980). Isolation to remove heat sources or sinks proximal to a themal element is typically achieved through bulk micromachining of single-crystal silicon from the frontside or backside of the silicon substrate, or alternately by surface micrmachining of a thin film sacrificial layer from underneath the thermal element. Thermal isolation structures have been extensively used for high performance thermal sensors. For example, thermoelectric infrared detector arrays have been realized using a commercial 1μm-linewidth CMOS process, with the polysilicon/aluminum interface acting as a thermopile detector, and backside bulk micromachining used to provide thermal isolation (Munch et al., 1997). In general, direct temperature measurement is performed using integrated metal film or polysilicon thermoresistors, or thermopiles formed from dissimilar metals, metal/polysilicon, or p/n polysilicon films. Less commonly, integrated thermodiodes or semiconducting thermistors have also be used. Thermal sensing has also been applied to the measurement of mechanical signals such as acceleration. Examples include a thermal microbeam accelerometer with integrated thermopiles to convert thermal changes due to beam vibration to an electrical signal (Hiratsuka et al., 1991), and a related thermal convection accelerometer design (Milanovic et al., 1998). Beyond passive thermal isolation and measurement, active thermal control has also been used for applications such as frequency tuning for microresonators, which often exhibit high temperature sensitivities which can limit their performance (Remtema & Lin, 2000).

Active thermal control has been used to perform direct mechanical actuation through thermal expansion. With relatively low voltage requirements and high force output, electrothermal actuation has important advantages over traditional electrostatic actuation. Both in-plane and out-of-plane actuators have been developed that rely on resistive heating and thermal expansion of silicon, polysilicon, nickle, or related materials. Thermal microactuators based on polymers such as polyimide have also been investigated (Eberfors et al., 1998). Thermal bimorphs (Riethmüller et al., 1998), pseudo-bimorphs (Comtois & Bright, 1996; Comtois et al, 1998), buckling beams (Lin & Lin, 1998; Que et al., 2001), compliant structures (Milanovic et al., 1998; Jonsmann et al., 1999; Moulton & Ananthasuresh, 2000), and high-aspect-ratio structures (Keller & Howe, 1997; Maloney et al., 2000) have all been demonstrated. Additionally, applications such as grippers (Pan & Hsu, 1997) and micromotors (Comtois et al., 1995; Comtois et al., 1998; Pai & Tien, 1999; Park et al., 2001; Maloney et al., 2001; Schreiber et al., 2001) have been described. Recently, work has been reported on the development of models to accurately predict actuator behavior (Reid et al., 1996; Butler et al., 1997; Huang & Lee, 1999; Maloney & DeVoe, 2000).

Thermal expansion for controlled mechanical work has also been applied to micropumps and microvalves for gas and liquid manipulation. Rather than using bubble expansion to expel fluid from the micromachined chip, as in the case of ink jet print head technology, a bubble pump for fluid mixing in a stationary silicon micromachined chamber has been demonstrated (Evans et al., 1997). Due to high surface tension forces, thermally-generated microscale bubbles tend to be highly stable and useful for generating mechanical work. An extension of fluid displacement via thermally-generated bubbles has also been used to generate bulk pumping in microchannel networks (Yuan & Prosperetti, 1999). A valveless micropump was demonstrated by combining a series of addressable microheaters to create a travelling sequence of bubbles capable of precise fluid pumping with a 0.5 nL/min flow rate (Kun & Kim, 1998). A larger scale thermal bubble micropump based in bulk silicon capable of delivering a 6.5mL/min flow rate (Tsai & Lin, 2001), and thermal bubble micropumps for drug dispensing from a microneedle (Zahn et al., 2001) have also been fabricated. Other micropumps utilize alternate thermal actuation methods such as bimetallic membranes (Lisec et al., 1994; Yue et al., 1996), a combination of bimetallic and thermopneumatic elements (Zou et al., 1997), or TiNi shape memory alloy actuators with integrated thermal control (Makino et al., 2001). Using thermal actuation methods similar to those employed for micropumps, microfabricated valves have also been investigated in a variety of MEMS technologies. In thermopneumatic microvalves, the saturated vapor pressure of a resistively heated working fluid deflects a micromachined valve plate to seal the valve. The heater elements are typically isolated from the substrate with the expansion cavity only partially filled to increase thermal efficiency. Devices with dynamic ranges above 40,000 have been demonstrated (Barth et al., 1994). The same concept has been applied to combined microvalve/micropump technology (Freygang et al., 1996), and microvalves with integrated termperature and pressure sensors (Fitch et al., 1998; Rich & Wise, 2000). Thermally actuated bimetallic microvalves have also been demonstrated. Examples include a nickel-silicon bimetallic microvalve developed by Hewlett Packard to replace bulky solenoid valves for gas chromatography (Wang et al., 1999), and a micro-liquid handling system combining a bimetallic valve and calorimetric flow sensor (Franz et al., 1996). Microvalves which employ thermal buckling of a single-layer membrane have been investigated (Lisec et al., 1994; Goll et al., 1995; Popescu et al., 1995), and thermally-controlled shape memory allow microvalves using integrated TiNi films have also been explored (Krulevitch et al., 1996). Thermal microsystems have also been widely applied to sensing bulk fluid flow. A variety of micromachined thermal sensors for measuring liquid flow rates have been demonstrated, such as microscale hotwire anemometers fabricated from polysilicon elements (Neda et al., 1995; Ebefors et al., 1998), and anemometers based on p/n polysilicon thermopiles with integrated heater elements (Mayer et al., 1996). A recent development is the demonstration of flow measurement below the nL/min range (Wu et al., 2001). The heaters in these devices act as classical hotwire anemometers, with a feedback loop used to maintain the temperature of the heating element as heat is removed by the flow. The sensors have demonstrated a flow rate resolution of 0.4 nl/min. Similar devices have been demonstrated by a number of groups, for example as a CMOS-compatible integrated thermoelectric flow sensor chip employing a membrane sensor element (Mayer et al., 1997). Resistive Joule heating elements for anemometry can also be readily formed using resistive metal or selectively-doped bulk silicon.

An important application for thermally actuated pumping, valving, and flow sensing is lab-on-a-chip microsystems. The development of lab-on-a-chip technology is ultimately directed toward chemical and biochemical analysis tools based on the miniaturization and integration of liquid handling, sample processing, and sample analysis in a single chip. In addition to flow control applications previously discussed, other issues in lab-on-a-chip technology are directly related to thermal phonomena. For example, miniature devices for polymearase chain reaction (PCR) amplification of DNA have been developed by a number of groups. The PCR process involves sequential heating and cooling, with 30s at 95°C for denaturation of double-stranded DNA molecules, and 1-2min each for annealing and extension at 72°C to form a new set of double-stranded molecules. With 25-40 cycles per reaction, a traditional cycler requires about 90min for amplification. In contrast, microfabricated polysilicon heaters integrated in a bulk micromachined silicon reaction chamber for PCR amplification were reported by Northrop et al. as early as 1993 (Northrup et al., 1993), with a cycle time of 60s/cycle and heating rate of 15°C/s, for a total amplification time of approximately 22min. More recently, an integrated device that accomplishes cell lysis, PCR, and capillary electrophoresis on monolithic microchips has been demonstrated using bacterial samples (Waters et al., 1998). Rapid PCR thermal cycling rates as fast as 17 s/cycle have been demonstrated in silicon microchambers (Belgrader et al., 1998). The short cycle times are made possible by the relatively small thermal mass of the microstructures. This technology is based on temporal control over the temperature within a single reaction chamber. Similarly, spatial temperature control has been used for continuous flow PCR, based on a glass chip with a single microchannel that meanders the sample across three zones of constant temperature to provide 20 amplification cycles in times ranging from 1.5 to 18.5 min (Kopp et al., 1998).

Because MEMS fabrication takes advantage of the same techniques used for patterning submicron features in microelectonic applications, extremely high-resolution elements for thermal sensing and control are possible. While the relative tolerances which can be achieved using these fabrication techniques is poor compared with traditional machining technology, the high absolute accuracy and resolution enable thermal microsystems which would otherwise be impossible. Futhermore, the ability to combine multiple thin film materials with different properties in a single device enables the realization of true integrated thermal microsystems. For example, micromachined thermal radiation emitters fabricated in a commercial CMOS process (Parameswaran et al., 1991) combine polysilicon thermal detectors for local temperature measurement with resistive traces for ohmic heating. The entire platform is suspended above the silicon substrate to provide effective thermal isolation. Using this microhotplate technology, infrared pixel arrays have been demonstrated with time constants of 5ms and platform temperatures of approximately 500°C. Further developments based on the same technology have resulted in platforms with time constants as low as 0.6ms (Suehle et al., 1993). These microhotplates are a good example a true thermal microsystem, combining thermal actuation and sensing for closed-loop temperature control on a thermally isolated platform.

1.1 *Scaling issues in thermal systems*

At present, most thermal MEMS applications involve elements which are on the order of at least several microns in scale, and traditional thermal modeling techniques can be applied for their analysis. However, as dimensions are further reduced into the nanoscale regime, thermal behavior is ultimately limited by quantum effects. Mesoscopic phenomena in energy transport processes occur when the wavelength or the coherence length of the carriers (i.e. phonons in the case of thermal systems) becomes comparable to, or larger than, the sample dimensions. For example, the quantization of electrical conduction is a result of this phenomena, with particular varieties of carbon nanotubes exhibiting electrical conduction occurring through only two electronic modes. Similar quantization in thermal transport has recently been demonstrated using suspended insulating nanostructures at very low temperatures (Roukes, 1999; Schwab et al., 2000).

Scaling issues must also be considered for micron-size elements. Because heat flux scales with surface area, while heat capacity scales with volume, thermal time constants in microsystems scale linearly with the length scale, resulting in a specific heat flux which shrinks linearly with decreasing dimensions. Similarly, another scale-related issue lies in the inherent coupling between the thermal state of a system and resulting atomic and molecular agitation. Thermal noise limits the ultimate minimum detectible signal for any given sensor, and at the microscale it is often larger than other noise sources such as preamplifier electronics (Gabrielson, 1993). One important thermal noise mechanism results from random thermal vibration, creating a white fluctuation force given by $F^2 = 4k_BTR$ [N^2/Hz]. In this expression, R is the mechanical resistance of the system, i.e. the mechanical damping. This noise source affects any mechanical system which possesses a damping mechanism for energy dissipation. Similarly, another important source of noise is from thermal noise in resistors, or Johnson noise, which results from random scattering of electrons due to thermomechanical vibration in a conductor, and is directly related to noise in purely mechanical systems. In fact, Johnson noise results in a white fluctuation voltage given by $V^2 = 4k_BTR$ [V^2/Hz], where R is the electrical, rather than the mechanical, resistance. Johnson noise is important for many thermal microsensors, defining the detection limits of elements such as thermoresistive temperature sensors and piezoresistive strain sensors.

2 THERMAL MICROSYSTEM DEVELOPMENT

2.1 *Microhotplate chemical sensors*

Micromachined gas sensors, originally developed at the National Institute of Standards and Technology (Suehle et al., 1993; Semancik et al., 2001; Panchapakesan et al., 2001), use semiconducting SnO_2 films deposited on microhotplate platforms to detect gas species such as hydrogen and methanol. The highly localized thermal control makes microhotplates an excellent platform for these sensors. A micrograph of a typical microhotplate element is shown in Figure 1(a). This thermal platform has lateral dimensions of less than 100µm, and is suspended over a bulk-etched cavity for thermal isolation. Furthermore, microhotplates may be fabricated in large arrays, such as the 36-

(a) (b) (c)

Figure 1. Micrograph of (a) a single microhotplate element, (b) a 36-element microhotplate array, and (c) sensor output in air and methanol.

element array shown in Figure 1(b). These arrays are particularly useful for applications in high-selectivity chemical sensing and combinatorial synthesis of materials. The physical architecture of each microhotplate consists of a polysilicon heater, a thermoresistive film for temperature measurement, and a semiconducting SnO_2 film which exhibits a change in conductance with the adsorption of chemical species. Each layer is sandwiched between SiO_2 films for electrical isolation, and contacts underneath the SnO_2 layer are used to measure the conductance of the film during gas sensing. With aluminum as the temperature sensing film, the micro-hotplates can operate up to 500°C. Thermal response times for 100μm wide microhotplates have been measured at around 0.6ms, with a thermal efficiency of 8 °C/mW (Panchapakesan et al., 2001). Figure 1(c) shows an example response of a single microhotplate sensor in air and methanol. Each peak in the plot reflects the change in conductivity as the hotplate is rapidly cycled between room temperature and the maximum allowable temperature, e.g. 500°C. Differences in the adsorption kinetics of the gas as a function of the hotplate temperature cause the shape of the conductance response to vary, and signal processing may then be used to identify the gas.

We have also used the precise temperature control offered by microhotplates to selectively deposit nanostructured SnO_2 films for improved gas sensing (Panchapakesan et al., 2001). In this process, an ultra-thin noncontinuous seed layer of metal such as Ag, Fe, or Ni is deposited by evaporation on a 36-element microhotplate array. Self-lithographic deposition of SnO_2 is then achieved by placing the chip in a cold-wall CVD reactor, and using the integrated microelectrodes to electrothermally raise the temperature of a single microhotplate within a 36-element array while flowing precursor gas over the chip. Thermal decomposition of the gas and deposition of SnO_2 occurs locally to the heated element. By selecting seed metals with different melting points, the film growth kinetics can be modified to produce SnO_2 films with controllable grain sizes ranging from 38nm for Ni seed to 121nm for Ag seed. The reduced grain sizes result in substantially improved gas sensitivity over unseeded sensors, and sensor elements with different seed layers exhibit identifiable changes in sensitivity in the presense of different gases. This thermal microsystems technology is leading towards highly sensitive and selective gas identification in a single chip by utilizing a CMOS-compatible micromachined platform offering local and addressable closed-loop thermal control, excellent temperature uniformity, and high temperature operation.

2.2 *Thermal elements in polymer microfluidic systems*

A strong area of growth in lab-on-a-chip systems is polymer microfluidics for genomic and proteomic analysis. In these systems, microchannel networks are used to manipulate biomolecules for purification, amplification, separation, and/or identification. By using a rigid polymer such as polycarbonate for the microchannel substrate, they offer the potential for low cost, disposable DNA and protein analytical tools. In particular, there is a need for embedded elements in these systems for temperature sensing and control, for example in micropumps and microvalves, and microheaters for PCR. Other applications include thermal lysis for extracting DNA from cells, and thermal denatura-

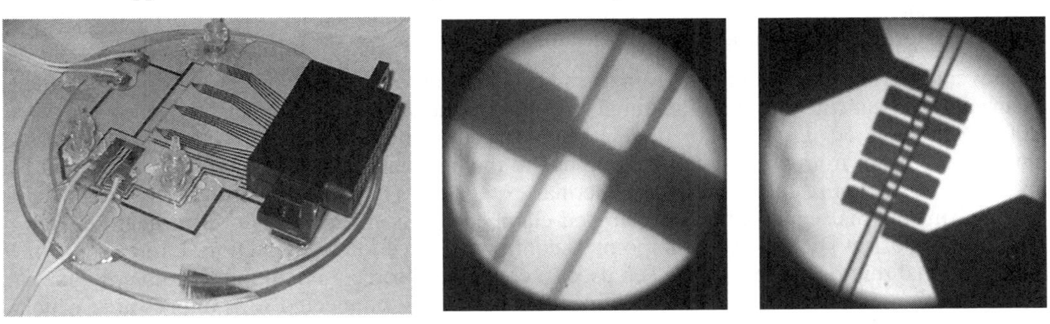

Figure 2. (a) Polymer microfluidic system with thin film heaters, thermoresistors with integrated wheatstone bridges, electrical interconnects, and 80μm wide microchannels, with (b) simple and (c) serpentine heaters.

tion of DNA for temperature-controlled separations. A polymer microfluidic test platform developed in our lab based on polycarbonate micromolding is shown in Figure 2(a). This device contains 80μm wide microchannels fabricated by bulk silicon micromachining to form inverse channel patterns, followed by thermal embossing of the patterns into a polycarbonate substrate. The channels are sealed by thermally bonding a second polycarbonate wafer onto the microchannel substrate. The test device includes thin film chromium thermoresistors and wheatstone bridge circuits for temperature measurement, and gold resistive heaters for local temperature control of fluids within the microchannels. Simple straight-line and sepentine microheaters are shown in Figure 2(b) and 2(c), respectively. Similarly, we have used microheaters for bubble pumping in polymer microfluidics, by sequentially growing and collapsing a series of bubbles to generate a net pressure head within a microchanne via peristaltic action.

2.3 *High-force linear micromotors*

Microactuators based on electrothermal transduction offer significantly larger energy density compared with electrostatic microactuators. Even recent piezoelectric microactuators based on high-performance piezoelectric films such as PZT can only provide a fraction of the energy density of electrothermal elements. The maximum electrostatic and piezoelectric energy density is given by $U=\varepsilon_r\varepsilon_o E^2/2$, where ε_r is the relative permittivity of the dielectric medium (air or PZT) and E is the applied electric field. On the other hand, electrothermal energy density is given by $U=V^2/\rho L^2$, where ρ is the resistivity, V the applied voltage, and L is the electrothermal element length. For typical configurations, electrothermal actuation provides 1 to 2 orders of magnitude higher energy density than piezoelectric actuation, and 4 orders of magnitude higher energy density than electrostatic actuation. We have fabricated electrothermal microactuators using silicon, polysilicon, and nickel structural components. For polysilicon, the maximum actuator height is typically 2 to 4 μm, limiting actuator force and risking stiction issues due to the relatively high out-of-plane compliance. In particular, stiction and bowing are a problem with long polysilicon U-beam actuators, limiting their maximum useful dimensions.

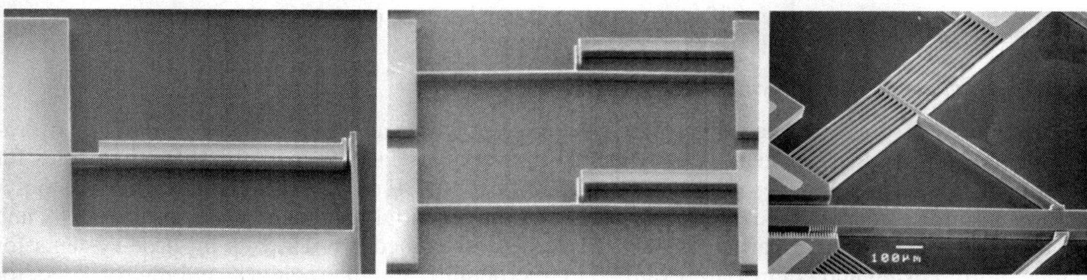

(a) (b) (c)

Figure 3. Individual (a) U-beam and (b) V-beam electrothermal actuators, and (c) a 10-element V-beam microactuator array used in a linear inchworm electrothermal micromotor.

Several thermal microactuator geometries have been investigated in our group, including the U-beam and V-beam geometries illustrated in Figures 3(a) and 3(b), respectively (Maloney et al., 2000; Schreiber et al., 2001). In the U-beam structure, deflection occurs as current is passed through the beam from asymmetric heating due to the unequal arm widths, with higher current flux in the thinner beam resulting in greater thermal expansion. The V-beam configuration deflects laterally from thermal expansion of a clamped-clamped beam undergoing uniform heat generation. The beam is offset slightly at its midpoint to control the direction of deflection. Large arrays of V-beam thermal microactuators have been used to create high-force thermal micromotors. Figure 3(c) shows a micrograph of a 10-element silicon V-beam array connected to a impact arm which is driven into the horizontal linear slider element near the bottom of the image. A symmetric impact arm can be seen on the other side of the slider. Using a total of 6 actuator arrays, the slider element can be positioned with submicron accuracy. Motor speeds as high as 1mm/s over a 2mm travel range have been achieved, with a slider a force of nearly 7mN (Maloney et al., 2001).

3 CONCLUSION AND FUTURE DIRECTIONS

Microscale systems are tightly coupled with a broad range of thermal phenomena. A subset of applications reviewed here include thermal inkjet print heads, micromachined thermal isolation structures, microscale elements for active thermal sensing and control, thermal microactuators, micropumps, microvalves, and thermal elements for lab-on-a-chip applications. Several critical thermal issues in microscale systems include the effects of thermally-induced noise on microsensor performance, and issues relating to device scaling into the micro and nano regimes. With ongoing developments towards functional devices at the nano-scale, managing thermal phenomena in these devices is becoming increasingly important for future progress in miniaturization. While there are a number of significant challenges facing miniaturization of mechanical elements below the micron scale, such as basic nanofabrication issues and sharply reduced coupling strength for most transduction mechanisms, thermal challenges are a key part of the overall problem. For example, at submicron dimensions it appears likely that noise due to temperature fluctuations, Johnson noise, and temperature-dependent adsorption/desorption effects will ultimately limit the applications of nanomechanical transduction elements (Vig & Yoonkee, 1999). For devices such as micromechanical and nanomechanical resonators, these thermal effects will result in frequency noise which complicates their use in high frequency filters (DeVoe, 2001) and molecular sensing applications (Vig et al., 1996). Exploring the limits in the miniaturization of sensors and actuators will require a full understanding of these thermal effects, and the device constraints they imply.

4 REFERENCES

[1] P. Barth, C. Beatty, L. Field, J. Baker, G. Gordon, “A robust normally-closed silicon microvalve,” Proc. Solid-State Sensor and Actuator Workshop (Hilton Head ’94), p.248-50, 1994.
[2] E. Bassous, H.H. Taub, L. Kuhn, "Ink jet printing nozzle arrays etched in silicon," Appl. Phys. Lett., v.31, p.135, 1977.
[3] P. Belgrader, W. Benett, D. Hadley, G. Long, R. Mariella, F. Milanovich, W. Nelson, J. Richards, P. Stratton, Clin. Chem., v.44, p.2191-2194, 1998.
[4] J.T. Butler, V.M. Bright, W.D. Cowan, “SPICE Modeling of Polysilicon Thermal Actuators,” Proceedings SPIE, v.3224, p.284-293, 1997.
[5] J.H. Comtois, M.A. Michalicek, C.C. Barron, “Electrothermal Actuators Fabricated in Four-Level Planarized Surface Micromachined Polycrystalline Silicon,” Sensors and Actuators, v.A70, p.23-31, 1998.
[6] J. Comtois and V. Bright, “Surface Micromachined Polysilicon Thermal Actuator Arrays and Applications,” Proc. Solid State Sensor and Actuator Workshop (Hilton Head '96)1996, p.174-177, 1996.
[7] J. Comtois, V. Bright and M. Phipps, “Thermal microactuators for surface-micromachining processes,” Proc. SPIE, v.2642, p.10-21, 1995.
[8] D.L. DeVoe, “Piezoelectric thin film micromechanical beam resonators,” Sensors and Actuators, v.A88, p.263-72, 2001.
[9] T. Ebefors, E. Kalvesten, G. Stemme, Sensors and Actuators, “Dynamic actuation of polyimide V-groove joints by electrical heating,” v.A67, n.1-3, p.199-204, 1998.
[10] T. Ebefors, E. Kalvesten, G. Stemme, “Three dimensional silicon triple-hot-wire anemometer based on polyimide,” Proc. IEEE MEMS, p.93-8, 1998
[11] J. Evans, D. Liepmann, A.P. Pisano, "Planar laminar mixer - MEMS fluid processing system," Proc. IEEE MEMS, p.96-101, 1997.
[12] J. Franz, H. Baumann, H.-P. Trah, “A monolithic silicon micro-liquid handling system with bimetallic valve and calorimetric flow sensor,” Proc. Actuator ’96, p.29-32, 1996.

[13] J.S. Fitch, A.K. Henning, E.B. Arkilic, J.M. Harris, "Pressure-based mass-flow control using thermopneumatically-actuated microvalves," Proc. Solid-State Sensor and Act. Work., p.162-5, 1998.
[14] M. Freygang, H. Glosch, H. Haffner, S. Messner, "Thermal actuation units for microvalves and micropumps," Proc. Actuator '96,, p.84-7, 1996.
[15] T.B. Gabrielson, "Mechanical-thermal noise in micromachined acoustic and vibration sensors," Trans. IEEE Elec. Dev., v.40, n.5, 1993.
[16] C. Goll, W. Bacher, A. Bustgens, D. Maas, W. Menz, W.K. Schomburg, " Microvalves with bistable buckled polymer diaphragms," Proc. Int. Conf. on Solid-State Sensors and Actuators, p.59-60, 1995.
[17] R. Hiratsuka, D.C. van Duyn, T. Otaredian, P. de Vries, "A novel accelerometer based on a silicon thermopile," Proc. Int. Conf. Solid-State Sensors and Actuators (Transducers '91), p.420-3, 1991.
[18] Q.-A. Huang, N.K.S. Lee, "Analysis and Design of Polysilicon Thermal Flexure Actuator," Journal of Micromechanics and Microengineering, v.9, p.64-70, 1999.
[19] J. Jonsmann, O. Sigmund, S. Bouwstra, "Compliant Thermal Microactuators," Sensors and Actuators, v.76, p.463-469, 1999.
[20] C.G. Keller, R.T. Howe, "Hexsil tweezers for teleoperated micro-assembly," Proc. MEMS, p.72-7, 1997.
[21] M.U. Kopp, A.J. deMello, A. Manz, Science, v.280, p.1046-1048, 1998.
[22] P. Krulevitch, A.P. Lee, P.B. Ramsey, J.C. Trevino, J. Hamilton, M.A. Northrup, "Thin film shape memory alloy microactuators," J. MEMS, v.5, n.4, p.270-82, 1996.
[23] T.K. Kun, C.-J. Kim, "Vaveless pumping using traversing vapor bubbles in microchannels," J. Appl. Phys., v.83, n.11, p.5658-64, 1998.
[24] L. Lin, S.-H. Lin, "Vertically Driven Microactuators by Electrothermal Buckling Effects," Sensors and Actuators, v.A71, p.35-39, 1998.
[25] T. Lisec, S. Hoerschelmann, H.J. Quenzer, B. Wagner, W. Benecke, "Thermally driven microvalve with buckling behaviour for pneumatic applications," Proc. IEEE MEMS, p.13-17, 1994.
[26] E. Makino, Mitsuya, T. Shibata, "Fabrication of TiNi shape memory micropump,"Sensors and Actuators, v.A88, n.3, p.256-62, 2001.
[27] J.M. Maloney, D.S. Schreiber, D.L. DeVoe, "Electrothermal linear micromotors fabricated by deep reactive ion etching of single crystal silicon," accepted for publication, Proc. ASME IMECE, 2001.
[28] J. Maloney, D.L. DeVoe, "Electrothermal microactuators fabricated from deep reactive ion etching of single crystal silicon," Proc. ASME IMECE, p. 233-240, Orlando FL, November 2000.
[29] F. Mayer, A. Haberli, G. Ofner, H.Jacobs, O. Paul, H. Baltes, "Single-chip CMOS anemometer," Proc. IEDM, p.895-8, 1997.
[30] F. Mayer, M. Hintermann, H. Jacobs, O. Paul, H. Baltes, "Thermoelectric CMOS anemometers," Proc. SPIE, v.2882, p.236-46, 1996.
[31] V. Milanovic, E. Bowen, N. Tea, J. Suehle, B. Payne, M. Zaghloul, M. Gaitan, "Convection-based accelerometer and tilt sensor implemented in standard CMOS," Proc. IEEE MEMS, p.487-90, 1998.
[32] T. Moulton, G.K. Ananthasuresh, "Micromechanical devices with embedded electro-thermal-compliant actuation," Sensors and Actuators, v.A90, p. 38-48, 2000.
[33] U. Munch, D. Jaeggi, K. Schneeberger, A. Schaufelbuhl, O. Paul, H. Baltes, J. Jasper, "Industrial fabrication technology for CMOS infrared sensor arrays," Proc. Transducers '97, v.1, p.205-8, 1997.
[34] T. Neda, K. Nakamura, T. Takumi, "A polysilicon flow sensor for gas flowmeters," Proc. International Solid-State Sensors and Actuators, p.548-5, 1995
[35] M.A. Northrup, M.T. Ching, R.M. White, R.T. Watson, "DNA amplification with a microfabricated reaction chamber," Proc. IEEE Int. Conf. on Solid-State Sensors and Actuators, p.924-6, 1993.
[36] P. O'Neill, "A monolithic thermal converter," Hewlett-Packard Journal, p.12, May 1980.
[37] M. Pai, N.C. Tien, "Current-Controlled Bi-Directional Electrothermally Actuated Vibromotor," Proc. IEEE Int. Conf. on Solid-State Sensors and Actuators, 1999.
[38] B. Panchapakesan, D.L. DeVoe, R. Cavicchi, M. Widmaeir, S. Semancik, "Nanoparticle engineering and control of tin oxide microstructures for chemical microsensor applications," Nanotechnology, v.12, p.336-49, 2001.
[39] B. Panchapakesan, D.L. DeVoe, R. Cavicchi, S. Semancik, "Stability and materials programmed selectivity of nanoparticle seeded SnO_2 microsensors," in press, Sensors and Actuators B, 2001.

[40] C.S. Pan, W. Hsu, "An Electro-Thermally and Laterally Driven Polysilicon Microactuator," J. Micromechanics and Microengineering, v.7, p.7-13, 1997.
[41] Parameswaran, M.; Robinson, A.M.; Blackburn, D.L.; Gaitan, M.; Geist, J., "Micromachined thermal radiation emitter from a commercial CMOS process," IEEE Elec. Dev. Let., vol.12, no.2 p. 57-9, 1991.
[42] Parameswaran, M.; Chung, R.; Gaitan, M.; Johnson, R.B.; Syrzycki, M., "Commercial CMOS fabricated integrated dynamic thermal scene simulator," Proc. IEDM, p. 753-6, 1991.
[43] J.-S. Park, L.L. Chu, A.D. Oliver, Y.B. Gianchandani, "Bent-beam electrothermal actuators - part II: linear and rotary microengines," J. MEMS, v.10, n.2, 2001.
[44] E. Peeters, S. Verdonckt-Vandebroek, " Thermal ink jet technology," IEEE Circuits and Devices Magazine, v.13, n.4, p.19-23, 1997.
[45] D.S. Popescu, D.C. Dascalu, M. Elwenspoek, T. Lammerink, " Silicon active microvalves using buckled membranes for actuation," Proc. Int. Conf. on Solid-State Sensors and Actuators, p.305-8, 1995.
[46] L. Que, J.-S. Park, Y.B. Gianchandani, "Bent-beam electrothermal actuators - part I: single beam and cascaded devices," J. MEMS, v.10, n.2, 2001.
[47] J.R. Reid, V.M. Bright and J. Comtois, "Force measurements of polysilicon thermal micro-actuators," Proc. SPIE, v.2882, p.296-306, 1996.
[48] T. Remtema, L. Lin, "Active frequency tuning for microresonators by localized thermal stressing effects," Proc. Solid-State Sensor and Actuator Workshop, p.363-6, 2000.
[49] C.A. Rich, K.D. Wise, "A thermopneumatically-actuated microvalve with improved thermal efficiency and integrated state sensing," Proc. Solid-State Sensor and Actuator Workshop, p.234-7, 2000.
[50] W. Riethmüller, W. Benecke, "Thermally Excited Silicon Microactuators," IEEE Transactions on Electron Devices, v.35, p.758-762, 1988.
[51] M.L. Roukes, "Yoctocalorimetry: phonon counting in nanostructures," Physica B, v.263-264 p.1-15, 1999.
[52] D.S. Schreiber, W.-J. Cheng, J.M. Maloney, D.L. DeVoe, "Surface micromachined electrothermal microstepper motors," accepted for publication, Proc. ASME IMECE, 2001.
[53] K. Schwab, W. Fon, E. Henriksen, J.M. Worlock, M.L. Roukes, "Quantized thermal conductance: measurements in nanostructures," Physica B, v.280, p.458-9, 2000.
[54] S. Semancik, R.E. Cavicchi, M.C. Wheeler, J.E. Tiffany, G.E. Poirier, R.M. Walton, J.S. Suehle, B. Panchapakesan, D.L. DeVoe, "Microhotplate platforms for chemical sensor research," Sensors and Actuators, v.B77, p.579-91, 2001.
[55] J.S. Suehle, R.E. Cavicchi, M. Gaitan, S. Semancik, "Tin oxide gas sensor fabricated using CMOS micro-hotplates and in-situ processing," IEEE Electron Device Letters, v.14, n.3 p.118-20, 1993.
[56] J.-H. Tsai, L. Lin, "Thermal bubble powered microfluidic mixer with gas bubble filter," Proc. Int. Conf. Solid-State Sensors and Actuators (Transducers '01), 2001.
[57] J.-H. Tsai, L. Lin, "A thermal bubble actuated micro nozzle-diffuser pump," Proc. IEEE MEMS, p.409-12, 2001.
[58] F.-G. Tseng, C.-J. Kim, C.-M. Ho, "A novel microinjector with virtual chamber neck," Proc. IEEE MEMS, p.57-62, 1998.
[59] F.-G. Tseng, C. Linder, C.-J. Kim, C.-M. Ho, Proc. ASME IMECE, v.59, p.183-7, 1996.
[60] J.R. Vig, K. Yoonkee, "Noise in microelectromechanical system resonators," IEEE Trans. Ultrasonics, Ferroelectrics and Frequency Control, v.46, n.6, p.1558-65, 1999.
[61] Vig, J.R.; Filler, R.L.; Kim, Y., "Chemical sensor based on quartz microresonators," J. MEMS, v.5, n.2, p.138-140, 1996.
[62] T.K. Wang, P. Barth, R. Alley, J. Baker, D. Yates, L. Field, G. Gordon, C. Beatty, L. Tully, "A production-ready silicon microvalve," Proc. SPIE, v.3876, p.227-37, 1999.
[63] L.C. Waters, S.C. Jacobson, N. Kroutchinina, J. Khandurina, R.S. Foote, J.M. Ramsey, Anal. Chem. v.70, p.158-162, 1998.
[64] S. Wu, Q. Lin, Y. Yuen, Y.-C. Tai, "MEMS flow sensors for nano-fluidic applications," Sensors and Actuators, vol.A89, no.1-2, p. 152-8, 2001.
[65] H. Yuan, A. Prosperetti, "The pumping effect of growing and collapsing bubbles in a tube," J. Micromechanics and Microengineering, v.9, n.4, p.402-13, 1999.

[66] Y. Yue, Z. Zhaoying, Y. Xiongying, J. Xiaoning, "A bimetallic thermally actuated micropump,"Proc. ASME IMECE, p.351-4, 1996.
[67] J.D. Zahn, A.A. Deshmukh, A.P. Pisano, D. Liepmann, "Continuous on-chip micropumping through a microneedle," Proc. IEEE MEMS, p.503-6, 2001.
[68] J. Zou, X.Y. Ye, Z.Y. Zhou, Y. Yang,"A novel thermally-actuated silicon micropump," Proc. Int. Symp. on Micromechanics, p.231-4, 1997.

ACKNOWLEDGEMENTS

The author wishes to thank Dr. Steve Semancik and Dr. Richard Cavicchi at the National Institute of Standards and Technology and Dr. Balaji Panchepakesan at the University of Deleware for their work in microhotplate chemical sensor development, Mr. John Maloney and Mr. David Schreiber with the University of Maryland for thermal micromotor development, and Mr. Chris Kimball and Ms. Zhou Yi with the University of Maryland for developments in thermal control for polymer microfluidic systems.

Thermal Challenges in Next Generation Electronic Systems, Joshi & Garimella (eds)
© 2002 Millpress, Rotterdam, ISBN 90-77017-03-8

Thermal challenges in MEMS applications: Phase change phenomena and thermal bonding processes

Liwei Lin
Department of Mechanical Engineering, University of California at Berkeley, Berkeley, CA, USA

Keywords: bubble, phase change, thermal bonding, MEMS

ABSTRACT: Two thermal challenges for current and next generation MEMS (Microelectromechanical systems) applications are discussed. The first topic is in the area of fundamental investigations of phase change phenomena in the microscale. Single, spherical and controllable thermal bubbles with diameter between 2 to 500μm have been generated by using micro resistive heaters for the first time. Both steady state and transient analyses that provides the scientific foundation of bubble nucleation and characterizations in the microscale are established. Specific device demonstrations are briefed including bubble-powered actuators, bubble-powered nozzle-diffuser micro pumps and bubble-powered micro mixers for applications in microfluidic systems. The second topic addresses key heat transfer issues during thermal bonding processes for MEMS fabrication and packaging applications. Basic thermal analyses on the microscale bonding processes are developed. Successful new thermal bonding processes are introduced, including localized eutectic bonding, localized fusion bonding, localized CVD (chemical vapor deposition) bonding, localized solder bonding and nanosecond laser bonding for encapsulating MEMS devices.

1 INTRODUCTION

Challenging thermal problems emerge as the size of micro mechanical devices is shrinking in MEMS applications. In the past twenty years, the application of microelectronic technology to the fabrication of mechanical devices stimulated many research in semi-conductor microsensors and microactuators. Micromachining technologies take advantage of batch processing to address the manufacturing and performance requirements of the sensor industry. The versatility of semiconductor materials and the miniaturization of VLSI patterning techniques promise new sensors with better capabilities and improved performance-to-cost ratio over those of conventionally machined devices. However, the shrinking of device size results in increasing surface-to-volume ratio. As a result, new devices and new manufacturing processes that were viewed as impossible previously in the macro world become feasible in the microscale.

In the area of heat transfer related topics, for example, two categories of thermal phenomena/processes are discussed in this paper. The first one deals with phase change from liquid to vapor in the microscale and its applications to microdevices. Although thermal bubbles can be generated during the boiling process that is routinely observed in our daily life, the generation and control of a single thermal bubble can only be accomplished in the microscale and this new phenomenon opens up many possible applications. Due to the high surface-to-volume ratio in the microscale, heating sources can be applied using tiny volumetric heat generation means and the generated heat flux can be balanced by the surface condensation of the vapor bubble. When the overall heat transfer process consisting of heat generation, condensation and heat losses to the environment reaches equilibrium, thermal bubbles becomes controllable [1]. In an effort to understand bubble formation processes in the microscale, the transient behavior of the phase change phenomena on microheaters is investigated [2]. These fundamental studies help the successful development of new micro bubble devices, including bubble-powered mechanical actuators [3], bubble-powered nozzle-diffuser pumps [4] and bubble-generated mixing effects [5].

The second microscale thermal challenge is in the thermal bonding process by using selective and localized heating. Again, the increase of volumetric heat generation rates and surface heat fluxes make several new thermal bonding process possible. In the microscale bonding process, the primary goal is to establish strong bonding with minimum thermal budget to reduce possible thermal damages to existing materials that were already placed on the wafer. This requirement happens often in the packaging of MEMS devices that a good hermetic sealing is required but existing MEMS materials or microelectronics can't be damaged during the bonding process. Localized heating provides the key advantage by utilizing thermal management schemes in the microscale to minimize the thermal budget and thermal damages since the energy is provided locally. At the same time, the bonding strength is maximized because the bonding temperature can be adjusted to be very high locally. Several successful bonding schemes are developed and will be discussed, including localized eutectic bonding [6], localized fusion bonding [7], localized solder bonding [8], localized CVD (Chemical Vapor Deposition) bonding [9], and nanosecond laser bonding [10].

2 PHASE CHANGE PHENOMENA IN THE MICROSCALE

In the phase change applications in the microscale, resistive heaters are commonly used as the heat generation sources in working liquids. In this paper, these resistive heaters are made of either heavily phosphorus doped polysilicon or metal layers with a thickness between hundreds of nanometer to 2μm. The width of these micro heaters varies from 1μm to 10μm and the length ranges from 50μm to thousands of μm. The following discussions start with the fundamental investigations of phase changes processes and continue with examples of successful device demonstrations.

2.1 *Fundamental Studies*

2.1.1 *Phase Change and Bubble Formation*

Micro heaters are designed and fabricated by using a standard micromachining process [11]. A one-dimensional micro heat transfer model is established based on the conservation of energy to estimate the temperature distribution on the micro heater [12]. For example, a 15-μm long microheater made of polysilicon has two widths of 3μm on the first 10μm portion and 2μm on the second 5μm part as shown in Fig. 1. The temperature prediction by using the established thermal model is plotted in Fig. 2. It can be observed that the temperature distribution is uneven and nonlinear. The highest temperature occurs at the 11.6μm spot as shown and it has great implication in the microscale phase change experiment. Due to the very intense local temperature distribution, strong Marangoni effect is expected. Shown in Fig. 1 is the experimental result by placing the fabricated micro heater in a pool of electrical-inert liquid of Fluorinert at room temperature. A thermal bubble of about 2μm in diameter is generated in this photo and it is controllable in size by adjusting the input power. Moreover, it is found that the bubble tends to stick to the hottest spot on the heater

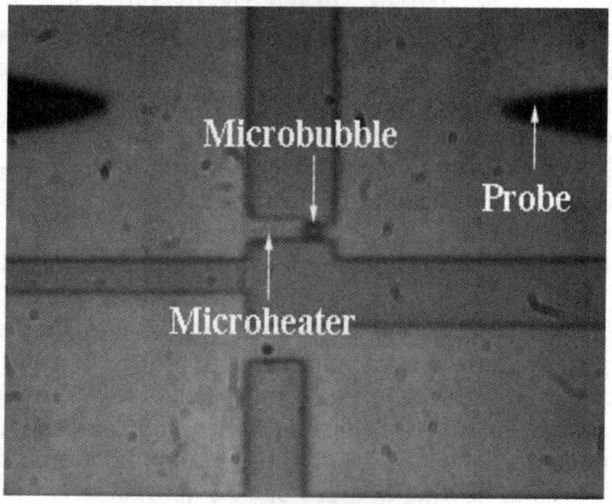

Fig. 1, A single thermal bubble of 2μm in diameter is generated on a micro resistive heater immersed in Fluorinert liquid [1].

due to the Marangoni effect. Comparing the experimental result with the simulation result, it clearly shows the consistency for the location of the micro bubble. Experimentally, individual, spherical vapor bubbles with diameters of 2 to 500μm have been generated with stable and controllable characteristics. Although initial thermal analyses were developed, including Hsu's analysis and conservation of energy to provide possible micro bubble formation mechanisms [1], further studies are required on phase change phenomena that may address the issues of in-situ temperature measurements down to 1μm, models of bubble incipience mechanisms by using microheaters under sub-cooled boiling conditions, transport process of vaporization and condensation and three-dimensional thermal analyses on micro bubble formation processes [13].

2.1.2 *Transient Phase Change Behavior*

The advanced study of bubble formation processes starts with the characterizations of transient phase change. It is observed that transient temperature responses can be classified into three groups by the level of input currents [2]. When the input current level is low, the wall temperature stays nearly constant and no bubble is generated. When the input current is set to be above a critical level, the wall temperature increases initially, quickly drops afterwards until a bubble is nucleated, and rises afterwards until stead state is reached. In the third group that has a high input current level, the wall temperature rises sharply as soon as the current is applied and a bubble is formed immediately. Overall, a major difference between micro and macro scale boiling experiments is observed in the second group of the input current level when the wall temperature drops before the nucleation of a micro bubble in the microscale experiments. Theoretical models are established to estimate the temperature of microheater during the transient phase change process. The time constant of the resistor is calculated in the microsecond range and that for the silicon substrate is calculated in the millisecond range. Although this transient phase change analysis only covers several very basic parameters, it does provide interesting and important foundations for future studies. Many challenging problems on transient behavior of phase change phenomena in the microscale, such as the evaporation and condensation processes, temperature fluctuations with respect to the input power and time require further investigations.

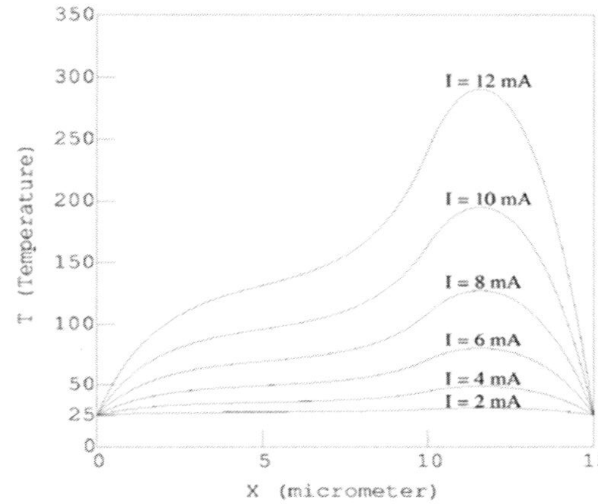

Fig. 2, Simulation results showing the isotherms of the resistive heater under various levels of input current. The width of the heater is 3μm from 0 to 10μm and 2μm from 10 to 15μm in the x-coordinate and the highest temperature occurs at the 11.6μm site [1].

2.2 *Applications*

The demonstration of phase change phenomena by using micro resistive heaters led to the successful implementation of several new micro fluidic devices based on thermal bubbles. These applications are different than the well-known bubble jet printers [14] but could be potentially made as commercial products as demonstrated by the success of bubble jet printers. Therefore, both fundamental phase change phenomena and application principles of these bubble-powered devices should be carefully studied.

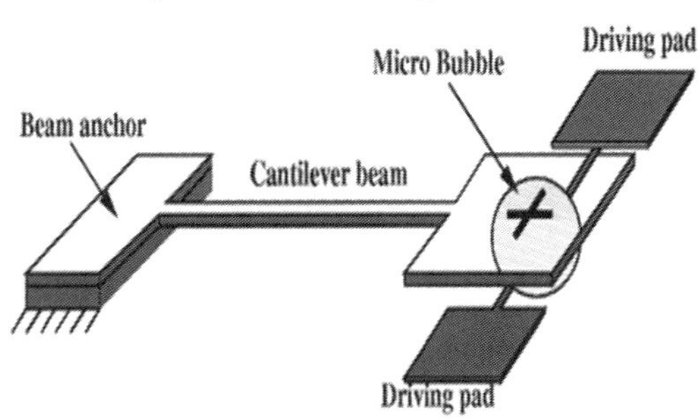

Fig. 3, Schematic illustration of a microbubble powered actuator [3].

2.2.1 *Micro Bubble-Powered Mechanical Actuators*

Micromechanical actuators based on the actuation of thermal bubbles have been demonstrated with the schematic diagram shown in Fig. 3. Surface micromachining is used to fabricate both the polysilicon resistors as microheaters and cantilever-type actuators as shown [11]. These actuators operate when they are entirely immersed in liquid. The heavily phosphorus doped polysilicon resistor works as the resistive microheater that can generate controllable, thermal bubbles by adjusting the input currents. Micro bubbles are generated directly underneath the actuator plate with either AC or DC inputs to control the height of the actuators. A maximum vertical deflection of 140μm and a maximum force of 2μN are achieved [3]. This new class of actuator is particular important in microfluidic systems that face difficult constraints imposed by fluid viscosity. However, challenging thermal issues such as the local temperature profile when the thermal bubble is in contact with mechanical constraints, the mechanical pressure-force relationship in conjunction with the thermal bubble geometry and the transient behavior of the actuation processes require further investigations.

2.2.2 *Micro Bubble-Powered Nozzle-Diffuser Pumps*

Micro pumps powered by the actuation of micro bubbles with the design of a pair of nozzle-diffuser flow regulator have been successfully demonstrated. The valve-less pump consists of a meander resistive heater, a pair of nozzle-diffuser flow controller, and a 1mm in diameter, 50μm in depth pumping chamber as shown in Fig, 4. Micro thermal bubbles are generated by pulsive electrical signals at a frequency up to 500 Hz. Liquid is actuated by this periodically expanding and collapsing thermal bubbles. The measured pumping flow rate can go up to 5μl/min and the highest pumping pressure is 377Pa with a power consumption of about 0.5 Watt as shown in Fig. 5 [4]. It is observed that the pumping rate increases as the driving frequency increases [15]. When residual bubbles begin to pile up, the pumping pressure reduces as shown. Many further investigations have to be conducted in order to fully understand the operation of this bubble pump, such as the heat transfer process during the periodically nucleating and collapsing thermal bubbles, the flow patterns and pressure head induced by the thermal bubbles as well as the effects of the nozzle-diffuser design.

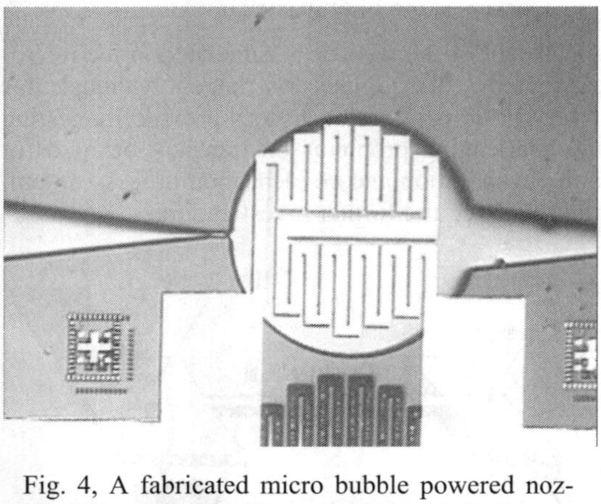

Fig. 4, A fabricated micro bubble powered nozzle-diffuser pump. The metal heater is built on a glass wafer that is transparent for the viewing of the pumping operations [4].

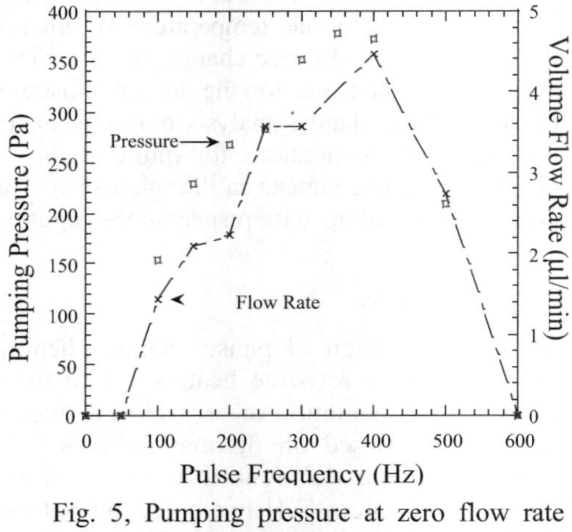

Fig. 5, Pumping pressure at zero flow rate and volume flow rate at zero back pressure of the micro bubble powered nozzle-diffuser pump under a 5% pulse duty and 0.5W average input power [5].

2.2.3 *Micro Bubble-Powered Mixers*

A microfluidic mixer powered by the developed bubble pump is demonstrated by using the generated oscillatory flow to induce wavy interfaces of flow in microchannels to increase the contact area of the mixing liquids and accelerate the mixing process. It if found that the mixing effect can be optimized when the pumping frequency reaches 200Hz in a 200μm wide, 50μm deep microchannel with a pumping volume flow rate of 6.5μl/min [5]. When the length of the mixing wave is shorter than the diffusion length of the mixing fluids, the mixing effect is optimized and this can be accomplished by adjusting the pulse frequency until the optimal mixing effect is achieved. Experimental results on normalized gray-scale measurements show that the gray-scale values that correspond to the mixing effect increase proportionally to the one-third power of the pumping pulse frequency. However, a good analytical model is lacking in the prediction and verification of the experimental data. This would be a challenging problem of thermal combined domain in the fields of thermal, fluidic and mechanics areas.

3 THERMAL BONDING PROCESSES IN THE MICROSCALE

3.1 *Fundamentals*

The development of thermal bonding processes is motivated by the fabrication and packaging needs for MEMS devices. In conventional IC industry, up to 95% of the manufacturing cost may be attributed to the packaging processes and the cost of packaging may go even higher for MEMS devices due to the stringent packaging requirements for microstructures. In the past, anodic bonding [16], fusion bonding [17], reactive gas sealing [18] and LPCVD sealing [19] processes have been used to encapsulate MEMS devices. However, none of them are suitable for the general purpose of MEMS packaging processes at low temperature and at low cost. Therefore, a new wafer-level approach that is insensitive to the MEMS fabrication processes is proposed based on the principle of localized heating and bonding as shown in Fig. 6 [20]. The basic idea is to increase the thermal bonding temperature locally but to keep the temperature low at the wafer level. As a result, high energy is supplied for strong bonding locally and low temperature environment is maintained at the wafer-level. Several new bonding processes are proposed and demonstrated based on this principle of localized heating and bonding.

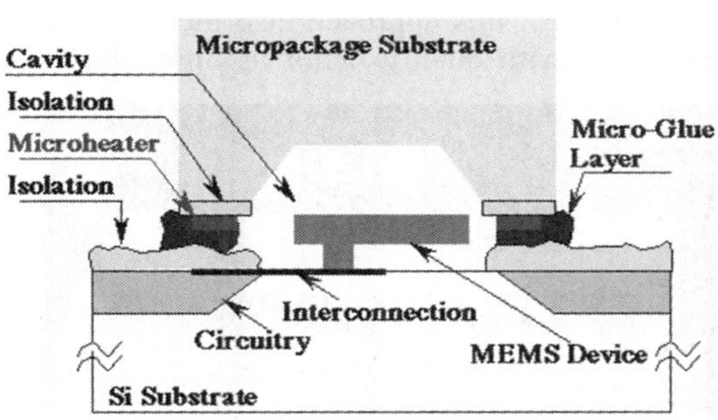

Fig. 6, The schematic diagram showing the principle of localized heating and bonding processes for MEMS packaging and fabrication applications. The bonding energy is locally generated such that the global temperature at the wafer-level can be kept low during the bonding process [20].

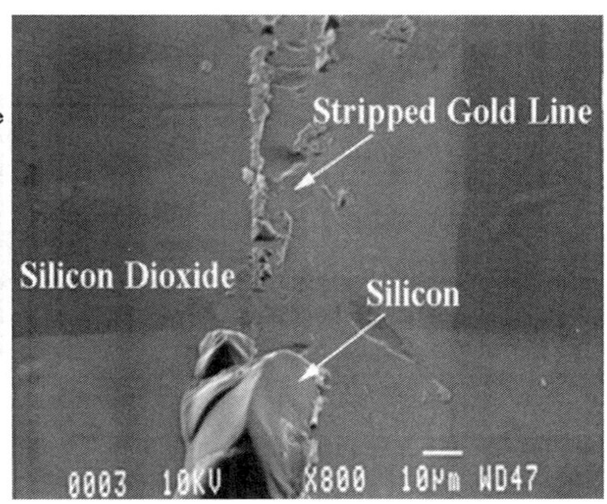

Fig. 7, A SEM microphoto showing the gold microheater after forcefully breaking the silicon-gold eutectic bond that was formed by using the localized heating method. The strength of the bond is comparable with the fracture toughness of the bulk silicon such that the fracture and separation occur either in the bulk silicon or at the gold-silicon dioxide interface [6].

3.2 *Applications*

3.2.1 *Localized Eutectic, Fusion and Solder Bonding*

The localized heating and bonding principle is first applied to conventional thermal bonding processes, including eutectic, fusion and solder bonding. In the localized eutectic bonding process, the silicon-gold system has been adopted for the demonstration because it can provide high bonding strength and good stability at a relatively low processing temperature (about 363°C). Experimentally, heating is achieved by applying a dc current through micromachined heater made of gold that serves as both the heating and bonding material. At the interface of silicon and gold, the formation of eutectic bond takes place in about 5 minutes [6]. After the bond is forcefully broken as shown in Fig. 7, the fracture and separation occur either in the bulk silicon or at the gold-silicon dioxide interface. Therefore, it can be concluded that the bonding strength is as strong as the silicon fracture toughness. The same scheme has been applied to the localized silicon-glass fusion bonding process with successful results. Moreover, the polysilicon-to-glass bonding activation energy is measured as 1.6eV and this number is close to those from silicon direct bonding [7]. The third localized heating and bonding scheme is to use metal solder in the bonding process. For example, a dew point sensor is designed and to be packaged inside the bonding cavity as shown in Fig. 8. The bonding solder is indium and the bonding system is indium-glass. A glass cap wafer is placed on top and sealed by localized heating and bonding and the dew point sensor is used to monitor the leakage of the package [8]. Several other solder-based bonding systems have been used, such as aluminum-glass and aluminum/polysilicon-glass system with improved results [21]. The advancement of this approach also includes a vacuum encapsulated micromechanical polysilicon resonators with quality factor as high as 10,000 [22]. The general thermal challenge for localized heating and bonding processes is from the existence of a strong local temperature gradient. As a result, several thermal related issues have to be reconsidered, for example, the residual thermal stress generated, the thermal material diffusion distance, the localized melting and solidification processes when the bonding solders are used.

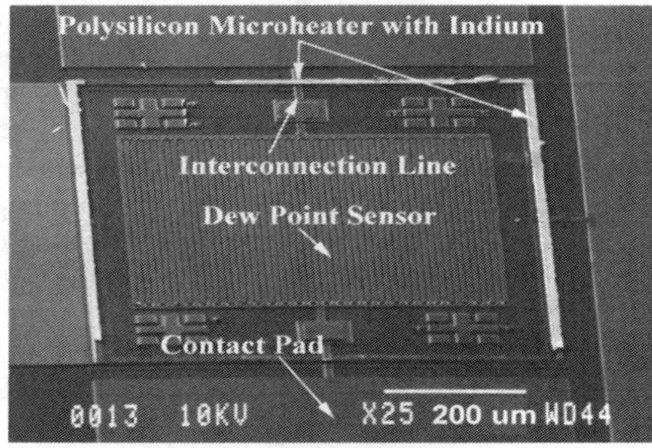

Fig. 8, A SEM micrograph of silicon substrate with a dew point sensor after breaking the indium-glass bond. The sensor is designed to monitor the hermeticity of the package [8].

3.2.2 *Localized CVD Bonding*

The localized CVD process is conducted by selectively growing polysilicon on a microheater in a low vacuum, silane environment. Experimental results show that silane will decompose at an elevated temperature of more than 800°C to form polysilicon locally on the microheater in contrast to the global CVD process that will have polysilicon grown all over the place. The top figure in Fig. 9 is the

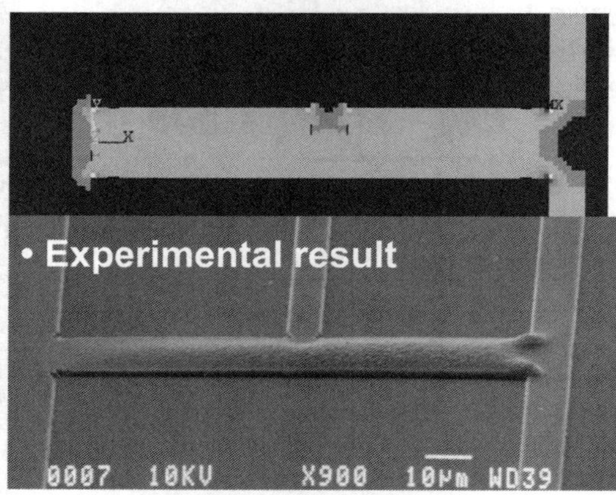

Fig. 9, (Top) Temperature simulation of a microheater under a current input. Hot spots can be identified. (Bottom) CVD polysilicon is deposited following the temperature distribution and surface-reaction limited process is identified.

electro-thermal simulation result (top) of the temperature distribution when an input current is applied to the experimental sample. The bottom photograph is the deposition profile. It is observed that the thickness of the deposited film seems to follow the temperature profile such that surface-reaction limited process is identified as the deposition process. This localized CVD process is applied to MEMS packaging to selectively encapsulate MEMS devices. It is found that an area of 1000 x 800 μm^2 can be enclosed in one hour of localized CVD process by filling up a gap of about 1 μm between the microheater and a top capping substrate [9]. This process is believed to be applicable for MEMS packaging and fabrication with the advantage of excellent bonding strength and low temperature processing at the wafer-level. Many thermal challenges have to be addressed before the process can be optimized. These will include thermal dynamics analysis on the gas flow passing through thin gap before and during the sealing process, the chemical vapor reaction during the deposition process under a strong local temperature gradient and other issues that may involve thermal analyses, gas flow as well as decomposition, recombination and solidification of materials in the microscale.

3.2.3 *Localized Nanosecond Laser Bonding*

In another effort to achieve localized heating and bonding, a nanosecond-pulsed laser bonding process with a built-in mask for MEMS packaging application has been successfully demonstrated. A YAG Surelit II laser with pulse duration of 4-6 nano-second, a wavelength of 355nm and a focal diameter of 1mm is used to provide the bonding energy for glass-to-silicon bonding with a 4μm-think indium as the intermediate bonding layer. The optimal bonding parameter is found experimentally when the laser energy is between 8 to 22 mJ and multiple laser shots are used [10]. With the assistance of a regular paper as the masking material with predefined patterns such as a ring as shown in Fig. 10, localized and selective heating and bonding can be achieved as shown in the bottom microphoto in Fig. 10. In addition to the thermal challenges suggested in the above sections, ultra fast laser bonding presents interesting issues in the time domain analyses and experiments to control and verification of the experimental results. Furthermore, the thermal problem of laser absorption at the mask, the proper masking material selection as well as the heat transfer processes dealing with heat conduction and phase change in such a small time scale and space scale may present very interesting challenges.

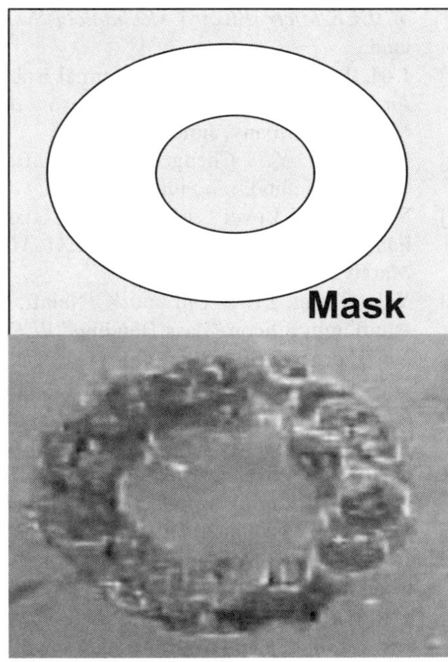

Fig. 10, Nano-second laser bonding with a built-in mask with (top) the mask and (bottom) the bonding result of indium to glass bond [10].

4 CONCLUSIONS

Thermal challenges in the area of phase change phenomena and thermal bonding processes in the microscale have been discussed. Although many interesting thermal challenges have been raised and some initial works have been conducted in issues related to these areas, it should be noted that these two subjects are merely two of the many scientific challenges in the microscale. There are plenty of rooms and fruitful of research directions in microscale thermal sciences that should be vigorously pursued. Another important observation as demonstrated by the above examples is that microscale thermal challenges will often combine with issues from other domains such as fluidic problems that have to be dealt with simultaneously and can't be overlooked.

ACKNOWLEDGEMENT

The author would like to thank Drs. J.H. Tsai, Y.T. Cheng, and C. Luo for contributions in several of the experiments presented in this paper. This work is supported in part by an NSF CAREER award and a DARPA/MTO/MEMS grant.

REFERENCES

[1] Liwei Lin, A.P. Pisano and V.P. Carey, "Thermal Bubble Formations on Polysilicon Micro Resistors," *ASME Journal of Heat Transfer*, Vol. 120, pp.735-742, Sep. 1998.

[2] Jr-Hung Tsai and Liwei Lin, "Transient Thermal Bubble Formation on Polysilicon Micro Resistors," *ASME Journal of Heat Transfer*, to be published.

[3] Liwei Lin and A.P. Pisano, "Thermal Bubble Powered Microactuators," *Microsystem Technologies Journal*, Vol. 1, No. 1, pp.51-58, 1994.

[4] Jr-Hung Tsai and Liwei Lin, "A Thermal Bubble Actuated Micro Nozzle-Diffuser Pump," *Proceedings of IEEE Micro Electro Mechanical Systems Conference*, pp. 409-412, Jan. 2001, Interlaken, Switzerland.

[5] J.-H. Tsai and Liwei Lin, "Thermal Bubble Powered Microfluidic Mixer with Gas Bubble Filter," *11th Int. Conference on Solid State Sensors and Actuators, Transducer's 01* , Technical Digest, pp. 966-969, Munich, Germany, June 2001.

[6] Liwei Lin, Y.T. Cheng and K. Najafi, "Formation of Silicon-Gold Eutectic Bond Using Localized Heating Method," *Japanese Journal of Applied Physics, Part II*, Vol. 11B, pp. 1412-1414, Nov. 1998.

[7] Y.T. Cheng, Liwei Lin and Khalil Najafi, "Localized Silicon Fusion and Eutectic Bonding for MEMS Fabrication and Packaging," *IEEE/ASME Journal of Microelectromechanical Systems*, Vol. 9, pp. 3-8, March 2000.

[8] Y.T. Cheng, Liwei Lin and K. Najafi, "A Hermetic Galss-Silicon Package Formed Using Localized Aluminum/Silicon-Glass Bonding," *IEEE/ASME Journal of Microelectromechanical Systems*, Vol. 10, No. 3, pp. 392-399, 2001.

[9] G.H. He, Liwei Lin and Y.T. Cheng, "Localized CVD Bonding for MEMS Packaging," *10th Int. Conference on Solid State Sensors and Actuators, Transducer's 99*, Technical Digest, Sendai, Japan, June 1999, pp. 1312-1315.

[10] C. Lu and Liwei Lin and M. Chiao, "Nanosecond-Pulsed Laser Bonding with a Built-in Mask for MEMS Packaging Applications," *11th Int. Conference on Solid State Sensors and Actuators, Transducer's 01* , Technical Digest, pp. 214-217, Munich, Germany, June 2001.

[11] D. Koester, R. Majedevan, A. Shishkoff and K. Marcus, "Multi-User MEMS Processes (MUMPS) Introduction and Design Rules, rev. 4, MCNC MEMS Technology Applications Center, Research Triangle Park, NC 27709, July, 1996.

[12] Liwei Lin, "Selective Encapsulations of MEMS: Micro Channels, Needles, Resonators and Microelectromechanical Filters", Ph.D. Dissertation, Mechanical Engineering Department, University of California, Berkeley, 1993.

[13] Liwei Lin, "Microscale Thermal Bubble Formation: Thermophysical Phenomena and Applications," *Microscale Thermophysical Engineering*, Vol. 2, No. 2, April-June, pp.71-85, 1998.

[14] N.J. Nielsen, "History of ThinkJet Printerhead Development," *HP Journal*, Vol. 36, No. 5, pp. 4-10, May, 1985.

[15] A. Olsson, G. Stemmme and E. Stemme, "A Valveless Planar Fluid Pump with Two Pump Chambers," Sensors and Actuators, Vol. 46-47A, pp. 549-556, 1995.

[16] G. Wallis and D. Pomerantz, "Filed Assisted Glass-Metal Sealing", *J. of Applied Physics*, Vol. 40, pp. 3946-3949, 1969.

[17] K.E. Peterson, P. Barth, et al., "Silicon Fusion Bonding for Pressure Sensors", *1988 Solid-State Sensor and Actuator Workshop*, Hilton Head, pp. 144-147, 1988.

[18] J.J. Sniegowski, H. Guckel and R.T. Christenson, "Performance Characteristics of Second Generation polysilicon Resonating Beam Force Transducers", *IEEE Solid-State Sensor and Actuator Workshop*, Hilton Head Island, 9-12, 1990.

[19] Liwei Lin, K. McNair, R.T. Howe and A.P. Pisano, "Vacuum Encapsulated Lateral Microresoantors", *7th Int. Conference on Solid State Sensors and Actuators*, pp. 270-273, Yokohama, Japan, June, 1993.

[20] Liwei Lin, "MEMS Post-Packaging by Localized Heating and Bonding," *IEEE Trans. on Advanced Packaging*, Vol 23, pp. 608-616, Nov. 2000.

[21] Y.T. Cheng, Liwei Lin and K. Najafi, "A Hermetic Galss-Silicon Package Formed Using Localized Aluminum/Silicon-Glass Bonding," *IEEE/ASME Journal of Microelectromechanical Systems*, Vol. 10, No. 3, pp. 392-399, 2001.

[22] Y.T. Cheng, W.T. Hsu, Liwei Lin, C.T. Nguyen and K. Najafi, "Vacuum Packaging Technology Using Localized Aluminum/Silicon-to Glass Bonding," *Proceedings of IEEE Micro Electro Mechanical Systems Conference*, pp. 18-21, Jan. 20 01, Interlaken, Switzerland.

Thermal Challenges in Next Generation Electronic Systems, Joshi & Garimella (eds)
© 2002 Millpress, Rotterdam, ISBN 90-77017-03-8

Computational techniques for simulation of sub-micron thermal conduction

Jayathi Y. Murthy
Department of Mechanical Engineering, Purdue University, West Lafayette IN, USA

Sanjay R. Mathur
Fluent Inc., Lebanon, NH, USA

Keywords: sub-micron, heat conduction, Boltzmann transport equation, numerical, finite volume, unstructured, coupled ordinates

ABSTRACT: Sub-micron heat conduction in dielectrics can be analyzed using the phonon Boltzmann Transport Equation (BTE) in the relaxation time approximation. The paper describes the development of unstructured finite volume numerical techniques to address the BTE. Spatial, angular, spectral and temporal discretization procedures are described. The performance of sequential solution procedures for the BTE is compared to that of a point-coupled algorithm called the coupled ordinate method (COMET). The sequential algorithm is shown to perform well for low acoustic thicknesses, but is quickly exceeded in performance by COMET as the acoustic thickness is increased. The proposed numerical schemes are applied to a variety of sub-micron conduction problems, both unsteady and steady. Favorable comparison is found with the published literature and with exact solutions.

1 INTRODUCTION

Heat transfer at extremely small length and time scales has come to play an increasingly important role in the design of microelectronics, superlattices and thin film structures, in nano-materials processing, and in short-pulse laser processing (Sverdrup *et al.* , 2001; Chen, 1997; Chen, 1996; Qui & Tien, 1993). In silicon-on-insulator (SOI) devices, for example, a thin silicon overlayer is located on top of a silicon dioxide layer of low conductivity. The high conductivity of the overlayer tends to conduct away the heat generated by the device so that temperatures are kept within acceptable limits. As devices become smaller and overlayer thicknesses become comparable to the phonon mean free path ($\approx$ 300 nm for silicon at room temperature) boundary phonon scattering is expected to cause substantial decreases in effective overlayer thermal conductivity, greatly exacerbating thermal problems. Sub-micron thermal effects also play a substantial role in electrostatic discharge (ESD) related failures in microelectronics. In thin film structures, effective thermal properties are a function of partially diffuse phonon reflection and transmission at the interfaces between layers, and understanding this behavior is critical to development of useful thermoelectric materials. In sub-picosecond laser processing, the speed of motion of the melting interface is critically dependant on sub-micron heat transfer. Understanding, predicting, and controlling sub-micron heat conduction is critical to a host of emerging technologies.

Heat transfer in dielectrics and semi-conductors is governed by phonon transport. When the domain length scale L becomes comparable to the phonon mean free path Λ, the Fourier heat conduction equation is no longer valid. When $L/\Lambda \approx O(1)$ but L is much greater than the phonon wavelength λ and phase effects are not important, heat conduction may be modeled using the phonon Boltzmann transport equation (BTE). A common and convenient practice is to use the relaxation time approximation. Over the last few years, there have been several attempts to numerically solve the phonon BTE. The similarity between the phonon radiative transfer equation (RTE) and the BTE in the relaxation time approximation has been noted for some time now. Majumdar (1993) derived a equation for phonon radiative transfer (EPRT) which established the correspondence between the photon radiative transfer equation (RTE) and the BTE. The primary difference is the existence of an unsteady term which lends the BTE a wave-like character and allows thermal phenomena to evolve with a finite wave speed; wave speed effects are normally not considered important in conventional thermal radiation problems.

Exploiting the similarity between the RTE and the BTE, Kumar et al (1990) developed a differential discrete ordinates method whereby the angular domain was discretized using techniques similar to the conventional discrete ordinates method (Modest, 1993); standard NAG and IMSL routines were used for the spatial operators. Majumdar and co-workers (Majumdar, 1993; Joshi & Majumdar, 1993) also used a discrete ordinates method with an explicit Euler scheme for unsteady terms and an upwind differencing of spatial derivatives. Sverdrup et al (2001) employed the discrete ordinates method with an explicit time-stepping scheme to numerically solve for the unsteady gray BTE in microelectronics device simulations. Murthy and Mathur (2001) developed an unstructured finite volume method for the BTE and applied it to the solution of gray and frequency-dependent steady and unsteady problems. As with the RTE, the solution methodology for the BTE typically involves a sequential procedure whereby the discrete equations for intensity in each ordinate direction are solved in turn, iterating between directions until convergence. At high acoustic thicknesses, inter-directional coupling through the collision term dominates, and sequential procedures become untenably slow. Mathur and Murthy (1999; 2001) developed the coupled ordinates method (COMET) to address this problem in the context of the steady gray RTE using an unstructured finite volume scheme. Here, temperature and the intensities in different directions at a cell were solved simultaneously to address coupling to the energy equation and to other directions; long-wavelength errors were eliminated through the use of a hybrid algebraic/ geometric multigrid scheme. The method was shown to substantially outperform sequential procedures at high optical thicknesses, and to exact little or no penalty at low optical thicknesses.

In this paper, we summarize our research in the development of finite volume schemes for the BTE in the relaxation time approximation. We present an unstructured solution-adaptive finite volume discretization for the BTE and discuss sequential and point-coupled solution procedures for solving it. We close with a discussion of the problems and challenges associated with devising numerical methods for micro- and nano-scale heat transfer and make recommendations for future research.

2 GOVERNING EQUATIONS

The phonon Boltzmann transport equation may be written in the relaxation time approximation in terms of the phonon distribution function $f_\omega(x, y, z, t, \omega, \mathbf{v})$ as (Joshi & Majumdar, 1993):

$$\frac{\partial f_\omega}{\partial t} + \mathbf{v} \cdot \nabla f_\omega = \frac{f_\omega^0 - f_\omega}{\tau_\omega} \tag{1}$$

Here, ω denotes a dependence on the phonon frequency, which ranges from zero to ω_d, the Debye cutoff frequency (Kittel, 1996). τ_ω is the relaxation time, which typically depends on the frequency and temperature. $\mathbf{v}$ is the velocity vector in the direction $\mathbf{s}$ and is given by:

$$\mathbf{v} = v\,\mathbf{s} = v\,(\sin\theta \sin\phi\,\mathbf{i} + \sin\theta \cos\phi\,\mathbf{j} + \cos\theta\,\mathbf{k}) \tag{2}$$

Here v is the phonon group velocity and (θ, ϕ) are the polar and azimuthal angles respectively. We define a phonon intensity as in Majumdar (1993)

$$I_\omega = \sum_p v \, f_\omega \; \hbar \; \omega \mathcal{D}(\omega) \tag{3}$$

Here the summation p is over the three phonon polarizations. $\mathcal{D}$ is the Debye density of states and $\hbar = h/2\pi$. Similar to Majumdar, we may write the equation of phonon radiative transfer (EPRT) as:

$$\frac{1}{v}\frac{\partial I_\omega}{\partial t} + \nabla \cdot (\mathbf{s} I_\omega) = \frac{I_\omega^0 - I_\omega}{v \tau_\omega} \tag{4}$$

I_ω^0 is given by

$$I_\omega^0 = \frac{1}{4\pi} \int_{4\pi} I_\omega d\Omega \tag{5}$$

where Ω denotes solid angle. The term on the RHS of Eq. 4 may be interpreted as an in-scattering term with a scattering coefficient $(1/v\tau_\omega)$ and an isotropic phase function.

For the purposes of this paper, we assume that the equilibrium temperature is known at the domain boundaries, and that the boundaries are perfect absorbers and emitters. Here, for directions outgoing from the boundary $(\mathbf{s} \cdot \mathbf{n} > \mathbf{0})$, the intensity I_ω in a frequency range $(\omega_1 - \omega_2)$ about ω is given by:

$$\int_{\omega_1}^{\omega_2} I_\omega (\mathbf{s}) \, d\omega = \left(F(\lambda_1 T_b) - F(\lambda_2 T_b)\right) \sigma T_b^4 / \pi \tag{6}$$

Here $\lambda = 2\pi v/\omega$. $F(\lambda T)$ is the emissive power fraction associated with the phonon equilibrium distribution function and is given by:

$$F(\lambda T) = \frac{15}{\pi^4} \int_{\frac{C_2}{\lambda T}}^{\infty} \frac{\xi^3 d\xi}{e^\xi - 1} \tag{7}$$

Here, $C_2 = hv/k_B$, where k_B is the Boltzmann constant. The Stefan-Boltzmann constant for phonons is taken to be (Majumdar, 1993)

$$\sigma = \frac{\pi^2 k_B^4}{40 \; \hbar^3 \, v^2}$$

2.1 *Recovery of Temperature*

An equivalent equilibrium temperature may be recovered from the phonon intensity field. If phonon frequencies in the range $(\omega_1 - \omega_2)$ are considered, the temperature is found from

$$\left[F(\lambda_1 T) - F(\lambda_2 T)\right] \frac{\sigma T^4}{\pi} = \frac{1}{4\pi} \int_{\omega_1}^{\omega_2} \int_{4\pi} I_\omega \; d\Omega \; d\omega \tag{8}$$

If the range $\omega_1 - \omega_2$ does not the cover the range $(0 - \infty)$, Eq. 8 represents an implicit equation in T and iteration is required to recover it.

3 NUMERICAL METHOD

A detailed description of the basic discretization is available in our previously published papers (Murthy & Mathur, 1998c; Murthy & Mathur, 1998a; Mathur & Murthy, 1998; Murthy & Mathur, 2001). We briefly summarize the main ideas behind the discretization of the BTE below.

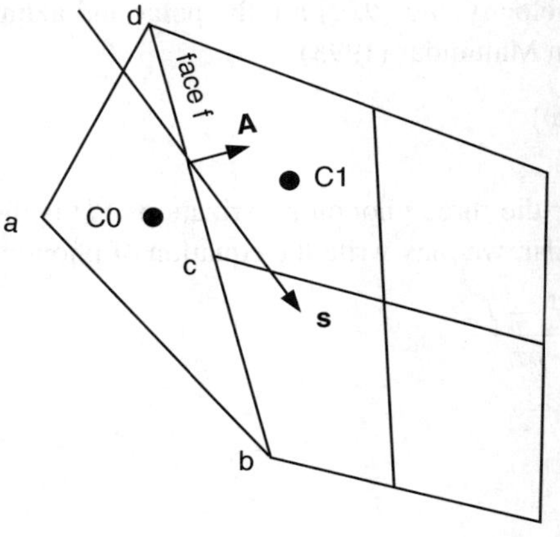

Figure 1: Control volume for phonon energy conservation in direction **s** in frequency band $\Delta\omega$

3.1 Discretization

The spatial domain is discretized into arbitrary convex polyhedral control volumes. The angular space 4π at any spatial location is discretized into discrete non-overlapping control angles Ω_l, the centroids of which are denoted by the direction vector $\mathbf{s}_l$, and the polar and azimuthal angles θ_l and ϕ_l. Each octant is discretized into $N_\theta \times N_\phi$ solid angles each with a constant angular extent of $\Delta\theta$ and $\Delta\phi$ to yield a total of $L(\equiv 8 \times N_\theta \times N_\phi)$ discrete directions. The angles θ and ϕ are measured with respect to the global Cartesian system (x, y, z). The angular discretization is uniform; the control angle extents are given by $\Delta\theta$ and $\Delta\phi$. The frequency domain (0-ω_H) is discretized in N_b frequency bands of extent $\Delta\omega$; these bands need not be equal in extent, but for simplicity, they are assumed so. All unknowns are stored at cell and boundary face centroids.

For each discrete direction i Eq. (4) is integrated over the frequency band $\Delta\omega$, the volume $\Delta\mathcal{V}_0$ of the control volume $C0$ in Figure 1, the solid angle Ω_i, and the time step Δt. The band-integrated cell centroid intensity I^l_{ik} in the cell i, direction l and frequency band k is used as the solution variable:

$$I^l_{ik} = \int_{\Delta\omega_k} I^l_{i\omega} d\omega \tag{9}$$

The variable I^0_{ik} is the angular average of I^l_{ik} at the cell centroid of cell i and is given by

$$I^0_{ik} = \frac{1}{4\pi} \sum_l I^l_{ik} \Delta\Omega \tag{10}$$

A step scheme for face intensity and a fully implicit time-stepping scheme are used as the baseline in this paper.

The final form of the discrete equation for a typical interior cell i in the frequency band k may be written as:

$$\sum_{j=1}^{N} \mu^l_{ij} I^l_{jk} + \left(\alpha^l_{ik} + \beta^l_i\right) I^l_{ik} - \gamma^l_{ik} \sum_{m=1}^{L} I^m_{ik} \Delta\Omega_m = \beta^l_i I^{n-1,l}_{ik} \tag{11}$$

Here, values at the previous time level are superscripted $n-1$; current time values do not carry a superscript for the time index. The coefficients are given by:

$$\begin{aligned} \alpha_{ik}^{l} &= \frac{1}{v\tau_k}\Delta\Omega_l \Delta\mathcal{V}_i \\ \beta_i^l &= \frac{\Delta\Omega_l \Delta\mathcal{V}_i}{v\Delta t} \\ \gamma_{ik}^{l} &= \frac{1}{4\pi}\frac{\Delta\Omega_l \Delta\mathcal{V}_i}{v\tau_k} \end{aligned} \tag{12}$$

The coefficients μ_{ij}^l in Eq. (11) result from the discretization of the $\nabla\cdot(\mathbf{s}I_\omega)$ term in Eq. (4). They denote the influence of neighboring cell intensities in the l direction on the intensity in cell i in the same direction (l) in the same frequency band (k). The α_i^l term results from out-scattering and depends on the band scattering coefficient $(1/(v\tau_k))$. The relaxation time τ_k is evaluated at the frequency band centroid as a second-order approximation. The β_i^l term results form the unsteady term and is independent of the frequency band. The γ_{ik}^l results from the discretization of the in-scattering term, and contains the contributions of other directions to the current direction l in the same band. It varies with the band because of the dependence on the band scattering coefficient $(1/(v\tau_k))$. Other terms appear at thermalizing boundaries because of boundary emission. Thus, the frequency bands are essentially independent of each other except at thermalizing boundaries where phonon energy is redistributed across bands in keeping with the equilibrium phonon distribution function.

3.2 Solution Techniques

Having obtained a discretization of the BTE and associated boundary conditions, it now remains to obtain the solution to the discrete equation set. The simplest method is to solve the discrete set for each direction and frequency band sequentially, cycling through all directions and bands until convergence. This approach requires little storage and is easy to implement. We have found it to be generally successful except when the acoustic thickness $L/(v\tau_k)$ is large. A large acoustic thickness increases the influence of the in-scattering term, coupling the directional intensity more strongly to other directions in the cell than to spatial neighbors in the same direction. Coupling between different directions is also introduced due to reflection and refraction at boundaries as well as due to geometric factors such as axisymmetry and periodicity. We have devised a solution technique called the coupled ordinate method (COMET) which addresses this issue. This is described next in brief. Details are available in Mathur and Murthy (1999; 2001).

3.3 Coupled Ordinate Method (COMET)

One approach to account for inter-directional coupling is to solve the entire equation system (all cells, directions and bands) simultaneously at each time step, but this too expensive to be tenable. Instead, COMET seeks to solve all the discrete equations at a point in a given band simultaneously. To this end, we write Eq. (11) at a cell i in the band k as

$$\mathbf{P}_{ik}\mathbf{q}_{ik} + \mathbf{r}_{ik} = 0 \tag{13}$$

where the solution vector $\mathbf{q}_{ik}$ for the band k is given by

$$\mathbf{q}_{ik} = \begin{bmatrix} I_{ik}^1 \\ I_{ik}^2 \\ \vdots \\ I_{ik}^L \end{bmatrix}$$

$\mathbf{P}_{ik}$ is a $L \times L$ matrix given by

$$\begin{bmatrix} M^1_{ii,k} - \gamma^1_{ik}\Delta\Omega_1 & -\gamma^1_{ik}\Delta\Omega_2 & \cdots & -\gamma^1_{ik}\Delta\Omega_L \\ -\gamma^2_{ik}\Delta\Omega_1 & M^2_{ii,k} - \gamma^2_{ik}\Delta\Omega_2 & \cdots & -\gamma^2_{ik}\Delta\Omega_L \\ \vdots & & & \\ -\gamma^L_{ik}\Delta\Omega_1 & -\gamma^L_{ik}\Delta\Omega_2 & \cdots & M^L_{ii,k} - \gamma^L_{ik}\Delta\Omega_L \end{bmatrix}$$

The matrix $M^l_{ij,k}$ is given by

$$M^l_{ij,k} = \begin{cases} \mu^l_{ij} & i \neq j \\ \mu^l_{ii} + \alpha^l_{ik} + \beta^l_i & i = j \end{cases} \tag{14}$$

The vector $\mathbf{r}_{ik}$ is given by

$$\mathbf{r}_{ik} = \begin{bmatrix} \sum_{j=1, i\neq j}^{N} \mu^1_{ij} I^1_{jk} - \beta^1_i I^{n-1,1}_{ik} \\ \vdots \\ \sum_{j=1, i\neq j}^{N} \mu^L_{ij} I^L_{jk} - \beta^L_i I^{n-1,L}_{ik} \end{bmatrix}$$

Thus the vector $\mathbf{r}_{ik}$ contains the influence of spatial neighbors in the same band in the same direction, as well as the influence of the previous time step. An equation set similar to Eq. (13) can be written for each band at the point i and solved assuming spatial neighbors temporarily known.

3.3.1 *Optimization of Point Solution*

The matrix $\mathbf{P}_{ik}$ is fully populated and is not always diagonally dominant, and is best solved using direct methods. However, the matrix size is proportional to the number of control angles L, and the effort for direct solution scales as $O(L^3)$. Consequently, COMET would be very expensive unless more efficient point solution procedures are devised. Under the relaxation time approximation, scattering is isotropic. Following Mathur and Murthy (2001), it is possible to define auxilliary variables on a band basis to obtain a direct solution procedure that scales as $O(L)$. We introduce the net irradiation I^a_{ik} at cell i and band k which is defined as

$$I^a_{ik} = \sum_{m=1}^{L} I^m_{ik}\Delta\Omega_m \tag{15}$$

We can then write Eq. (13) as

$$\tilde{\mathbf{P}}_{ik}\tilde{\mathbf{q}}_{ik} + \tilde{\mathbf{r}}_{ik} = 0 \tag{16}$$

where $\tilde{\mathbf{q}}_{ik}$ is the augmented vector

$$\mathbf{q}_{ik} = \begin{bmatrix} I^1_{ik} \\ I^2_{ik} \\ \vdots \\ I^L_{ik} \\ I^a_{ik} \end{bmatrix}$$

and $\tilde{\mathbf{P}}_{ik}$ is a $(L+1) \times (L+1)$ matrix given by

$$\begin{bmatrix} M^1_{ii,k} & 0 & \cdots & 0 & -\gamma^1_{ik} \\ 0 & M^2_{ii,k} & \cdots & 0 & -\gamma^2_{ik} \\ \vdots & & & & \\ 0 & 0 & \cdots & M^L_{ii,k} & -\gamma^L_{ik} \\ -\Delta\Omega_1 & -\Delta\Omega_2 & \cdots & -\Delta\Omega_L & 1 \end{bmatrix}$$

Matrix $\tilde{\mathbf{P}}_{ik}$ is easily upper-triangularized and the equation system solved in $O(L)$ operations per cell per band.

3.3.2 Geometric Multigrid Procedure

The point-coupled solution is embedded as a relaxation sweep in a multigrid procedure. It is not feasible to store the entire coefficient matrix for all cells in all directions in a band since the number of discrete directions is large. Consequently a *geometric* multigrid procedure is used. Coarse mesh levels are formed by agglomerating fine-level cells as described in Mathur and Murthy (1997), resulting in arbitrary (and possibly non-convex) unstructured polyhedra at coarse levels. The first-order step scheme is easily implemented on these polyhedra, as are all the other discretization procedures outlined in Mathur and Murthy (1999). Since these computations are relatively inexpensive, coefficients are recomputed for each coarse level cell when it is visited.

One iteration of COMET consists of the following steps:

1. Create the hierarchy of coarse levels by agglomeration For each frequency band k:
   (a) Perform one V cycle consisting of the following:
      i. Perform ν_1 relaxation sweeps at the finest level. This requires visiting each cell and constructing and solving Eq. (16).
      ii. Compute the residuals at the finest level. This also requires visiting each cell and constructing Eq. (16)
      iii. Inject the residuals to the next coarse level.
      iv. Recursively repeat steps (1(a)i) through (1(a)iii) for each coarse level.
      v. Using the solution at the coarsest level, correct the solution at the next finer level. Perform ν_2 relaxation sweeps.
      vi. Repeat (1(a)v) recursively for each finer level.
   (b) Repeat (1a) till the desired reduction of residuals is obtained.

With the coupled multigrid procedure multiple iterations are required within a time-step only if bands or directions are coupled due to boundary conditions or if insufficient V-cycles are used per iteration. The unsteady problem is solved by time-marching, with one or more V-cycles being applied within each time step until convergence within the time step is obtained.

4 RESULTS

In this section, we first present results to two micro-scale heat transfer problems to illustrate the underlying physics and to demonstrate the versatility of the unstructured finite volume formulation.

We then present results comparing the performance of the sequential solution procedure to that of COMET. All computations are done using double precision on a 32-bit Pentium III personal computer. The solution is judged to have converged when the scaled residuals for all the equations have reduced to 10^{-10}. At any global iteration, the linear equation set is solved to a tolerance of 0.1 reduction in the residual.

4.1 Unsteady 1-D Conduction in Thin Diamond Film

We consider here the case of unsteady heat conduction in a diamond film of width L, similar to the problem considered by Majumdar and co-workers (Majumdar, 1993; Joshi & Majumdar, 1993). The film is initially at $T_2 = 300$ K. At $t = 0$, the left boundary is raised to $T_1 = 301$ K. The objective is to compute the time evolution of the equivalent temperature defined in Eq. 8. Two phonon scattering processes are considered, impurity scattering and Umklapp scattering (Joshi & Majumdar, 1993). For impurity scattering, the relaxation time τ_i is taken to be:

$$\tau_i = \frac{1}{\alpha\Phi\eta v}$$

Here, R is the radius of the lattice imperfections, α is a constant that is approximately unity and η is the number density of impurities. The scattering cross-section Φ is taken to be:

$$\Phi = \pi R^2 \frac{\xi^4}{\xi^4 + 1}$$

where

$$\xi = \frac{\omega R}{v}$$

Values of the constants used are given in Table 1. For Umklapp scattering, the relaxation time is taken to be

$$\tau_u = A \frac{T}{\theta_d \omega} \exp\left(\frac{\theta_d}{\gamma T}\right)$$

Values of the constants A, γ and the Debye temperature θ_d are given in Table 1. Mathiessen's rule is used to obtain the combined time constant of the two scattering processes:

$$\frac{1}{\tau_\omega} = \frac{1}{\tau_i} + \frac{1}{\tau_u}$$

Frequency-dependent calculations are done using three different film thicknesses, L=0.1, 1.0 and 10.0 μm. At the frequency corresponding to the dominant energy($\omega_{dom} = k_B T/\hbar$), the acoustic thickness $\kappa = L/v\tau_{dom}$ for these three values ranges from 0.224 to 22.4. A mesh of 80 cells is used. An angular discretization $N_\theta \times N_\phi$ of the octant of 2×1 is used for all but the lowest acoustic thickness, for which an angular discretization of 4×1 is used. The frequency domain (0-ω_d) is divided into 29 equal frequency bands and the relaxation time τ_ω in each band is evaluated at the mid-point frequency. A dimensionless time step $\Delta t/(L/v)$ of 0.01 is chosen. Numerical experiments varying these discretizations were done for $\kappa = 2.24$ and indicate that the maximum error in the equivalent temperature with respect to finer discretizations is well under 1%.

Results for unsteady state for the three acoustic thicknesses are shown in Figures 2(a)-(c). Corresponding plots for Fourier conduction are also shown. The latter are computed using the methods

Table 1: Physical properties of diamond

Property	Symbol	Value
Impurity number density	η	0.154×10^{26} $(\frac{1}{m^3})$
Radius of lattice imperfection	R	1.785 Å
Phonon velocity	v	1.2288×10^4 m/s
Umklapp scattering constant A	A	163.94
Umklapp scattering constant	γ	1.58
Debye temperature	θ_d	1860 K
Stefan-Boltzmann constant	σ	50.47 W/m^{2}K^4

outlined in Murthy and Mathur (1998b). For acoustic thickness $\kappa = 0.224$, the wave nature of transport is clearly visible, though scattering tends to spread the wavefront somewhat. The Fourier solutions tend to reach steady state too quickly for low acoustic thicknesses, and do not exhibit the temperature jump at the boundaries either during the time-evolution or at steady state. For larger κ, the discrepancies between the BTE and Fourier solutions decrease, but substantial differences in the evolution speed exist even at κ=22.4.

4.2 Effective Conductivity of Array of Micro-Rods

To demonstrate the geometric flexibility of our scheme, we consider a periodic array of rods embedded in a matrix of interstitial material as shown in Figure 3. The array is four rods deep, and has lateral periodicity as shown. Adjacent rows of rods are displaced randomly with respect to each other; we present here results from a single realization where the rows starting from the top are displaced by $0.236928l$, $0.633978l$ and $0.139087l$ respectively with respect to the bottom row where l is the width of one module. The top and bottom boundaries are held at $T_1 = 300$K and $T_2 = 301$K respectively, yielding an emissive power ratio of 1.013; this has been kept deliberately small to yield thermal conductivities that are constant with respect to temperature. Both boundaries are assumed to be diffuse emitters and are completely absorbing.

The Fourier law is assumed to hold within the rods; this is valid if $(d/\Lambda) >> 1$. The interstitial material is considered to consist of a material which may be acoustically thin. Thus, the BTE is solved in the interstices, while Fourier conduction is solved in the rods. The interface is assumed diffuse and completely absorbing. The aspect ratio of the bed depth to the rod diameter (d/L) is assumed to be 5.333. All computations are gray and steady. Mixed BTE/Fourier computations are done for a range of acoustic thicknesses $l/(v\tau)$. Details of the treatment of translationally periodic boundary conditions used here may be found in Mathur and Murthy (1998) In addition, computations are also done for the case when Fourier conduction is valid in both the interstices and in the rods. The interstitial conductivity is deduced from kinetic theory (Kittel, 1996).

$$k_i = \frac{1}{3}\bar{C}v^2\tau$$

A dimensionless conductivity ratio $k_s/\frac{1}{3}\bar{C}v^2\tau = 0.1$ is chosen, corresponding to a Planck number of $k_s/(v\tau)/(4\sigma T_1^3)$ of 0.1364.

A fine mesh of 18,592 triangular cells is used in the calculation, with a 4×4 angular discretization of the octant with 1×10 pixelation to address the non-orthogonal mesh. A detail of a typical mesh is

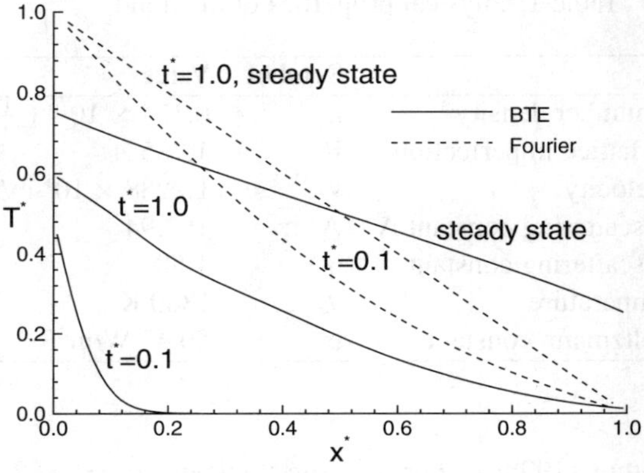

(a) $\kappa = 0.224$

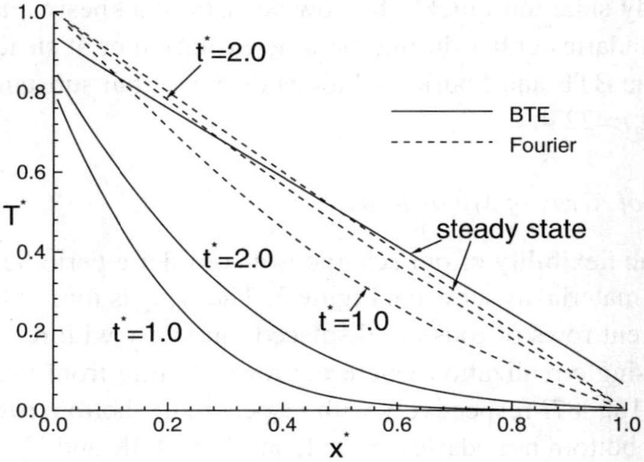

(b) $\kappa = 2.24$

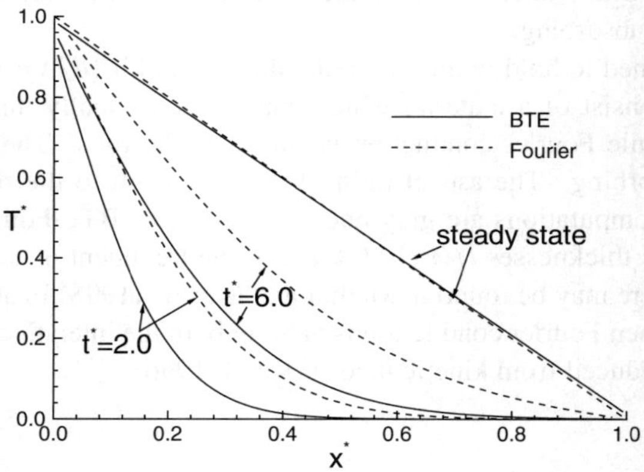

(c) $\kappa = 22.4$

Figure 2: Unsteady Conduction in diamond thin film

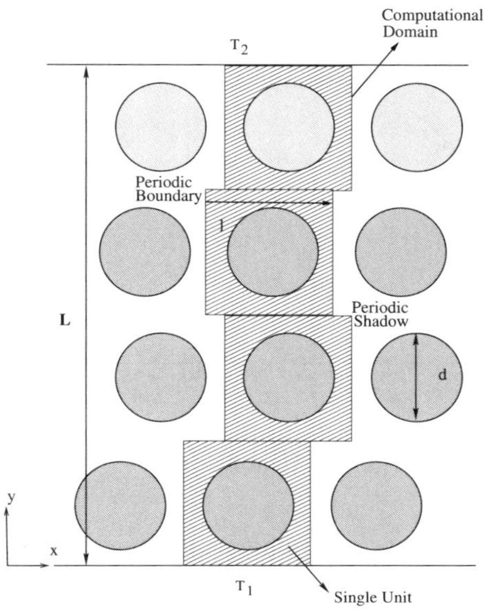

Figure 3: Rod array embedded in matrix

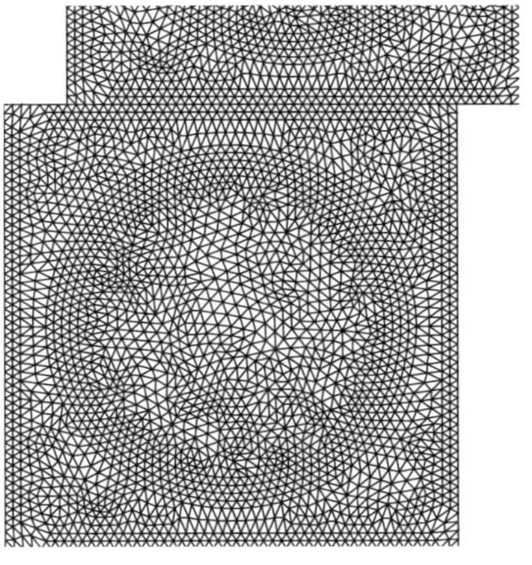

Figure 4: Detail of mesh

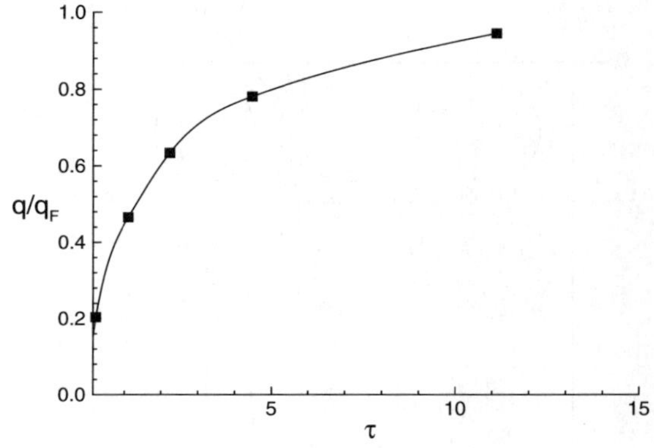

Figure 5: Heat flux ratio

shown in Figure 4. We also did computations with both finer and coarser triangular meshes; for the mesh used here, the overall boundary heat flux is accurate to better than 1%.

Figure 5 shows the ratio of the boundary heat flux obtained from the mixed BTE/Fourier computations and the pure Fourier case; this ratio is also the ratio of equivalent bed conductivities for the two cases. The dimensionless Fourier flux $qL/k_i\Delta T$ is an invariant. For high acoustic thicknesses, the Fourier and the mixed BTE/Fourier computations yield the same heat flux, and the ratio tends to unity. For lower acoustic thicknesses, however, the BTE/Fourier computations yield a lower heat flux. This is primarily a result of absorption and diffuse re-radiation in the rods; since the emission is diffuse, phonon energy is radiated back to the cold boundary at lower acoustic thicknesses.

4.3 Comparison of Sequential and COMET procedures

We consider here the case of unsteady heat conduction in a trapezoidal domain shown in Figure 6; geometric details may be found in Mathur and Murthy (1999). A 640-cell unstructured triangular mesh is considered for all calculations, with an angular discretization $N_\theta \times N_\phi$ of 4×4 directions per octant and a 1×10 pixelation to account for control angle overhang (see Murthy and Mathur (1998c)). The domain is initially at $T_2 = 300$ K. At $t = 0$, the bottom boundary is raised to $T_1 = 301$ K. The domain is considered to have the properties of diamond, as in the first example. Impurity and Umklapp scattering are considered as before.

Tables 2 and 3 present performance comparisons for unsteady conduction using the sequential and COMET procedures. The first column lists the acoustic thickness associated with the dominant frequency. Table 2 presents CPU time and iterations for a time step $\Delta tv/L = 10^{-2}$ for three different discretizations of the frequency spectrum. Here L is the length of the bottom wall of the trapezoid. Table 3 examines the effect of time step and presents performance comparisons for $\Delta tv/L = 10^{-3}$ and 10^{-4} for a single frequency discretization of ten bands. We see that the performance of the COMET procedure is essentially independant of acoustic thickness, both in terms of CPU time and the number of iterations to convergence. The performance of the sequential procedure, on the other hand, depends strongly on the acoustic thickness. For low acoustic thicknesses, the sequential and COMET procedures are comparable, and the sequential procedure may even outperform the COMET procedure in the transparent limit. However, as the acoustic thickness increases, the time to conver-

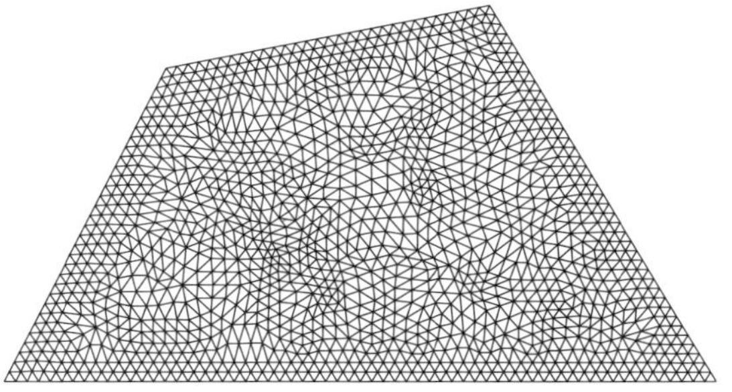

Figure 6: Trapezoidal cavity

gence of the sequential procedure increases rapidly, and quickly exceeds that of COMET by factors of 3-50, depending on the time step. The disparity between the sequential and COMET procedures is most pronounced for larger time step sizes ($\Delta tv/L = 10^{-2}$) than for smaller ones ($\Delta tv/L = 10^{-4}$) because smaller time step sizes keep the solution closer to that at the previous time instant. These trends are independant of the number of frequency bands, as see from Table 2; this is reasonable since the BTEs in the different frequency bands are independant of each other except at thermalizing boundaries.

5 FUTURE DIRECTIONS

We have described an unstructured finite volume scheme for solving both the steady and unsteady Boltzmann transport equation (BTE) for micro-scale heat conduction. The basic methodology for discretization has been developed and solution procedures for solving the discrete set have been shown to perform satisfactorily. The methodology is consistent with commonly used finite volume methods for flow and heat transfer. As a result, BTE calculations can be coupled in a conservative fashion with continuum calculations of heat transfer and fluid flow. We now consider extensions to the methodology and improvements to our understanding of the underlying physics.

The computations presented here model heat conduction assuming a single phonon branch and a simple dispersion relationship using a single phonon velocity. More realistic modeling of thermal transport at the microscale must distinguish between different phonon branches. In unsteady heat conduction, for example, the capacitative contribution of transverse acoustic phonons must be considered in addition to transport through longitudinal acoustic phonons (Ju & Goodson, 1999). In many modern microelectronics devices, electron-phonon energy exchange occurs through a transfer of energy to high-energy optical phonon modes; ultimately, this energy must find its way into acoustic modes. These different phonon branches are associated with different dispersion curves (Kittel, 1996) and hence, different phonon velocities. One approach is to also discretize the velocity space and to devise conservative discretization schemes for the transfer of energy between phonon branches. Other approaches, such as the Monte Carlo technique (Jacoboni & Reggiani, 1983), may also become profitable as the discretization space becomes very large. For complete consistency and conservation, it would be useful to consider electron transport within the finite volume framework presented here

Table 2: Unsteady sub-micron conduction in a trapezoidal cavity: CPU time and iterations to convergence for different acoustic thicknesses for $\Delta t v/L = 10^{-2}$

$L/(v\tau_{dom})$	Sequential		COMET	
	CPU s	Iter	CPU s	Iter
2 Frequency bands				
0.0484	98.68	5	79.52	5
0.484	98.85	5	69.62	5
4.84	97.72	6	68.54	5
48.4	162.25	10	67.95	5
484.0	529.4	33	61.22	5
10 Frequency bands				
0.0484	484.02	6	424.36	6
0.484	476.23	6	338.19	5
4.84	772.17	9	338.38	5
48.4	2922.96	29	337.03	5
484.0	19006.72	191	354.74	5
20 Frequency bands				
0.0484	970.04	6	958.01	7
0.484	960.8	6	882.56	6
4.84	1609.13	10	754.18	5
48.4	5701.64	34	828.82	6
484.0	39333.18	225	921.48	6

as well. As we resolve not only spatial, temporal and frequency spaces, but also include different phonon branches and the velocity space, the size of the problem to be solved increases multifold. Large-scale parallel computations are critical to obtaining realistic and reliable predicitons.

Ultimately, the assumptions underlying the Boltzmann transport equation must themselves be called into question. When physical length scales become comparable to the phonon wavelength or if phase coherence effects are important, it will become necessary to perform atomistic simulations to delineate the physics. Approaches of this type have begun to appear in a variety of contexts (Volz & Chen, 2000; Etcheverry & Mesaros, 1999). However the results of atomistic simulations depend substantially on the modeling involved, particularly the form of the potential function, number of particles, and domain boundary conditions; careful examination and validation of results is necessary. Again, massively parallel computations are necessary to obtain sufficiently large-scale atomistic simulations to address these questions. Conservative methodologies to couple atomistic, BTE and continuum calculations are another fruitful area for future research.

6 ACKNOWLEDGMENTS

The numerical procedures described here were implemented in Fluent 5.4 from Fluent Inc. We wish to acknowledge its use, and also the use of Fluent Inc.'s mesh generators, PreBFC and TGrid, in this work.

Table 3: Unsteady sub-micron conduction in a trapezoidal cavity: CPU time and iterations to convergence for different acoustic thicknesses for $\Delta tv/L = 10^{-3}$ and $\Delta tv/L = 10^{-4}$ for ten frequency bands

$L/(v\tau_{dom})$	Sequential		COMET	
	CPU s	Iter	CPU s	Iter
	$\Delta tv/L = 10^{-3}$			
0.0484	367.47	4	234.96	3
0.484	408.62	4	171.4	3
4.84	488.12	5	171.34	3
48.4	769.59	8	171.38	3
484.0	2803.55	25	171.41	3
	$\Delta tv/L = 10^{-4}$			
0.0484	159.92	2	163.33	2
0.484	159.88	2	99.56	2
4.84	211.13	3	98.72	2
48.4	271.06	3	91.67	2
484.0	364.19	4	91.66	2

REFERENCES

Chen, G. 1996. Nonlocal and Nonequilibrium Heat Conduction in the Vicinity of Nanoparticles. *Journal of Heat Transfer*, **118**, 539–545.

Chen, G. 1997. Size and Interface Effects on Thermal Conductivity of Superlattices and Periodic Thin Film Structures. *Journal of Heat Transfer*, **119**, 220–229.

Etcheverry, J.I., & Mesaros, M. 1999. Molecular Dynamics Simulation of the Production of Acoustic Waves by Pulsed Laser Irradiation. *Physical Review B*, **60**, 9430–9434.

Jacoboni, C., & Reggiani, L. 1983. The Monte Carlo Method for the Solution of Charge Transport in Semiconductors with Applications to Covalent Materials. *Reviews of Modern Physics*, **55**(3), 645–705.

Joshi, A., & Majumdar, A. 1993. Transient Ballistic and Diffusive Phonon Heat Transport in Thin Films. *Journal of Applied Physics*, **74**, 31–39.

Ju, Y.S., & Goodson, K.E. 1999. *Microscale Heat Conduction in Integrated Circuits and Their Constituent Thin Films*. Kluwer Academic Publishers. Chap. 5, pages 58–79.

Kittel, C. 1996. *Introduction to Solid State Physics*. John Wiley.

Kumar, S., Majumdar, A., & Tien, C.L. 1990. The Differential Discrete-Ordinates Method for Solutions of the Equation of Radiative Transfer. *Journal of Heat Transfer*, **112**, 424–429.

Majumdar, A. 1993. Microscale Heat Conduction in Dielectric Thin Films. *Journal of Heat Transfer*, **115**, 7–16.

Mathur, S.R., & Murthy, J.Y. 1997. A Pressure Based Method for Unstructured Meshes. *Numer. Heat Transfer*, **31**(2), 195–216.

Mathur, S.R., & Murthy, J.Y. 1998. Radiative Heat Transfer in Periodic Geometries Using a Finite Volume Scheme. *Journal of Heat Transfer*, **121**, 357–364.

Mathur, S.R., & Murthy, J.Y. 1999. Coupled Ordinates Method for Multigrid Acceleration of Radiation Calculations. *Journal of Thermophysics and Heat Transfer*, **13**(4), 467–473.

Mathur, S.R., & Murthy, J.Y. 2001. Acceleration of Anisotropic Scattering Computations Using Coupled Ordinate Method (COMET). *Journal of Heat Transfer*, **123**(June), 607–612.

Modest, M. F. 1993. *Radiative Heat Transfer.* Series in Mechanical Engineering. McGraw Hill.

Murthy, J. Y., & Mathur, S. R. 1998a. Radiative Heat Transfer in Axisymmetric Geometries Using an Unstructured Finite Volume Method. *Numerical Heat Transfer B: Fundamentals*, 397 – 416.

Murthy, J.Y., & Mathur, S.R. 1998b. A Conservative Numerical Scheme for the Energy Equation. *J. Heat Transfer*, **120**(November), 1081–1085.

Murthy, J.Y., & Mathur, S.R. 1998c. Finite Volume Method for Radiative Heat Transfer Using Unstructured Meshes. *J. Thermophys. Heat Transfer*, **12**(3), 313–321.

Murthy, J.Y., & Mathur, S.R. 2001 (November). Computation of Sub-Micron Thermal Transport Using an Unstructured Finite Volume Method. *In: Proceedings of IMECE 2001*. ASME, New York, NY.

Qui, T.Q., & Tien, C.L. 1993. Heat Transfer Mechanisms During Short-Pulse Laser Heating of Metals. *Journal of Heat Transfer*, **115**, 835–841.

Sverdrup, P.G., Ju, Y.S., & Goodson, K.E. 2001. Sub-Continuum Simulations of Heat Conduction in Silicon-on-Insulator Transistors. *Journal of Heat Transfer*, **123**(February), 130–137.

Volz, S.G., & Chen, G. 2000. Molecular-Dynamics Simulation of Thermal Conductivity of Silicon Crystals. *Physical Review B*, **61**(4), 2651–2656.

Thermal Challenges in Next Generation Electronic Systems, Joshi & Garimella (eds)
© 2002 Millpress, Rotterdam, ISBN 90-77017-03-8

Hot topics for cool devices – Thermal challenges for a small planet

J. Rantala
Nokia Research Center, Helsinki, Finland

Keywords: telecommunications, industrial needs, third generation, fourth generation, thermal management

ABSTRACT: Coming third generation and still undefind fourth generation standards for telecommunication devices, together with the market demand, push the designers to solutions bringing along still higher heat intensities. The technical development of telecommunication devices is discribed together with the demands set for thermal management in the product design. Some further trends in the applicable cooling and design solutions are discussed about.

1 INTRODUCTION

Telecommunication is probably that sector of industry that has most changed our daily life during the last decade of 20th century. The increasing speed of data transmission together with mobile applications have made our planet smaller so that many of us can be counted to be accessible whenever and where ever. The demand of ever increasing performance of ever smaller devices has led into situation where the newest innovations in semiconductor and electronic design are first taken into use just in telecommunications applications. In many cases these high demands have caused the thermal management of the components and systems being the limiting factor in reliable design. Even though the tough market situation in 2001 has momentarily decreased the speed of the technological development, it is seen that the challenges set for the thermal design are not going to decrease in the coming years.

2 TRENDS IN TELECOMMUNICATIONS

Some clear megatrends are visible in the telecommunications: in 1990's the wireless voice set through in the customers' mind and wireless phones became the preferred ones. This happened first in Europe where the more unified standardization and roaming between different networks supported the free mobility. Some years later the same trend became true also in North America. Simultaneously internet found its way into homes of ordinary people. When these both are combined, the next development can be easily forecasted: wireless internet, first with narrowband and then with broadband data transfer. This brings then more attractive features for consumers, as transfer of images. It might be easily thought, with the current economical slow down, that the consumers will never be ready to pay for image phones, but let us think what happened when television was invented. Many people were saying they will never buy a television set because they are already having a radio, totally satisfying their needs…

Additionally to higher speed and increased mobility, telecommunication is also coming more involved in normal daily life, even though we are not necessarily noticing it. Bluetooth concept makes possible to have telecommunication devices communicating with several peripherals and sensor systems when the user is walking from one room to another one in his home. Ubiquitous computing can have several applications from smart clothing to smart car tyres moniroting the pressure constantly. It has been forecasted that this device-to-device telecommunication is the next large boom in this industry.

Also with conventional voice call telephones the impact is growing due to the user circle widening. Even younger children are going to have their own phones, and on the other hand the phones of elderly and handicapped people are going to be an essential asset for their health care.

The digital convergence – the difference between phones and computers decreasing- brings along easy-to-carry personal trusted devices (PTD). These contain the features of today's personal digital assistents added with features needed for fluent mobile commerce and management of daily life. Through these one can securely do money transmissions for internet purchasing, check the saldo of the bank account; and with the security check system they also work as ID cards and keys for home, office, and car.

All this development looks nice and rosy, but having all our daily life dependent on higky sophisticated electronic devices sets huge demands on their user friendliness and reliability. Just consider a situation where so simple thing as an empty battery prevents you to use your PTD when you are outdoors in the late night. Unfortunately you can not enter neither your home nor your car to charge the battery because the keys do not work. You can not take a taxi to home because you can not pay it through your PTD and nobody any more uses cash money or credit cards. You can not also call home because your phone is not working and there are no more phone booths becase everybody is having a PTD.

To handle the networks for local needs, and on the other hand to prevent the interfering signals in densely populated area, there will be short range radio links and base stations. These might be locating even on the walls of office buildings, which sets huge demands on their size, noise, and temperature creation.

To fulfill the demands of these new ways of use, the devices should be light and small to carry with, be safe and confortable to use, have no noise, and they should be extremely reliable.

3 CHALLENGES SET FOR THERMAL MANAGEMENT

It is currently seen that the future miniaturization in telecommunication terminals goes in two directions: either there are more functions in the current size products, or the size of the products will decrease. The history and current situation in the terminals size and weight can be seen in Figure 1. Similar kind of development has happened also in network elements.

3.1 *Higher demand through the new standards*

The UMTS standard used for third generation telecommunication devices brought along the decrease in the efficiency in the transmitter when compared to GSM. Additionally, in UMTS the transmitter power amplifier is having a duty cycle of 100% whereas in GSM it was 12.5%. This makes the same nominal power device to produce eight times as much heat which is a comparable challenge for thermal management. Simultaneously, more processing power and memory is needed when phones will have features previously regarded as those for PC's. On the other hand, to satisfy the market needs and industrial competition, the device size should not be bigger than nowadays. This definitely leads to more dissipated power in shorter time with the same or smaller volume as before. Fourth generation devices standard is not yet known, but definitely it will increase the data transmission speed still, meaning even higher power consumption.

In component level the maximum heat intensities will start approaching 150-200 W/cm^2, the most challenging components being RF power amplifiers, and some processors and ASICs. Additionally to the older thermal management tasks to transfer the heat from the component and to dissipate the heat from the system, a third one has become essential: to keep the enclosure surface cool enough.

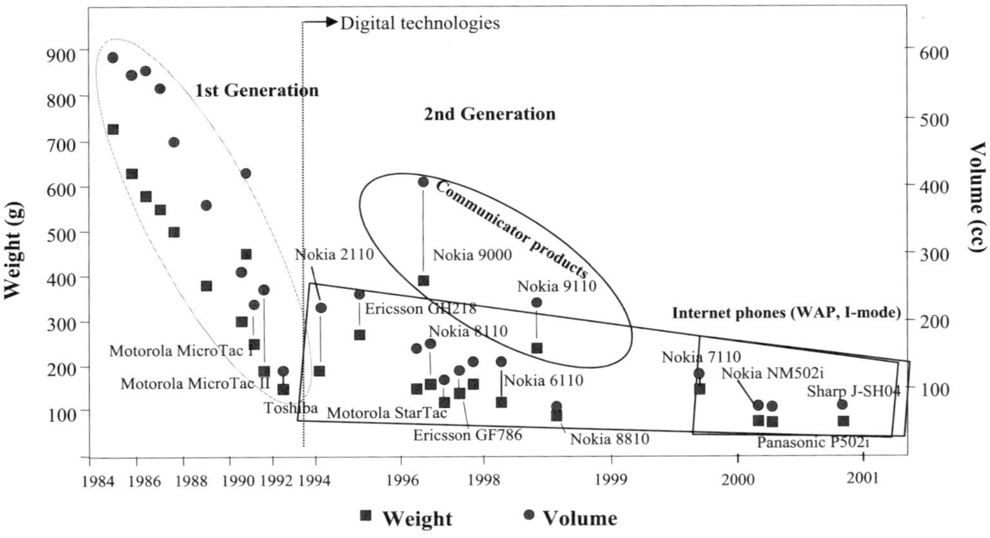

Figure courtesy by Dr. Petri Savolainen, Nokia Mobile Phones

Figure 1. Development in phone size and weight

3.2 *Changes in design approach*

In the telecommunications sector the thermal management used to be pretty straightforward: when problems in cabinets were encountered in the later phase of product development, the cooling fluid velocity and volume was increased by using more fans with higher rotational speed, or the size of heat sinks was increased. Nowadays the whole heat transfer cycle has to be optimized already at the early design phase. There is no time to redesign the system when problems occur. Additionally, the design should have a modular approach allowing reuse of certain design components for product variants.

In the optimized design, it is cheapest to tackle the problems where they are generated. This brings the component and even circuit design more into attention from the thermal design point of view. There are already programs simulating the physical circuit layout performance. The main application has been on finding the electrical hazards in the design, but thermal behavior is definitely getting here more importance. This will have an effect also in ASIC design guidelines, and more interdisciplinarity will be needed in circuit design teams. Coupling of thermal effects into electrical performance will become more important. This far this have been done for RF design, but with the increasing data transmission the whole physical, and many times temperature dependent, behavior has to be taken into account also in digital design. However, currently these circuit level design tools are very heavy to use and the use of memory and CPU capacity is excessive – thus improvements are awaited in this area.

In package design the longer term development is clearly going towards more integrated cooling systems and packaging. This allows both effective cooling and product miniaturization. Reliability issues are still becoming more important, and more fundamental knowledge is needed in material science. Package selection and PWB layout design will be more linked because of the thermal optimization of the layout.

In the design of enclosures the thermal issues have to be taken more into account. It is not enough just to have good thermal conductivity to spread the heat, also the heat transfer to the ambient shoud be maximized, and the effusivity should be suitable to make the surface feel comfortable.

As many of the local network elements will be located inside offices and homes, the acoustic noise will become an essential factor. To decrease the noise, both heat sink and fan optimization and development will have increasing importances. This includes also methods to make concurrent design for both thermal and acoustical issues.

The big question in the thermal management remains as how to define the design tolerances. On one hand, the more challenging design would tend to push the temperature limits upwards, but there is not enough information on the total effect of temperature on long term reliability. Especially the role of thermal transients and gradients is vaquely known, at least for practical industrial design purposes. At the moment, even though attractive, the change of design guidelines is very difficult because of the unknown risk.

4 CONCLUSIONS

There are lots of challenges for thermal management in telecommuncations sector. As the fourth generation standard is not defined yet, we are not even aware what is waiting for us. Both fundamental scientific and applied technological work are needed. There will be increasing interest toward component and circuit level work, and for small systems the heat dissipation from the whole system starts to be the main problem. Interdisciplinary approach is needed, both for heat transfer and research and industrial thermal design. And, to give the basis for the thermal management work, we still do not know too much what is the real link between temperature and reliability.

In all these thermal management developments, it should be remembered that the good thermal performance should not hazard other product properties. And, as always, cost still remains a very important factor. Unfortunately, the customers are not paying for sophisticated thermal management solutions but for high-performance reliable products.

Thermal Challenges in Next Generation Electronic Systems, Joshi & Garimella (eds)
© 2002 Millpress, Rotterdam, ISBN 90-77017-03-8

New possibilities in the thermal evaluation, offered by transient testing

Marta Rencz
MicReD, Budapest, Hungary

Keywords: transient testing, structure functions, die attach quality, board thermal conductivity

ABSTRACT: After giving the theoretical summary of the structure function method for the evaluation the thermal transient measurement results, the broad applicability of the method is demonstrated by practical examples. The first example shows how to control the quality of the die attachment or soldering of packages with fast transient measurements and subsequent evaluation. The second example presents how can transient testing be used to determine thermal material parameters; e.g. the effective thermal conductivity of printed circuit boards.

1 INTRODUCTION

It is well known for decades already [1] that thermal transient measurements that is, measuring the temperature increase inside the system resulting from a step function excitation (see Figure 1) carry a large amount of information about the thermal behavior of the structure and via this about the structure itself. Transforming however the measured data into thermal resistance, thermal capacitance or conductivity values or further into geometrical dimensions is not easy. Many authors deal with the evaluation of the measured heating or cooling curves, since the early paper of [1], like [2-4]. The paper [2] was exceptional in the sense that it has shown for the first time a direct transformation method that is capable to deliver even material structural data from the measured thermal transient curves.

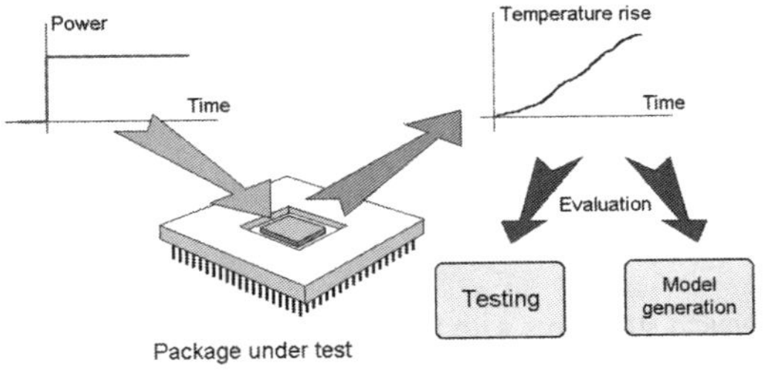

Figure 1. Thermal transient testing of a package

The base of the evaluation method is a relatively recent result in electrical network theory, providing a new treatment of distributed parameter networks [5], the network identification by deconvolution (NID) method. Based on [5] the time constant density, that is the intensity of the time constants in the function of the time of any distributed parameter system can be determined from the response function of the system. Since the heating curve obtained by the thermal transient measurement is the response function of the thermal system, with the NID method the time constants of the thermal system may be determined with arbitrarily fine resolution, in case if the heat flow is mainly into one direction. This enables accurate modeling of the heat flow path with partial thermal resistances and capacitances, and from these values further characteristic features of the system can be calculated.

In the rest of the paper first the theoretical background of the evaluation method is summarized (Section 2), then in Section 3 the applicability is presented with practical examples.

2 THEORETICAL BACKGROUND

The simplest representation of a thermal system is presented in Figure 2 – it is characterized by one thermal resistance and one thermal capacitance.

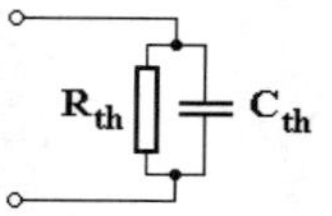

Figure 2. The simplest representation of a thermal system

If we apply a P power step to this model the temperature will rise on its ports according to

$$T(t) = P R_{th}(1 - \exp(-t/\tau)) \quad , \tag{1}$$

where $\tau = R_{th}C_{th}$ is the time constant of the system.

Real physical structures have usually more time constants, in this case the temperature response function is the sum of the appropriate exponential functions, as

$$T(t) = P\sum_{i=1}^{N} R_{thi}(1 - \exp(-t/\tau_i)). \tag{2}$$

To such response functions two different model networks may be ordered. These are not independent, one may be calculated from the other: a Foster and a Cauer type network, see Figure 3. Thermal systems are to be represented by Cauer networks, since the thermal capacitances are always connected to the ground, but as every Cauer network has its Foster equivalent, the Foster approximation of thermal networks exists as well.

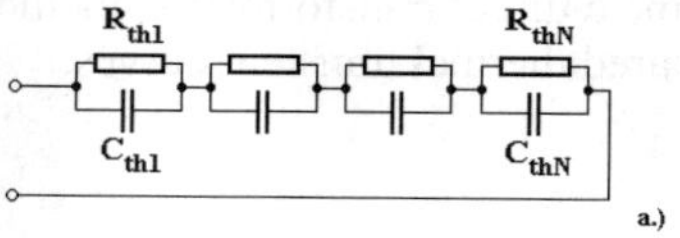

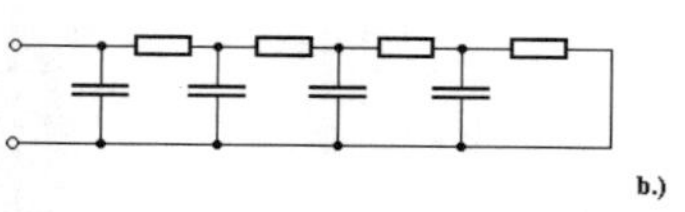

Figure 3. Foster(a) and Cauer (b) type representation of physical structures with finite time constants

Real physical structures however are represented with an infinite number of time constants, since any infinitesimal cube of a matter shows a certain thermal resistance and capacitance value, see Figure 4, resulting in as many number of time constants in the response function as the number of its capacities.

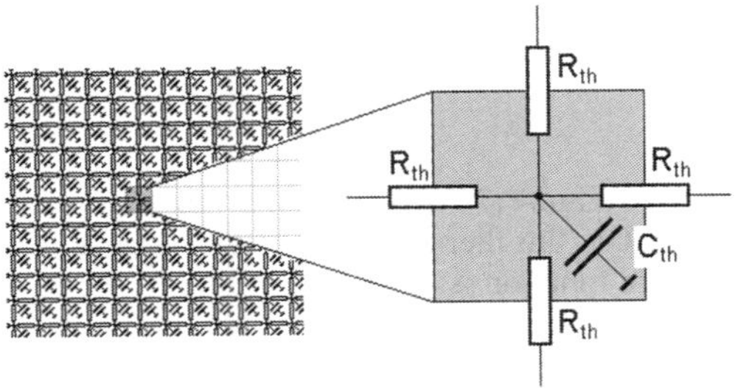

Figure 4. Real physical structures have infinite number of time constants

These amount of time constants may be best represented with their density function. See Figure 5.

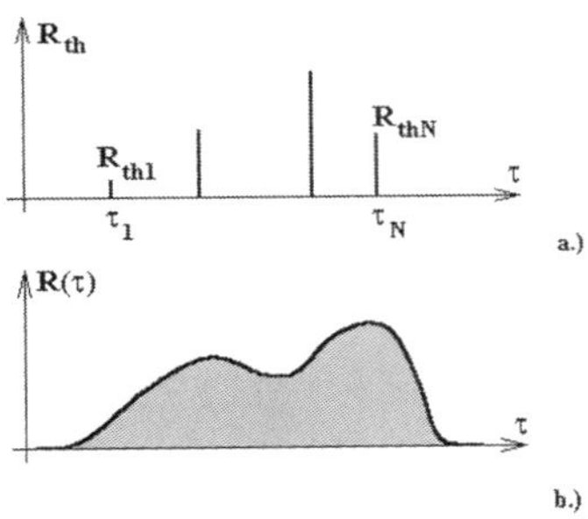

Figure 5. Time constants in a lumped element system (a) and in a distributed parameter system (b). In the latter case the time constants form a continuous curve, the time constant density function

Knowing the time constant density of a system an arbitrarily well approximating Foster equivalent circuit may be constructed by approximating the $R(\tau)$ function by infinitesimally narrow boxes, determining each a parallel RC pair in the Foster chain. Finding the Cauer equivalent of this network by textbook transformations we obtain a true physical equivalent of the heat transport in thermal systems. From this circuit we can draw up the so-called *cumulative structure function* [6], that gives the sum of the thermal capacitances C_Σ in the function of the sum of the thermal resistances R_Σ of the thermal system, measured from the point of excitation towards the ambient, see Figure 6.

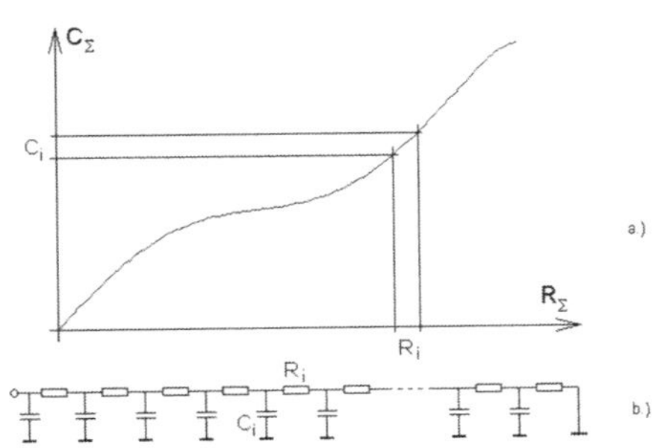

Figure 6. The cumulative structure function and the related Cauer equivalent circuit

The *differential structure function,* referred usually as the *structure function* [2]. is defined as

$$K(R_\Sigma)=\frac{dC_\Sigma}{dR_\Sigma}. \tag{3}$$

As the capacitance of a *dx* wide slice of a matter (see Figure 7) is $dC_\Sigma=cAdx$, the resistance is $dR_\Sigma=dx/\lambda A$, where c is the volumetric heat capacitance, λ is the thermal conductivity and A is the cross section area of the heat flow, the value of the structure function is

$$K(R_\Sigma)=\frac{cAdx}{dx/\lambda A}=c\lambda A^2. \tag{4}$$

This value (frequently designated also with *S)* is proportional to the *c* and λ material parameters, and to the square of the cross sectional area of the heat flow, consequently it is related to the structure of the system. In other words: this function provides a map of the square of the heat current-flow cross section area as a function of the cumulative resistance.

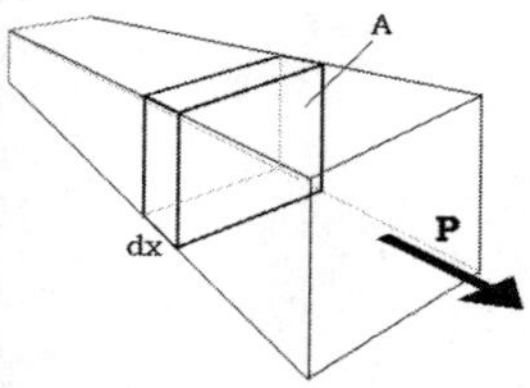

Figure 7. To the interpretation of the Structure function, P is the heat flux

Measured structure functions of power transistors on a cold plate are presented in Figure 8. In this functions the local peaks represent usually reaching new surfaces (materials) in the heat flow path, and their distance on the horizontal axis gives the partial thermal resistances between these surfaces. If we know the λ and *c* parameters for the used materials, even the cross section area vs. distance map can be constructed for the examined structure [7].

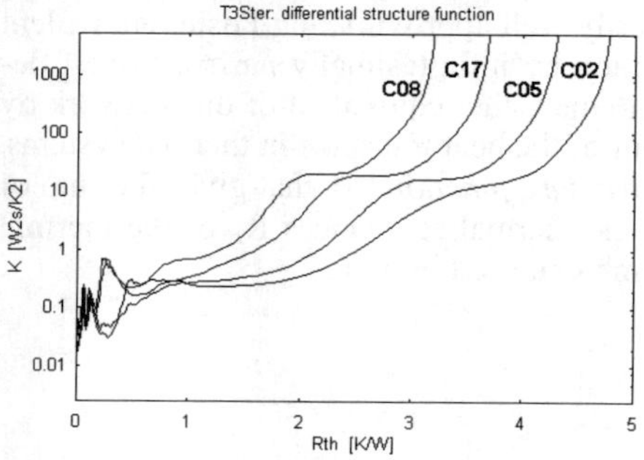

Figure 8. Structure functions of power transistors on a cold plate, measured and evaluated by T3ster® [9]

Obviously this simple interpretation of the structure function is possible only in the case if the heat is streaming along a single main path. In such cases the derived model corresponds directly to the physical structure, enabling the reconstruction of this structure. In case of complex, 3D streaming the derived model may be considered as an *equivalent physical structure* providing the same static or pulse thermal resistance as the original structure. This equivalent structure can not be considered however as a reconstruction of the physical structure to be modeled.

3 PRACTICAL EXAMPLES

The structure functions are extremely useful in the evaluation of the thermal behavior of various structures, especially packages. Now two of these possibilities will be presented.

3.1 *Determining partial thermal resistances, e.g. to control die attach quality*

Die attach failures are very dangerous packaging problems, the increased thermal resistance between the die and the platform may result in locally increased temperatures and eventually in serious reliability problems. In order to detect the samples with die attach failures early enough *in-line testing* of the die attach quality would be rather advantageous. Since a steady state R_{th} measurement for in-line testing is out of question for the reason of the relatively very long time needed to reach the thermal steady state, only transient measurements may be contemplated.

3.1.1 *Theory*

Die attach failures and soldering failures manifest as increased thermal resistances between the die and the platform or between the platform and the board, respectively. The differential structure function offers the possibility of locating the material transitions in the heat flow path, characterized by the cumulative thermal resistance, and finding the partial thermal resistances between the different locations. Comparing the structure function of the measured device with that of a known good device the location and the value of the increased partial thermal resistance may be easily determined.

3.1.2 *Measurement example*

To demonstrate the method we present the results of measuring a series of power transistors, each mounted on a copper base plate and fixed on a larger aluminum mounting plate, see Figure 9, that we measured on a water cooled cold plate in order to assure faster transients. The measurement was done by T3ster®[9], with the resolution of 1µs and 0,012 °C. [10]

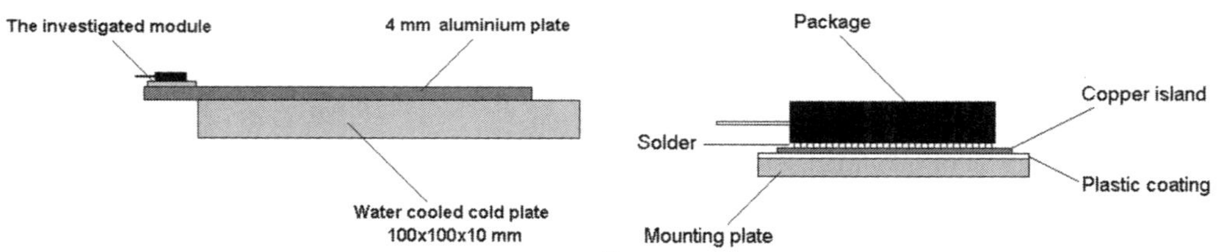

a. b.

Figure 9. The measurement arrangement, on b. the package mounting area is enlarged

The measured thermal transient curves are presented in Figure 10.

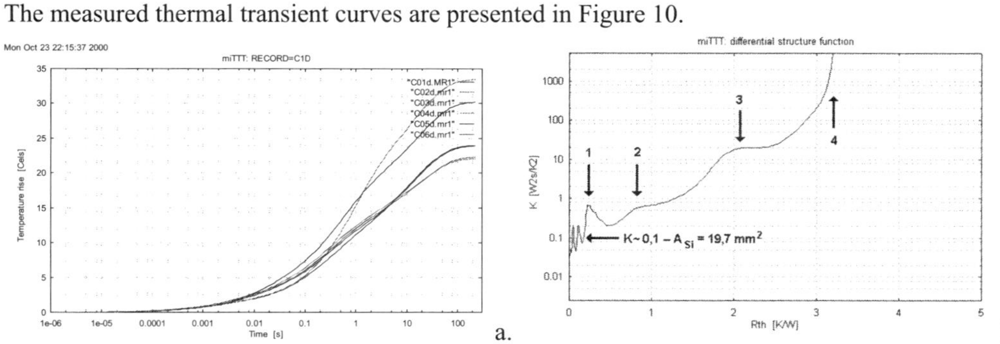

a. b.

Figure 10 a. Measured thermal transient curves of good and bad devices, b. The differential structure function of the C08 reference device. The arrows point to characteristic locations of the structure.

Examining the measured transient curves we can notice differences, but the evaluation of these curves is rather difficult. The evaluation of the *structure functions* is much easier, see Figure 8. It can be noticed that there are characteristic differences between the presented functions. To understand these differences let us discuss first the differential structure function of C08, the known good reference device, presented in Figure 10b. The left-hand side of this curve refers to the chip, the right hand end to the cold plate, arrow 4 shows this point. The value read on the horizontal axis gives the thermal resistance between the chip and the cold plate, it is 3.2 K/W. The zigzagged beginning of the curve shows the presence of some noise, but an average K=0.1 value can be considered. In case of silicon material this is equivalent to a 19.7 mm^2 cross sectional area, which equals in fact the area of the chip. The next peak, 1 refers to the heat capacitance of the transistor case, determined by the dominant heat capacitance of the copper base plate of the case. The next peak 2 refers to of the copper island of the mounting plate, peak 3 is the heat capacitance of the mounting plate itself.

After locating these characteristic points, the partial thermal resistance values can be read from the figure. The thermal resistance between the 1-2 points is about 0.6 K/W, this is the thermal resistance component of the transistor soldering. Between points 2-3 the thermal resistance of the plastic coating can be read, in our case this is about 1.3 K/W. The thermal resistance between the mounting plate and the cold plate determines the distance between the points 3 and 4.

Comparing the structure function of C02 to the reference function of C08 (Figure 11.) we notice that at C02 a characteristic minimum is visible at the right hand side of peak 2, and the thermal resistance to the next plateau is much (2.5 times) higher. This suggests a soldering problem at C02.

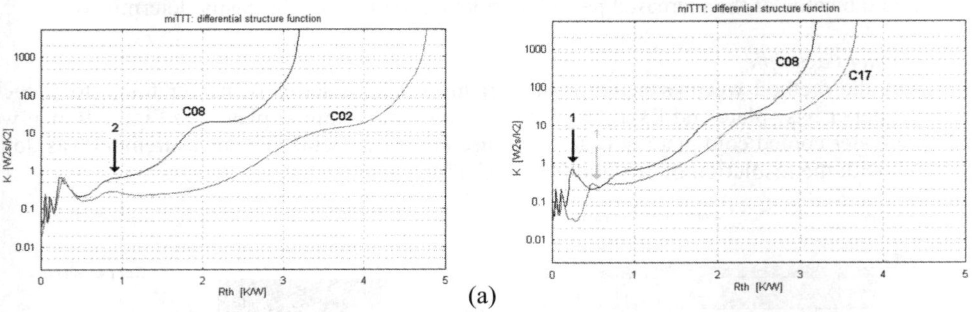

Figure 11 (a).Comparison of the differential structure functions of C02 and C08.The shift in peak 3 suggests soldering error at C02 . (b). The differential structure function of C17 referred to the structure function of C08, the known good device. The shift of peak 1 suggests die attach failure in C17.

The differential structure function of the C17 device is presented in Figure 11b. In case of the C17 device peak 1 is shifted to the right with a value of 0.4 K/W and the rest of the curve shows the same right shift. This means the presence of an extra thermal resistance between the chip and the copper platform of the case, which indicates that the chip is not attached to the platform appropriately.

On the measured transient curves (Figure 10a.) we can notice that the curves of both these devices are running above the nominal one with about 20-25% in the 0.1-0.2 sec range of the transient measurements. This is a very important experience, suggesting that die attach failures can be detected by short transient measurements, offering the possibility of using the method even for in-line testing.

3.2 *Measuring effective board thermal conductivity*

The second example demonstrates how can material parameters be determined by transient testing. Our example is the effective thermal conductivity measurement of patterned printed circuit boards.

Knowing the effective thermal conductivity values of patterned printed circuit boards can be very important both for thermal simulations and design. Thermal transient measurements offer a simple and fast method to determine these values.

3.2.1 *Theory*

In this measurement we exploit that in case of radial heat propagation the structure functions are linear in lin-log scale, and the slope is proportional to the thermal conductivity [10]. Radial heat propagation can be forced in the board by forcing isothermal boundary condition around the area that we wish to measure, and considering the thermal transients in the middle of this region, e.g. by mounting a transistor in the center of the region, see Figure 12., serving both as dissipator and sensor. From the recorded thermal transients the parameters of the board can be extracted as follows.

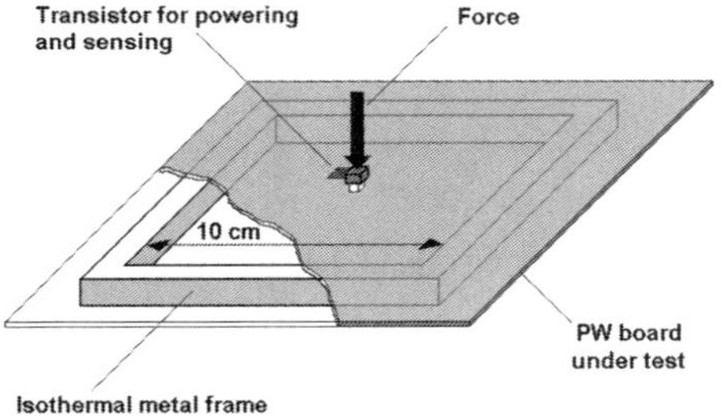

Figure 12 Setup to measure effective board thermal conductivity

For radial heat propagation in a disc of w thickness and λ thermal conductivity the thermal resistance can be written as

$$R_\Sigma = \frac{1}{2\pi w \lambda} \ln(r / r_0) \tag{5}$$

where r_0 is a reference location (e.g. the radius of the contact area under the dissipator/sensor transistor), and R_Σ is the thermal resistance related to the reference location r_0. The thermal capacitance is given by

$$C_\Sigma = wc_v \pi (r^2 - r_0^2) \cong wc_v \pi r_0^2 (r / r_0)^2 \tag{6}$$

where c_v is the volumetric heat capacity. This value is related again to the location r_0.
The r location can be expressed in terms of R_Σ as

$$r = r_0 \exp(2\pi w \lambda R_\Sigma) \tag{7}$$

Substitution of this equation into (3) results in

$$C_\Sigma = wc_v \pi r_0^2 \exp(4\pi w \lambda R_\Sigma) \tag{8}$$

and the derivative of this function is

$$K(R_\Sigma) = (2\pi r_0)^2 wc_v w\lambda \cdot \exp(4\pi w \lambda R_\Sigma) \tag{9}$$

Both $C_\Sigma(R_\Sigma)$ and $K(R_\Sigma)$ appear as a straight line in the lin-log coordinate system, with the slope of $4\pi\lambda w$ where λ is the thermal conductivity. This means that based on this slope, the λw product can be easily determined. A very interesting feature is that this slope is independent of the heat capacity.

3.2.2 *Measurement example*

The use of the method is demonstrated by measuring a board, where the patterning influences the effective thermal conductivity. The sample is shown in Figure 13 a. In two subsequent experiments the probe head was positioned to the points A and B. The results are shown in Figure 13 b.

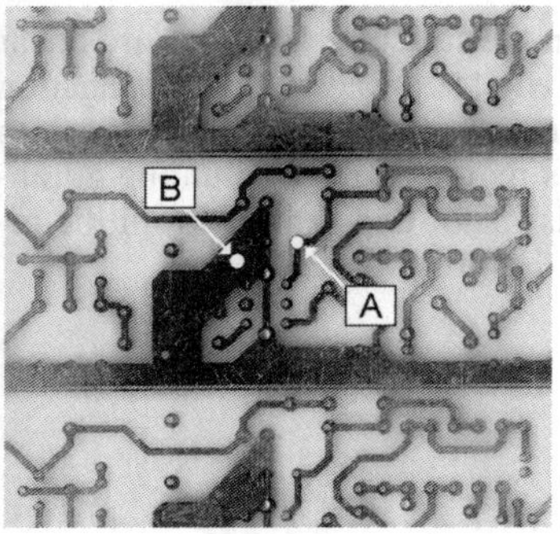

(a)

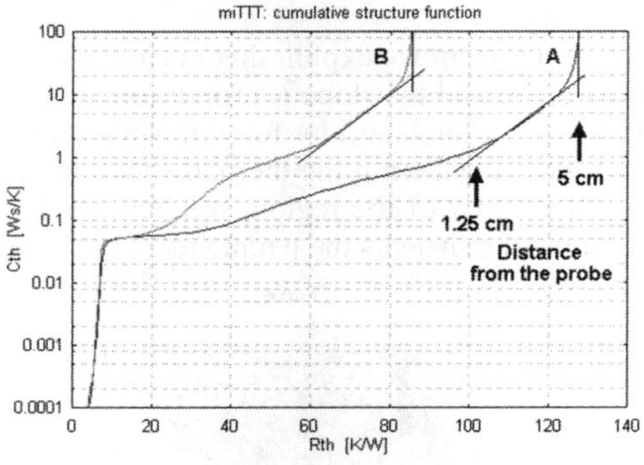

9b)

Figure 13 Pattern of the investigated board with the two positions of the probe (a)and the cumulative structure functions for the board at the A and B locations

Although the curves start with the same slope, that refers to the measuring transistor itself, near the probe the actual pattern strongly influences the heat flow, see the 1st and 2nd parts of the curves. The distance data show clearly that most of the board is represented by the rather short section of the curve, above the 1.25 cm mark. Thus, although the near thermal field of the probe produces a confused picture, the "far-field" shows the regular behavior. It is important to observe that the right-hand side of both functions can be approximated with about the same slope and same capacitance values. This shows that it is possible to calculate average or effective material parameters for a patterned board using the structure function, and the actual position of the probe-head is not critical. During the measurements we have to assure however that the shunt thermal resistance of the probe towards the ambient be much larger than that of the panel. Therefore a special probe has to be used that fulfils this requirement [11].

4 CONCLUSIONS

In this paper we intended to demonstrate that the NID method based evaluation of the measured heating curves might reveal a large amount of information about the physical structure, that was measured. Although the evaluation of the time constant spectrum may have similar practical importance in this paper we limited ourselves to the structure function based evaluation of the transient measurements.

Because of the limited length of this paper only two examples could be presented. The first example demonstrates the possibility to determine partial thermal resistances in a heat flow path with the help of transient measurements and the subsequent structure function evaluation. This feature may be used very broadly to measure interfacial thermal resistance values, and among others to detect die attach failures or soldering inaccuracies of packages. The presented other example demonstrated how to measure effective thermal conductivity values of patterned printed circuit boards.

Examples showing how to use the method for structure restoration may be found e.g. in [7].

Because of the length limitation we could not present how to use the *structure function* method e.g. to measure very small steady state thermal resistance values, but the presentation will give some examples also for this and some further applications.

5 REFERENCES

1. B.S. Siegal: "Measuring thermal resistance is the key to a cool semiconductor", Electronics, Vol. 51. Pp. 121-126, 1978.
2. V. Székely and Tran Van Bien: "Fine structure of heat flow path in semiconductor devices: a measurement and identification method", *Solid-State Electronics*, V.31, pp. 1363-1368 (1988)

3. W. Sofia: “Analysis of thermal transient data with synthesized dynamic models for semi-conductor devices”, *IEEE Trans. on Comp. Pack. & Manuf.* V.18, No.1. pp. 39-47 (1995)
4. G. Oliveti, A. Piccirillo and P.E. Bagnoli: "Analysis of laser diode thermal properties with spatial resolution by means of the TRAIT method", *Microelectronics Journal*, V.28, No.3. pp.293-300 (1997)
5. V. Székely, “On the representation of infinite-length distributed RC one-ports”, IEEE Trans. on Circuits and Systems, Vol.38. pp.711–719, 1991.
6. E.N.Protonotarios, O.Wing: "Theory of nonuniform RC lines", *IEEE Trans. on Circuit Theory*, V.14, No.1, pp. 2-12 (1967)
7. V. Székely, M. Rencz: "Thermal Dynamics and the Time Constant Domain", *IEEE Trans. On Components and Packaging Technologies,* Vol.23, No.3, pp 687-594 (2000)
8. V. Székely: “A New Evaluation Method of Thermal Transient Measurement Results”, *Microelectronics Journal*, Vol. 28, No. 3, pp. 277-292, 1997
9. http://www.micred.com/index1.htm
10. M. Rencz, V. Székely: “Determining partial thermal resistances in a heat flow path with the help of transient measurements”, *Proc. of the 7th THERMINIC Workshop,* 27-27 Sept. 2001 Paris, France, pp 250-256
11. V. Székely, M. Rencz, S.Török, S.Ress, B.Vizy: Experiments on effective thermal conductivity measurements, *Proc. of the 6th THERMINIC Workshop,* 24-27 Sept. 2000, Budapest, pp 26-30

Thermal Challenges in Next Generation Electronic Systems, Joshi & Garimella (eds)
© 2002 Millpress, Rotterdam, ISBN 90-77017-03-8

Thin film solid state refrigerators for spot cooling of integrated circuits

Ali Shakouri, Daryoosh Vashaee & James Christofferson
Jack Baskin School of Engineering, University of California at Santa Cruz, Santa Cruz, CA, USA
Tel: (831) 459-3821, Fax: (831) 459-4829, ali@cse.ucsc.edu

Xiaofeng Fan & Chris Labounty
Electrical and Computer Engineering, University of California Santa Barbara, CA, USA

EXTENDED ABSTRACT: The on-chip temperature of the packaged VLSI circuits not only can reach as high as 100C on average, but also can vary by as much as a few tens of degrees from one location to another (see Fig. 1). Because the failure rate of microelectronic devices depend heavily on the localized operating temperature, hot spots due to high local-power dissipation have become a long-term IC reliability concern in diverse applications such as high-performance microprocessors and digital signal processing chips. As integration level and clock speeds increase and device feature size decreases, overheating and hot spots is becoming more of a problem in conventional metal-oxide-semiconductor (MOS) circuits as well. One of the key reliability issues, electromigration, is exponentially dependent on temperature. So cooling by as little as 10C can increase the mean time to failure by a factor of 2 or 3.

In this talk we will review the performance of conventional BiTe thermoelectric coolers and describe recent advances in SiGe and III-V thin film solid-state refrigerators. We will also examine how thin-film coolers could be integrated with IC chips in order to remove the hot spots and make the chip temperature more uniform. Current commercial thermoelectric devices range is size from 1.8x3.4x2.4 up to 62x62x5.8 mm^2 and they can cool by as much as 70C at room temperature. However they are fabricated out of BiTe compounds in a bulk manufacturing technology that is not compatible with IC processing. In addition the cooling power density is limited to 10's of W/cm^2 (see Fig. 2). On the other hand, recently SiGe-based superlattice devices have demonstrated 4C cooling at room temperature and cooling power densities exceeding 500W/cm^2 (see Fig. 3). These refrigerators can be grown directly on a silicon substrate and they can be used to control the temperature on a short time scale (tens of micro seconds) over small areas down to hundreds of microns square.

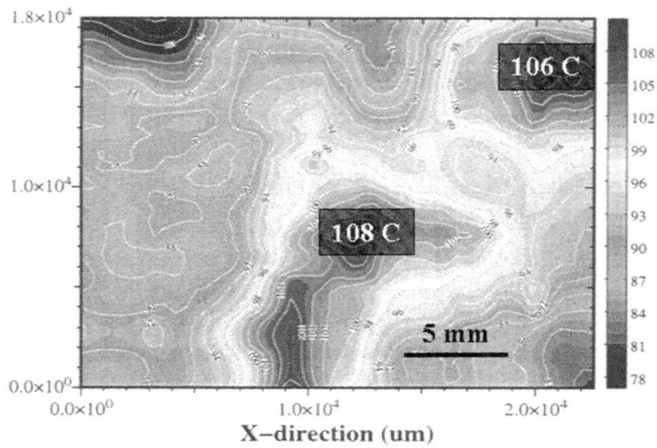

Fig. 1 Simulated temperature distribution in a VLSI Chip showing hot spots (Y-K Cheng et al. "Electrothermal analysis of VLSI Systems," Kluwer 2000).

Typical values for Bi_2Te_3 Seebeck=200 mV/K
resistivity=0.001 Ωcm
thermal conductivity=0.015 W/(cmK)

Device diameter=40□m, T_0=300K, device length=1mm

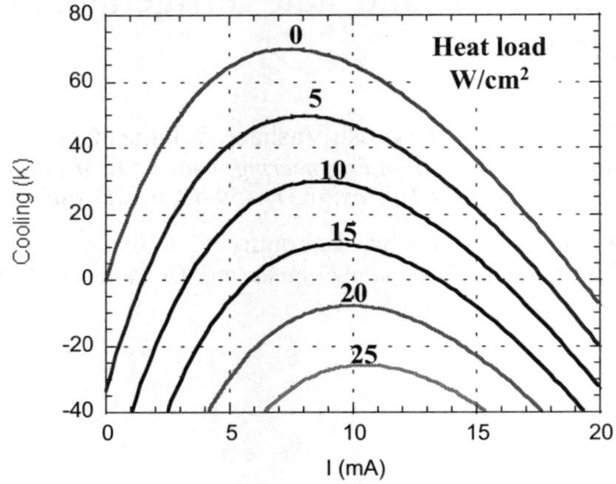

Fig. 2 Cooling versus current for different heat load power densities for of a conventional BiTe thermoelectric cooler.

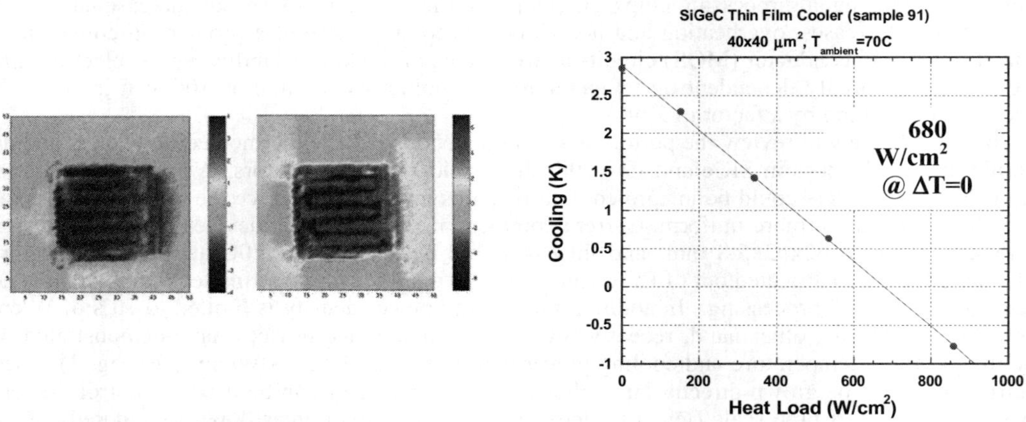

Fig. 3 Maximum cooling of a $40x40\mu m^2$ n-SiGeC refrigerator versus heat load power density.

CHALLENGES IN MICRO AND NANO SCALE TRANSPORT

Thermal Challenges in Next Generation Electronic Systems, Joshi & Garimella (eds)
© 2002 Millpress, Rotterdam, ISBN 90-77017-03-8

Anisotropic thermal conductivity of aligned carbon nanotube arrays

T. Borca-Tasciuc
Department of Mechanical, Aerospace, and Nuclear Engineering, Rensselaer Polytechnic Institute, Troy, NY, USA

R. Vajtai, B. Q. Wei, & P. M. Ajayan
Department of Materials Science and Engineering, Rensselaer Polytechnic Institute, Troy, NY, USA

M. S. Shur
Electrical, Computing, and Systems Engineering Department, Rensselaer Polytechnic Institute, Troy, NY, USA

J. Deng & R. Gaska
Sensor Electronic Technology, Inc., Latham, NY, USA

Keywords: carbon nanotubes, anodic alumina, thermal conductivity, GaN transistors

ABSTRACT: By improving the thermal conductivity of both the substrates and the package materials, carbon nanotube technology is expected to address the self-heating problem, which seriously degrades the device performance in high-power electronic and optoelectronic devices. In this work we present our ongoing efforts in the fabrication and thermal properties characterization of structures with expected high thermal conductivities based on aligned carbon nanotube arrays. The arrays are grown on selective sites on silicon substrates and into self-organized anodic aluminum oxide films with deep submicron periodic honeycomb structures. The anisotropic thermal conductivity measurement of the carbon nanotube filled anodic alumina samples is based on a differential 2-wire 3ω method. Heaters and temperature sensors instrumented onto patterned mesa structures are employed for the thermal characterization of the aligned carbon nanotube films grown on silicon substrates.

1 INTRODUCTION

The controlled growth of high purity, aligned, carbon nanotube (CNT) arrays (Zhang et al. 2000) paves the way for the possible use of CNTs as field emission sources, interconnects for 2D and 3D IC architectures, and heat transport media for the thermal management of electronics, optoelectronics, and solid state cooling devices. CNTs structures with large thermal conductivities are highly desirable, and theoretical predictions indicate that the thermal conductivity along the length of a single CNT may be equal or even larger than the thermal conductivity of diamond or graphite (Berber et al. 2000, Osman & Srivastava 2001). However, reported values of the experimentally measured thermal conductivities for several carbon nanotube structures are typically much lower than the theoretical predictions (Hone at al. 1999, Yi et al. 1999, Hone at al. 2000). Growth defects, entanglements of the carbon nanotubes ropes, and imperfect alignment, have been proposed as possible causes for the observed thermal conductivity reduction. Experimental measurements of the thermal conductivity in CNT structures have been reported for high-purity "mats" of tangled nanotube bundles (Hone at al. 1999), millimeter long bundles of aligned multiwall CNTs (Yi et al. 1999), and magnetically aligned single wall CNTs films (Hone at al. 2000). Removal of the CNTs from the deposition substrate was required to perform the alignment of the nanotubes (Hone et al. 2000) or the thermophysical characterization (Yi et al. 1999). Furthermore, the previous experimental reports are focused on thermal characterization along, or in the plane, of the alignment direction. However, the anisotropic configuration of the aligned CNT structures is expected to induce

Figure 1. SEM micrograph of an array of aligned CNT bundles grown perpendicular to the silicon substrate (shown on the left). A thin layer of amorphous carbon is present onto the top surface of the CNT film.

a highly anisotropic thermal conductivity, which could be important in the thermal management applications of CNT structures.

In this work we present our ongoing efforts in the fabrication and characterization of the anisotropic thermal properties of high purity, aligned carbon nanotube arrays grown on selective sites on silicon substrates and into self-organized anodic aluminum oxide films with deep submicron periodic honeycomb structures. A differential 2-wire 3ω method is proposed for the anisotropic thermal conductivity measurement of the carbon nanotube filled anodic alumina samples. Heaters and temperature sensors instrumented onto patterned mesa structures are employed for the thermal characterization of the aligned carbon nanotube films grown on silicon substrates.

2 SAMPLES AND MEASUREMENT TECHNIQUE

Figure 1 shows a SEM micrograph of an aligned CNT film selectively grown over a 100nm SiO_2 film deposited onto the silicon substrate. The CNT bundles grow perpendicular to the substrate and are self-standing. The total thickness of the CNT film is ~ 90µm. A thin layer of amorphous carbon is present on the top surface of the film. A more mechanically stable structure could be obtained by incorporating carbon nanotubes into self-organized aluminum anodic oxide films with deep submicron periodic honeycomb structures. Figure 2 show a SEM picture of an anodic alumina sample, which serve as a template for CNT growth. The individual pores are ~40nm in diameter. The aluminum oxide with carbon nanotubes will be an alternative substrate, which may have superior thermal performance to SiC. For example assuming the voids are filled with carbon nanotubes with a thermal conductivity of 30 W/K-cm along the tube and a 40% initial porosity of the template (the ratio of the void area compared to the whole area), the thermal conductivity of CNT filled alumina substrates can be as high as 12W/cm-K, two times higher than SiC substrate. By improving the thermal conductivity of both the substrates and the package materials, carbon nanotube technology is expected to address the self-heating problem, which seriously degrades the device performance in high-power electronic and optoelectronic devices. Figure 3 shows a proposed example for using such arrays for improved heat sinking of high power GaN-based transistors.

The anisotropic thermal conductivity measurement of the carbon nanotube filled anodic alumina samples is based on a differential 2-wire 3ω method (Borca-Tasciuc et al. 1998, Borca-Tasciuc et al. 2000). A similar method was used by Kumar et al. (2000) for the thermal conductivity charac-

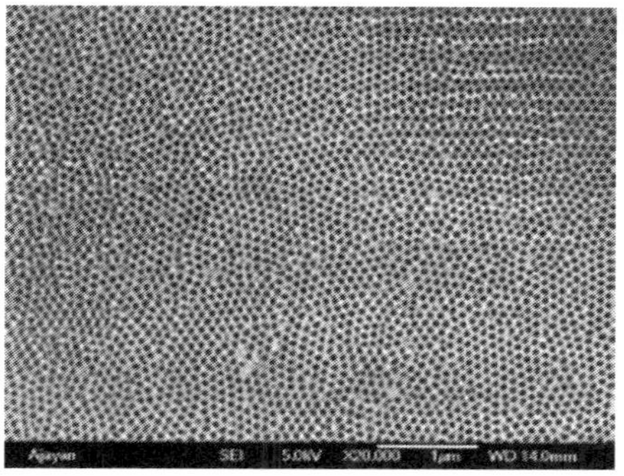

Figure 2. SEM micrograph of the top surface of a self organized aluminum anodic oxide films with deep submicron periodic honeycomb structures, which is used as a template for CNT growth.

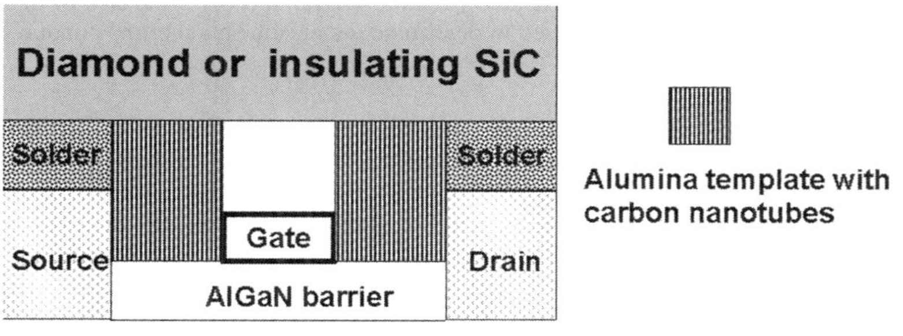

Figure 3. Possible heat sinking scheme for high power AlGaN/GaN HFETs using high thermal conductivity alumina templates with carbon nanotubes.

terization of anodized alumina templates. In the 3ω method (Cahill 1990), a long metallic wire deposited onto the sample acts as both heater and temperature sensor when modulated electrical current is passed through it. Figure 4 illustrates in cross-section the configuration of the samples necessary to perform the measurement. Metallic heaters/temperature sensors with wire widths between 2 - 40 μm are simultaneously patterned on the alumina templates and carbon nanotube filled samples using standard cleanroom microfabrication techniques such as lithography, e-beam evaporation and lift-off. For a given wire width, the same power is passed through the heaters and the frequency dependent temperature rise is recorded.

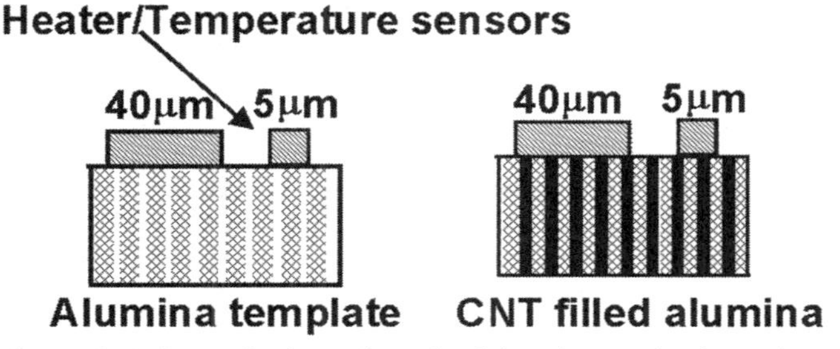

Figure 4. Anisotropic thermal conductivity characterization using microfabricated heaters/temperature sensors with different widths deposited on the alumina templates and carbon nanotube filled samples.

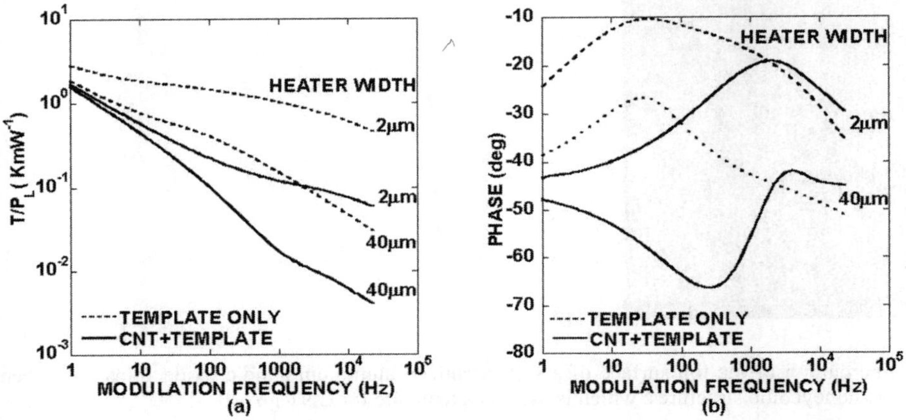

Figure 5. (a) The temperature rise normalized to power dissipated per unit length, and (b) the phase of the simulated thermal response of heaters/temperature sensors used to determine the anisotropic thermal conductivity of the empty and CNT filled alumina templates.

The anisotropic thermophysical properties of the samples are determined by fitting the experimental temperature rise collected for different wire widths with predictions of heat transport models based on 2D heat conduction in a multilayer-film-on-finite-substrate structure (Borca-Tasciuc et al. 2001). Figure 5 shows the predicted thermal response of heaters with different widths deposited over both empty and CNT filled alumina templates. The simulations consider free-standing 75μm thick alumina templates with porosity 35%, 40nm pore diameter, and assumes the channels are only partially filled (20%) with multi-wall CNT of thermal conductivity 1000W/mK. The effective thermal conductivity of the empty template was taken as 1W/mK along the direction of the channels and 0.7W/mK along the direction perpendicular to the channels. The contribution of the CNT to heat conduction in the direction perpendicular to the channels was neglected. As shown in Figure 5a, the temperature signal is more sensitive to the thermal transport along the channels as frequency increases and heater width decreases. At low frequencies and for large width heaters, the temperature signal is more sensitive to transport in the plane of the alumina template film. Figure 5b shows the phase of the thermal signal. The shift of the phase peak to larger modulation frequencies is due to the increase in thermal diffusivity along the channels of the simulated CNT filled template.

The same technique could be employed in principle to obtain the anisotropic thermal conductivity of the aligned CNT films shown in Figure 1. However, typical cleanroom microfabrication of metallic heaters and temperature sensors on the CNT film is difficult, partially due to the mechanical fragility of the films. Therefore, we used a different approach to pattern heaters and temperature sensors on the top surface of the film. Since the CNT films grow selectively over the SiO_2 surface, we patterned "heaters" on the SiO_2 film before the growth. The result after the growth of the CNT is shown in Figure 6. Millimeter long mesa structures with different heater widths and the same thickness as the CNT film are easily obtained. The amorphous carbon layer on the top surface could be used as a heater or a metallic film could be evaporated on the top surface. A differential 3ω method is employed to measure the temperature difference between same width heaters on top of CNT "mesas" with different thickness. Figure 7 shows a schematic of the differential measurement technique used to determine the thermal conductivity along the CNT growth direction. After etching the SiO_2 layer underneath the long "mesas", the structures can be used to determine the thermal conductivity of the CNT film in the direction perpendicular to the CNT growth.

Figure 6. SEM micrograph of the top view of a sample with a selectively grown CNT film forming long "mesas" on the surface. The top surface of the "mesa" is electrically conductive due to the presence of a thin amorphous carbon layer. Evaporation of a thin metal layer could be performed to increase the electrical conductivity. The numbers indicate the width (in microns) of the wire between the inner contact pads.

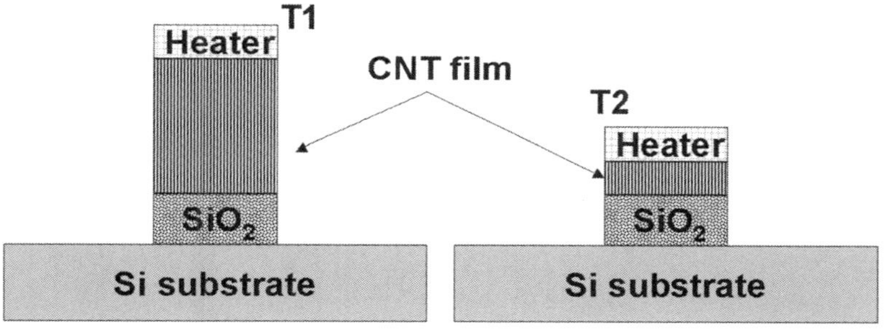

Figure 7. "Mesas" of aligned CNT films with different thickness are employed in a differential 3ω technique to determine the thermal conductivity of the film along the direction of CNT growth. After etching the SiO_2 layer underneath the long "mesas", the structures can be used to determine the thermal conductivity of the CNT film in the direction perpendicular to the CNT growth.

3 CONCLUSIONS

We present our efforts in the fabrication and thermal properties characterization of structures based on aligned carbon nanotube arrays with expected high thermal conductivities. CNT arrays are grown on selective sites on silicon substrates and into self-organized anodic aluminum oxide films with deep submicron periodic honeycomb structures. The differential 2-wire 3ω method will be used for the anisotropic thermal conductivity measurement of the carbon nanotube filled anodic alumina samples. Heaters and temperature sensors instrumented onto patterned mesa structures are proposed for the thermal characterization of the aligned carbon nanotube films grown on silicon substrates. Thermal conductivity characterization of aligned CNT arrays is currently under investigation.

4 ACKNOWLEDGEMENTS

The financial support of Office of Naval Research (00014-00-1-0250) and the Interconnect Focus Center at RPI, funded by MARCO and New York State are acknowledged by PMA. The work at

SET, Inc. was supported by the Ballistic Missile Defense Organization under Navy SBIR contract N00174-01-C-0048 monitored by Dr. J. P. Letellier of Naval Research Laboratory.

REFERENCES

Berber S. et al. 2000. Unusually high thermal conductivity of carbon nanotubes. *Phys. Rev. Lett.* 84: 4613-4616.

Borca-Tasciuc, T. et al. 1998. Anisotropic thermal conductivity of a Si/Ge superlattice. *Mater. Res. Soc. Symp. Proc.* 545: 473-478.

Borca-Tasciuc, T. et al. 2000. Anisotropic Thermal Conductivity of a Si/Ge Quantum-Dots Superlattice, *ASME-HTD* 366-2:381-384.

Borca-Tasciuc, T. et al. 2001. Data Reduction in 3ω Method for Thin-Film Thermal Conductivity Determination. *Rev. Sci. Instrum.* 72: 2139-214.

Cahill, D. G. 1990. Thermal Conductivity Measurement from 30-K to 750-K – The 3ω Method. *Rev. Sci. Instrum.* 61: 802-808.

Hone, J. et al. 1999. Thermal conductivity of single-walled carbon nanotubes. *Phys. Rev. B* 59: R2514-R2516.

Hone, J. et al. 2000. Electrical and thermal transport of magnetically aligned single wall carbon nanotube films. *Appl. Phys. Lett.* 77: 666-668.

Kumar, A. R. et al. 2000. Thermal conductivity of nanochanneled alumina. *ASME-HTD* 366-2: 393-398.

Osman, M. A. & Srivastava, D. 2001. Temperature dependence of the thermal conductivity of single-wall carbon nanotubes. *Nanotechnology* 12: 21-24.

Yi, W. et al. 1999. Linear specific heat of carbon nanotubes. *Phys. Rev. B* 59: R9015-R9018.

Zhang, Z. J. et al. 2000. Substrate-site selective growth of aligned carbon nanotubes, *Appl. Phys. Lett.* 77: 3764-3766.

Thermal Challenges in Next Generation Electronic Systems, Joshi & Garimella (eds)
© 2002 Millpress, Rotterdam, ISBN 90-77017-03-8

Chip integrated micro cooling system for high heat flux electronic cooling applications

S. Chowdhury, J. Darabi* & M. Ohadi
University of Maryland at College Park, College Park, MD, USA

J. Lawler
Advanced Thermal and Environmental Concepts. Inc., College Park, MD, USA

Keywords: electronic cooling, Electrohydrodynamics (EHD), MEMS, micropump, polarization

ABSTRACT: Rapid development in VLSI Technology anticipates an urgent need for new micro cooling strategies for high heat flux electronic chips. This paper presents a novel high efficiency micropump/evaporator, based on electrohydrodynamic (EHD) forces, high heat flux electronic sensors. The device incorporates a MEMS-based active micro-evaporative surface and temperature sensors into a single package that can be directly attached to the heat source. The pump is designed to supply and maintain an ultra-thin liquid film on a 10 × 15 mm evaporative surface, which was a quartz or sapphire wafer with an array of EHD electrodes having gaps varying between 10 and 100 μm. A pumping pressure of up to 300 Pa and a cooling capacity of 25 W/cm2 were obtained at electric field strength of 4 V/μm using a 3M electronic coolant (HFE-7100) as the working fluid. The EHD power consumption in this case was only 2.5 mW for the highest heat flux. Electrical field simulations were performed using FemLab-2D software to understand the effects of electrode geometry on the electric field and on the fluid pumping effect. This work is part of an ongoing effort to develop a complete self-controlled two-phase micro cooling system for cooling single and multiple electronic components.

1 INTRODUCTION

The miniaturization of electronic components and the rapid increase in power density of advanced microprocessors and electronic components have created a need for improved cooling technologies to achieve high heat dissipation rates. Some of future microprocessors and power-electronic components have been projected to dissipate over 100 W/cm^2. The traditional methods of electronic cooling such as conduction and natural/forced convection have proved to be insufficient for such high heat fluxes. The need for new technologies capable of dissipating high heat fluxes and requiring low input power is of critical importance to many industries. Coupled with these are requirements of low volume/weight addition, high portability, high reliability and large-scale fabricability of such future cooling devices.

Among various microfluidic devices, EHD pumping is a promising option, since it offers the unique advantages of no moving parts, low cost and minimal power consumption. The use of electrohydrodynamic forces for pumping of fluids at the micro-scale has been widely investigated (Bryan et al, 1992; Wong et al., 1996; Bart et al., 1996; Darabi et al., 2001). Also, evaporation of thin liquid films has long been recognized as a means of achieving high heat transfer coefficients. The devices developed in the present study uniquely combine fluid pumping concept based on polarization electrohydrodynamics of dielectric coolants with rapid evaporation of thin liquid films

* Currently with University of South Carolina, Columbia, SC, USA-20298

for high heat removal rates. The resistance to heat flow during the evaporation process in many circumstances is due to conduction and/or convection through the liquid film on the heated surface. One method to reduce the heat transfer resistance of the film is to control the film thickness. The electric field offers an alternative means to control the film thickness in thin-film evaporation.

One of the most critical issues in the electronic industry is the thermal management of electronic packaging and maintaining specific temperature uniformity across the chip. The device developed in this study presents a promising alternative that will provide a uniform temperature distribution and heat removal at high heat flux levels. These factors directly affect the performance, cost, and reliability of electronic devices.

2 THEORETICAL BACKGROUND

The mechanism by which liquids rise between a pair of parallel electrodes if subjected to an electric field can be understood by considering the simple case shown in Figure 1 in which a pair of parallel electrodes is dipped into a dielectric liquid (Melcher, 1981).

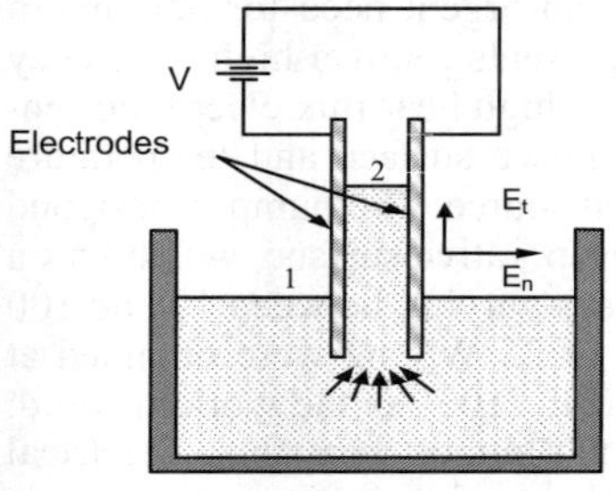

Figure 1. Liquid pumping with a pair of parallel electrodes dipped in a dielectric liquid

If a potential difference V is applied between the electrodes, the dielectric liquid rises between the electrodes due to a net electrostatic force acting on dipoles in the liquid. This force arises from the local electric field gradient at the end of the electrode pair. In a polarized medium, each individual dipole is subjected to an electric force. A dipole can be represented by a pair of oppositely signed charges of (+q) and (–q) separated by a distance $\vec{r}$ (Fig. 2). The net force on each dipole can be given by:

$$\vec{f} = \vec{p} \cdot \nabla \vec{E} \tag{1}$$

where $\vec{p} = q \cdot \vec{r}$ is the dipole moment. Melcher [Melcher, 1981] assumed that the dipoles were subject to an average or macroscopic electric field.

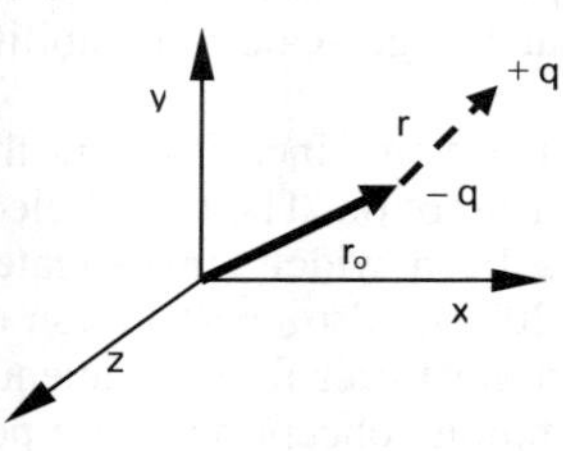

Figure 2. Dipole moment

This assumption ignores the distortion of the electric field at one dipole because of neighboring dipoles. The total polarization force can be found by averaging these polarization forces over all dipoles within the dielectric medium. One simple approach would be to multiply the net force on each dipole by n, the total number of the dipoles per unit volume. Defining the polarization density as $\vec{P} = n\vec{p}$, the polarization force density can be written as:

$$\vec{F} = \vec{P} \cdot \nabla \vec{E} \tag{2}$$

Thus the relation between the polarization force and electric field, $\vec{E}$ is (Melcher, 1981):

$$\vec{P} = \varepsilon_0 (\kappa - 1) \vec{E} \tag{3}$$

where κ is the dielectric constant. Substituting Equation (3) into Equation (2) and assuming constant permittivity, the polarization force density becomes:

$$\vec{\mathrm{F}} = \nabla \left(\frac{\varepsilon_0}{2} (\mathrm{k} - 1) \vec{\mathrm{E}} \cdot \vec{\mathrm{E}} \right) \tag{4}$$

The polarization force given by Equation (4) does not take into account the interaction between dipoles. Moreover, it is an average over a spectrum of dipole moments. Now, generalized equation of motion for a viscous incompressible fluid is given by the Navier-Stokes equation as:

$$\rho_m \frac{D\vec{V}}{Dt} = \rho_m \vec{F}_b - \nabla p + \mu \nabla^2 \vec{V} \tag{5}$$

and $\nabla . \vec{V} = 0$

where ρ_m is the fluid density, μ, the dynamic viscosity, p, the pressure and $\vec{F}_b$ the generalize body force. The first term on the right is due to the body force, the second due to pressure, and the last term is due to viscosity or fluid friction. In case of a fluid at rest in an electric field, we can write:

$$\nabla p = \vec{F}_e \tag{6}$$

If p is the pressure in the liquid, the mechanical volume force $-\nabla p$, which is set up as a result of the pressure gradient, is equal and opposite to $\vec{F}_e$.

By a comparison of Equations (4) and (6), we obtain:

$$p_1 - p_2 = \frac{\varepsilon_0 (k-1)}{2} E^2 \tag{7}$$

Equation (7) gives only the net difference in pressure from Level 1 to Level 2 in Figure 1. It does not describe the detailed pressure behavior in the liquid between these two interfaces. The height to which the liquid rises against the gravity between the electrodes may be expressed as:

$$h = \frac{\varepsilon_0 (k-1)}{2\rho g} E^2 \tag{8}$$

where ρ is the density of the liquid, and g is the acceleration due to gravity. The force that pushes the liquid up against the gravity is exerted at the lower edges of the plates where the field is inhomogeneous and the electric field density varies rapidly. This mechanism can also be explained by the fact that the energy of the dipoles in a field-free space is higher than their energy in the electric field. Thus, the dipoles in the field-free region are pulled into the region of higher electric field between the plates, where their potential energy is reduced. In the present device, the electrodes are planar and there are no capillaries or microchannels.

3 DESCRIPTION OF DEVICE

3.1 *Electrode design*

A conceptual design of the device is shown in Figure 3.

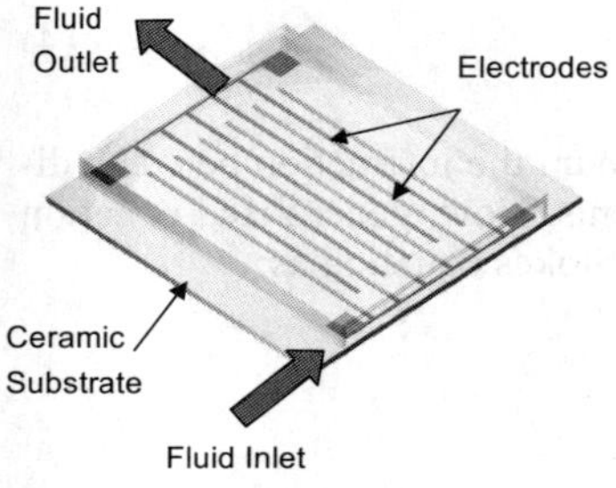

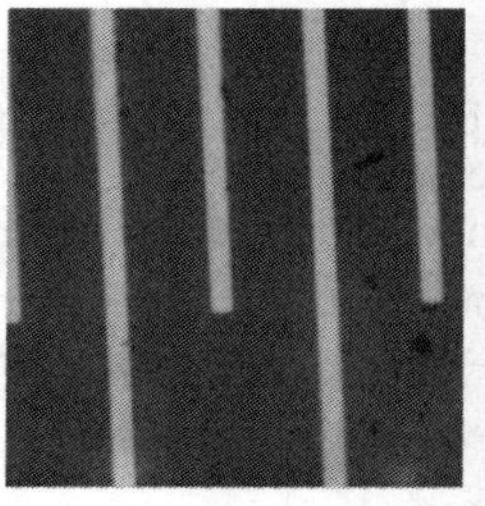

Figure 3. Schematic of polarization micropump and close-up view under microscope (50×)

The device incorporates an active evaporative cooling surface, a polarization micro-pump, and temperature sensors into a single package. On one side of the substrate wafer an array of electrodes were created from a 0.25 μm thick Niobium metal layer that wads sputter-deposited, patterned by photolithographic process and finally etches by DRIE process. The electrode geometry and dimensions have been selected to allow a wide range of tests conditions. The basic designs of the electrode array is shown in Figure 4.

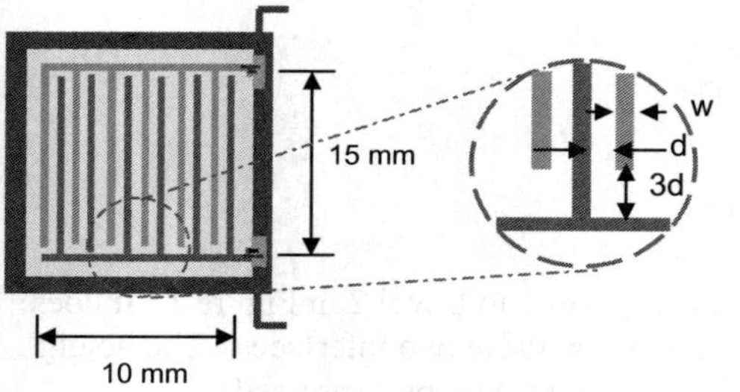

Figure 4. A Schematic diagram of the two electrode patterns

The mask layouts for these designs are drawn with *Cadence* software. Four designs with constant gap electrodes have gaps d = 10, 20, 50, and 100 μm between electrodes. The width of each electrode is w = 10 μm, and the thickness is 0.25 μm. Each electrode pattern is 10 mm × 15 mm centered on a 19 mm × 21 mm polished wafer of sapphire or quartz.

3.2 *Fabrication and packaging*

The fabrication sequence uses several processes and begins with wafer pre-metallization cleaning, followed by a deposition of a layer of Niobium using an e-beam evaporator. Pattern transfer is achieved by standard photolithography process followed by deep reactive ion etching (DRIE) to give the electrode patterns. The sapphire/quartz wafer is then diced into 19 mm × 21 mm individual devices. Each device is mounted on a 16 mm × 21 mm enclosure with integrated inlet and outlet ports for the fluid (Fig. 5(a)), produced by curing photosensitive resin with computer-aided-manufacturing (CAM) rapid prototyping equipment.

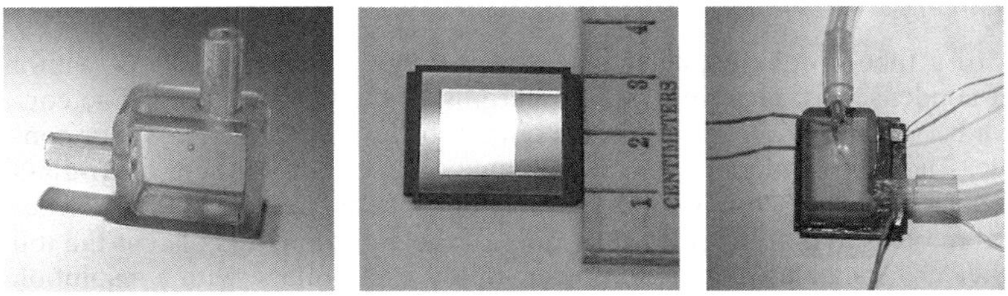

(a) (b) (c)

Figure 5. (a) Photo-cured resin enclosure; (b) A photograph of the actual device without enclosure. The shiny region consists of closely spaced electrodes; (c) Final Packaged device with thermocouples and lead wires

A lead-less ceramic chip carrier (LCC) of dimensions 25 mm × 25 mm × 1mm is used to package the device. The device (wafer) is mounted on the flat surface of the chip-carrier using a thin layer (0.2 mm) transparent epoxy. Figure 5(b) shows a photograph of the module without the resin enclosure. The shiny region on the device contains the patterned electrodes. The other face of the chip carrier contains an 11.5 mm × 11.5 mm cavity to place the actual silicon chip. To simulate the actual chip as a heat source, a palladium foil heater of dimensions 5 mm × 5 mm with a resistance of 10 Ω is mounted into the cavity using a thin film (~ 0.1mm) of pressure sensitive adhesive as backing. After the curing process, two small copper tabs are placed on the edge of the LCC. The copper tabs are then wire bonded to the bond pads/leads on the wafer. External lead wires are soldered to these copper tabs. The enclosure covers the electrode surface and is bonded to the substrate (wafer) using transparent epoxy. To facilitate the temperature measurement, two small holes were drilled at the inlet and outlet ports of the enclosure and tips of two very thin thermocouples were inserted in the holes. The holes were then sealed using epoxy glue. The connections to the cooling loop were made using vinyl tubing. The final packaged form of the device in Figure 5(c).

3.3 *Working fluid*

The criteria for the selection of test fluids were quite challenging, since a moderate dielectric constant with good heat transfer and electrical properties are required. A new fluids from 3M, tailored for electronic cooling purpose, namely HFE-7100, was selected. This fluid has low vapor pressure, and hence, do not require a high-pressure closed-loop system. Also, it has a good dielectric constant, k ~7.4, a relatively low viscosity high thermal conductivity. The boiling point of HFE-7100 fluid is 61°C at atmospheric pressure. Thermo-physical properties of the fluid are listed in Table 1.

Table 1. Thermo-physical properties of HFE-7100

Thermo-physical Property	HFE-7100
Boiling Point (°C)	61
Liquid density (kg / m^3)	1402
Liquid kinematic viscosity (cSt)	0.38
Liquid specific heat (J / kg / K)	1253
Dielectric strength (kV / mm)	11
Dielectric constant	7.4

3.4 *Cooling loop design and instrumentation*

The cooling loop consists of a tube-in-tube copper condenser of cooling capacity 100 W with a needle valve at the inlet of condenser for charging and discharging of the working fluid. The condenser is a copper tube-in-tube. Chilled water flows in the outer tube as the refrigerant vapor condenses inside the inner tube. The loop components were mounted on a 15 cm × 20 cm plexi-glass plate. Temperatures were measured using thin T-type thermocouples calibrated with a high accuracy RTD. Temperatures were measured at the inlet and outlet of the enclosure and also at the top and side edges on the foil heater. A standard 1 kV DC power supply with voltage with a resolution of 1 V and 1μA was used to provide necessary voltage to the EHD device. The power to heater was provided by a separate 30V/0.5A DC source. The power to heater was measured by a 4-wire method for better accuracy. A Hewlett-Packard (HP) data acquisition system with associated peripherals was used to monitor and acquire data for temperatures, voltages and currents. The schematic of the test loop is shown in Figure 6.

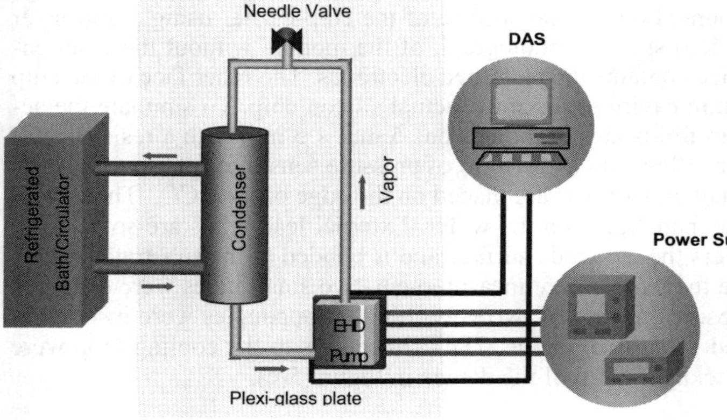

Figure 6. Schematic of the test loop

4 EXPERIMENTAL PROCEDURE

4.1 *Pumping effect visualization*

Testing and flow visualization using a video camera was performed to determine the rise of liquid film on the active surface due to the action of EHD polarization voltage applied to the pump. The experimentation was done in open atmosphere by placing the device without enclosure upright on a groove in a Teflon block inside a beaker. The beaker was partly filled with HFE-7100 upto a level that was approximately 2 mm above the bottom edge of the electrodes. The rise of the liquid film pulled up by the pump was determined by enlarging the video frames containing the device and a set of calipers. Measurements were performed using three different modules: Module I with 100 μm electrode gap, Module II with 50 μm gap and Module III with 20 μm.

4.2 *Heat transfer experimentation*

Experiments to measure heat transfer capabilities of the device is done with the packaged device connected to the complete loop as depicted in Figure 6. HFE-7100 is injected into the system through the needle valve. The heater is put on and vapors of the fluid are generated. The pressure is released after a while so that the vapor of coolant drives out the non-condensable gases (air). This process is repeated several times to make sure that there is only negligible amount of air present in the system. The needle valve is then closed, and an experiment is run by increasing the input voltage to the heater. At each heat input, the system was allowed a sufficient amount of time to stabilize and temperatures of the heater surface, liquid inlet and vapor outlet are recorded. As the heat input was increased, the pump voltage was also increased to supply more liquid to the evaporative surface.

5 RESULTS AND DISCUSSION

Figure 7 shows the pumping head results of three devices as compared with the theoretical value. The pumping curves was repeatable and showed parabolic nature similar to the theoretical curve. No capillary effect was observed visually. The results clearly indicate that with smaller gaps lower voltages are required to get the same pumping effect. An interesting observation was that the device tolerated electric fields at and even beyond macroscopic breakdown potential of air indicating that behavior at microscopic level is vastly different than that which is observed on large scale.

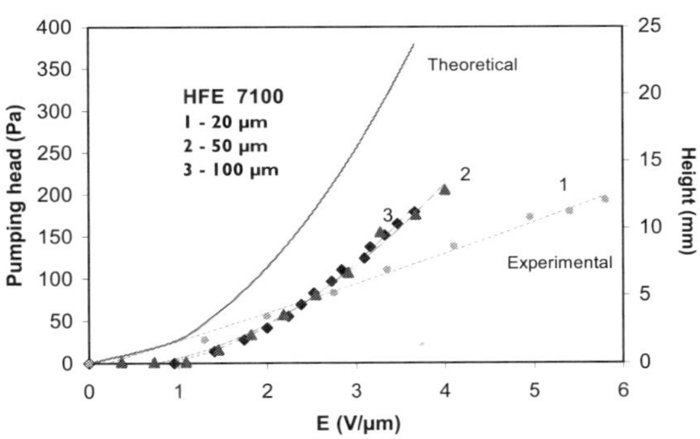

Figure 7. EHD Pumping results

Selected heat transfer results are shown in Figure 8. The heat flux was calculated based on nominal heater area. The results indicate a maximum heat flux of 25 W/cm^2. The chip temperature at this maximum heat flux was 97.5 °C and maximum power consumed by the pump was only 2.5 mW, which is only 0.01% of the cooling rate in the test device.

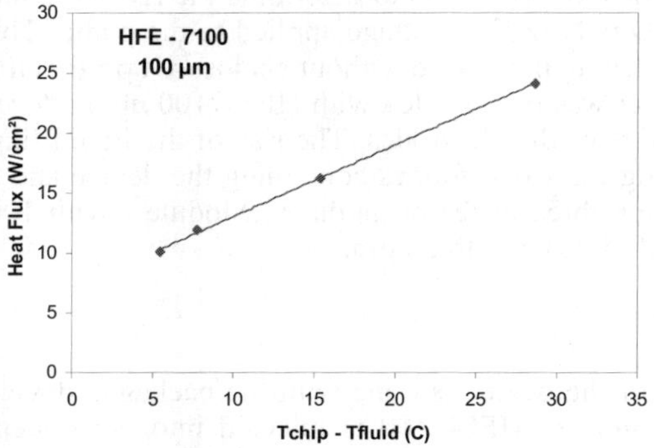

Figure 8. Heat flux versus ΔT for a device with a 100 μm

Several exercises were also made with enclosure-less devices placed in a beaker to observe the modes of heat transfer. It has been found that at very low heat fluxes, the heat transfer was based on convection mode; liquid was seen to be flowing up along the surface partly evaporating and partly overflowing from the top end on the electrodes. For higher heat fluxes, it was observed that the liquid level at a particular voltage dropped with increase in the heat input. A dry out region existed above the top liquid film. However increasing the pumping voltage had an effect of pulling up of the liquid film and thereby reducing the dry out region. At zero heat input the pumping head measured from the height of the liquid film achieved was 180 Pa. However, this pressure head is exceeded as we increase the heat input and increase the pumping voltage to reduce the dry out region. The pressure head thus achieved was calculated not on the basis of height but on the theoretical equation giving values over 450 Pa. For intermediated to high heat fluxes, the dominant mode was still observed to be convective with tiny bubbles appearing at the top edge of the liquid film which is the thinnest part of the film on account of evaporation. These bubbles indicate that some 1 boiling at the top edge of the film.

For very high heat fluxes, beyond 25 W/cm^2 the dry out region was large and the liquid film showed unstable oscillations. The chip temperature also drastically increased on account of the large dry out region. Reasonable increase in pumping voltage failed to restore the fluid film because of excessive demand for fluid. However it must be noted that this device utilizes minimal active heat transfer area, which is even less that the area of the LCC. Much higher heat fluxes can be achieved with an expanded design of the active area and improved electrode geometry.

The situation can also be mitigated to a certain extent, by an inclination of the pumping surface at a certain angle to the horizontal so that at a given voltage (pumping head) a greater region of wetting can be achieved. Such effects of inclination have been observed to lessen chip temperature. However, in order to determine detailed effects and optimum angle of inclination a series of experiments at different orientations needs to be performed. Near horizontal orientations will result in flooding of the surface by bulk liquid thereby preventing formation of the EHD-induced thin film and hence reducing the situation to one of pool boiling.

Simulation of electric field for a simplified case of two electrodes is shown in Figure 9. An explanation of the discrepancy between theoretical and observed pumping curves can be explained from this simulation. It can be seen that at the sharp corners of the electrodes a steep gradient of electric field exists, whereas there is a moderate gradient of electric field in the central region. These two gradients are in opposition. The gradient due to the sharp corner is acting only around the corner region, thereby reducing the effect of the central gradient only in this region. The conflict of these two gradients is more acute at higher electric field and smaller gaps, thus explaining the deviant behavior of a 20 μm gap device from the theoretical curve.

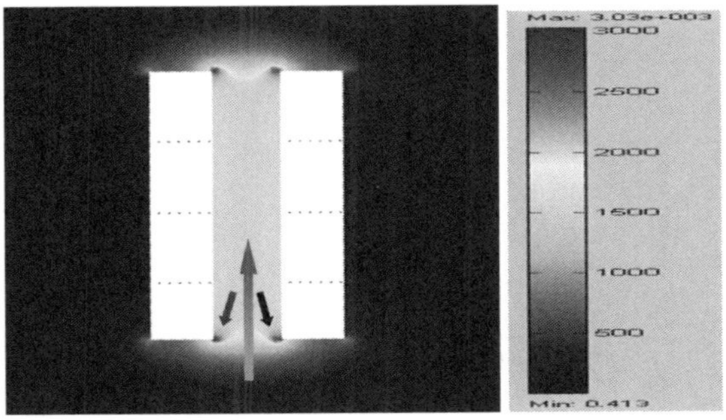

Figure 9. Numerical simulation of simplified parallel electrode geometry using FEMLab2D

6 CONCLUSIONS

An experimental investigation was conducted to study the feasibility of combining a novel EHD-induced ultra thin film evaporation technique with MEMS-based fabrication methods to create an efficient heat rejection unit for the purpose of cooling of high heat flux electronic sensors and chips. Several integrated active micro pump/evaporator devices with different electrode geometries have been fabricated and tested. Cooling rate of 25 W/cm^2 was achieved with a minimal EHD power consumption of 2.5 mW. Much higher heat fluxes should be achievable with proper design improvements. This technology shows great promise for commercialization due its to low voltage requirements and compact geometry.

REFERENCES

Bart, S.F., Tavrow, L.S., Mehregany, M., & Lang, J.H., 1996, Microfabricated electrohydrodynamic pumps, Sensors and Actuators, A. Physical. A21-23: 193-197.

Bryan, J., Seyed-Yagoobi, J. 1992. An experimental investigation of ion-drag pump in a vertical and axisymmetric configuration. IEEE Transactions of Industrial Applications, 28 (2): 310-316.

Castellano, A. 1998. Electrohydrodynamics, New York: Springer Wein.

Darabi, J., Ohadi, M.M., De Voe, D. 2001. An electrohydrodynamic polarization micropump for electronic cooling", Journal of Microelectromechanical Systems, 10, (1) : 98-106.

Melcher, J.R. 1981. Continuum Electromechanics, The MIT Press.

Panofsky, W.K.H. & Phillips, M. 1962. Classical Electricity and Magnetism. Reading : Addison-Wesley.

Richter, A., Plettner, A., Hofmann, K.A. & Sandmaier, H. 1991. A micromachined electrohydrodynamic (EHD) pump. Sensors and Actuators, 29 (2): 159-168.

Wong, C.C., Adkins, R.D., Chu, D. 1996. Development of a micropump for microelectronic cooling. Microelectromechanical Systems (MEMS). 59: 239-244. ASME.

Thermal Challenges in Next Generation Electronic Systems, Joshi & Garimella (eds)
© 2002 Millpress, Rotterdam, ISBN 90-77017-03-8

Simultaneous, spatially-resolved temperature and velocity measurements in microchannel flows

Steve Wereley
School of Mechanical Engineering, Purdue University, West Lafayette, IN, USA

Vince Hohreiter & Jacob Chung
Department of Mechanical Engineering, University of Florida, Gainesville, FL, USA

Keywords: micro particle image velocimetry, PIV, temperature, Brownian, microchannel

ABSTRACT: This paper demonstrates an extension to the micro-particle image velocimetry (μPIV) technique that allows simultaneous velocity and temperature measurement. The technique is based on the observation that Brownian motion will cause width-wise broadening of the cross-correlation peak. A standard μPIV algorithm measures the deterministic component of the tracer particle velocity while a new algorithm measures the increase in correlation peak width caused by the random, uncorrelated Brownian particle motion. The Brownian motion can be related directly to the ratio of dynamic viscosity to absolute temperature. Experimental results were obtained using fluorescent (rhodamine 542/612) spherical 700nm diameter polystyrene-latex particles in water. Temperature changes up to 25°C were determined with an experimental uncertainty of ±3°C. Modeling results show that this uncertainty can be reduced to less than ±1°C. Finally, guidelines are established for choosing experimental parameters anywhere in the temperature/velocity phase space.

1 INTRODUCTION

Responding to the thermal challenges imposed by the next generation of electronic systems will require experimental techniques capable of fully and accurately characterizing flows in of small-scale devices. In the current context, fully characterizing a flow includes measuring its velocity field as well as its temperature field, preferably simultaneously. Regardless of the length-scale of the experimental work, fluid dynamics and heat transfer intertwine to govern the physics of mechanical processes. Therefore the ability of a micron-resolution particle image velocimetry (μPIV) system to produce high-resolution velocity data naturally begins the search for methods that provide temperature measurements with the same high spatial resolution. The small scales of such MEMS devices often make physically probing a flow impractical, as even the smallest probe would likely be on the order of the size of the device itself. This condition puts minimally invasive measurement techniques, such as PIV, at a premium.

Early literature on μ-PIV noted that for small seed-particles (<1μm) and low speed flows (<10μm/s) Brownian particle motion was measurable and could be significant enough to introduce considerable error into velocity measurements (Santiago et al., 1998). Because it was viewed as an undesirable effect on velocity data, the effect of Brownian motion on the PIV correlation function—namely a width-wise spreading of the correlation peak which added to the uncertainty of locating the peak center—was often intentionally and substantially reduced through ensemble averaging over multiple images for a given flow field. However, Olsen and Adrian (2000a) postulated that such spreading of the correlation peak could be used to deduce fluid temperature since Brownian motion has a direct temperature dependence. Furthermore, since peak location (which yields velocity information) and peak width are independent parameters it is conceivable that PIV may be used to simultaneously measure both temperature and velocity.

The following work experimentally applies the latter approach to the creation and analysis of μ-PIV data. Instead of imaging a moving flow, measuring its velocity field, and ensemble-averaging away the effects of Brownian particle motions (which take the form of correlation peak broadening), this work—in implementing PIV-based thermometry—images stagnant flows and measures the variations in correlation peak broadening with variations in temperature. The end-goal, which is theoretically possible but not experimentally treated in this work, is a μ-PIV algorithm which interprets both peak-location and peak-width data from the same image set to yield both velocity and temperature information for the flow field.

2 THEORY OF PIV-BASED THERMOMETRY

2.1 *Particle diffusion*

The theoretical basis for the hypothesis of the current work—that temperature measurements can be deduced from imaging particles in a flow—rests in the theory of particle diffusion. Constant random bombardment by millions of fluid molecules causes a random resultant displacement in the seed-particles, whether or not there is a net fluid velocity. The expression

$$D = KT / 6\pi\mu d_p \tag{1}$$

gives the diffusivity D of particles of diameter d_p immersed in a liquid of temperature T and kinematic viscosity μ; K is Boltzmann's constant. Noting the dimension of D is [length^2/time], the square of the expected distance traveled by a particle with diffusivity D in some time window Δt is given by

$$< s^2 > = 2D\Delta t \tag{2}$$

Combining equations (1) and (2) it is observed that an increase in fluid temperature, with all other factors held constant, will result in a greater expected particle displacement, $\sqrt{< s^2 >}$. However, absolute viscosity, μ, is a strong function of temperature and $< s^2 > \propto T/\mu$. The effect of this relationship depends on the phase of the fluid. For a liquid, increasing temperature decreases absolute viscosity—so the overall effect on the ratio T/μ follows the change in T. For a gas, however, increasing temperature increases absolute viscosity meaning that the effect on T/μ , and hence $< s^2 >$, would need to be determined from fluid specific properties.

2.2 *PIV considerations*

Relating changes in fluid temperature to changes in the magnitude of seed-particle motion is the first step in the analysis. To qualify the use of PIV for temperature measurement, the following section relates the magnitude of seed-particle motion to changes in the cross-correlation peak width or, alternately, cross-sectional area. Brownian motion necessitates that, in a given time window, all particles immersed in a fluid undergo slight random variations in displacement, resulting in a correlation function with a predictable and measurable peak width increase over the case of no Brownian motion. Figure 1 demonstrates this peak width broadening phenomenon.

Following Keane and Adrian (1992) the spatial cross-correlation of images $I_1(\mathrm{X})$ and $I_2(\mathrm{X})$ can be locally estimated by the convolution integral

$$R(\mathrm{s}) = \int I_1(\mathrm{X})I_2(\mathrm{X} + \mathrm{s})d\mathrm{X} \tag{3}$$

Of principal concern in this work is the randomness of the shift, s, over each of the regions whose sum comprise the correlation, $R(\mathrm{s})$—the more random the shift of each particle between images $I_1(\mathrm{X})$ and $I_2(\mathrm{X})$, the more broad the resulting correlation peak.

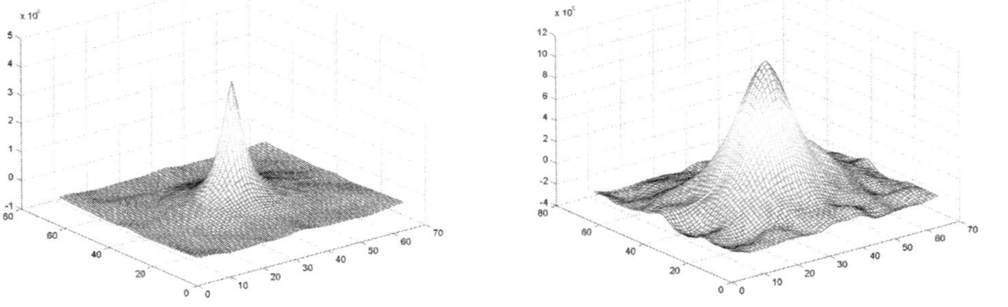

Figure 1. A pair of correlation functions demonstrates possible variations in peak width. Note also that the height of the correlation peak reduces from 5×10^6 to 1.2×10^6.

The relationship between a particle's actual cross-sectional area and that of its image is governed by the characteristics of the imaging optics (Adrian, 1991). The image of a particle with finite diameter is the convolution of two quantities: the geometric image of the particle and, given a diffraction-limited lens, the point response function of the imaging system. The point response function is an Airy function with diameter given by

$$d_s = 2.44(1 + M)\, f^{\#}\, \lambda \tag{4}$$

where M is the magnification of the lens, $f^{\#}$ is the f-number of the lens, and λ is the wavelength of light reflected, scattered, or fluoresced by the particle.

If both the geometric image and the point response function are approximated by Gaussian functions, the following approximate formula for particle image diameter is obtained

$$d_e = (M^2 d_p^2 + d_s^2 + d_z^2)^{1/2} \text{ with } d_z = M\, z\, D_a / (x_o + z) \tag{5}$$

where d_p is the actual particle diameter, z is the distance of the particle from the object plane, D_a is the aperature diameter of the microscope objective, and x_o is the object distance (Olsen and Adrian, 2000b). It should be noted that d_z changes throughout the interrogation volume making d_e difficult to calculate for volume illumination PIV.

Following the theoretical analysis of Olsen and Adrian (2000a), the effects of Brownian motion on cross-correlation PIV are seen in the correlation peak width, Δs_o—taken as the 1/e diameter of the Gaussian peak. Following an extensive derivation, Δs_o can be approximated as

$$\Delta s_{o,c} = \sqrt{2}(d_e^2 + 8M^2\beta^2 D\Delta t)^{1/2}/\beta, \tag{6}$$

where β is a parameter arising from the approximation of the Airy point-response function as a Gaussian function. Adrian and Yao (1985) found a best fit to occur for β^2=3.67.

The difficulty in calculating d_e for volume illumination PIV can be avoided by strategic manipulation of equation (6). The difference in area between the auto-correlation (Δt=0) and cross-correlation (arbitrary Δt) peaks can be directly related to the temperature-viscosity ratio as

$$\Delta A = \pi/4\, (\Delta s_{o,c}^2 - \Delta s_{o,a}^2) = C_0\, (T/\mu)\, \Delta t \tag{7}$$

where C_0 is the parameter $2M^2K/3d_p$.

The expected particle displacement, $\sqrt{< s^2 >}$, of classical diffusion theory can now be tied to the peak width of the correlation function, Δs_o, or the change in peak area , ΔA, through the diffusivity, D.

3 SIMULATIONS

Before any measurements were made using the μ-PIV system, an experimental model and feasibility analysis were designed using simulated particle images and the cross-correlation PIV algorithm. The initial analysis allowed for the characterization of the algorithm's sensitivity to particle displacement and anticipate the likelihood that good experimental images would result in measurable trends.

The particle image simulation code creates randomly spaced white particle images of similar size and number density (particles / pixel area) to an expected experimental particle image. The particle images are given a Gaussian intensity distribution. After an initial particle image was generated, a random number generator (mean-zero, normal distribution) was used to impart a random shift to the x and y position of each particle. The particle shift routine was repeated several times, each time imparting a random shift from the previous particle positions, in order to produce an image set. These successive shifts approximated the spatial shifts that the experimental particles undergo between frames of video.

The simulated particle images were then cross-correlated and the trends in peak broadening and peak location observed. Assuming truly random shifts, it was expected that the average correlation indicate a net-zero velocity. Furthermore, because the shifts were done in an additive fashion, a simulated sequence of images,1-2-3, should yield an increase in peak-width from a correlation of 1-2 to a correlation of 1-3, and so on.

4 EXPERIMENTS

An Olympus BX50 system microscope with BX-FLA fluorescence light attachment (housing the dichroic mirror/prism and optical filter cube) was used to image the particle-laden solution, see Figure 2. All experiments were carried out with a 50X objective (NA=0.8). A Cohu (model 4915-3000) 8-bit CCD video camera was used—the CCD array consisted of 768 (horizontal) x 494 (vertical) pixels and a total image area of 6.4 x 4.8 mm. The particles—700nm diameter polystyrene latex microspheres (Duke Scientific, Palo Alto, CA)—had a peak excitation wavelength at 542nm and peak emission at 612nm. A variable intensity halogen lamp was used for illumination—optical filters were used to isolate the wavelength bands most applicable to the particles—520-550nm for the incident illumination, and >580nm for the particle fluorescence.

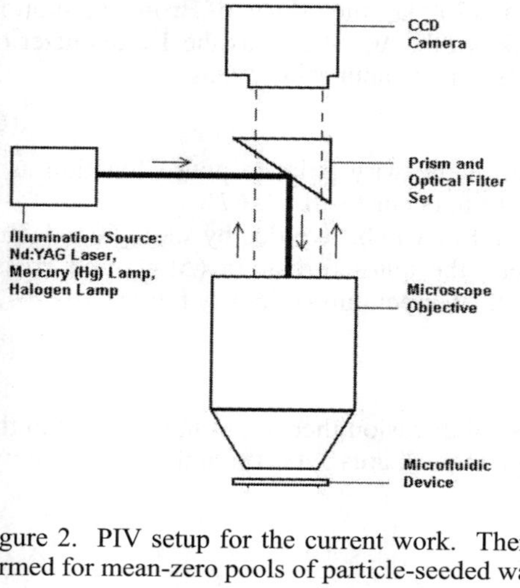

Figure 2. PIV setup for the current work. There is no flow actuation device as all experiments were performed for mean-zero pools of particle-seeded water.

In the experiments voltage supplied to a patch heater was adjusted incrementally and video of particle motion was captured at each of several temperature steps. At each temperature, the system was allowed 5 minutes to come to a steady temperature which was recorded by a thermocouple. The gating between frames was controlled by the camera frame rate of 30 frames per second. In order to get numerical data, the individual frames were cross-correlated using the PIV algorithm (Wereley et al., 1998). The first frame in each sequence was correlated with itself (i.e. autocorrelated) to provide a reference value of correlation peak width. Then the first frame was cross-correlated with the second frame, and then with the third, and the fourth, and so on—time between cross-correlated frames linearly increased. Each individual cross-correlation of such successive pairs of images produced one correlation peak (the average over each interrogation region peak). The width of each peak was recorded and trends in the measurements of peak widths were observed. Temperatures from 20°C to 50°C were measured in this way.

5 RESULTS

5.1 *Experimental uncertainty*

The results of the simulated particle images are shown in Figure 3(left) while the experimental results are shown in Figure 3(right) for a certain fixed temperature. In both figures, the PIV-determined peak area data are plotted as the rhombus-shaped points while the line is a best fit linear profile. The salient feature of these two plots is that the area increase is very nearly proportional to time, as predicted by equation 7. From a set of such figures, it is possible to extract a calibration constant representing the particle diameter, Boltzmann's constant, pi, etc. The quality of the linear fit gives an estimate of the uncertainty of the technique. The root-mean-squared (RMS) difference between experimental and theoretical values of peak area increase (at each data point) was taken for many such curves. This RMS difference in square-pixels was converted to an equivalent temperature difference, which is the nominal value of temperature resolution given above. The average error over all the test cases is about ± 3°C. By combining many such data sets, the temperature at many different values can be assessed and compared to the thermocouple measurements, as in Figure 4(left).

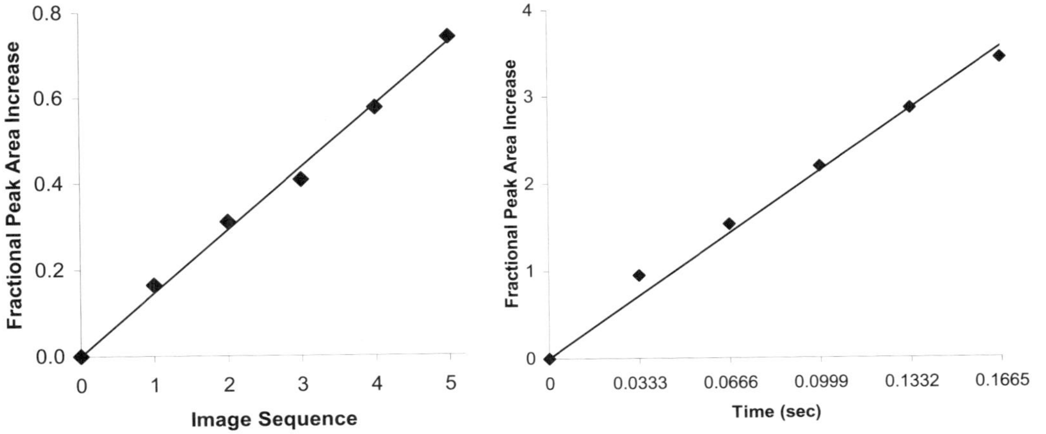

Figure 3. Measured versus theoretical peak broadening trend for particle diffusion in simulated images (left) and actual experimental images (right).

To further explore the uncertainty of the measurements, Equation 7 will be used in a standard uncertainty analysis to find

$$\delta T = \pm T \cdot \left[1 - \frac{T}{\mu}\frac{d\mu}{dT}\right]^{-1} \cdot \sqrt{\left(\frac{\delta(\Delta A)}{\Delta A}\right)^2 + \left(2\frac{\delta M}{M}\right)^2 + \left(\frac{\delta(\Delta t)}{\Delta t}\right)^2 + \left(\frac{\delta d_p}{d_p}\right)^2} \tag{8}$$

As an estimate of the uncertainty in area difference $\delta(\Delta A)$, we can use a representative value from both the experiments and simulations. For the simulations, $\delta(\Delta A)$=0.95 pixels^2 and ΔA=42 pixels^2 while for the experiments, $\delta(\Delta A)$=5.0 pixels^2 and ΔA=126 pixels^2. Using these values we can plot in Figure 4(right) the expected uncertainty of the measurements and simulations over the range for which water is normally a liquid. The measurement uncertainty of the simulations is about half as large as that of the experiments. There are several possible explanations for this. One is that the experiments were performed using volume-illuminated micro-PIV while the simulations were done more in keeping with light sheet PIV, wherein the particles have uniform brightness and size within the light sheet and no particles are visible in front of or behind the light sheet. Both of these factors tend to lower signal-to-noise ratio and raise the uncertainty of the measurements. The second reason is that the statistics may not have been stationary for the experimental results. Further experiments and simulation using more particle images and better simulation approximations will shed further light on the best possible results achievable.

The one unusual result in this uncertainty calculation is the term raised to the negative one power that comes from the temperature dependence of the viscosity combined with the assumption that the viscosity is a known property of the fluid dependent only on the temperature. The values of the parameters in Equation 7 used in the current work can be substituted into Equation 7 to find the expected accuracy of the technique. At this stage we will assume that the particles are monodisperse (δd_p=0), that the laser pulse separation is exactly known ($\delta(\Delta t)$=0), and that the magnification of the microscope is exactly known (δM=0). These assumptions are reasonable because the uncertainty in the area difference $\delta(\Delta A)$ dominates the right hand side of Equation 7.

5.2 *Simultaneous velocity and temperature measurement*

In the introduction section of this work, the potential for the simultaneous measurement of both ve-

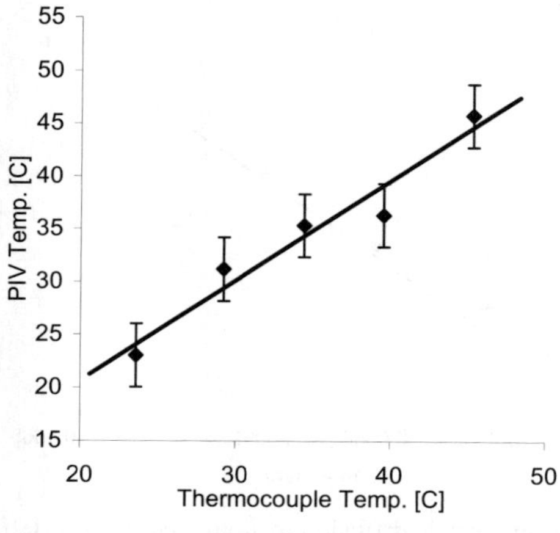

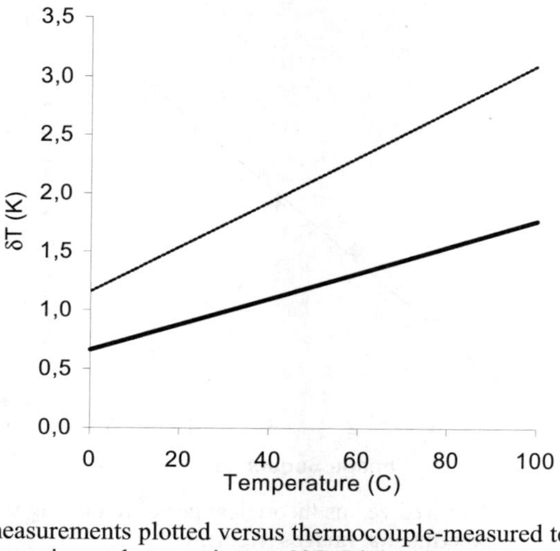

Figure 4. Left: Temperature calculated from PIV measurements plotted versus thermocouple-measured temperature. Error bars indicate the range of average experimental uncertainty, ± 3°C. Right: Measurement uncertainty over the range of temperatures for which water is a liquid.

locity and temperature was noted. While it is true that all experimental measurements reported in the current work were made using fluids with a net-zero velocity, it is also true that the correlation peak location (from which the velocity measurement is derived) and the correlation peak width (which yields the temperature measurement) are independent measurements whose convolution results in the correlation function. Therefore, it is not mere speculation to say that the two may be simultaneously measured. The key is finding the range of flow parameters over which both may be measured.

For a control surface, take a standard *n x n* pixel interrogation region (IR) in a larger image of a particle-seeded fluid at some temperature *T* and some steady, bulk velocity *V*. A set of images is recorded—with some time delay between each image—for the purposes of measuring *T* and *V*. The time delay, Δt, must satisfy two criteria: (a) it needs to be sufficiently long to allow measurable particle motion to take place, yet (b) short enough that a significant fraction of the particles in the first image also be present in the second. The former criterion is imposed by Brownian particle motion, the latter by bulk fluid motion.

The first criterion establishes a minimum Δt and requires that particles move beyond a measurable threshold—a property of the optical system—in order to be accurately resolved. With a precision micrometer and stationary fluorescing particles (dried onto a microscope slide) under the experimental (50X) microscope objective, the threshold of the current system was found to be about 100nm.

The second criterion establishes a maximum Δt and requires that particles not leave the interrogation region between frames. While there are two possible ways a particle can leave an IR—either by convection, following a streamline of the flow or by diffusion, following a Brownian path—for bulk flows with velocities greater than 10μm/s (using particles on the order of 100nm), bulk motion will dominate diffusion in violating the second criterion (Santiago et al., 1998). Furthermore, judicious window shifting can allow particles to move a distance bounded only the size of the CCD array—much farther than the size of an interrogation region, which is the bound that exists without a window shifting scheme.

Combining equations (1) and (2) yields

$$\sqrt{< s^2 >} = (KT\Delta t / 3\pi\mu d)^{1/2} \tag{9}$$

which gives the average particle displacement due to Brownian motion in terms of experimental parameters. This value is bounded below by the measurement threshold of the system, about 100nm, and above by the size of the CCD. Constraining the analysis to the current experimental parameters—the minimum Δt required to get a $\sqrt{<s^2>}$=100nm, for 700nm diameter particles in liquid water, ranges from 0.016s for 20°C to 0.008s for 50°C. Furthermore, using the size of the CCD array (6.4mm) as the limiting factor for bulk flow, a Δt of 0.016s under 50X magnification limits velocities to 8mm/s. A Δt of 0.008s allows for velocities up to 16mm/s. Extending the calculation to the limit of liquid water, a 100°C pool would allow velocity measurements to approach 3.5cm/s.

6 CONCLUSIONS

The random component of Brownian particle motion is directly related to the temperature of a fluid through the self-diffusion mechanism. Changes in fluid temperature result in changes in the magnitude of Brownian particle motions. Such fluctuations can be detected by analyzing a spatial cross-correlation of successive particle images.

A standard PIV algorithm provides a reliable spatial cross-correlation. Where standard PIV looks at correlation peak location to measure velocity, PIV thermometry focuses on changes in correlation peak area—the cross-sectional peak area being a measure of the average magnitude of Brownian motion. Experimentally, temperatures were measured for zero-velocity pools of particle-seeded water over the range 20-50 °C using a standard PIV system. Results for experimentally measured temperatures, when compared with thermocouple-measured temperatures, agreed within a range of ±3°C.

Experimental error for PIV thermometry is tied closely to the optical resolution of the PIV system being used—the higher the optical resolution the more accurately fluid temperatures can be measured using this technique. Increasing optical resolution also has effects on the potential for PIV to simultaneously measure both velocity and temperature. The results of the current work indicate that, using water as the fluid, measurements of fluid velocity less than or equal to 8mm/s can be made while simultaneously measuring temperatures greater than 20°C.

REFERENCES

Adrian, R.J. 1991. Particle-imaging techniques for experimental fluid mechanics. Annual Review of Fluid Mechanics, 23: 261-304.

Adrian, R.J. & Yao, C.S. 1985. Pulsed laser technique application to liquid and gaseous flows and the scattering power of seed materials. Applied Optics, 24: 44-52.

Keane, R.D. & Adrian, R.J. 1992. Theory of cross-correlation analysis of PIV images. Applied Scientific Research, 49: 1-27.

Meinhart, C.D., Wereley, S.T., & Gray, M.H.B. 2000. Volume illumination for two-dimensional particle image velocimetry. Measurement Science and Technology, 11: 809-814.

Olsen, M.G. & Adrian, R.J. 2000a. Brownian motion and correlation in particle image velocimetry. Optics and Laser Technology, 32 (7-8): 621-627.

Olsen, M.G. & Adrian, R.J. 2000b. Out-of-focus effects on particle visibility and correlation in microscopic particle image velocimetry. Experiments in Fluids, 29: S166-S174.

Santiago, J.G., Wereley, S.T., Meinhart, C.D., Beebe, D.J., & Adrian, R.J. 1998. A particle image velocimetry system for microfluidics. Experiments in Fluids, 25: 316-319.

Wereley, S.T., Meinhart, C.D., Santiago, J.G., & Adrian, R.J. 1998. Velocimetry for MEMS applications. Proc. ASME/DSC, Anaheim, CA: 453-4.

Thermal Challenges in Next Generation Electronic Systems, Joshi & Garimella (eds)
© 2002 Millpress, Rotterdam, ISBN 90-77017-03-8

Nonequilibrium electron and phonon transport and energy conversion in heterostructures

Taofang Zeng
Mechanical Engineering Department, North Carolina State University, Raleigh, NC, USA

Gang Chen
Mechanical Engineering Department, Massachusetts Institute of Technology, Cambridge, MA, USA

Keywords: nonequilibrium, transport, heterostructures, energy conversion

ABSTRACT: We establish a unified model dealing with the transport of electron and phonon in double heterojunction structures with the coexistence of three nonequilibrium processes: (1) nonequilibrium among electrons, (2) nonequilibrium among phonons, and (3) nonequilibrium between electrons and phonons. Using this model, we investigate the energy conversion efficiency based on concurrent thermoelectric and thermionic effects on electrons and size effects on electrons and phonons. It is found that heterostructures can have an equivalent figure of merit higher than the corresponding bulk materials.

1 INTRODUCTION

Size effects and large deviation from equilibrium often characterize transport of electrons and phonons in nanostructures. In very thin films with a thickness less than one electron mean free path (MFP), for example, electrons going through the film experience little or no scattering and the electric current is controlled by the barrier heights of the heterojunction, as described by the Richardson's thermionic emission theory [1]. Cooling and power generation based on such thermionic processes in heterostructures and superlattices have been studied [2-5]. On the other hand, scattering dominates the transport process of electrons flowing in thick films and bulk materials. Electrons and phonons are at quasi-equilibrium states and their transport is diffusive, as described by the thermoelectric theory [6]. Enhanced thermoelectric energy conversion along thin films based on quantum effects on electron energy states but diffusive quasi-equilibrium electron transport has been predicted and confirmed [7,8]. In addition, phonon size effects in thin films and superlattices are also under consideration as a means to enhance thermoelectric and thermionic energy conversion efficiency [9-11]. Some studies on nonequilibrium transport of electrons and phonons in thin films have been reported based on hot carrier assumption [12]. In these studies, however, the subsystems of electrons and phonons individually are still at quasi-equilibrium. In the ballistic limit, the quasi-equilibrium assumption for electrons as a subsystem or for phonons as a subsystem is clearly not valid. Although electronic thermoelectric and thermionic effects and phonon size effects in heterostructures all have been evaluated individually as potential mechanisms for improving the efficiency of solid-state coolers and power generators, a general theory that could encompass the effects of nonequilibrium among electrons, among phonons, and between electrons and phonons, has been missing. In this study we fill in this gap and establish a unified model including all these nonequilibrium processes such that the relative importance of size effects on electrons and phonons, and thermionic and thermoelectric energy conversion processes can be evaluated.

Our model treats the nonequilibrium transport of nonequilibrium electron and phonons subsystems and their interactions. It is based on the Boltzmann transport equations (BTEs) for nonequilibrium electrons and phonons, and their nonequilibrium coupling. The BTE is appropriate under the particle description of electrons and phonons, which means that the film thickness must be larger than the thermal de Broglie wavelength of electrons and phonons. Approximate solutions of BTE are obtained, leading to the distribution of the electron and phonon temperatures and Fermi level inside heterostructures. We emphasize that the temperature and Fermi level cannot be understood in the conventional sense of an equilibrium quantity since we are dealing with nonequilibrium processes. Rather, they are considered as a measure of the local electron and phonon energy for temperature and the local particle number for the Fermi level. The coexistence of discontinuities in the electron and phonon temperatures, as well as the Fermi levels at the interfaces are demonstrated for the first time, as a consequence of nonequilibrium transport when the film thickness is much smaller than the electron and phonon mean free path (MFP). The coexistence of thermoelectric and thermionic effects may increase the power factor. Phonon size effects can enhance the equivalent figure of merit of heterostructures.

2 PHYSICAL MODEL

Figure 1 shows the objective in this study. A semiconducting thin film (II) is sandwiched within another semiconductor or metal, which are used as substrates I and III maintained at uniform temperature T_c and T_h respectively. The two substrates act as two reservoirs. This is a very simplified schematic, whereas it allows us to focus on the main physical insights on nonequilibrium propensities inside the film and at the interfaces. We further limit our attention to the films with a thickness larger than the tunneling length and the coherent length or matter wavelength, so that we can neglect the quantum effects. In this regime, the BTE is applicable for describing nonequilibrium transport of nonequilibrium electrons and phonons. We begin with electrons: $v_x \partial f / \partial x + F_x \partial f / \partial p_x = (\partial f / \partial t)_c$, where f is the distribution function, v_x the electron velocity component in the x direction, F_x the local electric force, and p_x the electron momentum component in the x direction. The right hand side represents the scattering. In this work, we will use the relaxation time approximation, $(df/dt)_c=-(f-f_0)/\iota$, where f_0 is the Fermi-Dirac distribution for electrons, and ι the relaxation time of electrons. This relaxation time approximation gives us a way to calculate the effective temperature and Fermi level, which are embedded in f_0. Since the major concern here is the deviation far from equilibrium in space, we assume that the deviation from equilibrium in the k-space is small such that $\partial f / \partial p_x \approx \partial f_0 / \partial p_x$. The assumption can be justified by assuming quick adaptation of electrons to the new electronic structure of the thin film [13]. In the BTE, $F_x = -q\boldsymbol{\varepsilon}$, where $\boldsymbol{\varepsilon}$ is the electric field, $\boldsymbol{\varepsilon} = dE_c / dx$ for electrons. By neglecting the space charge and assuming a Boltzmann distribution at the interface for the electrons, which can be rationalized, for example, with doping inside the film and a reasonable barrier height, we can obtain the following solution to the BTE:

$$f^+(\xi,\mu)=f^+(0,\mu)\bullet e^{-\xi/\mu}+\int_0^{\xi} S^0(\xi_1)e^{-(\xi-\xi_1)/\mu}d\xi_1, \quad \text{(+ means positive direction, or } 0<\mu<1) \quad (1a)$$

$$f^-(\xi,\mu')=f^-(\xi_h,\mu')\bullet e^{(-\xi+\xi_h)/\mu'}-\int_{\xi}^{\xi_h} S^0(\xi_1)e^{(\xi_1-\xi)/\mu'}d\xi_1, \quad \text{(- means negative direction, or } -1<\mu'<0) \quad (1b)$$

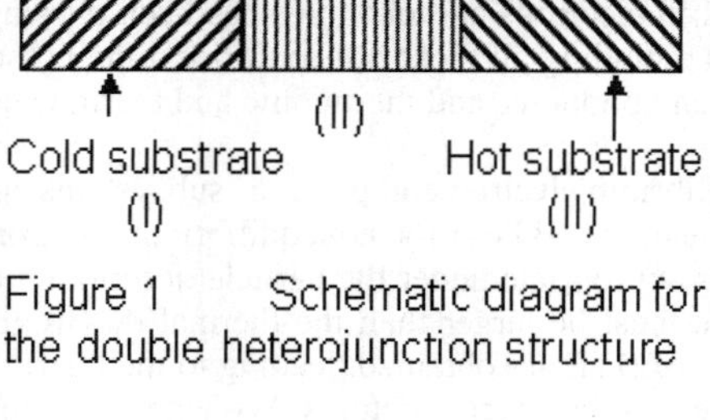

Figure 1 Schematic diagram for the double heterojunction structure

where, $\mu=\cos\theta$ is the directional cosine of the electron random velocity with the positive x-direction, $\xi = x/\Lambda$, the dimensionless coordinator, Λ the electron MFP, and $S^0 = \partial f_0/\partial x + q\boldsymbol{\varepsilon}\,\partial f_0/\partial E$. The electric current can then be obtained by integrating f^+ and f^- [14]:

$$J(\xi)\bullet\pi^2\hbar^3/qm^* = 2J_1E_3(\xi) - 2J_2E_3(\xi_h-\xi) + 2\int_0^{\xi} H(\xi-\xi_1)E_3(\xi_1)d\xi_1 + 2\int_0^{\xi_h-\xi} H(\xi+\xi_1)E_3(\xi_1)d\xi_1\,, \tag{2}$$

where $H(\xi) = -\int_0^{\infty}\Lambda(\partial f_0/\partial x + q\varepsilon\bullet\partial f_0/\partial E)EdE$, and $E_n(x)=\int_0^1\mu^{n-2}e^{-x/\mu}d\mu = \int_1^{\infty}e^{-xt}t^{-n}dt$ the exponential integral function of n-th order. We have made the parabolic band assumption in Eq. (2) and m^* is the effective mass. The boundary terms J_1 and J_2 are determined by the thermionic emission boundary conditions, $J_1=j(\phi+E_{cI}-E_{fI},T_c)-j[E_c(0)-E_f(0),T_e(0)]$, $J_2=j(\Delta E_c+\phi+E_{cIII}-E_{fIII},T_h)-j[\Delta E_c+\phi+E_c(h)-E_f(h),T_e(\xi_h)]$, and the j function is define by $j(x,T)\equiv\exp(-x/K_BT)\bullet(K_BT)^2$. E_c and E_f are conduction band edge and the Fermi energy, $\Delta E_c = E_c(0) - E_c(h)$, $T_e(x)$ or $T_e(\xi)$ is the electron temperature distribution inside the film, and K_B the Boltzmann constant. Similarly, the electronic heat flux can be expressed as

$$C\bullet Q_e = 2P_1E_3(\xi) - 2P_2E_3(\xi_h-\xi) + 2\int_0^{\xi} R(\xi_1)E_3(\xi-\xi_1)d\xi_1 + 2\int_{\xi}^{\xi_h} R(\xi_1)E_3(\xi_1-\xi)d\xi_1\,, \tag{3}$$

where, $C=\pi^2\hbar^3/m^*$, $R = -\int_0^{\infty}\Lambda(\partial f_0/\partial x + q\varepsilon\bullet\partial f_0/\partial E)E(E-E_f)dE$, and E_f is relative to E_c. The thermionic heat fluxes at the two interfaces can be expressed as $P_1 = j(\phi+E_{cI}-E_{fI},T_c)\bullet[\phi+(E_{cI}-E_{fI})+2K_BT_c]-j[E_c(0)-E_f(0),T_e(0)]\bullet[E_c(0)-E_f(0)+2K_BT_e(0)]$, and $P_2 = j(\Delta E_c+\phi+E_{cIII}-E_{fIII},T_h)\bullet[(\Delta E_c+\phi+E_{cIII}-E_{fIII}+2K_BT_h] - j[\Delta E_c+E_c(h)-E_f(h),T_h]\bullet[\Delta E_c+E_c(h)-E_f(h)+ 2K_BT_e(h)]$.

The phonon Boltzmann transport equation, under the relaxation time approximation is, $v_x dg/dx = -(g-g_0)/t_{ph}$, where g is the phonon distribution function, g_0 is the Bose-Einstein distribution for phonons, and t_{ph} phonon relaxation time. The equation can be solved, leading to the following expression for the local phonon heat flux [9]

$$Q_{ph}(\tau) = 2I_1E_3(\tau) - 2I_2E_3(\tau_h-\tau) + 2\pi\int_0^{\tau} I^0(\tau-\tau_1)E_2(\tau_1)d\tau_1 + 2\pi\int_0^{\tau_h-\tau} I^0(\tau+\tau_1)E_2(\tau_1)d\tau_1\,, \tag{4}$$

where, τ ($= x/\Lambda_{ph}$) is the non-dimensional coordinate, I is the total phonon intensity, I^0 the equilibrium phonon intensity expressed as $I^0 = C_Pv_{ph}/4\pi\bullet(T_{ph}-T_{ref}) + I^0{}_{ref}$ under the temperature independent specific heat(C_p) assumption [9,15], T_{ph} phonon temperature, v_{ph} phonon group velocity, and Λ_{ph} the average MFP of phonons in the corresponding bulk material. We assume the substrates acting as two reservoirs and emit phonons such that $I_1 = C_Pv_{ph}/4\pi\bullet(T_c-T_{ref}) + I^0{}_{ref}$, and $I_2 = C_Pv_{ph}/4\pi\bullet(T_h-T_{ref}) + I^0_{ref}$ [15].

Equations (2)-(4) contain the local electron and phonon temperatures, and the Fermi level inside the heterostructure that must be determined by satisfying the conservation of charge and energy. These equations are

$$dJ(x)/dx = 0,\ dQ_e(x)/dx = -G(T_e-T_{ph}),\ \text{and}\ dQ_{ph}/dz = G(T_e-T_{ph})\,, \tag{5}$$

where, $G = 3/2K_B/\iota$, the coupling factor describing the energy exchange rate between electrons and phonons. Exact solution of Eq. (5) will need complicated numerical approach. Our interest is to obtain analytical solutions so that the nonequilibrium electron and phonon and their transport mechanisms can be clearly illustrated. To keep the problem amendable for analytical solutions, we use an approximation solution technique in thermal radiation for the photon BTE [16]. The optically thick approximation in thermal radiation, when combined with appropriate boundary conditions, leads to results in excellent agreement with the exact solution even for the films with a thickness smaller than the photon MFP. We infer that the same technique can be applied to the current problem with reasonable accuracy. The idea is to assume a large electronic distance and acoustic distance away from the interfaces, i.e., $\tau>>1$, $\tau_h-\tau>>1$; $\xi>>1$, and

ξ_h-ξ>>1. Under these assumptions the influence of the boundaries becomes negligible and Eqs. (2)-(4) can be simplified into the ordinary diffusion form,

$$J(x)=e/C\bullet H(x),\ Q_e(x)=1/C\bullet R(x),\ Q_{ph}(x)=-1/3\bullet C_P v_{ph} dT_{ph}/dx. \quad (6)$$

The boundary conditions are derived from the Deissler's jump boundary conditions [16] as follows:
$J_1=1/2H_x(0)$, $J_2=-1/2H_x(\xi_h)$, $P_1=1/2R_x(0)$, $P_2=-1/2R_x(\xi_h)$,

$$I_1-I(0)=1/2Q_{ph}(0),\ I_2-I(\tau_h)=-1/2Q_{ph}(\tau_h), \quad (7)$$

where $H_x=L_{xx}^{(0)}(\varepsilon+\nabla E_f/q)+L_{xx}^{(1)}(-\nabla T_e)$ and $R_x(x)=L_{xx}^{(1)}(\varepsilon+\nabla E_f/q)+L_{xx}^{(2)}(-\nabla T_e)$ are standard thermoelectric expressions, except in the current case, the electron temperature and its gradient are unknown and must be solved with the boundary terms. Based on the parabolic band approximation and the Fermi-Dirac statistics, the transport coefficients can be expressed as $L_{xx}^{(\alpha)}=q^2\tau(8m^*)^{1/2}(K_BT_e)^{3/2+\alpha}/3\pi^2\hbar^3\bullet\int_0^\infty x^{3/2}(x-\varsigma)^\alpha e^{x-\varsigma}/\left(e^{x-\varsigma}+1\right)^2dx$, where $\varsigma=E_f/K_BT_e$ is the dimensionless Fermi level.

3 RESULTS AND DISCUSSION

We now focusing on the open-circuit condition, J(x) = 0, and derive the corresponding Seebeck coefficient. By introducing the approximation of $(T_h-T_c)/T_c << 1$, and $\Delta E_f=q\alpha[T_e(0)-T_e(h)]\langle\langle[E_c-E_f]$, where α is the Seebeck coefficient of the corresponding material, which can be realized by maintaining a very small temperature difference at the two substrates, Eq. (7) can be solved analytically. Figure 2(a) shows the distributions of electron and phonon temperature inside the film and Fig. 2 (b) the corresponding Fermi level distribution. The temperature profiles are very similar to those of photons between two parallel plates in thermal radiation [16] and phonon heat conduction across a thin film [9,15]. When the film thickness is very small compared to the electron MFP, electrons transport inside the film is purely ballistic. Due to the lack of scattering, electrons are not close to an equilibrium state as these in the diffusive transport regime. Temperature thus calculated is only a measure of the local average energy if these nonequilibrium electrons were to interact and to thermalize with each other. The temperature discontinuities (jumps) at the boundaries are the consequence of using this equilibrium concept to represent nonequilibrium transport. As the film thickness becomes larger, the temperature jumps at the interface become smaller relative to the temperature drop inside the film. When the thickness is very large, the electrons reach quasi-equilibrium states. Phonons follow similar behavior. The Fermi level distribution can be explained following similar logic for the temperature distribution. At the very thin film limit, because electrons are not at equilibrium states, they do not obey the Fermi-Dirac distribution. The equivalent Fermi level is a measure of the total number of electrons at each location.

The equivalent Seebeck coefficient of the structure, defined as $S = -(E_{fI}-E_{fIII})/[q(T_c-T_h)]$, is shown in Fig. 3(a). This Seebeck coefficient is determined by two mechanisms. One is thermionic emission at the interfaces and the other is the diffusive transport inside the film. When the film is thick compared to the electron MFP, the Seebeck coefficient is control by the thermoelectric process inside the film since most temperature drop occurs there. Under the given doping level, the Seebeck coefficient approaches 250 μV/K, which is the value for bulk Si under the given doping concentration [17]. When the film is thin compared to the electron MFP, the thermionic effect dominates and the barrier heights determine the Seebeck coefficient, because the electron temperature drop inside the film is small. Not surprisingly, when the barrier height is small, S is very close to the bulk material. And when the barrier height is large, S is much larger than the bulk material, because S is determined by the barrier height plus $2K_B/q$ in the purely thermionic process [2].

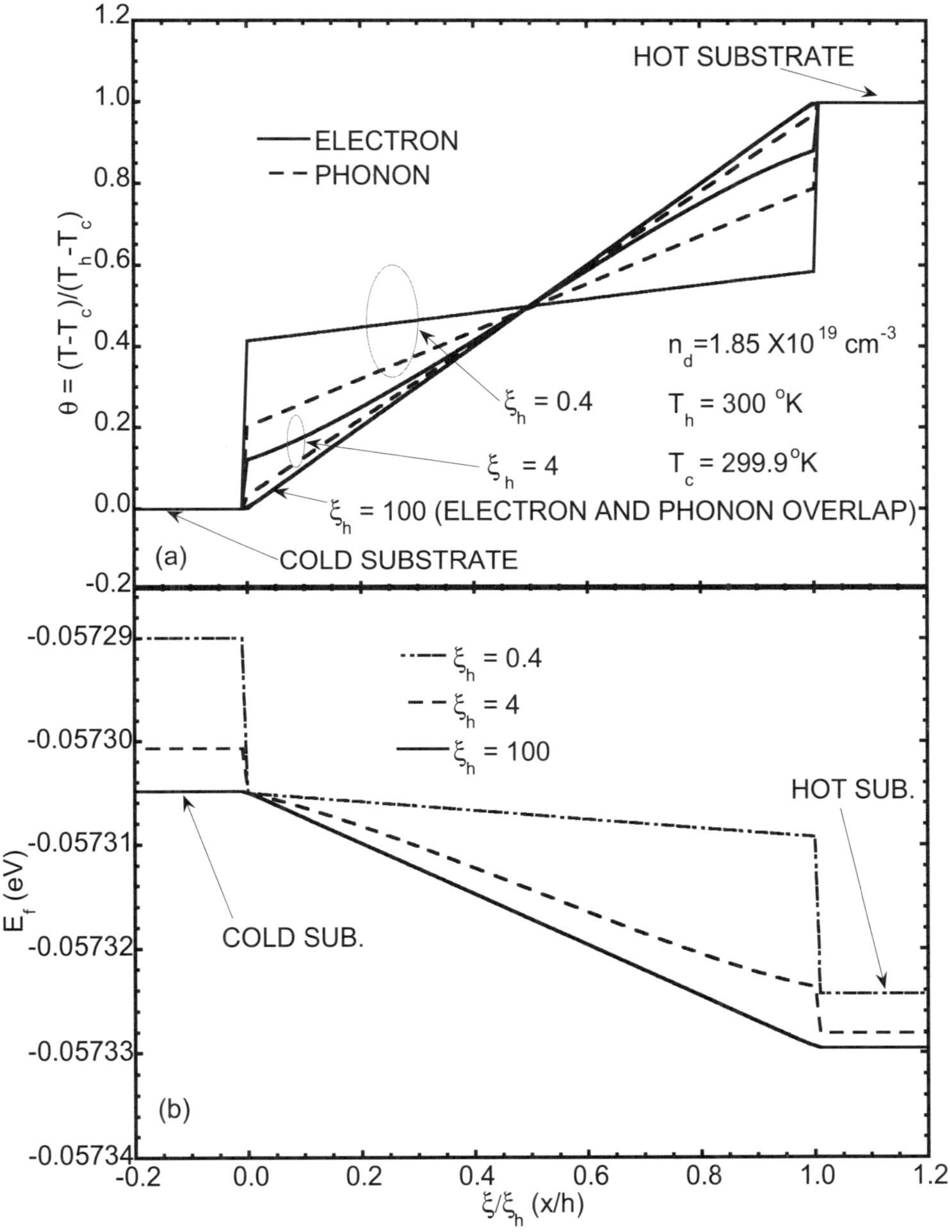

Figure 2(a) Electron and phonon temperature distributions inside the double heterojunction structure under open circuit condition, and (b) corresponding Fermi level distributions. Discontinuity occurs at the interface due to the nonequilibrium nature of electron and phonon transport.

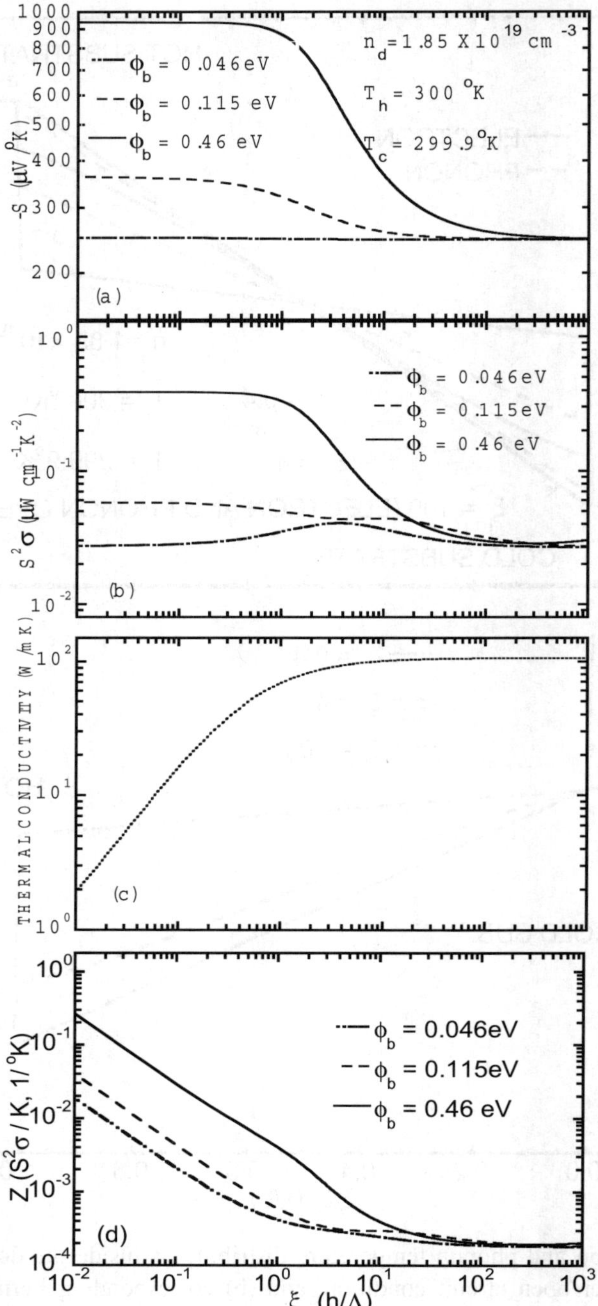

Figure 3 Equivalent thermoelectric properties as a function of the film thickness for different barrier heights, (a) Seebeck coefficient, (b) power factor, (c) thermal conductivity, and (d) equivalent figure of merit. Calculations based on properties of silicon.

The equivalent electrical conductivity can be calculated from: $\sigma = (h + h_{sub})/(h/\sigma_{equ\_film} + h_{sub}/\sigma_{sub})$, where, h is the film thickness, h_{sub} is the thickness of the substrate [14], $\sigma_{equ\_film} = \sigma_{bulk}(E_{fI} - E_{fIII})/(E_f(0) - E_f(h))$ is the equivalent electrical conductivity of the film. The physical meaning of the equivalent electric conductivity can be understood by the fact that for purely ballistic transport, there is no scattering and voltage-drop in the film. Therefore, the electrical conductance inside the film is infinity. In this case, the excess electron energy is lost inside the substrate. The equivalent thermal conductivity of the structure is made of contributions from electrons and phonons, which are estimated from $K_{e-film} = K_e(T_e(0) - T_e(h))/(T_c - T_h)$ and $K_{ph\_film} = K_{ph}(T_{ph}(0) - T_{ph}(h))/(T_c - T_h)$, where K_e and K_{ph} are the thermal conductivity of the bulk materials corresponding to the film. Upon determination of S, σ, and K, we can estimate the power factor $S^2\sigma$ and the figure of merit, $Z = S^2\sigma/K$, of the heterostructure, as shown in Figs. 3(b)-3(d). When the film thickness is less than 10 percent of electron MFP, Seebeck coefficient is determined by the barrier height. As the film becomes thicker, the power factor approaches that of corresponding bulk materials. For low barrier heights, the coexistence of thermoelectric and thermionic effects creates a peak in power factor when the film thickness is comparable to electron MFP. Finally, size effects on thermal conductivity reduction represents another significant advantage on the overall figure of merit of low-dimensional structures, as shown in Fig. 3(c). The interface thermal boundary resistances on phonons and to a lesser extend, on electrons, leads to high figure of merit.

4 CONCLUSIONS

In summary, we have established a unified approach to deal with electron and phonon energy transport in heterostructures under highly nonequilibrium conditions existing within the electron and phonon subsystems and between the two subsystems. We apply this model to treat the concurrent thermionic and thermoelectric processes in a semiconductor heterostructure, combined with consideration of size effects on phonons and electrons. Our calculation shows that when the film thickness is small compared to the electron MFP, the transport of electrons is controlled by thermionic emission because no temperature gradient develops inside the film. When the film is thick, the transport is controlled by thermoelectric process. When the thickness is on the order of MFP, both thermionic emission and thermoelectrics are important. The coexistence of both mechanisms can create peak in the thermoelectric power factor. The thermal conductivity reduction due to interface scattering can significantly increase the figure-of-merit.

ACKNOWLEDGMENTS

This work is supported in part by a DOD/ONR MURI grant on thermoelectrics and the DARPA HERETIC project.

REFERENCES

1. S. M. Sze, Physics of Semiconductor Devices, 2nd Ed. (John Wiley & Sons, New York, 1981).
2. G. D. Mahan and L. M Woods, Phys. Rev. Lett. 80, 4016 (1998).
3. A. Shakouri, E. Y. Lee, D. L. Smith, V. Narayanamurti, and J. E. Bowers, Microscale Thermophysical Engineering 2, 37 (1998); A. Shakouri, C. LaBounty, J. Piprek, P. Abraham, and J. E. Bowers, Appl. Phys. Lett. 74, 88 (1999).
4. L. W. Whitlow and T. Hirano, J. Appl. Phys. 78, 5460 (1995).
5. R. J. Radtke, H. Ehrenreich, and C. H. Grein, J. Appl. Phys. 86, 3195 (1999).

6. For example in H. J. Goldsmid, Thermoelectric Refrigeration (Plenum Press, New York, 1964) and T. C. Harman and J. M. Honig, Thermoelectric and Thermomagnetic Effects and Applications (McGraw-Hill, New York, 1967).
7. L. D. Hicks and M. S. Dresselhaus, Phys. Rev. B 47, 12727 (1993).
8. L. D. Hicks, T. C. Harman, X. Sun, and M. S. Dresselhaus, Phys. Rev. B 53, 10493 (1996).
9. G. Chen, Phys. Rev. B 57, 14958 (1998).
10. P. Hyldgaard and G. D. Mahan, Phys. Rev. B 56, 10754 (1997).
11. R. Venkatasubramanian, Phys. Rev. B 61, 3091 (2000).
12. V. S. Zakordonets and G. N. Logvinov, Semiconductors 31, 265 (1997).
13. C. M. Wu and E.S. Yang, Solid-State Electronics 22, 241 (1979).
14. T. Zeng and G. Chen, to be published.
15. A. Majumdar, J. Heat Transfer 115, 7 (1993).
16. M. F. Modest, Radiative Heat Transfer (McGraw-Hill, New York, 1992).
17. A. W. Van Herwaarden, Sensors Actuators 6, 245 (1984).

NOVEL THERMAL MANAGEMENT CONCEPTS

Thermal Challenges in Next Generation Electronic Systems, Joshi & Garimella (eds)
 ISBN 90-77017-03-8

Dynamics and topology optimization of piezoelectric fans

Philipp Bürmann[1], Arvind Raman[2] & Suresh V. Garimella
School of Mechanical Engineering, Purdue University, West Lafayette, IN, USA

Keywords: piezoelectric fans, resonators, portable electronics cooling

ABSTRACT: Piezoelectric fans are very low power, small, very low noise, solid-state devices that have recently emerged as viable thermal management solutions for a variety of portable electronics applications including laptop computers, cellular phones and wearable computers. Piezoelectric fans utilize piezoceramic patches bonded onto thin, low frequency flexible blades to drive the fan at its resonance frequency. The resonating, low frequency blade creates a streaming airflow directed at key electronics components. The optimization of a piezoelectric fan with two symmetrically placed piezoelectric patches is investigated through an analytical Bernoulli-Euler model as well as a finite element (FE) model of the composite piezo-beam. The closed form analytical solution is used to demonstrate that different optimal piezoceramic-to-blade length ratios and piezoceramic-to-blade thickness ratios exist for maximizing the electromechanical coupling factor (EMCF), tip deflection and rotation, and stroke volume rate. Such optimization procedures provide simple design guidelines for the development of very-low power, high flow rate piezoelectric fans.

1 INTRODUCTION

Piezoelectric resonators are used as actuators (Pan *et al.* 1992, Rivory *et al.* 1994, Tiersten 1969, Seeley & Chattopadhyay 1993, Seshu & Naganathan 1998) and sensors (Lee & Saravanos 1996, Aldraihem & Kheir 2000, Main *et al.* 1994) for a wide variety of applications including piezoelectric fans, optical beam choppers, ultrasonic motors, microelectromechanical density sensors, and resonating viscometers. For these applications, little seems to be known about an optimal actuator-beam configuration including patch location, actuator patch-to-beam length ratio, and patch-to-beam thickness ratio. Moreover, the optimization criterion is application dependent and can vary between maximal electromechanical coupling factor (EMCF), optimal mode shapes and frequencies, and tip deflections. In light of this, we focus in this article on analytical solutions of the dynamic response optimization of a piezoelectric fan modeled as a partially electroded (on both faces) elastic beam near resonance.

Several authors (Pan *et al.* 1992, Wang & Wang 2000, Khdeir & Aldraihem 2001, Brennan *et al.* 1997, Lee & Saravanos 1996, Crawley & Anderson 1990, Aldraihem & Kheir 2000, Seeley & Chattopadhyay 1993, Seshu & Naganathan 1998, Yang 1997) have investigated the modeling and optimization for *static* deflection/actuation of piezoelectric-elastic composite structures.

Analytical, experimental and computational investigations of the *dynamic* response of such structures are presented in Rivory *et al.* (1994), Tiersten (1969), Barboni *et al.* (2000), Main *et al.*

[1] Currently at the Department of Solid Mechanics, Technical University-Dresden, Germany
[2] Corresponding author, raman@ecn.purdue.edu

(1994), Lobontiu *et al.* (1999), Shen (1995), Pan & Hansen (1994), Abramovich (1998), Crawley & de Luis (1987), and Strambi *et al.* (1995). Rovory *et al.* (1994) and Tiersten (1969) present an analytical dynamical model. However, the free vibration problem is not considered and thus the electromechanical coupling factor is not computed. In Barboni *et al.* (2000) a similar approach to a solution for the forced mode shape is derived after the time-dependent terms have been eliminated. In Main *et al.* (1994) dynamic response of a beam that is fully covered by the patch is considered. Lobontiu et al (1999) presents a beam that is partially covered by two piezoelectric patches. However, the additional mass and stiffness of the piezoelectric patches are not considered, while the actuation is included as two external point bending moments applied on the beam. The coupling effects between curvature and electric field are neglected. The forcing from the piezoelectric patches in Shen (1995) is modeled as an internal bending moment distribution in the beam. The resulting equations are written in state equation form and the control issues are discussed. However, no coupling between piezoelectric patches and beam is taken into account. A lumped parameter approach is followed in Pan & Hansen (1994). Mass, damping, stiffness and forcing are lumped as parameters in a one degree-of-freedom model. The forcing term is included explicitly and an extensive parametric study is carried out. This is however not based on a continuum model. The wave response in beams and plates with attached piezoelectric patches is considered in Abramovich (1998). However, the excitation is included through Dirac delta and Heaviside step functions implying the neglect of coupling between piezoelectric patch and the substrate. In Crawley & de Luis (1987) an infinitely long beam attached to piezoelectric patches is considered and is treated as a wave generator. The forcing is included as external point bending moments neglecting the coupling between the piezoelectric patch and substrate. Strambi *et al.* (1995) present a static and dynamic model of a beam with two piezoelectric patch actuators, incorporating true coupling between the piezoelectric patch and beam. However, optimization issues are not discussed and their approach requires *a priori* knowledge of the phase difference between forcing and response. Wolf (2000) presents a comprehensive analysis of optimal electromechanical coupling factors in the context of a piezoelectric patch attached to an *infinitely long* beam. A similar approach is followed in the present paper, however a focus is placed on the analysis for a *finite* length beam. This leads to substantially different analysis techniques and optimization issues.

Finite element approaches for the optimization of piezoelectric structures are presented in Benjeddou *et al.* (1999), Chandrashekhara and Varadarajan (1997), Silva and Kikuchi (1999), Yang (1997), and Zhai *et al.* (2000). In Chandrashekhara and Baradarajan (1997) a cantilever beam with segmented shear piezoelectric actuators is analyzed using sandwich beam finite elements. In Yang (1997) the determination of the optimal thickness of a single piezoelectric actuator is addressed. In Zhai *et al.* (2000) a topology optimization of a geometrically complex, stack actuator is performed. Finite element based optimization techniques are very expensive computationally. The focus of the present paper is to develop a closed-form analytical solution for a resonating, composite piezobeam and utilize this solution to develop guidelines for optimal fan configurations. Finite element models of the fan are used to confirm the validity of the analytical solution.

2 MODELING AND ANALYSIS

Consider the beam shown in Figure 1. The beam is of total length L_3, while the piezoelectric patches are bonded perfectly to the beam and are of length $L_2 - L_1$. The thickness of the beam and the actuators is given as $2h$ and h_p, respectively, where the coordinate x_3 is measured from the neutral axis of the beam. The widths of the beam and patches are $2b$. The part of the fan from the clamp up to the beginning of the beam will be referred to as *region 1*, the section covered by the actuators as *region 2* and the section from the end of the patches to the tip of the beam as *region 3*.

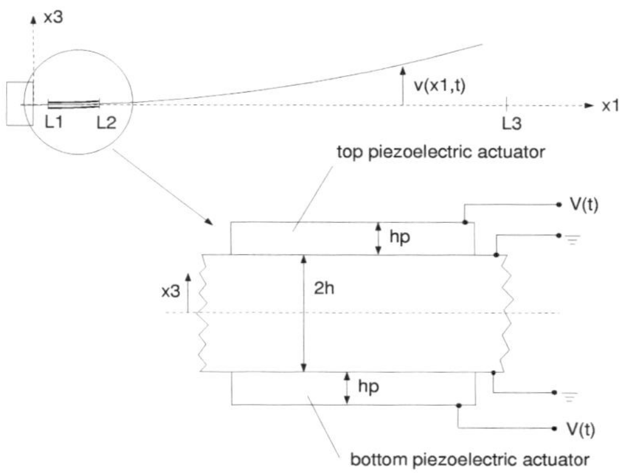

Figure 1. Piezoelectric fan geometry. The fan is driven at resonance by piezoelectric patches that are operated out-of-phase from an electronic circuit.

2.1 *Analytical model*

The transverse displacements of the neutral axis of the substrate in each region $v_i(x_1, t)$, $i=1,2,3$, are measured with respect to the longitudinal coordinate x_1 along the neutral axis of the substrate. The composite piezoelectric fan is modeled as a composite Bernoulli-Euler beam. The equations of motion governing the transverse deflection of the piezoelectric fan can be derived through the use of linear piezoelectric constitutive relations, the simplifying Bernoulli-Euler assumptions and application of Hamilton's principle inclusive of strain energy and electromechanical energy terms in the formulation. The resulting equations of motion for the three regions and the corresponding boundary conditions can be shown to be

$$\begin{aligned}
&EI\, v_1'''' + m_s \ddot{v}_1 = 0, \qquad K\, v_2'''' + m_p \ddot{v}_2 = 0, \qquad EI\, v_3'''' + m_s \ddot{v}_3 = 0\\
&v_1(0,t) = 0, \quad v_1'(0,t) = 0, \quad v_1(L_1,t) = v_2(L_1,t), \quad v_1'(L_1,t) = v_2'(L_1,t),\\
&EI\, v_1''(L_1,t) = K\, v_2''(L_1,t) + a_4(t)\, A_p, \quad EI\, v_1''(L_1,t) = K\, v_2''''(L_1,t),\\
&v_2(L_2,t) = v_3(L_2,t), \quad v_2'(L_2,t) = v_3'(L_2,t), \quad K\, v_2''(L_2,t) + a_4(t)\, A_p = EI\, v_3''(L_3,t),\\
&K\, v_2'''(L_2,t) = EI\, v_3'''(L_2,t), \; v_3''(L_3,t) = 0, \; v_3''(L_3,t) = 0
\end{aligned} \tag{1}$$

where the primes are derivatives with respect to x_1

$$m_s = \rho_s\, A_s, \; EI = E_s\, I_s, \; m_p = \rho_s\, A_s + 2\, \rho_p\, A_p, \; K = E_s\, A_s + 4\, a_2\, I_p + 4\, a_3\, A_p \tag{2}$$

and $\rho_s, \rho_p, E_s, A_s, A_p, I_s, I_p$ denote respectively the mass density of the beam and the piezoelectric patch, the Young's modulus of the substrate, the areas of cross section of the substrate and patch, and the area moment of inertias of the substrate and patch. The coefficients a_2, a_3, a_4 are related to the piezoelectric-structure coupling and are given by

$$a_2 = \frac{1}{2}\left(E_p + \frac{e_{31}^2}{\varepsilon_{33}}\right), \; a_3 = \frac{-e_{31}^2 (2h + h_p)^2}{8\, \varepsilon_{33}}, \; a_4(t) = -\frac{e_{31}(2h + h_p)\Delta\Phi(t)}{2h_p}, \; a_5 = \frac{-\varepsilon_{33}(\Delta\Phi)^2}{2h_p^2} \tag{3}$$

where $E_p, e_{31}, \varepsilon_{33}$ are respectively the Young's modulus of the piezoelectric patch, the product of E_p and the piezoelectric material coefficient d_{31}, and the product of the relative dielectric constant

K_{33} and the free space permitivity ε_0. $\Delta\Phi(t)$ is the potential difference between the top and bottom of the electrodes of the piezoelectric patches.

It must be noted here that no explicit forcing term appears in equation (1). The piezoelectric excitation is only contained in the boundary conditions because $a_4(t)$ depends on the potential difference $\Delta\Phi(t)$ (equation 3). In addition to the boundary conditions at the clamp and the free end of the beam, there are matching conditions at each interface between two regions. The deflection and slope, and the internal bending moment and the internal shear force are related to each other. Note that there is a time-dependent jump in bending moment and that the stiffness term K consists of the stiffness of the underlying substrate E_sI_s, the added mechanical stiffness of the piezoelectric actuators $4a_2I_p$ and an electromechanical coupling term $4a_3A_p$. For free vibrations of the piezoelectric fan, $\Delta\Phi(t)$ can either be 0 (short-circuited or SC electrodes) or be generated naturally by the bending vibrations (open circuited or OC electrodes). For forced vibrations, $\Delta\Phi(t)$ is a prescribed, harmonic waveform with a frequency matching the short-circuited fan frequency.

The free vibration eigenvalue problem is solved piecewise in each region by use of expressions for the prescribed potential difference $\Delta\Phi(t)$ and through separation of variables in equation (1):

$$v_i(x_1,t)=\varphi_i(x_1)e^{i\omega t},\quad \varphi_i(x)=A_i\cos(\beta_i x_1)+B_i\sin(\beta_i x_1)+C_i\cosh(\beta_i x_1)+D_i\sinh(\beta_i x_1) \quad (4)$$

$i=1,2,3$ and the wavenumbers β_i are given by

$$\beta_1=\beta_3=\left(\frac{m_s}{E_sI_s}\omega^2\right)^{\frac{1}{4}},\ \beta_3=\left(\frac{m_p}{K}\omega^2\right)^{\frac{1}{4}} \quad (5)$$

It can be shown that the partial differential equations and boundary conditions (1) are satisfied exactly with the above solution forms. Substitution of (5) into the boundary conditions (1), casting the system in matrix form and setting its determinant to zero, the eigenvalues ω of the fan can be computed. For both the SC and OC cases, all unknown coefficients A_i, B_i, C_i, D_i are expressed in terms of any one amongst them. This last coefficient is found by normalizing the mode shapes in the standard way with respect to the mass operator Meirovitch (1997). This yields an exact closed form analytical solution for the normalized free vibration mode.

When the beam is driven by the piezoelectric actuators the matching conditions (1) for internal bending moments become non-homogeneous and non-autonomous. To homogenize the boundary conditions, the following transformations are introduced:

$$v_1(x_1,t)=w_1(x_1,t),\qquad v_2(x_1,t)=w_2(x_1,t)+h(x_1)\,a_4(t)\,A_p,\qquad v_3(x_1,t)=w_3(x_1,t) \quad (6)$$

The function $h(x_1)$ is determined such that the boundary conditions (1) are satisfied and rendered homogeneous. The resulting equations for $w_i(x_1,t)$ feature homogeneous partial differential equations with explicit forcing terms and can be solved using Galerkin's method:

$$w_i(x_1,t)=\varphi_i(x_1)\,q(t),\ i=1,2,3 \quad (7)$$

where $\varphi_i(x_1)$ are the eigenfunctions for the SC free vibration problem. Substitution of the previously computed $\varphi_i(x_1)$ into (7) and subsequently into the homogenized equations for $w_i(x_1,t)$, and taking inner products in succession with the $\varphi_i(x_1)$, leads to the following discretized equations near resonance:

$$\ddot{q} + 2\varsigma\omega_{SC}\dot{q} + I_5 q = -I_2 A_p \ddot{a}_4 - I_6 A_p a_4$$

$$I_2 = m_p \int_{L_1}^{L_2} \varphi_2(x_1) h(x_1) dx_1, I_6 = K \int_{L_1}^{L_2} \varphi_2(x_1) h''''(x_1) dx_1,$$

$$I_5 = EI \int_0^{L_1} \varphi_1(x_1)\varphi_1''''(x_1)dx_1 + K \int_{L_1}^{L_2} \varphi_2(x_1)\varphi_2''''(x_1)dx_1 + EI \int_{L_2}^{L_{31}} \varphi_3(x_1)\varphi_3''''(x_1)dx_1 = \omega_{SC}^2 \tag{8}$$

Note that linear viscous damping has been added through the damping ratio ς. Emergence of the short-circuited natural frequency ω_{SC} in equation (8) is a consequence of the use of SC eigenfunctions in Galerkin's method. The particular steady state solution of (8) is

$$q_p(t) = F(\Omega)\sin(\Omega t - \gamma(\Omega))$$

$$F(\Omega) = \sqrt{\frac{(I_6 - I_2)\hat{V}^2}{(\Omega^2 - \omega_{SC}^2)^2 + 4\zeta^2\omega^2}}, \quad \gamma(\Omega) = -\arctan\left[\frac{\omega_{SC}^2 - \Omega^2}{2\zeta\omega\Omega}\right], \quad \hat{V} = -2e_{31}(2h + h_p)b\overline{V} \tag{9}$$

and $\overline{V}$ is the amplitude of excitation in volts.

Table 1. Material Properties of a Piezoelectric Fan.

	Physical Quantity	Value
E_s	Young's Modulus of the Beam	$10 * 10^{10} \frac{N}{m^2}$
E_p	Young's Modulus of the Actuator	$6.2 * 10^{10} \frac{N}{m^2}$
ρ_s	Density of the Beam	$8400 \frac{kg}{m^3}$
ρ_p	Density of the Piezoelectric Actuator	$7800 \frac{kg}{m^3}$
2h	Height of the Beam	0.1016*mm*
h_p	Height of the Piezoelectric Actuator	0.127*mm* – 0.254*mm* (varied)
2b	Width of Beam and Piezoelectric Actuator	8.89*mm*
L_1	Length of Region 1	0.5*mm*
L_2	Length of Region 1 and 2	3*mm* – 27*mm* (varied)
L_3	Overall Length of the Structure	27.424*mm*
K_{33}	Relative Dielectric Constant	3800
ε_0	Permitivity of Free Space	$8.9 * 10^{-12} \frac{C}{Vm}$
d_{31}	Piezoelectric Strain Coefficient	$-320 * 10^{-12} \frac{m}{V}$
ζ	Damping Ratio	0.04

2.2 *Finite Element Model*

To confirm the validity of the analytical model the piezo-beam is modeled in ANSYS 5.7 using 3-D linear solid brick elements. The piezoelectric constitutive equations are utilized in their three-dimensional planar isotropic form and the properties chosen for the computation are listed in Table 1. Some three-dimensional piezoelectric constants are not retrievable from the manufacturer's data and are approximated from typical handbook values. A regular three-dimensional mesh was generated using 3D brick piezoelectric 20-node elements. Overall 16,008 nodes were used in the computation. Piezoelectric and elastic beam properties were assigned to the respective volumes. All nodes above and below the piezoelectric patches were assigned a voltage coupling corresponding to the SC/OC electrical boundary conditions.

2.3 *Comparison of analytical and FE model*

For typical actuator lengths and one value of thickness ($h_p = 0.178\,\text{mm}$) the SC and OC natural frequencies of the analytical and FE model corresponding to the first and second bending modes were compared and found to be in excellent agreement (Table 2). For a fixed length of patch $L_2 - L_1 = 6.5\,\text{mm}$ the mass-normalized first and second SC and OC mode shapes are compared in Figure 2. The results show excellent agreement, with the difference between the analytical and FE predictions increasing with increasing mode number.

Table 2. A comparison of the natural frequencies predicted by the analytical and the FE model. The FE model is used as a reference when computing the error.

Mode	L_2= 7mm	L_2= 14mm	L_2= 21mm
1st SC anal.	125.63 Hz	194.44Hz	162.35Hz
1st SC FEM	126.97Hz	192.22Hz	158.85Hz
% Error	−1.05%	1.16%	2.52%
1st OC anal.	125.86Hz	194.79Hz	163.49Hz
1st OC FEM	127.55Hz	195.27Hz	161.41Hz
% Error	−1.32%	−0.25%	1.29%
2nd SC anal.	614.86Hz	496.57Hz	1278.62Hz
2nd SC FEM	624.18Hz	509.825Hz	1246.00Hz
% Error	−1.49%	2.60%	2.62%
2nd OC anal.	616.06Hz	496.58Hz	1306.6Hz
2nd OC FEM	627.91Hz	510.09Hz	1307.60Hz
% Error	−1.89%	−2.65%	−0.08%

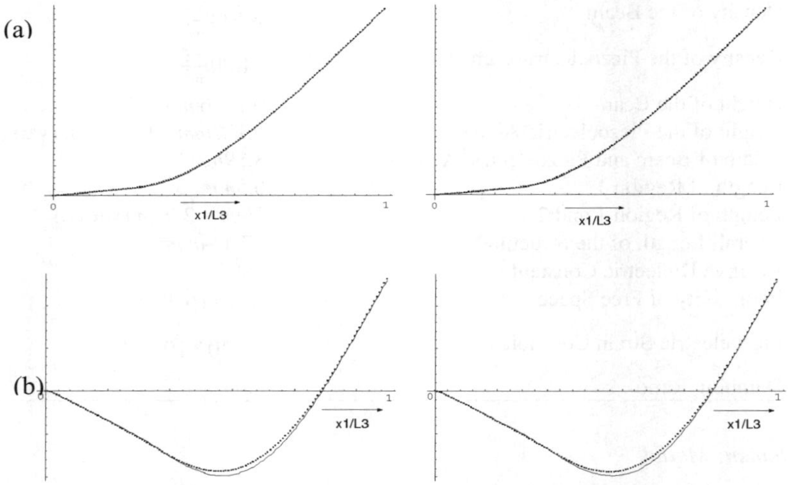

Figure 2. Normalized mode shapes for (a) 1st bending mode, short-circuited (left), and open-circuited (right), and (b) 2nd bending mode, short-circuited (left), and open-circuited (right). Solid-line: analytical model, dotted line: FEM model.

3 OPTIMIZATION

3.1 *Results & discussion*

Now two parameters of the system are allowed to vary: (i) the length of the patch, while L_1 is held fixed, and (ii) the height of the patch. All other geometric constants are held constant (Table 1). L_2 is varied from 3 mm to 27 mm in steps of 1 mm while h_p ranges from 0.127 mm to 0.254 mm in increments of 0.0254 mm. Only the first bending mode was considered, it being the most common resonant mode used in piezoelectric fans.

First the natural frequencies for SC and OC electrodes are calculated. These are used to compute the electromechanical coupling factors (EMCF), while for the forced response only the SC electrode mode shapes and natural frequencies are sufficient. The EMCF is a measure of the electromechanical energy conversion and is defined as

$$\mathrm{EMCF} = \sqrt{\frac{\omega_{OC}^2 - \omega_{SC}^2}{\omega_{OC}^2}}$$

The variation of EMCF's as a function of the patch-to-beam thickness ratio and patch-to-beam length ratio is plotted in Figure 3a. The EMCF for the first bending mode of the fan is seen to reach a maximum when the patch thickness is reduced to equal the beam thickness (hp/2h = 1). Further an optimum patch-to-beam length ratio exists (approximately 0.55) where the EMCF is maximized. This variation is closely connected to the variation with optimization parameters of the fan natural frequencies. Piezoelectric patches add distributed elastic and piezoelectric stiffness as well as mass to the fan. For small patch lengths, the added stiffness effect dominates and increases the natural frequencies while at longer patch lengths the added mass effect reduces the natural frequencies. Therefore the natural frequencies and consequently EMCF are maximal at an optimal value of patch-to-beam length ratio.

Using the SC mode shapes and the relation $v_3(L_3, t) = w_3(L_3, t) = \varphi_3(L_3)\, q(t)$, the maximum tip deflection can be given as $\bar{v}_3(L_3) = \phi_3(L_3)\, F(\omega_0)$ where ω_0 is the damped natural frequency of the fan with short-circuited electrodes. The variation of the maximum tip deflection as a function of the patch-to-beam thickness ratio and patch-to-beam length ratio is plotted in Figure 3b. Equal thickness of the patch and beam are seen to lead to maximal tip deflections while the optimal patch-to-beam length ratio (approximately 0.45) is lesser than that required for optimal EMCF. For a fixed damping ratio, the maximum tip amplitude is affected by variations in mode shape and modal forcing.

Maximizing the tip rotation of the fan is expected to lead to an increase in the centrifugal forces on the fluid under actuation. For this reason, the variation of the tip rotation defined as $\bar{v}_3'(L_3) = \varphi_3'(L_3)\, F(\omega_0)$ is also plotted as a function of the patch-to-beam thickness ratio and patch-to-beam length ratio in Figure 3c. Both tip deflection and tip rotation reach their maxima for the same fan configuration. However, the performance drop for off-optimal configurations is sharper if the optimization criterion is posed in terms of tip rotation as compared to tip deflection. Tip rotation may also be of interest when considering high tip velocities where vortex shedding can significantly affect the performance of the piezoelectric fan.

Finally, the variation of the stroke volume flow rate defined as $\dot{V} = \frac{4}{\pi}\, b\, F(\omega_0)\, \omega_0 \int_0^{L_3} \varphi(x_1)\, dx$ as a function of the patch-to-beam thickness ratio and patch-to-beam length ratio is plotted in Figure 3d. This quantity is a measure of the total volume per second of fluid displaced by the fan and is expected to correlate well to the axial flow rate of the fan. The stroke volume flow rate is also maximized when the patch thickness equals the blade thickness. However, the optimal patch-to-beam length ratio (approximately 0.6) is slightly greater than that required for maximal EMCF. Stroke flow rate is consists of both the natural frequency (EMCF) and the maximum tip deflection expressions. Therefore, its optimal configuration in general is different from optimal configurations based on each constituent term.

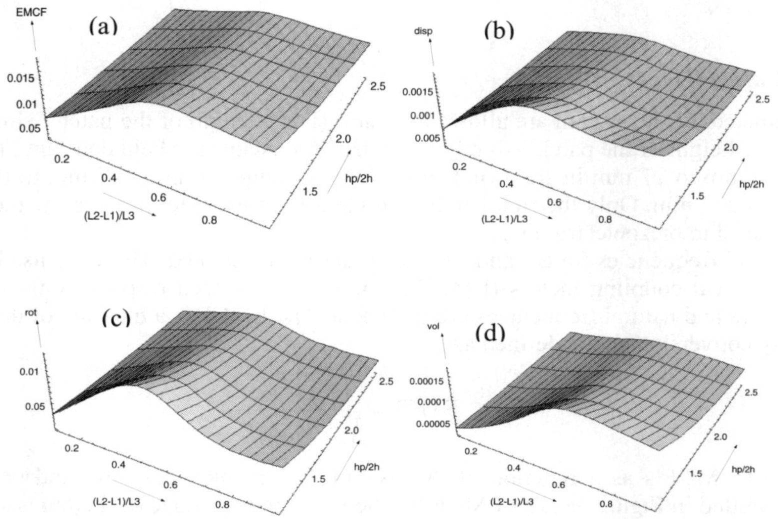

Figure 3: Variation with patch-to-beam length and thickness ratio of (a) Electromechanical coupling factor (EMCF), (b) Tip deflection, (c) Tip rotation, and (d) Volume stroke rate of the resonant fan.

4 CONCLUSIONS

A closed-form analytical solution of the dynamic response of a resonant, composite piezoelectric beam is developed for the optimization of piezoelectric fans. The predicted open- and short-circuited frequencies and mode shapes match excellently the results of finite element computations. The results indicate that the optimal values of patch-to-beam ratio and patch-to-beam thickness are different depending on whether the optimization criterion is posed in terms of (i) maximal electromechanical coupling factor (EMCF), (ii) maximal tip deflection and rotation at resonance, and (iii) maximal volume stroke rate. Such analytical and FE models are expected to provide simple design guidelines for the development of low power, high flow rate piezoelectric fans.

REFERENCES

Abramovich, H. 1998. Deflection control of laminated composite beams with piezoceramic layers - closed form solutions. *Composite Structures* 43: 217-231.

Aldraihem, O.J. & Kheir, A.A. 2000. Smart beams with extension and thickness-shear piezoelectric actuators. *Smart Materials and Structures* 9: 1-9.

Barboni, R., Mannini, A., Fantini, E. & Gaudenzi, P. 2000. Optimal placement of PZT actuators for the control of beam dynamics. *Smart Materials and Structures* 9: 110-120.

Benjeddou, A., Trinade, M.A. & Ohayon, R. 1999. New shear actuated smart structure beam finite element. *AIAA* 37 (3): 378-383.

Brennan, M.J., Elliot, S.J. & Pinnington, R.J. 1997. The dynamic coupling between piezoceramic actuators and a beam. *Journal of the Acoustical Society of America* 102 (4): 1931-1942.

Chandrashekhara, K. & Varadarajan, S. 1997. Adaptive shape control of composite beams with piezoelectric actuators. *Journal of Intelligent Material Systems and Structures* 8: 112-124.

Crawley, E.F. & Anderson, E.H. 1990. Detailed model of piezoceramic actuation of beams. *Journal of Intelligent Material Systems and Structures* 1: 5-25.

Crawley, E.F. & de Luis, J. 1987. Use of piezoelectric actuators as elements of intelligent structures. *AIAA* 25 (10): 1373-1385.

Khdeir, A.A. & Aldraihem, O.J. 2001. Deflection analysis of beams with extension and shear piezoelectric patches using discontinuity functions. *Smart Materials and Structures* 10: 212-220.

Kim, S.J. & Jones, J.D. 1996. Influence of piezo-actuator thickness on the active vibration control of a cantilever beam. *Journal of Intelligent Material Systems and Structures* 6: 610-623.

Lee, H.-J., & Saravanos, D.A. 1996. Coupled layerwise analysis of thermopiezoelectric composite beams. *AIAA* 34 (6): 1231-1237.

Lobontiu, N., Goldfarb, M. & Garcia, E. 1999. Achieving maximum tip deflection during resonant excitation of piezoelectrically actuated beams. *Journal of Intelligent Material Systems and Structures* 10: 900-913.

Meirovich, L. 1997. Principles and Techniques of Vibration, Prentice-Hall.

Main, J.A., Garcia, E. & Howard, D. 1994. Optimal placement and sizing of paired piezoactuators in beams and plates. *Smart Materials and Structures* 3: 373-381.

Pan, J., Hansen, C.H. & Snyder, S.D. 1992. A study of the response of a simply supported beam to excitation by a piezoelectric actuator. *Journal of Intelligent Material Systems and Structures* 3: 3-16.

Pan, X., & Hansen, C.H. 1994. Piezoelectric crystal vs point force excitation of beams and plates. *Journal of Intelligent Material Systems and Structures* 5: 363-370.

Rivory, J.F., Hansen, C.H. & Pan, J. 1994. Further studies of the dynamic response of a simply supported beam excited by a pair of out-of-phase piezoelectric actuators. *Journal of Intelligent Material Systems and Structures* 5: 654-663.

Seemann, W., Straub, A., Chang, F.-K., Wolf, K. & Hagedorn, P. 1997. Bonding Stresses Between Piezoelectric Actuators and Elastic Beams. *SPIE* 3041: 665-675.

Seemann, W., & Sattel, T. 1999. The influence of bonding on the efficiency of piezoceramic patches as actuators in smart structures. *SPIE* 3668: 677-687.

Seeley, C.E. & Chattopadhyay, A. 1993. The development of an optimization procedure for the design of intelligent structures. *Smart Materials and Structures* 2: 135-146.

Seshu, P. & Naganathan, N.G. 1998. Finite-element analysis of strain transfer in an induced strain actuator. *Smart Materials and Structures* 7: 76-88.

Shen, M.-H. H. 1995. A new modeling technique for piezoelectrically excited beams. *Computers & Structures* 57 (3): 361-366.

Silva, E.C.N. & Kikuchi, N. 1999. Design of piezoelectric transducers using topology optimization. *Smart Materials and Structures* 8: 350-364.

Strambi, G., Barboni, R. & Gaudenzi, P. 1995. Pin-force and Euler-Bernoulli models for analysis of intelligent structures. *AIAA* 33 (9): 1746-1749.

Tiersten, H.F. 1969. Linear piezoelectric plate vibrations. New York: Plenum Press.

Wang, Q. & Wang, C.M. 2000. Optimal placement and size of piezoelectric patches on beams from the controllability perspective. *Smart Materials and Structures* 9: 558-567.

Wolf, K.D. 2000. Electromechanical energy conversion in asymmetric piezoelectric bending actuators. Ph.D. Thesis, Technical University-Darmstadt, Germany.

Yang, J.S. 1997. Equations for the flexural motion of elastic plates with partially electroded piezoelectric actuators. *Smart Materials and Structures* 6: 485-490.

Zhai, B., Lim, S.-P., Lee, K.-H., Chew, C.-H., Dong, S. & Lu, P. 2000. A design of an ultrasonic linear motor based on theoretical analysis. *Smart Materials and Structures* 9: 774-779.

Thermal Challenges in Next Generation Electronic Systems, Joshi & Garimella (eds)
 ISBN 90-77017-03-8

Vibration-Induced Droplet Atomization heat transfer cell for high-heat flux dissipation

S.N. Heffington, W.Z. Black & A Glezer
Woodruff School of Mechanical Engineering, Georgia Institute of Technology, GA, USA

Keywords: two-phase heat transfer, spray cooling, microelectronics cooling

ABSTRACT: This paper describes a unique two-phase cooling method that includes a closed heat transfer cell, similar to a thermosyphon that can be used to cool microelectronic packages. The cooling method is based upon a Vibration-Induced Droplet Atomization, or VIDA, process that can generate small liquid droplets inside a closed cell and propel them onto a heated surface. The VIDA technique involves the violent break-up of a liquid film into a shower of droplets by vibrating a piezoelectric actuator and accelerating the liquid film at resonant conditions. The droplets continually coat the surface with a thin liquid film, which evaporates on the heated surface, and the vapor is condensed on the internal surfaces of the heat transfer cell as well as the liquid working fluid. The condensed liquid is returned via gravity to the piezoelectric actuator where it is again atomized.

A VIDA heat transfer cell 145 mm in diameter and 75 mm thick was constructed. Test data described in this study include the heat transfer characteristics and cooling capabilities for a small-scale cell that is suitable for cooling a desktop microprocessor. The VIDA process produces droplets of relatively uniform diameter, and the droplets have sufficient momentum to reach the remotely located heated source. Heat fluxes as high as 100 W/cm^2 have been measured when a single fan is used as the external heat removal device.

1 INTRODUCTION

In the microelectronics industry, advances in technology have brought about an increase in transistor density and faster electronic chips. As electronic packages increase in speed and capability, the level of heat flux that must be dissipated to maintain reasonable chip temperatures has also risen. Cooling levels are projected to reach the 100 – 150W range according to the SIA Packaging Technology Roadmap [SIA 1997]. To provide reliable cooling for this expected level, the single level integrated module under development by the Packaging Research Center at the Georgia Institute of Technology is expected to dissipate between 20 and 100W/cm^2 [Georgia Institute of Technology 1995].

Two-phase heat transfer, involving the evaporation of a liquid in a hot region and the subsequent condensation of the resulting vapor in a cold section, can provide the large heat fluxes needed for microelectronic packages to operate at acceptable temperature levels. By changing the phase of the working liquid, a two-phase heat transfer cooling scheme can transport high heat transfer rates across moderately small temperature differences. Heat pipes and thermosyphons are examples of efficient heat transfer devices that exploit the benefits of two-phase heat transfer [Peterson 1994, Kiewra & Wayner 1986, and Mudawar et. al. 1987]. Immersion cooling involving pool boiling of a dielectric working fluid on the surface of the package is another example of two-phase cooling technology useful for microelectronic applications [Arik & Bar-Cohen 1998]. A cooling module based on the VIDA principle is a new technology capitalizing on the benefits of two-phase cooling while improving on heat pipe performance by eliminating the wicking structure and employing a more active means to transport the liquid phase back to the heat source [Smith et. al. 1998]. Without a wick structure, the operation of a VIDA cell is not limited by the capability of the wick to continually supply the boiler section with liquid. Furthermore, the VIDA cell has the potential to improve upon the performance of a thermosyphon, because it creates a condition of thin film boiling in the evaporator section. The evaporation of a thin film prevents the formation of an insulating vapor blanket that can exist in pool boiling situations. Also, the atomized droplet momentum is sufficient to propel the liquid droplets through the vapor layer and to spread the impinging liquid into a thin film on the heated surface. While the VIDA cell has advantages over heat pipes and thermosyphons due to the production of a thin evaporating film, the VIDA cell, like a thermosyphon, has one limitation because it must be operated in a nearly horizontal orientation. Because the VIDA process creates a thin, evaporating liquid film, it can provide high heat transfer rates and can minimize temperature gradients that cause potentially dangerous thermal stresses [Denney et. al. 1996].

A schematic of one type of VIDA cell is shown in Figure 1. The cell consists of a vibrating driver that creates a shower of small diameter secondary droplets by breaking up a larger primary drop that forms on the driver. An optional orifice plate can be placed over the driver to regulate the thickness of the liquid film on the driver. In this type of design, the driver ejects small secondary droplets through the holes in the orifice plate. The secondary droplets are propelled toward the heated surface where they form a thin film of liquid on the heater. The liquid film evaporates and fills the interior cavity of the cell with vapor. The exterior surfaces of the cell are cooled via heat transfer to the ambient and the vapor condenses on these surfaces. The condensed liquid is then returned via gravity to the driver where it once again is atomized and the process is repeated. The entire cell can be very small and the driver requires only milliwatts of energy to operate.

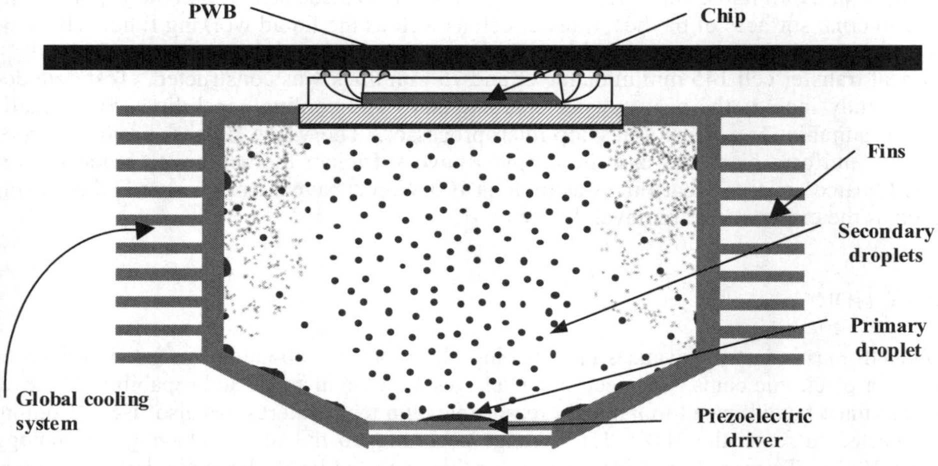

Figure 1. Schematic of a simple VIDA heat transfer cell

2 THE VIDA PROCESS

In order to design an efficient heat transfer cell, a major challenge is understanding the VIDA process. The VIDA cell includes a metallic disc coated with a piezoelectric material that is energized with a sinusoidal varying voltage. The resulting vibrations are capable of accelerating a single liquid droplet or liquid film to an extent that the fluid has sufficient velocity to break-up or atomize into numerous smaller secondary droplets. The secondary droplets are propelled upward and they are capable of impacting a remotely located heated surface. The frequency and amplitude of the voltage used to energize the piezoelectric transducer are important parameters that must be properly controlled in order to achieve the liquid breakup phenomena that is essential to the VIDA process.

The schematic and photograph in Figure 2 illustrate the VIDA atomization process used in this research investigation. An aluminum piezoelectric driver that was 31 mm in diameter was placed below a brass perforated plate. This perforated orifice plate contained holes 1.59 mm in diameter spaced on 3.18 mm centers in a square grid. The thin vertical spacing between the orifice plate created surface tension forces that are sufficient to pump water onto the driver surface from a reservoir surrounding the driver. This liquid layer was accelerated by the motion of the driver which resulted in atomization of the liquid through the holes in the orifice plate. Secondary droplets are produced that are much smaller than the diameter of the holes in the orifice plate. The orifice plate provides a robust continuous VIDA process by regulating the mass of water on the driver surface while not allowing the driver to become overloaded or starved for water. This droplet atomization procedure ensures a continual supply of liquid droplets to the heater surface.

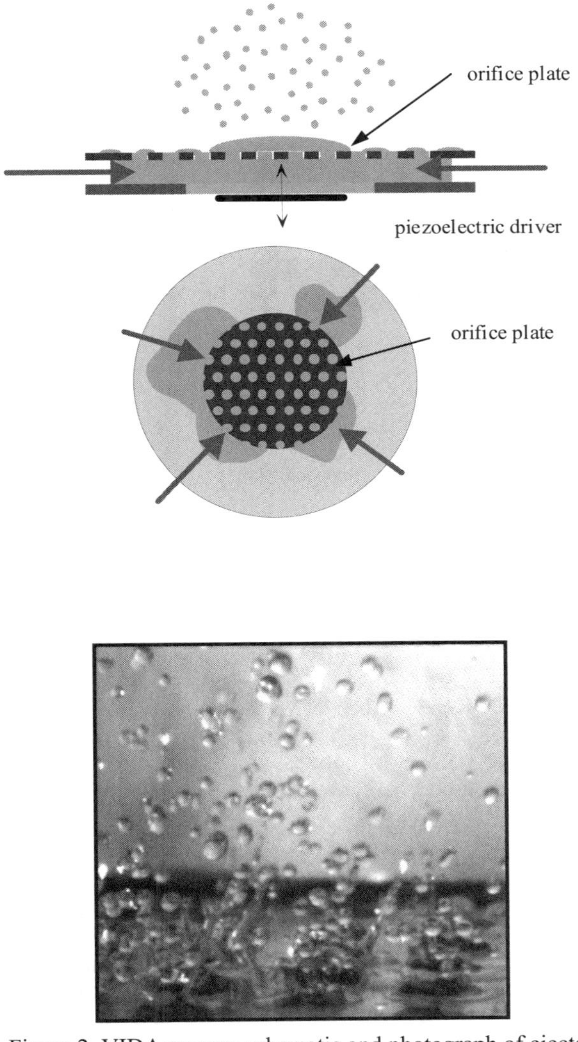

Figure 2. VIDA process schematic and photograph of ejected secondary droplets

3 VIDA HEAT TRANSFER CELL

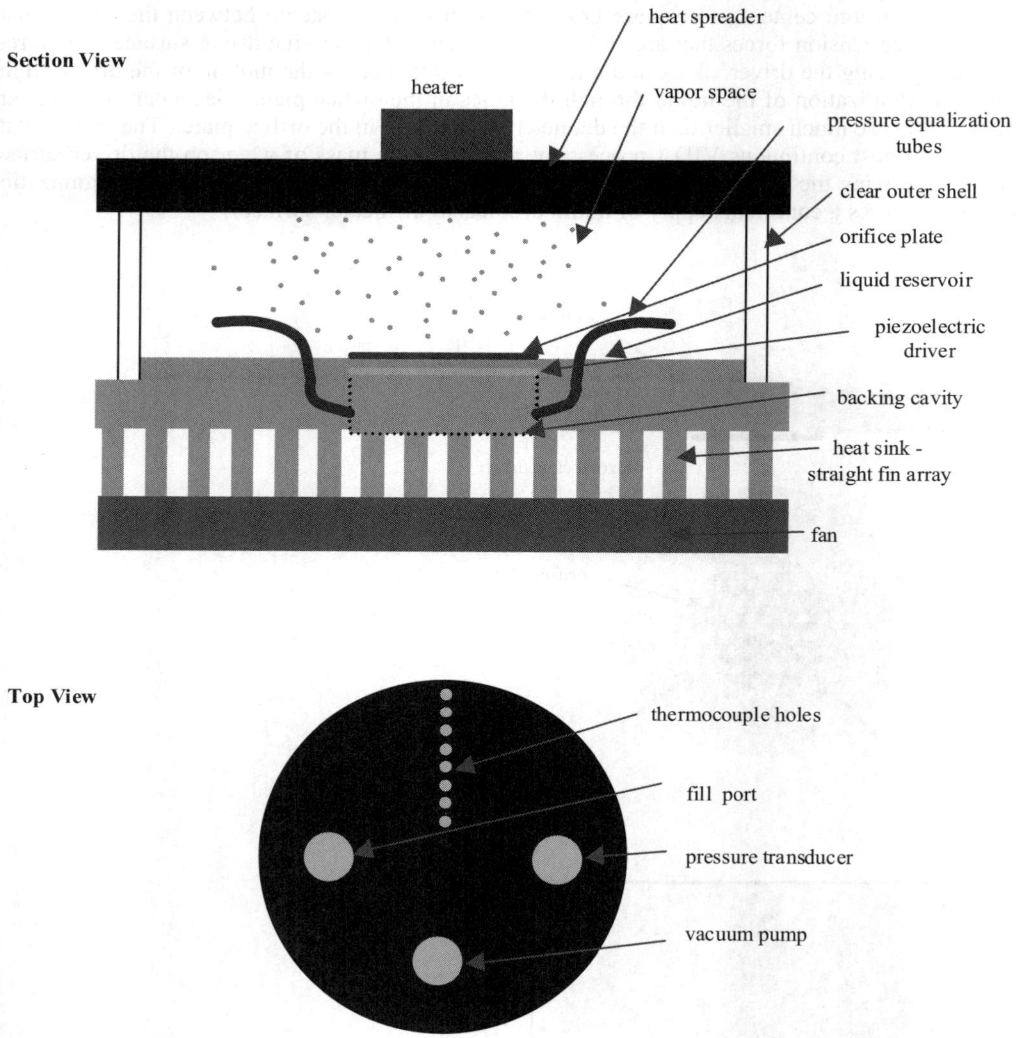

Figure 3. Schematic of VIDA heat transfer cell

Several different VIDA cells have been built and tested to evaluate their heat transfer characteristics. A schematic of one of the VIDA heat transfer cell designs is shown in Figure 3. This cell was designed around a Globe motors fan model # D47-B15A-04W3-100. This fan was 38.1 mm thick and 120 mm square. Wakefield aluminum extrusion fin array number 2002 was chosen to form the bottom of the cell. This aluminum extrusion was 141.2 mm wide and was purchased in a single 1.83 m length. To form the cell, a square piece of the fin array 141.2 mm on a side was cut from the long stock. The extrusion had 16 longitudinal fins that were 1.7 mm wide at the tip and 2.5 mm wide at the base and 25.4 mm long. The base of the extrusion was 7.9 mm thick. The driver was located flush with the bottom of the cell leaving a small backing cavity with a volume of 3618 mm^3

behind the driver. Two brass tubes with an outer diameter of 3.2 mm were used to connect the backside of the driver to the topside. These tubes were curved to prevent water from leaking through them to the backside of the driver and used to equalize the pressure on both sides of the driver when the cell was operated under vacuum conditions. An O-ring groove was machined in the aluminum extrusion to allow for a 139.7 mm outer diameter plastic tube. After the aluminum heat sink was machined, its upper surface was gold plated to prevent the aluminum surface from corroding where it was in contact with water.

Several lengths of plastic tube were cut to form the sides of the cell. These transparent pieces could be easily interchanged to vary the overall height of the cell. Also, by using clear plastic it was possible to view the VIDA process during operation with and without the heater energized. The top of the cell was machined from aluminum and it had several holes to accommodate a pressure transducer, two valves, several thermocouples, and the calibrated heater. The 9.5 mm thick aluminum heat spreader was black anodized to prevent the aluminum from reacting with water to produce noncondensible gases. One valve was connected to the vacuum pump, and the other valve was connected to a graduated buret that was used to charge the cell with water. The thermocouple holes made it possible to measure eight temperatures within the cell even under vacuum conditions.

The liquid reservoir is designed to provide a large condensation area so that the base of the fins is maintained at a temperature close to the saturation temperature of the water. The liquid reservoir is well mixed due to the vibrating action of the piezoelectric driver. This liquid pool also provides a uniformly thick layer of water over the top of the driver. The piezoelectric driver is 31 mm in diameter, and it is activated by a 25 V_{rms} sinusoidal source. A perforated orifice plate with uniformly spaced holes is centrally located over the driver. The small spacing between the driver and orifice plate assures a reliable flow of liquid to the surface of the driver. The frequency and amount of water inside the cell were varied until a continuous VIDA process was observed on the driver. For the VIDA cell shown in Figure 3, a strong VIDA process was experienced when the cell was charged with 50 to 90 mL of water and it was vibrated at a frequency of 0.7 to 1.2 kHz.

4 HEAT TRANSFER RESULTS

The first heat transfer experiment conducted with this cell was a series of transient tests at different pressure levels with no orifice plate present in the cell. These results are plotted in Figure 4. These three plots are individual runs at different pressure levels inside the cell. As the pressure was changed by introducing air into the cell, the optimum frequency of VIDA operation was also changed. These distinct curves indicate that the critical heat flux temperature increases as the pressure inside the cell increases as expected. However, the significant decrease in the level of heat removal as the pressure decreases was unexpected, since the driver was known to have the same accelerations under all pressures. It was also noted that as the pressure decreased, the VIDA process was visually noticed to decrease in robustness. Often it was impossible to generate a spray at extremely low pressures (below 13.8 kPa).

Further visual inspection of the VIDA process under reduced pressures revealed the cause of the diminished VIDA performance. At reduced pressures when the driver was excited, vapor bubbles were seen in the liquid layer located above the driver surface. The bubbles in the liquid layer above the driver were due to the localized low pressures that existed directly above the driver. The already low pressure within the cell combined with the lower pressure in the liquid layer that resulted due to the acceleration of the liquid by the driver. The resulting pressure was so low that the saturation temperature of the water was below room temperature, producing a pulsating cavitation in the liquid layer on the driver. The presence of the cavitation bubbles hindered the droplet atomization process. The bubbles caused the formation of a liquid dome as high as two mm above the reservoir surface, and the domed region moved chaotically with the random nature of the cavitation bubbles.

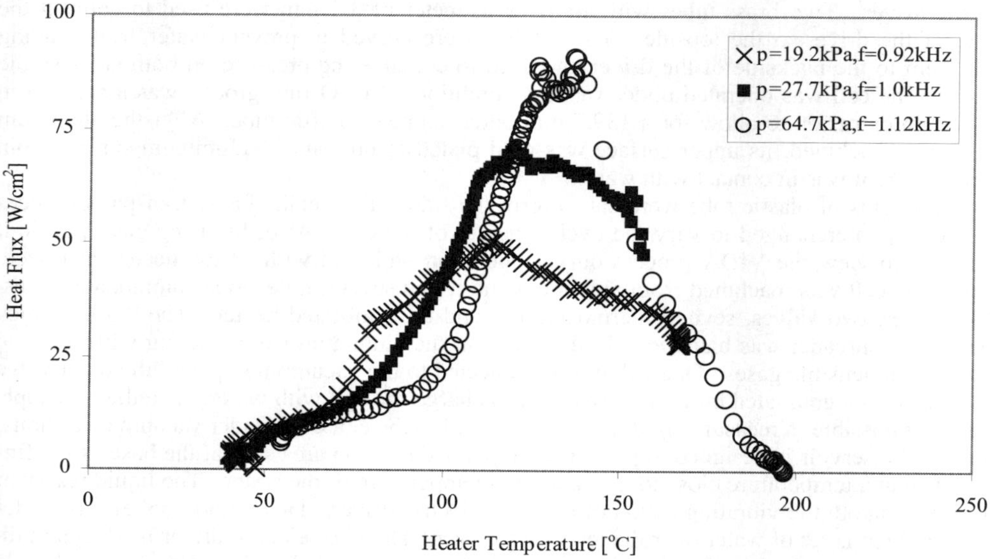

Figure 4. Heat transfer results for VIDA heat transfer cell without orifice plate

Generating a thin liquid layer above the driver surface rather than a thick pool was believed to be the answer to the cavitation problem. Since the surface of the driver is not easily wetted by water, a large liquid reservoir was required to overcome the surface tension forces and to completely wet the driver. The thick layer of water over the driver encouraged the formation of vapor bubbles and cavitation. It was thought that if the liquid layer over the driver was sufficiently thin, the vapor bubbles formed due to cavitation would simply escape into the cell and not hinder the atomization process. Therefore the release of the bubbles could in fact help propel the secondary droplets toward the heater. The first attempt to solve the cavitation problem involved coating the driver with a hydrophilic substance. RainX brand anti-fog solution was used to coat the driver. While it appeared to initially help, the RainX anti-fog solution did not last, and water quickly stopped wetting the surface.

The other solution to the cavitation issue involved using the perforated plates that had been used in the past. These perforated plates would not be used as a pump to drive liquid onto the driver surface, but more as a way of controlling the thickness of the liquid layer on the driver, which would limit the cavitation problem. The brass orifice plate with 1.59 mm holes on 3.18 mm square centers worked well. When the orifice plate was placed less than one mm above the driver, no visual difference in the VIDA effect could be noticed at all pressures, and the cavitation process was eliminated. The remainder of the results for VIDA heat transfer cell utilized the brass orifice plate located one mm above the driver.

During the one atmosphere tests, the characteristics of the liquid reservoir were studied. Thermocouples that were located in the pool of liquid measured only a slight temperature drop of less than 5°C from the center of the cell to the outside. These measurements indicated that the pool was well mixed by the vibration of the driver. In a separate test, small red aluminum flakes were added to the liquid pool after the driver was energized. The motion of the driver caused these flakes to distribute rapidly into the entire pool. By keeping the pool nearly uniform in temperature, the heat conduction to the fins located directly below is more efficient.

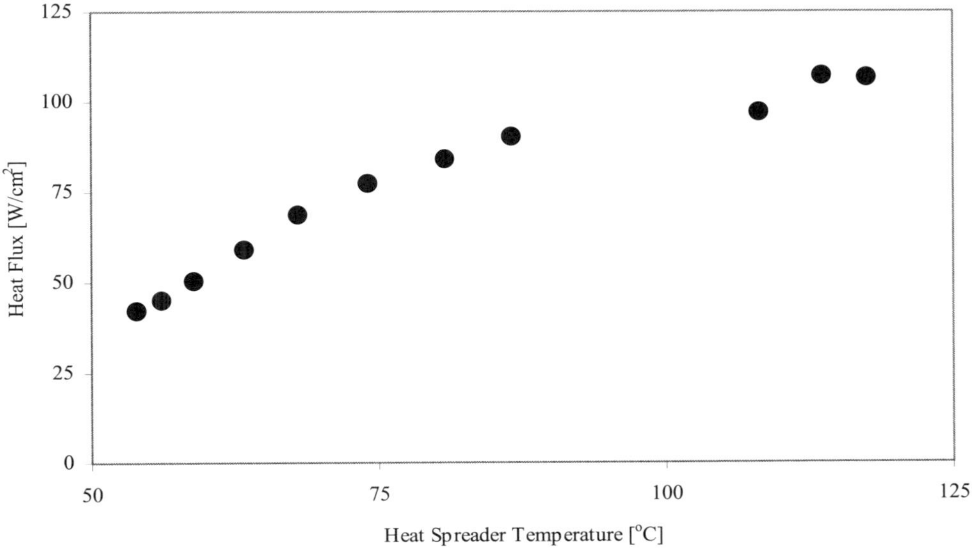

Figure 5. Heat transfer results for VIDA cell at 101.3 kPa with heat spreader

Several steady-state tests were completed using the VIDA heat transfer cell. The atmospheric pressure results are plotted in Figure 5. These data points were measured using VIDA heat transfer cell with the orifice plate present. The 19 mm plastic tube was used during these experiments, and cell was charged with 90 mL of distilled water. Figure 5 indicates that significant levels of cooling were seen below a saturation temperature of 100°C. By heating the entire surface of the cell, the secondary droplets are utilized more efficiently. In previous cases, a large percentage of the atomized droplets did not evaporate because they did not strike the heater. With the heat spreader in place, a vast majority of the droplets that strike the heat spreader evaporate because its surface was above the saturation temperature of the water.

The results of the reduced pressure tests are plotted in Figure 6. The average pressure inside the cell for these data points was 23.1 kPa. It was difficult to keep the pressure constant during the entire test, because the cell pressure tended to rise slightly as the heat flux to the heater increased. The results at lower pressures showed an increase in heat transfer at lower cell temperatures. With a heat spreader temperature of 75°C, the heat transfer cell dissipated nearly $100 W/cm^2$. This level of heat flux translates to a total heat removal rate of nearly 130 W. By better utilizing the VIDA process, the VIDA heat transfer cell was able to dissipate a high heat flux at a temperature that is compatible with microelectronics.

A benchmark was needed to evaluate the performance of the cell during initial runs for VIDA heat transfer cell. A test was conducted in which the VIDA portion of the cell was replaced with a solid metallic cylinder. Results from this test would allow comparison of the VIDA process with a pure conduction model in which the heat was simply conducted from the heater to the fin surfaces. The original benchmark test involved the aluminum heat sink mounted on top of the fan with a calibrated heater attached directly to the center of the heat sink. This heater was held tightly to the finned surface, and the thermal contact resistance between the heater and the heat sink was reduced by a thin layer of thermal grease. The heater was energized with a constant power supply, and the heat sink/heater combination was allowed to reach steady state. This process was completed for nine different power levels.

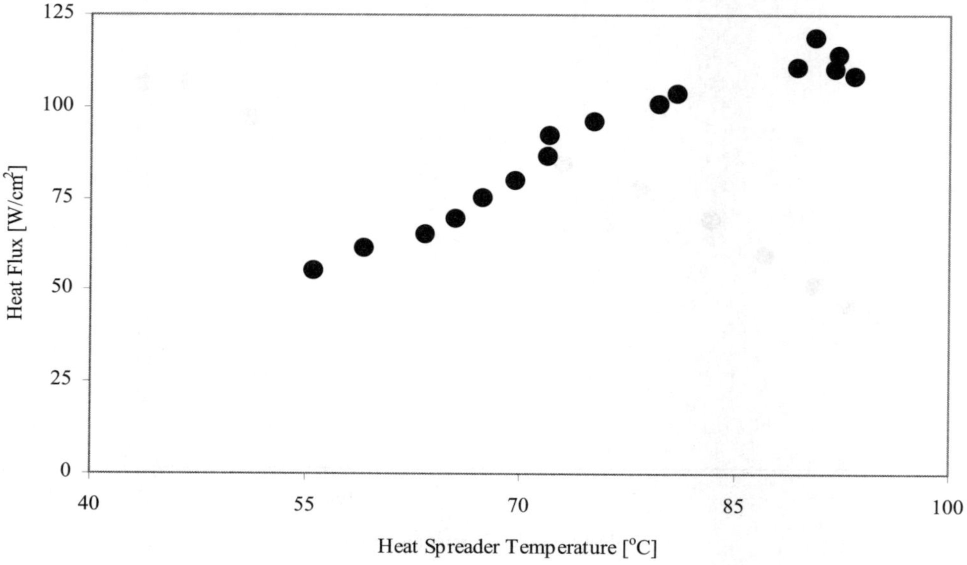

Figure 6. Heat transfer results for VIDA cell at 23.1 kPa with heat spreader

A second benchmark test was completed. For this test, a solid copper cylinder was placed between the heater and the aluminum fin array. The copper cylinder was 19 mm thick and 139.7 mm in diameter. The copper cylinder replaced the two-phase VIDA portion of the cell with a solid metallic conductor. A thin layer of thermal grease was applied to the interface between the heater and the copper cylinder as well as between the copper cylinder and the heat sink. The heater was energized to twelve different power levels when the copper cylinder was in place, and the heater temperature was recorded.

The cell consisting of the combination of the copper plus heat sink was capable of removing a significantly larger amount of heat than the cell with only the aluminum heat sink. The enhanced performance with the copper cylinder present is a result of improved heat spreading which increases the effectiveness of the fin array. Assuming one-dimensional conduction within the cell, the curves should be linear since the heat flux is proportional to the effective thermal conductivity of the cell. For these benchmark tests the fan was operated at the same power levels as the previous VIDA tests, and the heat sink was identical in all situations.

A comparison of the benchmark and VIDA heat transfer cell results is presented in Figure 7. This graph shows that the reduced pressures tests were able to lower the temperature of the heater without sacrificing the level of heat dissipation. Also, the VIDA cell outperforms a solid copper conductor of the same volume. By designing a cell to take advantage of the VIDA technology, a successful heat transfer device has been constructed and verified. In particular, the VIDA heat transfer cell is capable of removing over 100W/cm^2 while maintaining the heater temperature below 85ºC.

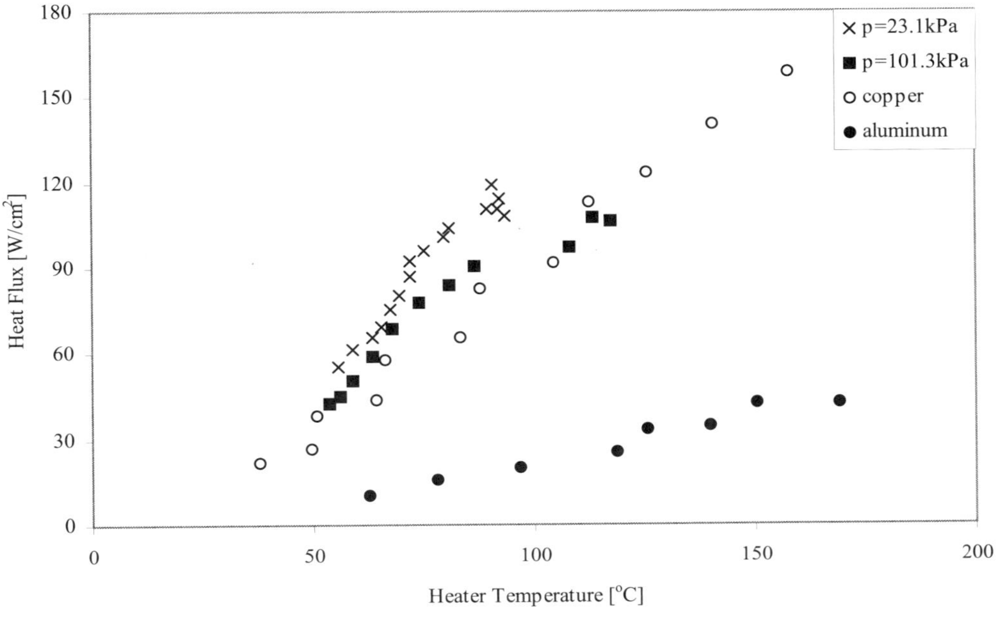

Figure 7. Comparison of heat transfer results for the copper conductor and cells #5B and #4B

5 CONCLUSIONS

A new heat transfer cell suitable for cooling small microelectronic packages is proposed which is based on a droplet atomization mechanism. A vibrating driver provides a continual generation of small diameter droplets that are propelled toward the heated surface where they form a thin liquid film. The liquid evaporates on the heater and the vapor condenses on the secondary droplets and the cool inner surfaces inside the cell. The condensed liquid returns to the driver where it is re-atomized.

Initial heat transfer results utilizing water and a simple cell design have shown that a cell based on the VIDA process can provide a heat flux over 100 W/cm^2. Reducing the pressure inside the cell increases the heat flux at a given heater surface temperature. In addition, for a given heat flux, the surface temperature of the heater is less at reduced pressures. These preliminary tests have indicated that a VIDA-based heat transfer cell has a potential to provide inexpensive cooling of small-scale electronic devices and the cooling capabilities are within the range required for most modern package designs.

6 ACKNOWLEDGEMENTS

This work was supported by DARPA through BAA 98-32.

7 REFERENCES

Arik, M. and A. Bar-Cohen, 1998, "Immersion Cooling of High Heat Flux Microelectronics with Dielectrid Liquids", 4th Int. Symposium and Exhibition on Advanced Packaging-Materials-Processes-Properties, Atlanta, pp. 229-247.

Denney, D. L., W. Z. Black, J. G. Hartley, and A. Glezer, 1996, "High Heat Flux Boiling with Monodisperse Water Droplets," Process, Enhanced, and Multiphase Heat Transfer, R. M. Manglik and A. D. Krause, eds., Begel House, Inc., New York, pp. 175-182.

Georgia Institute of Technology, 1995, Packaging Research Center, Third Annual Review, Atlanta Georgia, pp. 4.1-4.5.

Kiewra, E. W., and P. C. Wayner, 1986. "A Small Scale Thermosyphon for Immersion Cooling of a Disc Heat Source," *Heat Transfer in Electronic Equipment,* ASME Symposium HTD-Vol. 57, Bar-Cohen, A. (Ed.), American Society of Mechanical Engineers, New York, pp.77-82.

Mudawar, I., T. A. Incropera, and F. P. Incropera, 1987, "Microelectronic Cooling by Fluorocarbon Liquid Films," *Proc. Int. Symposium of Cooling Technology for Electronic Equipment.*

Peterson, G. P., 1994, An Introduction to Heat Pipes, John Wiley & Sons, Inc.,, New York, pp. 285-326.

SIA Semiconductor Industry Association, 1994, "The National Technology Roadmap for Semiconductor: Technology needs – 1997", Semiconductor Industry Association, Washington DC.

Smith, M. K., A. James, B. Vukasinovic, and A. Glezer, 1998, "Vibration-Induced Droplet Atomization," submitted for publication.

Thermal Challenges in Next Generation Electronic Systems, Joshi & Garimella (eds)
 ISBN 90-77017-03-8

Transient and sub-atmospheric performance of a closed-loop electroosmotic microchannel cooling system

L. Jiang, J. Mikkelsen, J.-M. Koo, L. Zhang, D. Huber, S. Yao, A. Bari, P. Zhou, J. Santiago, T. Kenny, & K. E. Goodson
Department of Mechanical Engineering, Stanford University, CA, USA,

J. Maveety & R. Prasher,
Intel Corporation, USA

S. Benning
AFRL/IFSC, USA

Keywords: electronic cooling, electroosmotic flow, microchannel heat exchanger, two-phase convection

ABSTRACT: The increasing power density in electronic systems has motivated research on aggressive, compact cooling technology. This paper explores the transient performance of a closed-loop microchannel cooling system based on electroosmotic pumping. The system is implemented at sub-atmospheric conditions to control the boiling temperature of the working fluid. The closed-loop system, at a loop pressure of 25 kPa, removes 38 W from a 1 cm × 1 cm chip with a pump power less than 1.5 W and junction temperature below 100 °C.

1 INTRODUCTION

Future microprocessors with power generation exceeding 200 W will pose challenges for thermal management and will require novel cooling technology. Furthermore, multi-chip integration and system miniaturization motivate reductions in the chip-backside volume occupied by any cooling solution. Conventional forced convection air-cooling is not practical for these levels of heat generation and integration. Heat pipes and vapor chambers have advantages over air cooling since they do not require an external electrical power supply and the working fluid is contained within a closed chamber. However, these capillary-driven devices are not optimal for chip powers exceeding several tens of watts because of the associated increases in the cross sectional area of the heat pipe or vapor chamber and the thermal resistance of the wicking structure [Prasher et al., 2001]. Fully-micromachined capillary-driven loops can eliminate on-chip hotspots but do not solve the problem of heat removal away from the chip [Greif, 2001]. A closed-loop two-phase thermosyphon takes advantage of gravity to generate flow head without using any pumping device [Joshi, 2001]. However, the demand on water flow rate, which is on the order of 100 mL/min necessary to remove 40 W/cm^2, could increase the size of the system significantly. While droplet impingement cooling loops are a promising replacement to conventional aluminum heat sink technology [Glezer, 2001], this technology requires substantial volume at the chip backside. Thermoelectric devices are attractive for refrigeration because they are extremely compact and may eventually be integrated with VLSI technology [Tamura et al., 1999]. Although recent advances have achieved high thermoelectric figures of merit for superlattice structures, more work is needed to reduce the additional power required to achieve a given temperature drop [Venkatasubramanian et al., 2001 & Fan et al., 2001].

Single and two-phase forced convection microchannel heat sinks have received attention because of their dramatic reduction in thermal resistance [Tuckerman & Pease, 1981; Bowers & Mudawar 1994]. Fundamental research has been conducted to understand the complex two-phase microchannel flow and heat transfer which might deviate from traditional theory [Peng et al., 1995; Jiang et al., 2001a; Koo et al., 2001 & Zhang et al., 2001]. A practical self-contained cooler based on convective cooling must be implemented with a pump within a closed loop. For space-critical desktop and portable applications, the pump must be compact, effective and reliable. Among the miniaturized pumps under development, electroosmotic pumps, which use electroosmosis to propel the liquid, are very attractive. They have advantages of no moving parts and on-demand operation and have potential to be compact and efficient. Our past work [Jiang et al., 2001c] demonstrated the feasibility of a closed-loop electroosmotic microchannel cooling system at powers up to 38 W. In the present paper, we report transient and sub-atmospheric performance of this system.

2 SYSTEM DESCRIPTION AND EXPERIMENTAL PROCEDURE

Figure 1 shows a schematic of the closed-loop system. The system consists of an electroosmotic (EO) pump, a microchannel heat exchanger and a heat rejecter. The EO pump propels the liquid through the loop at desired flow rates and pressures using an electric field. The two-phase microchannel heat exchanger, attached on the backside of the thermal test chip, removes heat from the chip and is designed to minimize chip temperature variations. Two-phase fluid exiting from the microchannel heat exchanger is condensed back to a liquid phase in a heat rejection device. Figure 2 is an image of the closed-loop cooling system.

The working principle of the EO pump is based on electroosmosis through an ultra-fine porous glass filter. The silanol groups on the glass surface deprotonate while in contact with an electrolyte and form an electric double layer. As an external electric field is applied through the structure, electroosmotic flow develops within the pores. Our current EO pump [Yao et al., 2001] is designed with a glass filter disk embedded in two acrylic chambers. The glass filter disk has a diameter of 30 mm and a thickness of 2 mm, with an effective pore size of 1 μm and a porosity (ratio of void to total volume) of about 0.2. The disk is sintered and chemically treated for stability and reliability. The two acrylic chambers serve as fluid reservoirs and provide the electric connections to the pump electrodes. A platinum catalyst recombiner is located at the top of the pump inlet chamber to recombine the gas generated during electrolysis. The EO pump with an active volume of 1 cm^3 delivers a maximum flow rate of 6.8 mL/min and a maximum differential pressure of 160 kPa under 200 V applied voltage with working fluid of 1 mM borate buffer in de-ionized water.

The microchannel heat exchanger is fabricated using plasma etching and anodic bonding technology. The parallel rectangular cross-sectional microchannels are etched using deep reactive ion etching (DRIE) in a standard <100> silicon substrate, with common inlet and outlet passages. The microchannels are sealed by anodic bonding of the silicon wafer to 500 μm thick Pyrex 7740 glass (Corning) with two pre-drilled holes serving as the inlet and outlet of the working fluid. The heat exchanger is bonded to a 1 cm × 1 cm Intel thermal test chip using thermally conductive

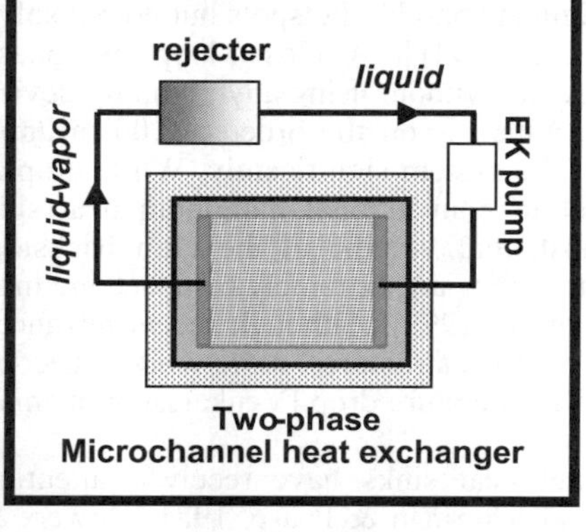

Figure 1. A schematic of the closed-loop cooling system, consisting of an electroosmotic pump, a microchannel heat exchanger and a rejecter.

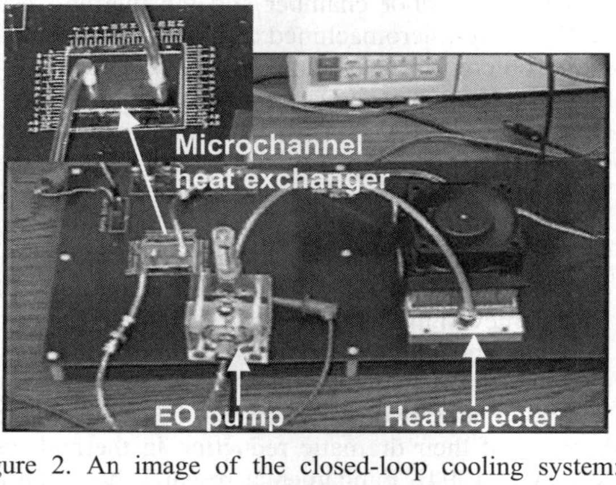

Figure 2. An image of the closed-loop cooling system: electroosmotic pump, a microchannel heat exchanger and a rejecter (Jiang et al., 2001b).

grease with a conductivity of 1.2 $Wm^{-1}K^{-1}$. The current microchannel heat exchanger consists of 40 microchannels, with channel dimensions of 20 mm in length, 100 μm in depth and 100 μm in width. Our open loop experiments showed that the microchannel heat exchanger provides a junction-to ambient thermal resistance of less than 1 K/W and a uniform temperature distribution within the chip under two-phase flow conditions [Jiang et al., 2001b]. The heat rejecter is designed as another heat exchanger with an array of minichannels. The channels are machined in an aluminum plate with hydraulic diameter of 2 mm and sealed by another thin aluminum plate. A commercial fin-array with fan is thermally bonded to the minichannel heat exchanger. The heat rejecter condenses the two-phase fluid, exiting from the microchannel heat exchanger, back to liquid phase and further cools the liquid down to the inlet fluid temperature. The cooled liquid is pumped into the microchannel heat exchanger forming a closed flow loop. The present heat rejecter has a thermal impedance of 0.35 K/W and a water flow impedance on the order of 10^{-6} atm/(mL/min). Optimal design of the minichannels and elimination of the thermal interface between the minichannel heat exchanger and the fin-array can further improve the performance of the heat rejecter.

The cooling system is experimentally studied using 1 mM borate buffer as the working fluid. A schematic of the experimental set-up is presented in Fig. 3. The thermal chip has integrated heater and temperature sensors. The embedded heaters in the thermal chip provide uniform heat flux within the entire chip and the embedded temperature sensors provide junction temperature of the chip. A thermocouple is located inside the pump to monitor the fluid temperature. A vacuum chamber is connected at the pump upstream with a valve to control the system operating pressure. A differential transducer is connected, in parallel, between the inlet and outlet of the microchannel heat exchanger, providing the pressure drop across the microchannel heat exchanger. The applied voltage and the current of the heater and the pump are recorded. The output of the pressure transducer and the temperature sensors are collected using a data acquisition system with a rate of 10 Hz. The output from the temperature sensor located in the center of the 1 cm × 1 cm chip is used to represent the junction temperature due to the uniform temperature distribution. The uncertainty in the experiments is estimated to be 0.1 °C in chip temperature, 3.2% in chip power, and 0.25% in pressure.

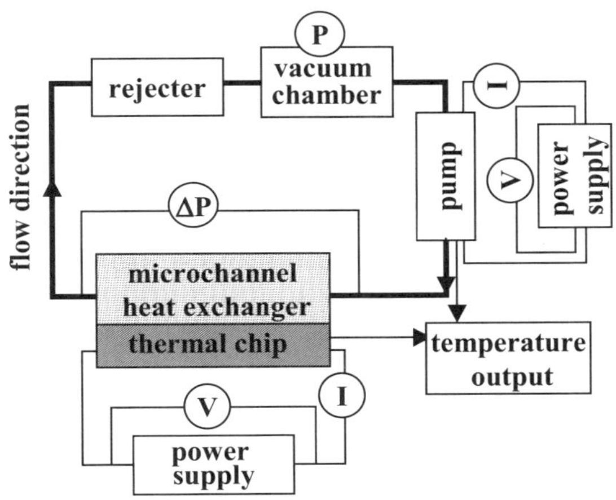

Figure 3. A schematic of the experimental set-up.

3 TRANSIENT BEHAVIOR

We report here the transient performance of the system under various transient power input profiles. Figure 4 is a typical transient response for a "start-up" situation, in which both the EO pump and the heater are turned on simultaneously. The junction temperature shows a response time of about 1 minute, and no dry-out is observed for start-up powers up to 20 W. The time scale for the EO pump to establish a desired flow is measured to be less than 1 second. This is much smaller than the time scale for the heater chip to build up the temperature, on the order of several tens of seconds. Thus, this electroosmotic pumping system is believed to be able to prevent the system from dryout under "start-up" situations. Figure 5 shows the transient response behavior for an "out-of-sleep" situation as the chip operating temperature approaches a new steady state due to a step increase in power while the pumping system maintains the same operating condition. The system time response for "out-of-sleep" situation is shown to be less than 1 minute.

Gas bubbles, resulting from dissolved gas or electrolysis in the EO pump are observed. The bubbles are collected at the pump outlet and form a gas plug with a volume of about 2 mm^3, traveling with the liquid. The total volume of the microchannels in evaporator is 8 mm^3, 4 times of the gas plug volume. Thus there is always liquid in microchannels being evaporated even at the time the gas plug is passing through the evaporator. This indicates that the gas plug with such volume cannot lead to burnout of the thermal chip. However, it suggests the volume of the gas plug must be controlled to avoid thermal failure. The gas plug results in a temporal temperature spike of 3 ~ 7 $^{\circ}C$, which is due to the poor heat capacity and the reduced convection performance of the gas phase. For the present closed-loop system, the time for each individual spike to pass through the microchannel heat exchanger is observed to be less than 1 second with a temperature recovering time of about ten seconds, and the time interval between two spikes is about several minutes.

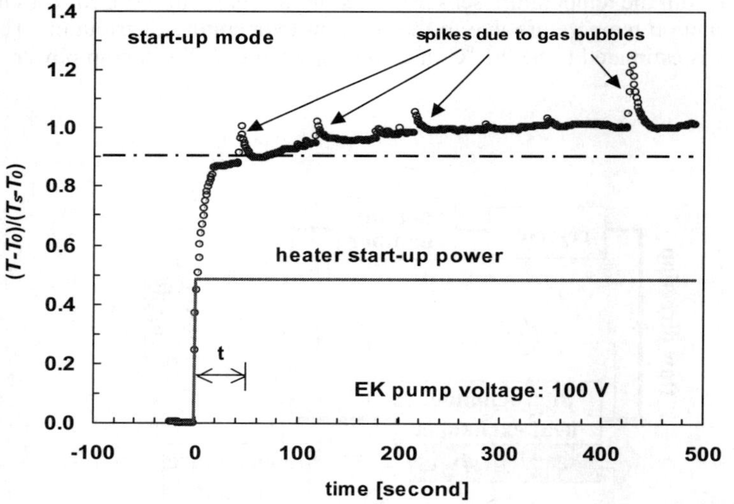

Figure 4. Transient response of a start-up mode. At time zero, the processor is powered up to 20 W and the EO pump is started simultaneously.

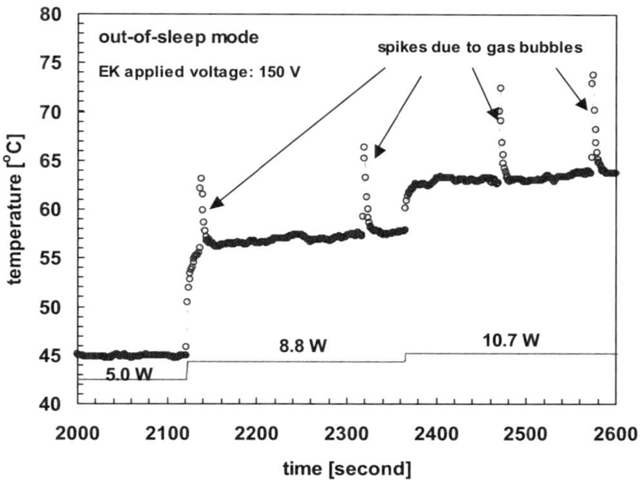

Figure 5: Transient response of an out-off-sleep mode. The processor has a step increase in power while the EO pump maintains the same operating condition. Settling time is shown to be less than 1 minute.

The gas bubble generation needs serious attention for system stability and reliability. There are two sources for the bubble generation. One is dissolved air in the water. The solubility of gas in water generally decreases with increasing temperature. The water temperature increases due to joule heating and thus the gas is released out of the water. Degassing the water right before filling it into the loop can minimize the gas generation from this source. The other source is the electrolysis of water at the pump electrodes. Oxygen is generated by the anode at the upstream of the pump and stays at the upstream reservoir. Hydrogen is generated by the cathode at the pump downstream and travels with the working liquid. The applied power to the pump affects the gas generation rate since more bubbles are observed under higher applied voltages to yield larger flow rate. We designed a recombiner to recombine the hydrogen and oxygen back to water. The recombiner chamber is located at the top of the upstream reservoir of the pump and a platinum spring inside the chamber functions as catalyst. The chamber collects the hydrogen and oxygen and recombines them back to water. Optimal design of the catalyst to increase the surface area exposed to the gases is expected to make the recombiner more effective.

4 SUB-ATMOSPHERIC OPERATION

The optimum operating condition of the processors requires the junction temperature to be lower than 85 °C. In order to achieve this goal, the closed-loop system is operated under sub-atmospheric pressure for lower water boiling temperature. The system was vacuumed, using a vacuum pump located at the upstream of the EO pump, to an absolute pressure of 15 kPa. Neglecting the pressure drop across the heat rejecter, the pressure at the inlet of the microchannel heat exchanger is estimated to be around 25 kPa and is defined as loop pressure. The voltage applied to the pump was set to 150 V, providing an estimated flow rate of ~ 3 mL/min.

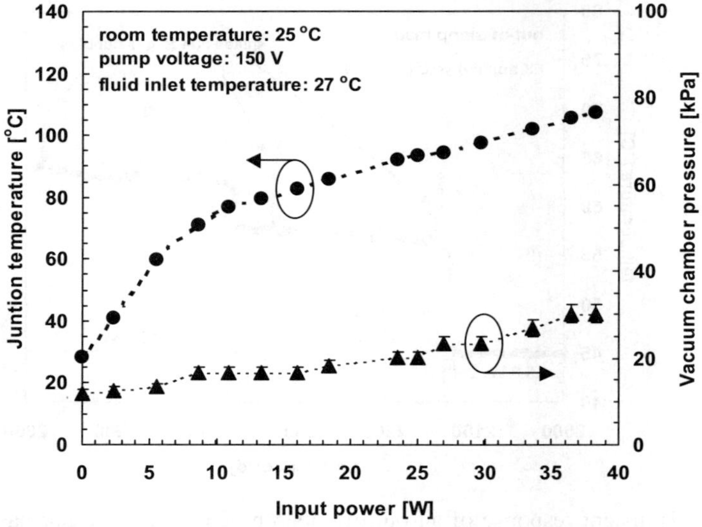

Figure 6. System sub-atmospheric performance.

Figure 6 shows the junction temperature and the corresponding system pressure for various input power. The temperature-input power curve has similar profile with that obtained under atmospheric pressure [Jiang et al., 2001c]. For power settings less than 7 W, a linear relation between temperature and input power was obtained as expected because the working fluid was in liquid phase. The slope of the curve decreases for higher powers indicating the fluid enters two-phase regime with increased heat removal capacity due to latent heat. The corresponding junction temperature at the onset of two-phase is about 75 °C, several degrees higher than water saturation temperature water under this pressure. In the two-phase regime, the fluid mixture quality increases with the increasing input power, resulting in increasing pressure drop across the microchannel heat exchanger. The continuing increase in pressure drop causes the fluid saturation temperature to rise, and consequently, increases the junction temperature. The present system achieves a junction-to-ambient thermal resistance of less than 2 K/W under system loop pressure of 25 kPa, half of the thermal resistance obtained under atmospheric pressure with similar flow rate [Jiang et al. 2001c]. The pressure at the upstream of the EO pump was recorded during the continuous operation of several hours of the experiment. The result suggests that there was 15 kPa increase in the loop pressure over the entire experiment. This pressure increase is believed to be from the gas generation due to electrolysis, which can be reduced by using an efficient recombiner.

5 SUMMARY AND CONCLUSIONS

The transient and sub-atmospheric behavior of a closed-loop cooling system with a microchannel heat exchanger and an electroosmotic pump are reported. The system time responses for both start-up situation and out-off-sleep situation were less than 1 minute. The system is believed to be able to avoid dryout under these transient power-increase situations. The closed-loop system, under a sub-atmospheric pressure of 25 kPa, is capable of removing 38 W from a 1 cm × 1 cm chip with a pump power less than 1.5 W. The system yields a junction temperature below 100 °C and even lower temperature is achievable by further reducing the system loop pressure. Gas generation due to electrolysis raises big concern for system stability and reliability. Further research is ongoing to improve the pump design that minimizes gas generation and to realize a more effective recombiner. Future work also targets optimal design of the microchannel heat exchanger to further reduce pressure drop and achieve lower thermal resistance.

ACKNOWLEDGMENT

This research is sponsored by the DARPA HERETIC Program through contract F33615-99-C-1442. We also appreciate support from Intel Corporation.

REFERENCES

Bowers M.B., & Mudawar I. 1994. High flux boiling in low flowrate, low pressure drop mini-channel and micro-channel heat sinks. *Int. J. Heat Mass Transfer* 37: 321-332.

Fan X., Zeng G., Labounty C., Bowers J.E., Croke E., Ahn C.C., Huxtable S., Majumdar A. & Shakouri A. 2001. SiGeC/Si supperlattice microcoolers. *Applied physics Letters* 78 (11): 1580-1582.

Glezer A. 2001. Microfluidic technologies for integrated thermal management: micromachnined synthetic jets. *DARPA HERETIC Principle Investigator's Meeting, Heat Removal by Thermo-Integrated Circuits* Atlanta, GA, May 23-25, 2001.

Greif R. 2001. Integrated micro-cooler module for high thermal flux removal. *DARPA HERETIC Principle Investigator's Meeting, Heat Removal by Thermo-integrated Circuits* Atlanta, GA, May 23-25, 2001.

Jiang L. Wong M. & Zohar Y. 2001a. Forced convection boiling in a microchannel heat sink. *J. Microelectromechanical Systems* 10: 80-87.

Jiang L., Koo J.-M., Zeng S., Mikkelson J., Zhang L., Zhou P., Maveety J., Tran Q., Kenny T.W., Santiago J.G., & Goodson K.E. 2001b. Two-phase microchannel heat sinks for an VLSI cooling system. *SEMI-THERM XVII*, San Jose, CA USA, March 21.

Jiang L., Mikkelson J., Koo J.-M., Hubber D. Yao S., Zhang L., Zhou P., Maveety J., Prasher R. Kenny T.W., Santiago J.G., & Goodson K.E. 2001c. Closed-loop electroosmotic microchannel cooling system for VLSI circuits. Submitted to *IEEE Transactions on Components, Packaging & Manufacturing Technology*.

Joshi Y. et al. 2001. Microfabrication alliance for innovative cooling of microelectrnics. *DARPA HERETIC Principle Investigator's Meeting, Heat Removal by Thermo-Integrated Circuits* Atlanta, GA, May 23-25, 2001.

Peng X.F., Peterson G.P. & Wang B.X. 1995. Experimental investigation of heat transfer in flat plates with rectangular microchannels. *Int. J. Heat and Mass Transfer* 38: 127-137.

Prasher R.S., Shipley J. & Devpura A. 2001 A simplified modeling scheme for design sensitivity study of thermal solution utilizing heat pipe and vapor chamber technology. *Proceedings of International Mechanical Engineering Congress and Exposition*, Nov. 11-16, 2001, New York: New York.

Tamura S., Tanaka Y., & Maris H.J. 1999. Phnon group velocity and thermal conduction in superlattices. *Physics Revue B* 60: 2627-2630.

Venkatasubramanian R., Siivola E. Colpitts T. & O'Quinn B. 2001. Thin-film thermoelectric devices with high room-temperature figures of merit. *NATURE* 413: 597-602.

Yao S., Huber D., Mikkelsen J. & Santiago J.G. 2001. A large flowrate electroosmotic pump with micron pores. *Proceedings of the International Mechanical Engineering Congress and Exposition, Sixth Micro-Fluidic Symposium* New York: New York.

Zhang L., Koo J-M, Jiang L., Goodson K.E., Santiago J.G., Kenny T.W. 2001. Measurements and modeling of two-phase flow in microchannels with nearly-constant heat flux boundary conditions. *J. Microelectromechanical Systems*: in press.

Thermal Challenges in Next Generation Electronic Systems, Joshi & Garimella (eds)
 ISBN 90-77017-03-8

A novel SOI CMOS compatible thermal device technology

Ty McNutt, Alex Lostetter, & Alan Mantooth
University of Arkansas, USA

Mohammad Mojarradi
Nasa's Jet Propulsion Laboratory, Pasadena, CA, USA

Keywords: data isolation, SOI, systems-on-a-chip, thermal transducer

ABSTRACT: A radiation-hardened SOI CMOS compatible thermal transducer intended for on-chip data isolation for mixed-signal systems-on-a-chip (SoC) is presented. This device technology is much smaller in size than on-chip inductors and MEMS devices, and requires no post-processing. Further, the device is capable of faster switching speeds (up to two orders of magnitude) than equivalent thermal-based MEMS structures due in part to the smaller thermal mass present in the transducer structure. Structure performance was verified using two- and three-dimensional finite-element simulators. Results show that a thermal-based data isolator can propagate a data bit in under a microsecond, which meets the targeted I2C bus specifications thus proving viability in select SoC applications.

1 INTRODUCTION

Electrical isolation is utilized to improve the robustness, reliability, and noise immunity of electrical systems. An important requirement of systems-on-a-chip (SoC), with combinations of analog, digital and RF circuitry, is on-chip electrical isolation. Isolation improves the noise performance of the overall system by reducing the coupling between the various technologies. Further, it helps to reduce the chances of cascading failures in a system.

Currently available on-chip isolation approaches are either not CMOS compatible (MEMS, optical), not viable for radiation filled environments (optical) or have not yet performed as desired (on-chip transformers, MEMS) [Streetman 1980]. On-chip transformers require significant die area, on the order of 100,000 μm^2, and the circuits used to drive them consume a considerable amount of power [Nguyen & Meyer 1990]. Even in specialized processes allowing MEMS, devices such as these are typically much larger and slower to respond than required [Parameswaran et al. 1991], [Wojciak et al. 1997].

One solution to a reliable source of data isolation is the use of an on-chip, CMOS compatible electrothermal device. The advantage of using thermal, as opposed to magnetics, as a means of isolation is the current availability of on-chip devices that are able to transmit and receive thermal signals. Also, the thermal based transducers described here can be made much smaller than their magnetic counterparts, typically around 1,600 μm^2.

The results presented here are based on isolator designs fabricated in 0.8 μm and 0.35 μm silicon-on-insulator (SOI) CMOS processes. SOI technology performs much better in a radiation filled environment compared with bulk CMOS technology [Colinge 1997]. Bulk processes are prone to single-event latchup (SEL), single-event upset (SEU), and other effects caused by particle bombardment. SOI technology is immune to SEL and has proven to be ten times less sensitive to SEU than bulk technology. Furthermore, SOI technology offers superior temperature performance over bulk, including functionality up to 800K.

2 THERMAL TRANSDUCER DEVICE STRUCTURE

Figure 1 shows a cross sectional view followed by a top view of a thermal transducer structure. The transducer structures presented here are approximately 40 μm on a side. The transducer can be decomposed into two sides, the input, or primary side, consisting of a resistor (polysilicon in this illustration) and the output, or secondary side, consisting of a thermal detector (a diode in this illustration). The polysilicon resistor acts as a heater that converts the incident electrical signal to a thermal signal. The diode converts the thermal signal back to an electrical signal. This is accomplished by biasing the detection diode at a fixed forward voltage. When the temperature in the diode structure rises, the detected electrical result is an increase in the diode's forward current. The coupling between the primary and secondary side is enhanced through the use of a thermal lens (aluminum in Fig. 1).

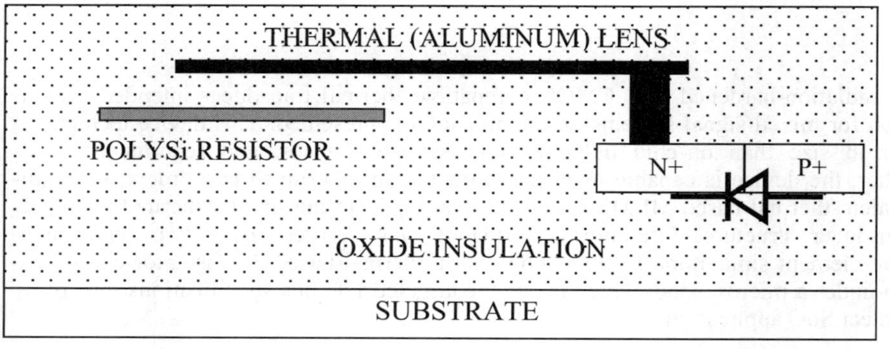

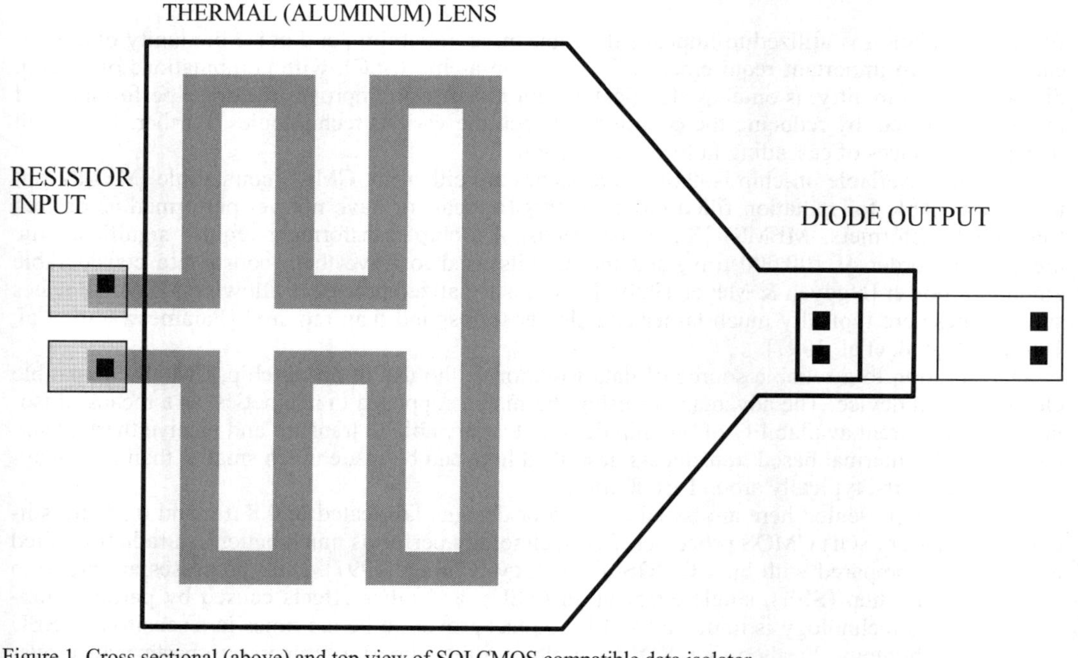

Figure 1. Cross sectional (above) and top view of SOI CMOS compatible data isolator.

The thermal signals generated by the heater propagate outward in all three dimensions. The thermal signal produced by the heater is gathered by the thermal lens and directed toward a set of vias that are connected directly to the cathode of the diode. The thermal signal also propagates directly toward the sensing diode providing two paths for the heat flux. The spacing between the polysilicon resistor and the aluminum thermal lens, as well as the spacing of the sensing diode from the resistor, determine the response time and the electrical isolation characteristics between the primary and secondary sides of the device.

Typically, the thermal transducer structure is chosen such that the realized isolation voltage will meet or exceed the specified isolation voltage. The fundamental criterion in choosing a transducer structure for a specific application is the above mentioned oxide thickness between the primary and the secondary sides. Due to process layout rules, generally the oxide thickness between the resistor and thermal lens is the determining factor in the achievable isolation of a structure.

In Fig. 1 the oxide layer thickness between the polysilicon and the first level of metal in the process is ~0.5 μm. The spacing of the diode from the poly resistor is governed by the process layout rules to be about 1 μm. These distances provide enough information to compute the electrical isolation achievable (~330V maximum for these dimensions) as well as simulate the thermal response time of the device.

Other novel structures can also be built in a typical SOI process. A slightly thicker oxide is achievable using a diffused silicon resistor with an aluminum (Metal 1) lens. The SOI process used here allows a linear capacitor option that consists of a polysilicon top plate with an $n{+}{+}$ doped diffused bottom plate. The resulting oxide thickness is about an order of magnitude thinner than the poly to metal 1 oxide, which reduces the amount of required input power and/or provides faster response times, but at the same time yields a lower electrical isolation rating (~60V).

3 THERMAL TRANSDUCER OPERATION PRINCIPLES

3.1 *Thermal transducer primary side*

Basically, the primary side of the transducer consists of a resistor. The resistor material (e.g., Si or polySi) can vary depending on the transducer structure selected. Once a transducer structure has been chosen according to isolation and speed specifications, a 3D finite element thermal simulator, such as Flotherm® [Flomerics, Ltd.], is employed to determine the power dissipated in the resistor that yields a target ΔT (1–1.5 °C for Fig. 1) at the diode such that a detectable electrical signal (ΔI = 1 μA) is created.

The power dissipated in the resistor must create enough thermal energy to cause the diode temperature to increase by ΔT in the specified amount of time, e.g., the transducer in Fig. 1 is able to pass a data bit in less than 1 μs to meet I^2C bus specifications. The figure of dissipated power, P_{dis}, is found from the 3D thermal simulation, as shown in Figure 2. Fig. 2 demonstrates the transient thermal analysis of the transducer, where 25 mW of power (a figure that is found to produce the desired response) was dissipated in the resistor. The top graph is a 3D outline of the transducer with a cutaway of the temperature distribution after 1 μs superimposed to demonstrate the ΔT at the diode. The bottom graph is a plot of the resistor and diode layer temperatures as a function of time. The dissipated power is then used to find the input current, I, needed according to $P_{dis} = IV$, where V is the input voltage. The input current dictates the minimum resistor width that can be realized according to the process current ratings.

The value of the input resistance, R, is also calculated from the dissipated power according to $P_{dis} = V^2/R$. The final layout of the resistor cannot exceed the size constraints of that simulated in 3D, otherwise the effective power density under the thermal lens and thus the change in the sensing diode temperature are reduced (i.e., the simulated structure is too small). If this violation occurs, the 3D simulation process must be repeated to achieve a new power figure, because the original power figure is insufficient for the resistor's resulting thermal mass.

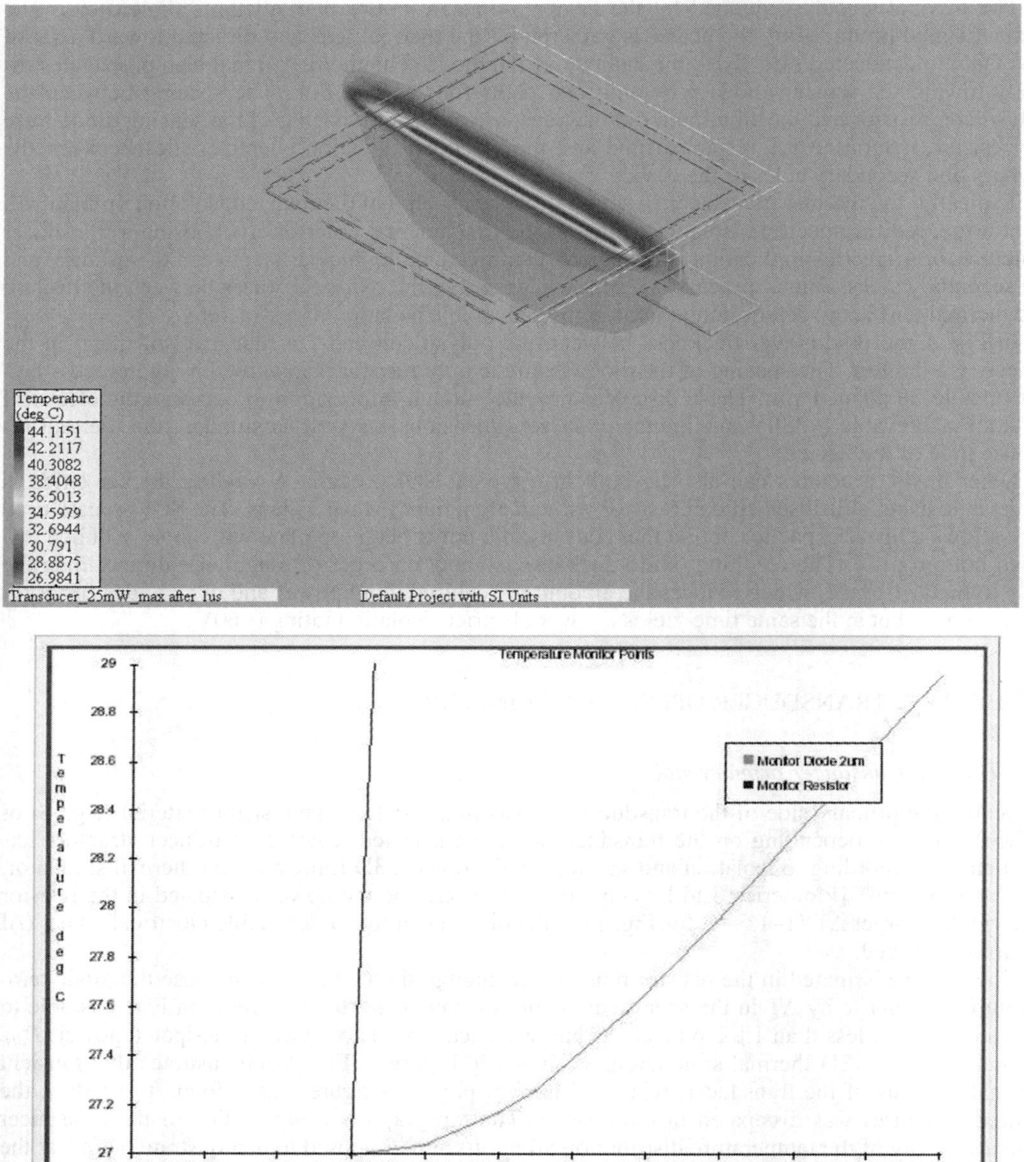

Figure 2. Three-dimensional transient thermal temperature distribution for transducer in Fig. 1 after 1 μs (above) and plot of the resistor and diode temperature vs. time.

On the other hand, if the original structure analyzed was found to be larger than necessary, then the power dissipation figure is also larger than necessary. Therefore, another iteration with a smaller thermal mass is required to optimize the power efficiency and switching speed of the device. The transducer shown in Fig. 1 was found to work acceptably well in the range of 20 μm to 40 μm per side.

3.2 *Thermal transducer secondary side*

The thermal signal produced at the primary side is sensed on the secondary side by a forward biased diode. This implementation detects a change in diode current as a function of changing temperature. The change in diode current, i.e., electrical analog signal, can be used to reconstruct the original digital data signal with additional circuitry.

The additional circuitry is designed such that a current reference value is established against which changing device temperatures are monitored. The resolution of this signal reconstruction circuitry determines the specifications of the diode sensor. The circuit utilized in testing the transducer in Fig. 1 was found to operate adequately with a diode biased at 0.7 V. The detectable change in current, ΔI, of the diode was 1 μA.

The diode used to detect the thermal signal is sized through device analysis using the Medici® finite element simulator package [Avanti Corp.]. Given the doping levels and layer thicknesses for the selected process, the diode is constructed in Medici and a sweep of the bias voltage is performed to obtain the current per micrometer. This figure is then used to calculate the diode area that yields the desired current for the bias voltage chosen previously (0.7 V), e.g. the diode used in Fig. 1 yields 129.3 μA at 0.7 V. Utilizing the same diode structure, an electrothermal simulation is then performed in Medici to obtain the minimum temperature change, ΔT, that the diode can sense and transition into a detectable ΔI (1 μA).

4 CONCLUSION

Two- and three-dimensional finite element analysis tools were used to analyze a thermal transducer design. The results have demonstrated acceptable performance. The transducer in Fig. 1 can achieve a propagation delay time on the order of 1μs, which meets I^2C bus specifications. Also, the transducer developed with the linear capacitor option, providing less electrical isolation, can meet these same specifications dissipating only 5 mW of power.

Logic dictates that a thermal-based transducer would suffer from environmental effects created by ambient temperature swings and on-chip localized heating due to circuitry and/or neighboring transducers. Also, of concern would be thermal saturation of the transducer rendering it insensitive to incoming signals. All of these issues are valid concerns. However, each one can be addressed through appropriate external circuitry. Using transducers in dual configurations, where the second transducer is used as a means of isolation between a sensor and control circuitry for shutting down the initial transducer, can address thermal saturation effects. Further, appropriate digital logic on the primary side of the initial transducer propagates signals through the device based on edge detection rather than data values. Likewise, other environmental concerns can be dealt with through additional diode sensors and circuitry and layout rules. Future work will involve testing fabricated transducers along with circuitry to address these environmental concerns.

ACKNOWLEDGEMENTS

The work described in this paper was performed for the Jet Propulsion Laboratory, California Institute of Technology under contract with the National Aeronautics and Space Administration.

REFERENCES

Streetman, B.G., *Solid State Electronic Devices*, Englewood Cliffs, NJ, Prentice Hall, 1995.

Nguyen, N., Meyer, R., "Si IC-Compatible Inductors and LC Passive Filters," *IEEE Journal of Solid State Circuits*, 1990, SC-25, pp.1028-1031.

Parameswaran, M., Robinson, A., Blackburn, D., & Gaitan, M., "Micromachined Thermal Radiation Emitter from a Commercial CMOS Process," *IEEE Electron Device Letters*, vol. 12, no. 2, February 1991.

Wojciak, W., Orlikowski, M., Zubert, M., Napieralski, A., "An Electro-thermal Converter in CMOS Compatible, Front Side Micromachined Technology," *Proceedings of the Mixed Design of Integrated Circuits and Systems Workshop*, Poznan, Poland, June 1997.

Colinge, J.P., *Silicon-On-Insulator Technology*. Norwell, MA: Kluwer, 1997.

Flomerics, Ltd, Surrey, England, *Flotherm® User's Manual v. 2.2*.

Avanti Corp., Fremont, CA, *Medici®User's Manual v. 2000.2.*

Thermal Challenges in Next Generation Electronic Systems, Joshi & Garimella (eds)
 ISBN 90-77017-03-8

Spray cooling of IGBT electronic power modules

G. Mitic, W. Kiffe, G. Lefranc, S. Ramminger
Siemens AG, Corporate Technology Department, CT MS 4, Munich, Germany

D.E. Tilton, B.A. Smetana, T.D. Weir
Isothermal Systems Research, Clarkston, WA, USA

Keywords: spray cooling, power electronics, IGBT module

ABSTRACT: Removing the dissipated heat directly from the surface of the chip by means of spray cooling increases the efficiency of cooling and reduces the temperature gradients at the chip surface. Both together lead to a significant increase in the reliability of the whole system. A demonstration system with a eupec BSM 150 IGBT module was built to show the efficiency of spray cooling for high power dissipation in electronic power modules. The maximum power dissipation of the module is rated at 1250 W. IGBTs with heat dissipation of almost 200 W/cm^2 were successfully cooled by spray cooling using Fluorinert. The transient electrical measurements showed that the thermal resistance between the junction and the surroundings was 0.051 K/W in the case of spray cooling in contrast to 0.075 K/W for direct flow convection between water and baseplate.

1 INTRODUCTION

Power semiconductors are used as switches in static converters, for instance to generate a sinusoidal current by means of pulse width modulation (Hierholzer 1996, Sommer 1997). Six such switches can generate a three-phase current whose frequency - and therefore motor speed - is variable. The increasing operating voltages of IGBT modules result in high heat dissipation (up to 200 W/cm^2) so that both mechanical and thermo-mechanical requirements must be satisfied. Special attention must also be paid to the increasing thermal resistance caused by thermal cycles. Figure 1 shows the set-up of a typical IGBT module. The waste heat is generated in the cell structure of the IGBTs and has to dissipate through the solder layers, the metallized substrate and the baseplate to the cooling system. In the event of high power losses, water-cooled plates to which the module is attached directly act as a heat sink. A thermally conductive grease or foil is used to reduce the thermal resistance of the interface between the module and the water-cooled plate (Khatri 2001, Rauch 1999). Despite this, the thermal resistance of the interface is comparatively high. To reduce the thermal resistance of the set-up, the Al_2O_3 ceramic is replaced by AlN with a high thermal conductivity of 180 W/mK (Mitic 1998). For cost reasons, however, this technological step is only feasible for modules with a high blocking voltage of 3.3 kV.

The heat generated during operation causes thermally induced stresses between the different materials due to their different coefficients of thermal expansion. Thermal fatigue of the soft solder layer is aggravated by temperature cycles, causing crack propagation which reduces the heat transfer between the IGBT chip and the heat sink and thus the electrical efficiency and operating life of the module (Mitic 1999, Mitic 2000). Isothermal IGBT temperatures are desirable during operation to maintain a degree of high reliability. In a conventional module set-up, the heat has to dissipate from the IGBT through many different thermally resistive layers. Table 1 shows the thermal conductivity and thickness of the different layers of the IGBT module used for the spray cooling measurement demonstrated here. As the module is cooled from the lower side of the baseplate, each layer increases the temperature of the chip.

Table 1. Thermal conductivity of the different layers in the IGBT module BSM 150 GT 120 DN2

Econo Module	Thermal Conductivity [W/m.K]	Thickness
Silicium	148	170 µm
Solder	34	100 µm
Cu (DCB)	395	0.3 mm
Al_2O_3	22	0.38 mm
Cu (DCB)	395	0.3 mm
Solder	34	200 µm
Cu (Baseplate)	395	3 mm

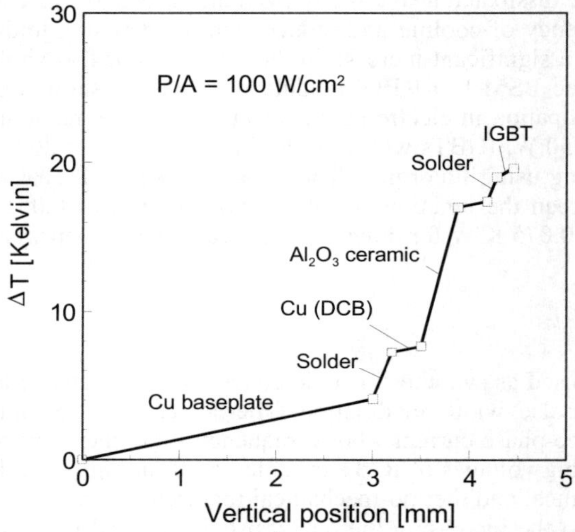

Figure 1. Temperature distribution in the IGBT module BSM 150 GT 120 DN2 with water cooler

Removing the dissipated heat directly from the surface of the chip by means of spray cooling increases the efficiency of cooling and reduces the temperature gradients at the chip surface. Both together lead to a significant increase in the reliability of the whole system.

A demonstration system with a eupec BSM 150 IGBT module was built to show the efficiency of spray cooling for high power dissipation in electronic power modules. The three-phase full-bridge has a nominal DC collector current of 150 A and the collector-emitter voltage is specified as 1200 V. At a gate voltage of 15 V, the maximum power dissipation of the module is 1250 W.

2 SPRAY COOLING BASICS

A very efficient technique to extract heat from a surface is to utilize the latent heat of evaporation of a liquid. A technique called spray cooling, in which liquid droplets sprayed onto hot electronic components, evaporates to remove excess heat. The hot vapor is recovered by rejecting the waste heat to the ambient in a heat exchanger where it is condensed back into a liquid. Thus, the fluid is continually recycled for reuse within a closed cycle system. This mechanically driven evaporation/condensation process is nearly isothermal, or "constant temperature," and can effectively maintain components at quite uniform temperatures to improve reliability. All surfaces exposed to the liquid/vapor environment remain very close to the saturation temperature of the fluid. This effectively reduces both "hot spots" and thermal cycling; these thermal mechanical stresses are the primary causes of failure in electronic systems. Spray cooling may therefore play a pivotal role in the design of high reliability, high performance electronic systems.

The vaporization process at the device level depends on the degree of superheat of the heated surface at temperature T_s. The superheat is defined by the temperature difference (ΔT) between the heated surface temperature, T_s, and the interface temperature, T_f, required to drive the flow of energy (heat) from the sprayed surface to the liquid/vapor interface where evaporation occurs. The efficiency of this process depends on many factors, including the thickness of the liquid film covering the surface, the spray characteristics, and the conditions of the surface itself. Figure 2 below illustrates a typical relationship between the heat flux (q') removed from the surface versus the superheat.

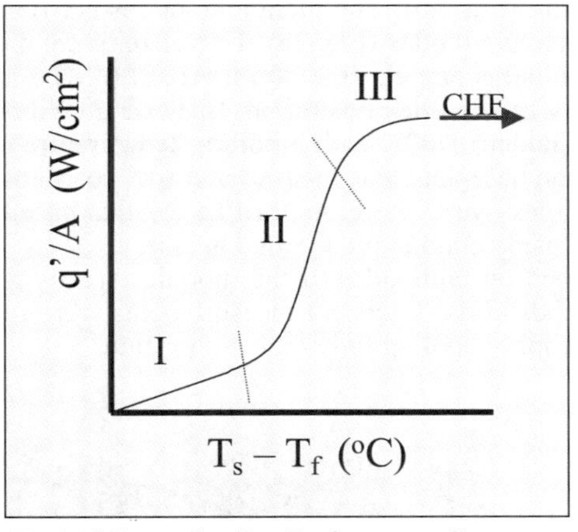

Figure 2. Typical Spray Cooling Performance Curve.

There are three distinct superheat regimes involved with heat from hot surfaces. These three regimes comprise the low, intermediate, and high superheat regimes and are illustrated in Figure 2. The portion of the curve labeled Region I, is the low superheat case. Here single phase forced convection is the dominant heat transfer mode. There is insufficient temperature difference to drive evaporation, or phase change. The portion of the curve labeled Region II has increased evaporation, resulting in an increase in the heat transfer efficiency. This region is characterized by large increases in heat flux achieved with only small increases in temperature difference. Depending on the fluid properties, surface characteristics, spray characteristics, and liquid film thickness, nucleation of vapor bubbles at the solid surface or within the liquid film may occur. The portion of the curve labeled Region III is where the surface begins to partially dry out, and any increase in heat flux results in larger increases in superheat. This region is considered relatively unstable since minor perturbations in heat flux can go beyond the Critical Heat Flux (CHF) point and result in a surface burnout condition. The preferred spray cooling system design will operate in the Region II and avoide operation in the unstable Region III. If CHF is exceeded, the surface temperature instantly jumps to a very high superheat regime. In this very high superheat regime, a vapor layer is formed between the droplet and the heated surface, reducing the heat transfer coefficient. This mechanism is most often referred to as the Leidenfrost phenomena.

The above discussion is useful for understanding the performance of spray cooling at the device level, as it relates to heat removal from the wetted surface. Of course, reliability and Mean Time

Between Failure (MTBF) is dependent on the junction temperature of the device. Thus, in applications where spray is applied to packaged devices, traditional methods of evaluating the conduction heat transfer from the junction to the package surface may be applied. However, it should be noted that spray cooling improves heat transfer from all wetted surfaces, where in conventional air-cooling, only the surface with the attached heat sink is effective. By directly spray cooing the device junction at the surface, the waste heat is removed directly from the heat source. Thus, thermal resistances may be substantially lowered with spray cooling.

In addition to understanding the heat transfer between the semiconductor junctions and the wetted surfaces, the heat transfer between the fluid and the environment (in the condenser or heat exchanger) will also impact the overall operating temperatures. Thus, spray cooling requires full system level modeling of the complete heat transfer paths.

3 SPRAY COOLING DEMONSTRATION SYSTEM

A demonstration system with an IGBT module designated as BSM 150 GT 120 DN2 was built to determine the thermal properties of spray cooling (Fig. 3b). The module contains six IGBTs each with dimensions of 11 x 11 mm and six free-wheeling diodes (Fig. 3a). To utilize direct die contact spray cooling, the module was packaged without silicone gel. PF5060 Fluorinert (3M) with a boiling temperature of 56°C was used for this spray cooling demonstration. Because a calibration has to be performed at temperatures up to 90°C, (Galden) D02 with a boiling temperature of 180°C was used for this purpose. The temperature in the baseplate was measured at two positions directly beneath the IGBTs as shown in Fig. 3a. Holes were therefore drilled in the baseplate and NiCr/NiAl thermocouples were fitted into them using conductive thermal grease.

The nozzles for spray cooling were positioned on both sides of the module directly above the IGBTs and beneath the baseplate at the IGBT positions. Transient measurements were performed with spray cooling on both sides as well as with spray cooling only on the chip side.

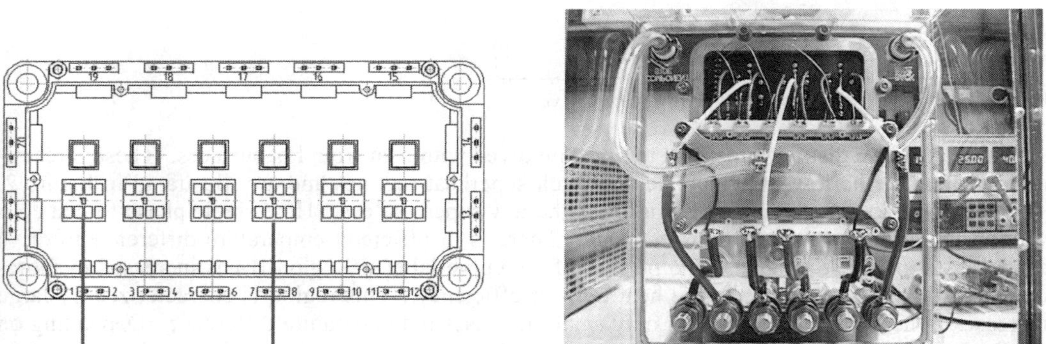

(a) (b)

Figure 3. Spray cooling demonstration system (b) with BSM 150 GT 120 DN2 IGBT module (a)

4 MEASUREMENT OF THERMAL RESISTANCE

Transient electrical or infrared measurements can be used to measure the chip temperature and the thermal resistance of power modules, including the cooling system. IR imaging allows the chip temperature to be measured during operation with an accuracy of one degree Kelvin. However, the surface of the unpackaged chip must not be covered by media which is optically absorbent in the infrared region. This means that IR imaging is not feasible when silicone gel is used for packaging.

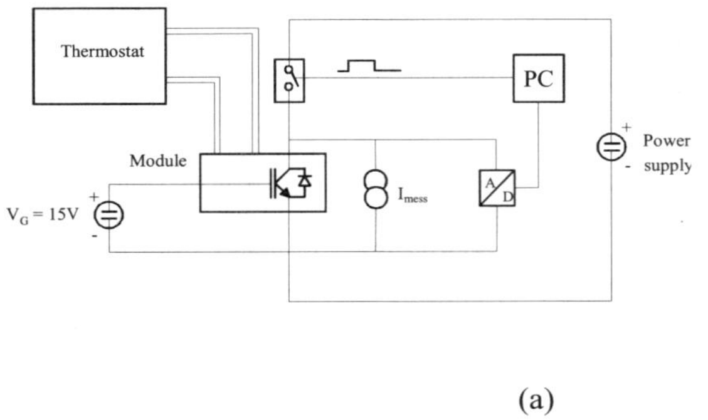

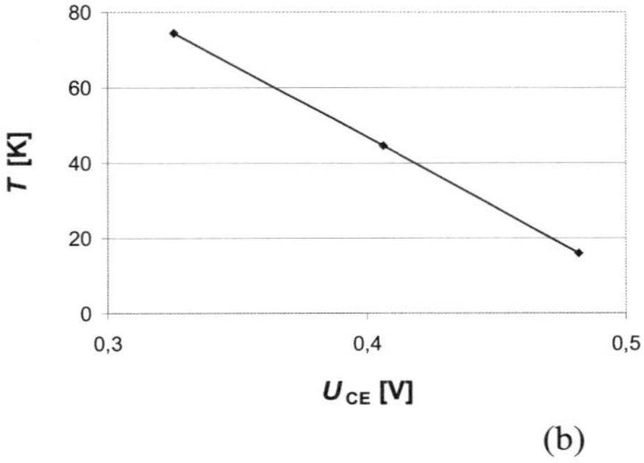

Figure 4. Set-up for transient measurements of the collector-emitter voltage (a) and temperature dependent collector-emitter voltage for calibration (b)

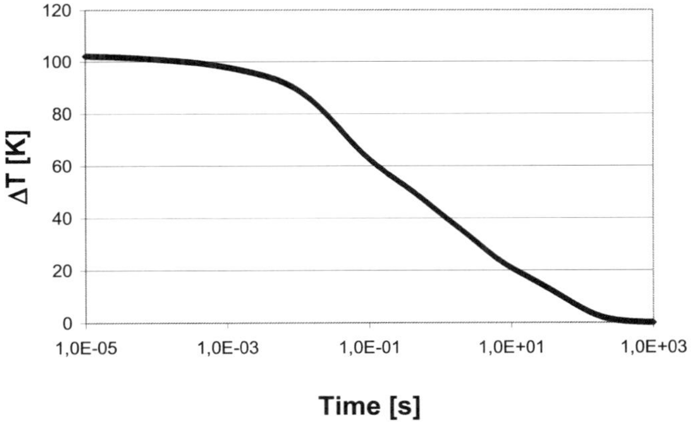

Figure 5. Measured transient temperature relative to the ambient temperature after switching off the IGBT module

To measure the chip temperature, we performed transient electrical measurements of the temperature-dependent collector-emitter voltage U_{CE}. A few microseconds after the device is switched off by a computer-controlled switch, a current $I_{CE} = 100\,\text{mA}$ is applied and U_{CE} is measured with an oscilloscope. The measured voltage can be assigned to a specific chip temperature after calibration at three different temperatures using a thermostat. From a comparison with infrared measurements, we know that the measured temperature corresponds to a median chip temperature of the active IGBT area. Figure 4 and 5 shows the set-up and basic results of transient electrical measurements.

5 RESULTS OF THE TRANSIENT ELECTRICAL MEASUREMENTS

The spray cooling experiment was performed at a maximum power dissipation of up to 1400 W with operation of all six IGBTs. This corresponds to a power dissipation of 190 W/cm^2. Table 2 summarizes the measured thermal data and compares it with that of an open water cooler.

The PF5060 Fluorinert had a temperature of approximately 40°C at the inlet and is sprayed onto the chip surface and onto the baseplate beneath the chips. Under these conditions, we found a temperature of 92.8°C (thermocouple 1) and 94.0°C (thermocouple 2) respectively in the baseplate beneath the chips as shown in Figure 3. The transient measurement yielded a chip temperature of 127°C. Because a large part of the IGBT surface is covered by a 15-µm thick polyimide layer with a low thermal conductivity of 0.25 W/mK, we anticipate high temperature gradients in that layer. When the polyimide is removed chemically from the IGBTs, the junction temperature decreases to 111.2°C and the temperatures in the baseplate are consequently reduced to 82.8°C and 82.2°C respectively.

These results from spray cooling are compared to those obtained by using a water-cooled plate. The best cooling efficiency is obtained by an open water cooler which avoids any interfaces with thermal conductive grease. To ensure similar thermodynamic conditions, the water flow was chosen to be equal to the Fluorinert flow of the spray cooling experiment, namely around 1.7 l/min. In the open water cooler, a rhombus structure of fins across the baseplate produces highly turbulent flow which increases the heat transfer. The ambient temperature of the water-cooled plate is 16.4°C and the transient electrical measurement yielded a junction temperature of 106°C. The temperature rise was thus 90°C in contrast to 71°C in the case of spray cooling.

Table 2. Comparison of the temperatures and the thermal resistance of spray cooling and water-cooled plates at the IGBT module BSM 150 GT 120 DN2

Type of cooling	Spray Cooling	Spray cooling without polyimide on chips	Direct liquid base-plate flow convection
Pressure [psi]	25	25	0.3
Power consumption [W]	108	108	
Flow rate Fluorinert / H_2O [l/min]	1.70	1.70	1.76
P_v power dissipation [W]	1371	1393	1198
T_c baseplate (thermocouple 1) [°C]	92.8	82.8	73.3
T_c baseplate (thermocouple 2) [°C]	94.0	82.2	70.8
T_a Fluorinert/H_2O (inlet) [°C]	38.8	40.2	16.4
T_j junction [°C]	127	111.2	106
$R_{th(j-c)}$ [K/W]	0.025	0.020	0.028
$R_{th(j-a)}$ [K/W]	0.064	0.051	0.075

6 DISCUSSION

The transient electrical measurements showed that the thermal resistance between the junction and the fluid was 0.051 K/W in the case of spray cooling in contrast to 0.075 K/W for direct flow convection between water and module baseplate. Although the experiment was performed with similar liquid flow rates, the thermal properties of water and Fluorinert are quite different. Under the demonstrated conditions, IGBTs with heat dissipation of almost 200 W/cm^2 were successfully cooled by spray cooling using Fluorinert with a low heat of evaporation of 89 KJ/kg compared with that of water, namely 2250 KJ/kg. In addition, the specific heat of water is four times higher than that of Fluorinert.

The tested module package is not optimized for spray cooling because of its many different material layers. These different layers in the IGBT module cause heat to spread across the baseplate. Direct water flow thus cools the module efficiently across the whole baseplate. Because the module is only sprayed with Fluorinert directly at the IGBTs and the corresponding area of the baseplate, further improvements in cooling efficiency could be achieved by spray cooling the whole module. Finally, the benefits of spray cooling could be increased even more by using an optimized package without a baseplate and ceramics made of aluminum nitride with a high thermal conductivity up to 180 W/mK instead of aluminum oxide with a typical value of 30 W/mK.

LITERATURE

Hierholzer, M. 1996. Application of high-power IGBT Modules. Proc. PCIM 96, 26

Khatri, P. & Ziemski, J. 2001. Dry-To-The-Touch Thermal Grease. 2001 International Symposium on Advanced Packaging Materials: 236-239

Mitic, G. et al. 1998. The thermal impedance of new power semiconductor modules using AlN subtrates. IAS, Industry Applications Society: 1026-1030

Mitic, G. et al. 1999. Reliability of AlN substrates and their solder joints in IGBT power modules. IEEE-Electron Device Society, ESREF 99: 1159-1164

Mitic, G. et al. 2000. AlSiC Composite Materials in IGBT Power Modules. IAS, Industry Applications Society: 3021-3027

Rauch, R.A. 1999. Test methods for characterizing the thermal transmission properties of phase-change thermal interface materials. Electronics Cooling, May

Sommer, K.H. et al. 1997. Multichip high-power IGBT modules for traction and industrial applications. Proc. EPE97

Thermal Challenges in Next Generation Electronic Systems, Joshi & Garimella (eds)
 ISBN 90-77017-03-8

Action of low and high gravity levels on cryogenic cooling with boiling of electronics

V.F. Prisniakov & K.V. Prisniakov
Inst. Geothecn. Mech. Nat. Acad. Science of Ukraine, Dnipropetrovsk, Ukraine

Keywords: superheat, density centres, withdrawal, frequency, heat flux, gravitational field

ABSTRACT: The report presents the fullest theoretical solution of a problem of heat and mass transfer at boiling. Known empirical formulae for heat flow by different authors are gotten from this solution as special case. Comparison of particular experiments with the calculations by the given the dependence coincide satisfactorily. The obtained analytical relations let estimate influencing a gravitational field $\bar{j} = j/g$ on the internal characteristics of boiling. The gravitation influences on growth rate of a steam bubbles includes the range $\bar{j}^{0}$- $\bar{j}^{-1/3}$, on the departure bubbles includes the range $\bar{j}^{-1/3}$ - $\bar{j}^{1/2}$, on frequency of bubbles withdrawal include the range $\bar{j}^{0}$ – $\bar{j}^{2/3}$ - $\bar{j}^{1}$, on the density of active nucleation sites $\bar{j}^{1/4}$ - $\bar{j}^{1/6}$, on the overheat at boiling $\bar{j}^{-1/4}$, on heat flux $\bar{j}^{1/3}$ - $\bar{j}^{-1/12}$ - $\bar{j}^{-1/3}$, on the heat transfer coefficient includes the range $\bar{j}^{-1/12}$.

1 INTRODUCTION

The cryogenic cooling with boiling in electronics systems are being used widely. The heat and mass transfer in boiling liquids is one of the most complicated phenomena. Usually the thermal operational modes with boiling in space systems are determined experimentally under normal gravitation so as to extend obtaining results for boiling under gravitational field, that is not correctly. Because of this it needs to have the analytical solutions for thermal- hydraulic characteristics of boiling, which give possibility to analyse the influence of gravitational field. It is well-known that, a vapor nucleus appears on a heating surface when a liquid heat up to certain temperature T'. The process of nucleation of the vapor phase provides the appearance of a certain number of vapor nucleuses n per m^2 of heating surface under conditions at certain superheating of the liquid as compared with the saturation temperature $\Delta T = T_w - T_s$, a vapor bubbles growth up to the moment of detachment $R(\tau = \tau_d) = R_d$. This process repeats itself with a certain frequency f.

2 SUPERHEATING OF LIQUID UPON BOILING

A vapor bubble in liquid may exist, provided a number of conditions of thermodynamic and mechanical equilibrium are valid. For a spherical nucleus, the condition of mechanical equilibrium is written as $p'' - p' = 2\sigma/R_*$. The external pressure in liquid p' is usually to be equal to the pressure of saturated vapour in a system, that is, $p' = p_s$. The value of vapor pressure in a bubble is defined by the saturation pressure at the temperature of surface heating T_w, that is, $p'' = p_w = p_s(T = T_s)$. Because, in practice, one manipulates temperature rather than pressure, the transition from equation Laplace, written in view of the above-identified assumptions as to its expression in terms of liquid super-heating is usually performed using the Clausius-Clapeyron relation. This approximate transition yields the widely employed formula ($\Im$ - heat evaporation, ρ - density; σ - surface tension, R_*- critical radius)

$$\Delta T = 2\sigma T_s/(\Im\rho'' R_*) \qquad (1)$$

If the value of the critical radius of the nucleus is substituted into this equation, the temperature difference will define the liquid superheating required for the beginning of boiling. For numerous liquids, formula (1) gives a tenfold understatement of the results. For hydrogen the difference of overheatings experimental and computational reaches 15 times (Verkin et al. 1987). This may be attributed, on the one hand, to the use of the Clausius-Clapeyron relation in the form of finite differences and, on the other hand, to the failure to include the specific features of formation and growth of the vapor nucleus as an elementary catastrophe. Obviously, a phase transition in liquid in principle cannot be regarded as a phenomenon characterised by continuous changes. However, it proved rather difficult to apply the apparatus of the catastrophe theory to description of the nucleation process. We will use Van der Waals' equation instead of the Clausius-Clapeyron relation, from which one can readily derive the values of p_w and p_s.. After some transformations, we have, in view of the known correlation between the critical parameters, the following formula for determining the superheating of liquid (**$\Re$ -** gas constant):

$$\Delta T = 2\sigma/(\Re\rho'' R_*) \qquad (2)$$

Compare this formula with (1). One can see that the proposed relation is distinguished by the presence of the factor $1/\Re$, which replaced the ratio $T_s/\Im$. Therefore, the new result ΔT_2 , derived by formula (2), is $B_* = \Im/\Re T_s$ times greater than the value of ΔT_1 (for hydrogen it is 5 times greater, for O_2 it is 7 times, for N_2 it is 8 times). The correction factor designed to make the results of calculation by formula (2) agree with the experimental data is provided by the coefficient B_*. The calculated values of the coefficient B_* are in adequate agreement with the experimental data.

3 CONCENTRATION OF NUCLEATION SITES

The theoretical solution of the density of vapour formation centres (DVFC) n is based on the connection of n with characteristics of surface roughness, particularly with normal distributions law of microuneveness height (Prisniakov 1970, 1988): $\underline{n} = n/n_c = 1/2 + \Phi_0[\varepsilon(1 - h/h_{rms})]$, where $\varepsilon = h_a/h_{rms}$; ($\varepsilon = 3.2$ or 4 depending on purity of processing[4]); Φ_0 is Frenel integral ; h_c, h_a, h_{rms} is current, average and root-mean-square microuneveness height. The existence of probability integral, in spite of its recognition was an obstacle for its wide application. Besides, a great number of parameters of surface roughness, their different standardisation in various countries made it difficult to calculate the density of vapour formation centres. One of the universal current characteristics of surface roughness is the so called standardised support roughness curve (known in the western countries as Abbot curve). This curve shows the connection between the summarised area F_h of bulge cross-section cut off by the plane on the height h (taken from the highest bulge h_c), related to the nominal base F_0 and relative height of micro-roughness $\underline{h} = h/h_a$: $\eta_s = F_h/F_0 = b\underline{h}^{\nu}$ .Here b and ν are parameters, depending on the kind of processing: $\nu = 0.2\, h_c/h_{rms}{}^{3/2}$; $b = 0.5(1 - k_p)$, where k_p is the coefficient of volumetrically filling the surface profile (at turning $k_p = 0.33$; at grinding 0.5; at polishing 0.66; at mechanical processing 0.2; at cutting, planing, milling 0.4). All parameters included here (as well as other expressions for Abbot curve formula) are well known to the designers, are widely presented in reference literature and can be determined for any specific kind of surface processing. It's a common knowledge, that vapour nuclei formation on the surface of boiling begins on the deepest hollow, that is at $h = h_c$, and with increasing of temperature difference ΔT covers still more shallows with the depth of h (estimated from the level of the deepest hollow). Taking into account a grate number of bulges and hollows on the surface ($\sim 10^8\ m^2$) and their equal quantity we can from "non-support" or ("dip") curve indicating the connection between the depth of the hollows h and the total area of hollow cross-sections at the distance h from the deepest hollow. If N_h is the number of hollows with the depth h, N_c is the total number of hollows on the base surface F, but N_{1h} and N_{1c} are their number on the surface unit, then the parameter η_s can be represented as follows: $\eta_s = F/F_0 = (N_h/N_{1h})/(N_c/N_{1c}) = (N_h / N_c)/(N_{1h} /N_{1c})$. The ratio of hollow densities as potential centres of vapor formation is proportional to the ratio of DVFC for the depth h n to the maximum possible value of DVFC n_c, that is $N_{1h}/N_{1c} = n/ n_c = \bar{n}$. Relation of the hollow numbers N_h/N_c may be connected with their distribution law $f(h)$ in the following way : $N_h/N_c = 1 - F(\bar{h})$, where $F(\bar{h})$ is integrated function of distribution. Here we took into account such peculiarity of vapour formation process that the depth h corresponds to the number of potential centres, which are their sum from the moment of boiling start (at the minimum overheating ΔT_0 at the depth h_c) up to the moment of acting the hollow with the depth h as the centre of formation (that corresponds to overheating ΔT). If we use connection (2) between the hollow depth h and liquid overheating ΔT we obtain formulae for DVFC $\bar{n}$:

$$\bar{n} = (1/b)(\Delta T/\Delta T_0)^{\nu}/[1 - F(\Delta T_0/\Delta T)] \tag{3}$$

According to summarised data at turning, planing and milling $\nu=1.2-2.2$; $b= 0.4-2$; at grinding $\nu=1.5-3$; $b= 0.6-4$; at polishing $\nu=1.5-3$; $b=0.7-2.5$; at honing $\nu= 1.6-1.8$; $b=0.7-2.5$ and surface shaping $\nu=1.1-1.4$; $b= 2.4-4.5$. Formulae (3) gives, as particular cases, the majority of the well-known dependence for DVFC both theoretical and experimental. When $\nu=2$ and the law of hollow distribution is not taken into consideration we have D.A. Labuntsov's dependence with constant multiplier accuracy, when $\nu=3$ we have K.A. Zhohov's formulae. If we take into consideration only microuneveness distribution on their depth then we have: with normal law a theoretical formulae for n author [4], with Maxwell's distribution – author's and N. Sitnikova's formulae [8]. The calculations carried out by us and their comparisons with the experiments have shown that the simplest and the most accurate case is being determined where by Reley's distribution where $\sigma=0.4$, that gives the following expression for calculation formulae

$$\bar{n} = \exp(-3.125\ \bar{h}^2)/(b\ \bar{h}^{\nu}) \tag{4}$$

For the purpose of received dependence control we took experimental data known in literature. We come across with satisfactory calculation and experimental data agreement that gives us then possibility to take a conclusion as to practical fitness of given dependence.

4 DEPARTURE OF BUBBLES FROM A HEATED SURFACE

The size of a bubble at detachment depends on many forces, which include in a general case: buoyant (lift) P_g, cohesion with a surface due to surface tension P_σ, inertial force of liquid (and vapor) P_f, drag force (due to molecular and whirling viscosity) P_μ, pressure forces determined by possible circulation inside a bubble etc. The analysis of these forces shows that those of P_g, P_σ, P_f, P_μ, are to be given primary importance. In the first case the design scheme of the bubbles departure allows to write down the condition of the equilibrium of forces, acting on the bubble in the moment of departure, in the following way: $[P_a + P_g \cos\alpha]^2 + [P_w + P_g \sin\alpha]^2 = [P_f + P_\sigma]^2$. This equality can be reduced to the following equations for the case of boiling in a large volume on a horizontal surface:

$$\bar{R}_d^3 - 3/2 \sin\theta\ \bar{R}_d \zeta_\sigma/\zeta_g - 7.2\ C^4 \nu \Phi^4 \mathrm{Ja}^4 = 0 \tag{5}$$

where $\mathrm{Ja} = c'\rho'\Delta T/\Im\rho''$.

Computations by these formulae are in good accord with well–know experimental data and formulas by other authors. In the case of a quasi-static regime of departure we have $\bar{R}_d = \sqrt[3]{3/2\bar{R}_c}$.

5 FREQUENCY OF VAPOUR BUBBLES DEPARTURE

Analytical solution of the problem on defining the frequency of departure vapor bubbles is possible to be found out proceeding from the following consideration. Let the bubble grow on the heated surface during time τ_d, achieving the value of departure radius R_d , after this it departs and goes into liquid. The latter goes into the volume occupied by the bubble, and because of this the heated surface comes into contact with colder layers of the liquid and during some period of waiting τ_w heat flows going into the liquid, overheat this liquid by value ΔT, which is enough for a new bubble formation. The frequen-cy of bubbles departure is equal to $f = 1/(\tau_w+\tau_d)$, that is , the problem is reduced to defining τ_w and τ_w . The expressions for τ_w and τ_d [5] allow to obtain the formula for the frequency of bubbles departure f:

$$f=\{\pi/4a'(\Delta\lambda'/q)^2+\ \bar{R}_d^2/(4\,C^2\,\mathrm{Ja}^2\,\phi^2 a')\}^{-1}=4/\pi(\,\mathrm{Pe}_*/\,\mathrm{Ja})^2\{1+\ \bar{R}\,\mathrm{Pe}_{*d}\,/\,\mathrm{Ja}]^2/(\pi\,C^2\,\phi^2)\}^{-1}\tau_* \qquad (6)$$

where $\mathrm{Pe}_*= qR_*/(\Im\rho''a')$, $R_*=\sqrt{\sigma/g(\rho'-\rho'')}$

Formula (6) gives most well-known empirical dependencies for f in the form of fR_d^m= const where m adopts particular values ½, 3/2, 2. The comparison of calculation, got through (6), is in accord with experimental data by various authors (Prisniakov 1975).

6 HEAT TRANSFER AT BOILING

A possible model for calculation of the heat transfer at boiling on the surface element being in-cluded the zone of activity of one vapour centre (and equal to $F_1= 1/n$). The quantity of the heat, removed from the surface of this surface element is equal to

$$Q = q_{av}F_1(\tau_w+\ \tau_d) = (q'_1+q''_2)F'_1(\tau_w+\ \tau_d)+\int_0^{\tau_d} q''F''d\tau+\pi\int_0^{\tau_d} q'_1\,(R_d^2-R^2)d\tau+\int_0^{\tau_w} q'_1F''d\tau .$$

Here, the first summand determines heat removal from the surface which is not included in the zone of active of the nucleation centre due to free convection q'_1 and agitation q'_2, the second summand is the sum of heat, removed by a forming bubble, the third is the heat, removed due to convection into the mass of liquid within the zone of vapour bubble activity (on a variable on account of its area growth $\pi(R_d^2- R^2)$), and the forth summand determines the heat removal due to convection during the waiting period. In the general case the formula for number Pe_* at natural convection can be presented in the fol-lowing way: $\mathrm{Pe}'_*= \kappa'_0 N_{T0}\mathrm{Ja}^{1+\gamma}$. The amount of the heat , transferred by vapour bubbles $q''=nfV_dL\rho''$ or in dimensionless form $\mathrm{Pe}''_*= 4/3\ \ \bar{n}\ \bar{f}\ \bar{R}_d^3$. That is why, passing finally on to dimensionless val-ues, we obtain $\mathrm{Pe}_*= \mathrm{Pe}'_*+\mathrm{Pe}''_*$ or, at boiling in a large volume

$$\mathrm{Pe}^3_* - \{\kappa'_0 N_{T0}\mathrm{Ja}^{4/3}+16/3\,\pi C^2\ \bar{n}\ \bar{R}_d\,\mathrm{Ja}^2\,\phi^2)\,\mathrm{Pe}^2_*++\pi C^2\ \bar{R}_d\mathrm{Ja}^4\ \phi^2\ \bar{R}_d^{-2}\,\mathrm{Pe}_*-\pi C^2\phi^2\kappa'_0 N_{T0}\mathrm{Ja}^{16/3}\ \ \bar{R}_d^2=0$$

Let as transpose this equation, introducing the following combinations of numbers:

$$\Lambda=(Pe_*\ \ \bar{R}_d/Ja^2);\ s=\kappa'_0 N_{T_0}\ \ \bar{R}_d/Ja^{2/3};\ \eta=nR^2_d=\ \bar{n}\ \bar{R}^2_d$$

Then we have

$$\Lambda^3-(s+(16/3)\,\pi\,c^2\,\eta\,\phi^2)\Lambda^2+\pi\,c^2\,\phi^2\,(\Lambda- s)=0 \qquad (7)$$

Complex s depends on the type of liquid, its temperature and number Ja; for water, for instance, s changes from 0.5 to 0.05 at the changes of Ja from 10 to 1000. At small values of the complex of nucleation centers density η (<0.5), the heat exchange due to free convection prevails, and at $\eta>0.5$ vapor bubbles make the main contribution into heat transfer. Empirical formulae by differ-ent authors are gotten from the obtained equation. The relationship between Λ and η, at a devel-oped nucleate boiling ($\eta>0.5$), is described accurately enough by the following dependence ob-tained while neglecting the last member in (7) and at $s<<16/3\ \pi C^2\phi^2\eta$: $\Lambda\cong 16/3\ \pi C^2\phi^2\eta$ or after substituting instead of Λ and η their values

$$\text{Pe}_* = 16/3\ \pi C^2 \phi^2\ \bar{n}\ \bar{R}_d\ \text{Ja}^2 . \tag{8}$$

At small numbers Ja (<40) R_d shows a weak dependence on Ja, and $n \approx \text{Ja}$, that is the proportionality analogous to W. Rohsenow formula: $\text{Pe}_* \cong \text{Ja}^3$. At large numbers Ja, when $\bar{R}_d = \text{Ja}^{4/3}$ and n shows a weak dependence on Ja, we get the relationship which is present in most empirical formulae on heat flux:

$$\text{Pe}_* \cong \text{Ja}^{10/3} \tag{9}$$

At small values of η complex Λ depends weakly on η, while approximately $\Lambda = 1.11s$ and

$$\text{Pe}_* = 1.11\ \kappa'_0 N_{T0} \text{Ja}^{4/3} \tag{10}$$

that is, at weak vapour formation heat removal is determined by natural convection and agitation. Thus it is seen that the range of changing exponents at number Ja, which is equal from 4/3 to 13/3, covers all its values obtained empirically.

7 EFFECT OF THE GRAVITY ON BOILING

7.1 *Introduction*

In many space power devices, which are situated, as a rule, in a local (and quite weak) gravitation field, the work mediums are either used to take off or give heat, changing their state of aggregation. Thus, the problem of liquid boiling in the conditions of the law of gravitation attract the attention of some rese-arch workers. On the other hand the intensification of cooling processes in the rotating elements of va-rious devices including space devices calls for knowledge of the acceleration effects exceeding the gra-vity acceleration. The above mentioned problems became of interest somewhat in the middle of 50-s. There are very few reports in this connection now.

The problems of heat and mass transfer in the acceleration conditions, which differ from the earth ones, have found their reflection in the reports (Prisniakov 1975, 1988; Verkin 1987). Rather detailed review of the reports on problems of boiling in the conditions of gravity differ from those of the earth, had been given in the papers of Siegel R. and Unterberg W., Congelliere J.. Thus, we shall not concern this problem. Let us consider the gravitational field influence on the main micro-characteristics of boiling R, R_d, n, f, q, α, ΔT. When analysing we shall already know the theoretical relation, allowing to predict the boiling parameters in the differing state from the earth gravitational fields.

7.2 *Velocity of steam bubbles growth*

The data influence j on velocity of steam bubbles growth are quite meagre and in somewhat contradicting (Prisniakov 1975). If the analysis of the well known theoretical formulae shows that j does not influence over R than the available experimental information proves the decreasing of growth of velocity when the acceleration increases. Let us show, the theoretical relation taken from (Prisniakov 1988) paper gives the same result. For the approximate of $\bar{j} = j/g$ on R we shall use the law of bubble growth taken in the most wide aspect

$$R = 2c\phi\ \text{Ja}\ \sqrt{a'\tau} \tag{11}$$

where c and ϕ - magnitude the terms which are determined differently in various reports (for example, in accordance with D. Labuntzov and V.V. Yagov $\phi = 1+\sqrt{(1+400\text{Ja}/3)}$; $c = 0.15$; Ja is

Jacob's number; a' is the thermal diffusivity of the liquid; τ is time.

In accordance with Prisniakov 1975

$$\phi = 1/[(1+\cos\theta)R] \int_{\delta_1}^{(1+cщыθ)R} (T'(y) - T)dy / \Delta T$$

where δ_l is the initial thickness of liquid microlayer under the bubble; θ is the angle of the moistening; $\Delta T = T_w - T$; T_w –is the temperature of wall; T' is the liquid temperature in accordance with the height of the thermal boundary layer; y is the coordinate; T is the temperature in a bubble.

If we consider proportional law of the temperature distribution in the height of the boundary layer for the approximate analysis

$$\Delta T(y) = \Delta T_0(1 - y/\delta) \qquad (12)$$

where δ is the thickness of heat boundary layer, we shall receive the next correlation between $\overline{R}_j = R_j/R_g$ and $\overline{j}$:

$$\overline{R}_j = [1 + (R_g/\delta_g)(\ \overline{j}^{1/3} - 1)]^{-1} \qquad (13)$$

Thus it may be proved that with the increasing of j R is decreasing, and on the contrary.

The comparison experimental data with theoretical one shows (Prisniakov 1975, 1988) that the region of a real gravity influence in our approximate analysis is limited with lines

$$\overline{R}_j = \ \overline{j}^{\,0} \ldots \ \overline{j}^{\,1/3} \qquad (14)$$

The region between lines $\overline{R}_j = \ \overline{j}^{\,0}$ and $\overline{R}_j = \ \overline{j}^{\,1/3}$ reflects quite well the influence of $\overline{j}$ over the velocity growth of the bubble.

7.3 *Radius of bubble departure R_d*

This parameter is much better investigated than others and the theoretical data are well confirmed by the experiments. In the case of great value boiling the solution results of (Prisniakov 1988) equation give two limit cases allowing to find the regions of the influence j over R_d:

$$R_{dj}/R_{dg} = \ \overline{j}^{1/2}\{[a\sqrt{j} + \sqrt{(a\sqrt{j} - b)}]^{1/3} + [a\sqrt{j} - \sqrt{(a\sqrt{j} - b)}]^{1/3}\}/\{[a\sqrt{g} + \sqrt{(a\sqrt{g} - b)}]^{1/3} + [a\sqrt{g} - \sqrt{(a\sqrt{g} - b)}]^{1/3}$$

and

$$R_{dj}/R_{dg} = \ \overline{j}^{1/2}\ [\cos(1/3 \text{ arc cos } d\sqrt{j})]/[\cos(1/3 \text{ arc cos } d\sqrt{g})]$$

where a, b, d are the coefficients, not depending from j.

Hence, for the case of great (>100) and small (<20) Jacob's numbers we receive the limits of the gravity influence over the radius of bubbles departure:

$$\overline{R}_{d\,j} = \ \overline{j}^{\,-1/3} \ \ldots \ \overline{j}^{\,-1/2} \qquad (15)$$

To control this result we have used the experimental data Beckman W., Siegel R. and Keshock E., Pomeranz M. (Prisniakov 1975). The comparison of experimental data with calculation according (15) give good result.

7.4 *Bubble nucleation site density n*

At present we practically know nothing about the gravity influence over n. Only two experiments of Judd R. and Merte H. (Prisniakov 1975) can be pointed out.

The theoretical bubble nucleation site density n is determined by $n = \bar{n}\, n_c$, where

$$n_c = 2\{\pi \sin^2\theta\,(1 + \pi c^2 \phi^2 / \Lambda_1^2)\}^{-1} R_{d1}^{-2} \tag{16}$$

$\Lambda_1 = 3.35$; R_{d1} is the radius of bubble departure at the moment of burnout, $\bar{n}$ is (4).
From (16) the following expression is received $n_j/n_g = \bar{n}_j = R^2_{d1g}/R_{d1j}^2$ or with (15) taken into account

$$\bar{n}_j = \bar{j}^{1/4} \ldots \bar{j}^{1/6} \tag{17}$$

The experimental points received in the accelerating power field[21] are rather contradictory:
$\bar{j} = 10$ $\bar{n}_j > 1$ and $\bar{j} = 100$ $\bar{n}_j < 1$. The first case is similar to that of the calculations on (17).

7.5 *Departure frequency of bubble f*

Using the formulae (6) of (Prisniakov 1988) for f it is not difficult to receive denoted by r and s independent from j value:

$$f_j/f_g = \bar{f}_j = (r + s R_{dg}^2) / (r + s R_{dj}^2) \tag{18}$$

With $r << sR_d^2$ we have

$$\bar{f}_j = \bar{j}^{1} \ldots \bar{j}^{2/3} \tag{19}$$

With $r >> sR_d^2$ we have

$$f_j = \bar{j}^0 = 1 \tag{20}$$

i.e. the boundary of departure frequency of bubble changes are determined by $\bar{j}$ powers from 0 to 1. This result agrees with the experimental data (Prisniakov 1975).

7.6 *Specific heat flux q*

The results of (Prisniakov 1988) give the possibility to determine theoretically the influence of j over the heat flux of boiling medium the calculating formula in criterion is given as:

$$Pe_* = 64/27\ \bar{n}\ R_d\, Ja^2$$

It shows that under the comparison with $\Delta T = const$ (Ja= const) and taken into account the results which were received above the range of the influence j over the heat flux are determined by the following

$$q_j/q_g = \bar{j}^{-1/12} \ldots \bar{j}^{-1/3} \tag{21}$$

In the case of the prevailing forced convection as followed the Pecle's number $Pe_* \approx Pe_*'$ (here and else-where the prime denotes the liquid phases and two primes denote the steam formed phase), i.e. we have

$$q_j/q_g = \bar{j}^{1/3} \tag{22}$$

Thus, it shows that the influence of j over q is sign –variable: under the developed boiling its increase decreases the heat flux, but insignificantly.

To control the dependences (21) and (22) we have been using the experimental results of Siegel R., Merte H., Klark I, Costello C., Tuthill W. (Prisniakov 1975). The theoretical conclusions are qualitatively confirmed by these experiments. Let us consider here that in accordance with the most of the experimental equations Katto $q \sim \sqrt{j}$.

7.7 *Temperature drop ΔT*

If we take the condition of q *=const,* the following expression determining the influence j over ΔT will be received:

-for developed boiling:

$$\Delta T_j / \Delta T_g = \bar{j}^{1/24} \dots \ \bar{j}^{1/6} \tag{23}$$

for weak boiling:

$$\Delta T_j / \Delta T_g = \bar{j}^{1/4} \tag{24}$$

This result agrees with the experimental data: under the weak boiling ΔT decreases with the increase of j by (23); under the developed boiling, the relation between ΔT and j is close to the theoretical curve based on (24).

7.8 *The heat transfer coefficient*

With ΔT =const, we have

$$\bar{\alpha} = \alpha_j / \alpha_g = \bar{j}^{1/12} \dots \ \bar{j}^{1/3} \tag{25}$$

The calculations according to: $\bar{\alpha} = \bar{j}^{1/12}$ good agrees with the experimental data (Verkin et al. 1987).

8 CONCLUSIONS

The analytical solutions for thermal-hydraulic characteristics of boiling are presented, which one give possibility to analyse the influence of microgravity. The obtained analytical relations let estimate influencing a gravitational field $\bar{j} = j/g$ on the internal characteristics of boiling.

REFERENCES

1. Prisniakov, V. F. 1988. *Kipenie (Boiling)*. Kiev: Naukova dumka,
2. Prisniakov, V. F. 1970. The density of vapour formation centers at boiling on the surface. *Automic energy* 29: 451-454.
3. Prisniakov, V. F. 1975. The Effects of the Gravity on Boiling of the Cooling Agents for the Space Power Systems. *Proc. of 11-th ISSTS,* Tokyo: 423-430.
4. Prisniakov, V.F. 1997. Density of Vaporization Centres. *Industrial Heat Engineering.* 19 (2-5): 20-27, 33 -39.
5. Rallis, C.J., Jawurek, H.H. 1964. Latent heat transport in saturated nuclear boiling. *Int. J. Heat and Mass. Transfer.* 7: 1051 – 1068.
6. Verkin, A., Kirichenko, Yu., Rusanov, K. 1987. *Heat exchange at the Boiling Cryogenic Liquids*. Kiev: Naukova dumka.

Thermal Challenges in Next Generation Electronic Systems, Joshi & Garimella (eds)
 ISBN 90-77017-03-8

Dynamics of a liquid plug in a capillary powered by vapor explosion

O. Suzuki
Mechanical Engineering Research Laboratory, Hitachi Ltd., Ibaraki, Japan

Y. K. Joshi
G. W. Woodruff School of Mechanical Engineering, Georgia Institute of Technology, Atlanta, GA, USA

W. Nakayama
ThermoTech International, Kanagawa, Japan

Keywords: vapor explosion, liquid plug, dynamics, microfluidics

ABSTRACT: A theoretical model of liquid plug dynamics powered by vapor explosion in a capillary duct has been proposed. The model can evaluate performance parameters such as the amount of useful work and displacement volume rate of liquid. The efficiency of energy conversion for water under atmospheric pressure and room temperature is 2.8 %. The prediction agrees well with existing experimental results and suggests the existence of optimum heat flux at which displacement volume rate is maximum.

1 INTRODUCTION

Significant progress of microfabrication technologies in the integrated circuit industry has inspired miniaturization of mechanical devices, and integration between electrical and mechanical devices, i.e. microelectromechanical systems (MEMS). More recently, microfluidic systems, integrated with microelectronics and micromechanics are becoming enabling technologies for biomedical and biochemical applications. Microfabricated fluidic loops also have a great potential as thermal management devices for next generation microsystems.

The fluid propulsion device; i.e. micro pump plays an essential role in a microfluidics system. One of typical fluid propulsion mechanisms utilized in microfluidics systems is membrane-actuated pump with inlet / outlet valves. Although various kinds of membrane-actuation mechanisms have been proposed and demonstrated (Koch et al. 2000), all of these have mechanical / moving parts, which cause reliability and durability concerns.

On the other hand, Tsai & Lin (2001) have recently demonstrated a "bubble actuated" micropump with nozzle / diffuser valves. They regarded the liquid-vapor interface generated by the bubble as a "virtual membrane" that drives the liquid in pumping chamber. Utilization of vapor explosions is already made with great commercial success in the inkjet printer. Asai (1991) has attempted to describe the bubble dynamics in such devices analytically. Huan at el. (1999a, b) and Ory at el. (2000) have investigated bubble formation and dynamics of liquid in a small tube numerically. But, there has been no published information about the useful work by vapor explosion and the efficiency of the energy conversion analytically.

In this study, we describe a theoretical model of liquid plug actuation by vapor explosion, in order to evaluate the efficiency of the actuation process, and to provide guidance for designing the actuator.

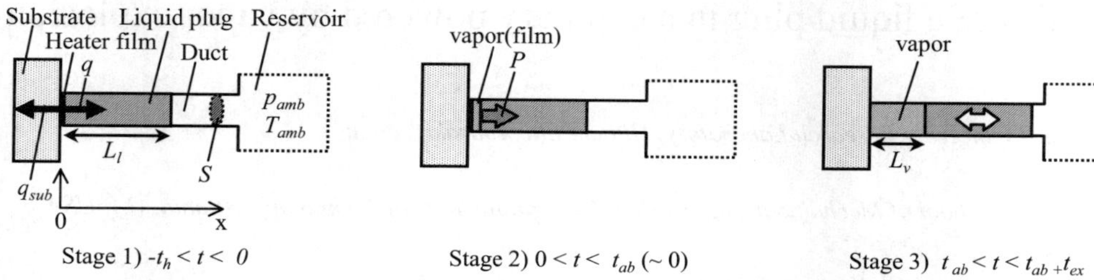

Figure 1. Schematic diagram of theoretical model

2 MODELING

2.1 *Physical configuration and dynamics of liquid plug*

Figure 1 shows physical circumstances of the theoretical model. The problem is treated as one-dimensional and the surface tension forces are neglected. A thin heater film, which covers the end of a small duct, is embedded on a substrate. Cross-sectional area of the duct S is sufficiently small to form liquid plugs by surface tension in the duct. The opposite end of the duct is connected to a large reservoir, in which the temperature and pressure are maintained at T_{amb} and p_{amb} respectively. A liquid plug, of length L_l, is initially located at the closed end of the duct. The initial temperature and pressure in the duct are as same as that of the reservoir. We assume that the decrease in mass of the liquid plug by vaporization is sufficiently small compared to its total mass; therefore we treat the length of the liquid plug L_l as constant.

We hypothesize three stages for the dynamics of the liquid plug by pulsed heating:

1) Rapid heating of liquid plug at constant heat flux q ($-t_h < t < 0$)

2) Vapor explosion or spontaneous nucleation of bubble at the heater surface and simultaneous generation of initial momentum P for the liquid plug ($0 < t < t_{ab}$)

3) Inertia-controlled bubble growth and collapse and the resulting liquid plug dynamics ($t_{ab} < t < t_{ab} + t_{ex}$)

Figure 2 shows profiles of the net heat flux q (as boundary condition) and temperature, where the profile between $-t_h$ and 0 shows the temperature on the heater surface and the profile between 0 and t_{ab} shows the temperature at liquid-vapor interface. The ratio of heat flux supplied to the liquid (q) to the heat flux into the substrate (q_{sub}) is expressed as follows:

$$\frac{q}{q_{sub}} = \left(\frac{k_l \rho_l c_l}{k_{sub} \rho_{sub} c_{sub}} \right)^{\frac{1}{2}} \tag{1}$$

where c, k and ρ are specific heat, thermal conductivity and density respectively, and subscript l and sub refer to liquid and substrate respectively. We assume that the thermophysical properties of liquid and substrate are constant, so that the ratio expressed by eq. (1) has constant value. As a result, we can evaluate q from the total heat flux supplied by the heater ($q + q_{sub}$) in advance. At the time $t = -t_h$, the liquid plug at uniform temperature T_{amb} is suddenly exposed to q, which is maintained from $t = -t_h$ to $t = 0$. At $t = 0$, the temperature of liquid in contact with the heater surface reaches its superheat limit T_{shl}, a bubble embryo forms spontaneously and drops the liquid interface temperature (at $x = 0$) to the saturation temperature at p_{amb}, $T_{sat}(p_{amb})$ suddenly. From $0 < t < t_{ab}$, explosive vaporization occurs at the interface, so that momentum P is generated towards the remaining liquid plug. After the momentum is imparted to the liquid plug, the plug begins to move by inertia ($t_{ab} < t < t_{ex}$).

As we assume that the time period in stage 2) is sufficiently small compared to that in stage 3), the model should be reasonable under the following condition:

$$\frac{t_{ab}}{t_{ex}} < 0.1 \tag{2}$$

In other words, we postulate inertia-controlled bubble growth (Carey, 1992) in the model.

2.2 *Superheat limit of the liquid*

There are two theoretical means of predicting the limiting superheat: the thermodynamic limit and the kinetic limit (Carey, 1992). In this study, we adopt the thermodynamic limit just because it gives more explicit threshold temperature than the kinetic limit. The threshold temperature of superheat limit of the liquid T_{shl} is shown as follow:

$$T_{shl} = \left[\frac{\left(3\left(\frac{v}{v_{cr}} \right) - 1 \right)^2}{4\left(\frac{v}{v_{cr}} \right)^3} \right]^{\frac{1}{(\lambda+1)}} \tag{3}$$

where λ is the exponent for the van der Waals ($\lambda = 0$) or Berthelot ($\lambda = 1$) equation of state, and v the specific volume. Subscript *cr* expresses critical condition of the liquid.

2.3 *Heating time*

The relation between the heating time t_h and the temperature is given by the analytical solution (Incropera & Dewitt, 1990):

$$T_{shl} - T_{amb} = T(0, t_h) - T_{amb} = \frac{2q\left(\frac{\alpha t_h}{\pi} \right)^{\frac{1}{2}}}{k} \tag{4}$$

where α is thermal diffusivity. Rearranging (4), one can expresses t_h :

$$t_h = \frac{\pi}{\alpha} \left[\frac{k(T_{shl} - T_{amb})}{2q} \right]^2 \tag{5}$$

The total input energy to the liquid in stage 1), E_{in} is expressed as follows:

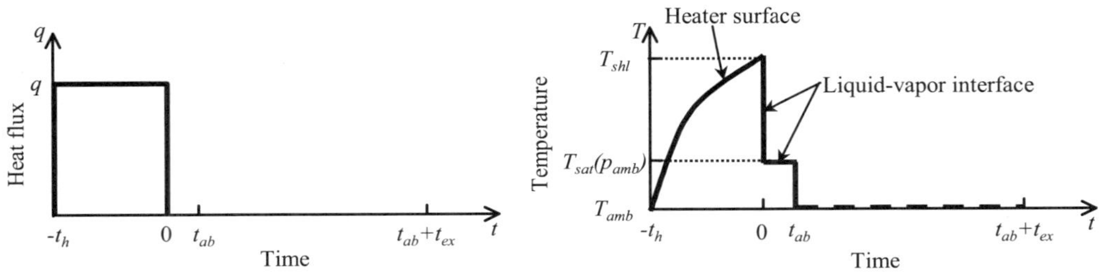

Figure 2. Profiles of heat flux and temperature

$$E_{in} = qSt_h = \frac{S\pi k^2}{4\alpha q}(T_{shl} - T_{amb})^2 \tag{6}$$

2.4 *Heat-transfer-controlled model of rapid vaporization*

Here we describe the rapid vaporization in a non-uniform temperature field based on the analysis developed by Mikic & Rohsenow (1969). The temperature distribution in the liquid plug is specified with the differential equation:

$$\frac{\partial T}{\partial t} = \alpha \frac{\partial^2 T}{\partial x^2} \tag{7}$$

The initial and boundary conditions for (7) are expressed as follows:

$$\begin{aligned} &T(x,-t_h) = T_{amb} && -t_h \le t \le 0 \\ &-k\left(\frac{\partial T}{\partial x}\right)_{x=0} = q && -t_h \le t \le 0 \end{aligned} \tag{8}$$

$$T(0,t) = T_{sat}(p_{amb}) \qquad 0 < t \tag{9}$$

One solves the problem by superposition (Myers, 1998) and gets the temperature distribution in the liquid by the following expression:

$$\begin{aligned} T(x,t) - T_{amb} = {} & \frac{2q\left[\frac{\alpha(t+t_h)}{\pi}\right]^{\frac{1}{2}}}{k} \exp\left[\frac{-x^2}{4\alpha(t+t_h)}\right] - \frac{qx}{k} erfc\left\{\frac{x}{2\left[\alpha(t+t_h)^{\frac{1}{2}}\right]}\right\} \\ & - (T_{shl} - T_{sat}(p_{amb}))erfc\left[\frac{x}{2(\alpha t)^{\frac{1}{2}}}\right] \qquad t \ge 0 \end{aligned} \tag{10}$$

where erfc expresses complimentary error function. The heat flux at the liquid-vapor interface ($x = 0$) q_{lv} follows from (10) as:

$$q_{lv} = k\left(\frac{\partial T}{\partial x}\right)_{x=0} = k\left[\frac{T_{shl} - T_{sat}(p_{amb})}{(\alpha\pi t)^{\frac{1}{2}}} - \frac{q}{k}\right] \tag{11a}$$

The positive value of (11a) expresses the heat flux from the liquid to the vapor. Although we assume a one-dimensional model in this study, the liquid-vapor interface is never plain but closer to-hemispherical, at the temperature of the super heat limit. Thus, the heat flux at the interface should be modified to evaluate the dynamics of the liquid plug precisely. Here we introduce the correction factor C ($=\sqrt{3}$) proposed by Mikic & Rohsenow (1969) for (11a):

$$q_{lv} = \sqrt{3}k\left[\frac{T_{shl} - T_{sat}(p_{amb})}{(\alpha\pi t)^{\frac{1}{2}}} - \frac{q}{k}\right] \tag{11b}$$

For Stage 2), we define t_{ab} as the time at which the evaporation is complete, and we postulated that t_{ab} is equal to the time when the heat flux q_{lv} becomes zero. Thus, t_{ab} is obtained from (11b):

$$t_{ab} = \frac{k^2 (T_{shl} - T_{sat}(p_{amb}))^2}{q^2 \alpha \pi} \tag{12}$$

It is noted that t_{ab} is the time at which the bubble volume is maximum, when the phase change is governed by heat transfer.
The energy transferred from the liquid to the vapor from $t = 0$ to t_{ab}, E_{lv} is expressed as follow:

$$E_{lv} = S \int_0^{t_{ab}} q_{lv} dt = \frac{\sqrt{3} S k^2 (T_{shl} - T_{sat}(p_{amb}))^2}{q \alpha \pi} \tag{13}$$

The mass of the liquid consumed by the vaporization from $t = 0$ to $t = t_{ab}$, m_l is expressed as follow:

$$m_l = \frac{E_{lv}}{h_{lv}} = \frac{\sqrt{3} S k^2 (T_{shl} - T_{sat}(p_{amb}))^2}{h_{lv} q \alpha \pi} \tag{14}$$

Finally, extractable work W, which will be done against the liquid plug by the vapor expansion , is given by

$$W = p_{amb} m_l \left(\frac{1}{\rho_v} - \frac{1}{\rho_l} \right) \approx \frac{p_{amb} m_l}{\rho_v} \tag{15}$$

2.5 *Inertia-controlled dynamic model of liquid plug*

We describe the entire dynamics of liquid plug based on the analysis developed by Yuan & Prosperetti (1999). They described the dynamics by neglecting the thermal aspects of the phenomenon and adding internal pressure of bubble as a boundary condition.

The dynamics of the liquid plug is described based on Newton's second law as follow:

$$\rho_l L_l S \frac{d^2 L_v}{dt^2} = (p_{sat}(T_{amb}) - p_{amb}) S \tag{16}$$

where L_v is length of generated vapor in the duct. Boundary conditions for (16) is given as follows:

$$\rho_l L_l S \frac{dL_v}{dt} = P, \qquad L_v = 0 \qquad \text{at } t = t_{ab} (\sim 0) \tag{17}$$

where P is initial momentum towards the liquid plug obtained from the conservation laws of momentum and energy as

$$P = (2 \rho_l L_l S W)^{\frac{1}{2}} \tag{18}$$

(16) is easily solved with the boundary condition (17):

$$L_v(t) = \frac{p_{amb} - p_{sat}(T_{amb})}{2 \rho_l L_l} \left[\left(\frac{P}{(p_{amb} - p_{sat}(T_{amb}))S} \right)^2 - \left(t - \frac{P}{(p_{amb} - p_{sat}(T_{amb}))S} \right)^2 \right] \tag{19}$$

From its formation, it is obvious that maximum vapor length L_{vmax}, the time at which the vapor length has its maximum t_{max}, and vapor residence time t_{ex} are expressed and rearranged using (19):

$$L_{v\max} = \frac{P^2}{2 \rho_l L_l S^2 (p_{amb} - p_{sat}(T_{amb}))} = \frac{p_{amb} m_l}{(p_{amb} - p_{sat}(T_{amb}))S} \left(\frac{1}{\rho_v} - \frac{1}{\rho_l} \right) \tag{20}$$

$$t_{ex} = 2t_{\max} = \frac{4}{p_{amb} - p_{sat}(T_{amb})}\left[\rho_l S L_l p_{amb} m_l \left(\frac{1}{\rho_v} - \frac{1}{\rho_l}\right)\right]^{\frac{1}{2}} \tag{21}$$

2.6 *Efficiency of energy conversion*

We define the efficiency of the energy conversion ε as the ratio of useful work W to the total energy input E_{in}. From (6) and (15) ε is obtained:

$$\varepsilon = \frac{W}{E_{in}} = \frac{4\sqrt{3}p_{amb}(T_{shl} - T_{sat}(p_{amb}))^2}{h_{lv}\pi^2(T_{shl} - T_{amb})^2}\left(\frac{1}{\rho_v} - \frac{1}{\rho_l}\right) \tag{22}$$

It is noted that ε depends on only thermal properties of the working fluid and boundary conditions. For example, it is computed as 2.8 % for water under atmospheric pressure and room temperature.

2.7 *Displacement volume rate*

The liquid plug volume displacement rate by the bubble is defined as follows:

$$D = \frac{SL_{\max}}{t_h + t_{ex}} \tag{23}$$

where we postulate that we can neglect t_{ab} or include it in t_{ex}, and the temperature distributions in the liquid and substrate come back to the initial condition just after the time period of t_{ex}.

3 RESULTS AND DISCUSSIONS

In order to compare the model predictions with available experimental results, here we extend the limit of validity of the model for the evaluation of V_{max} as follow:

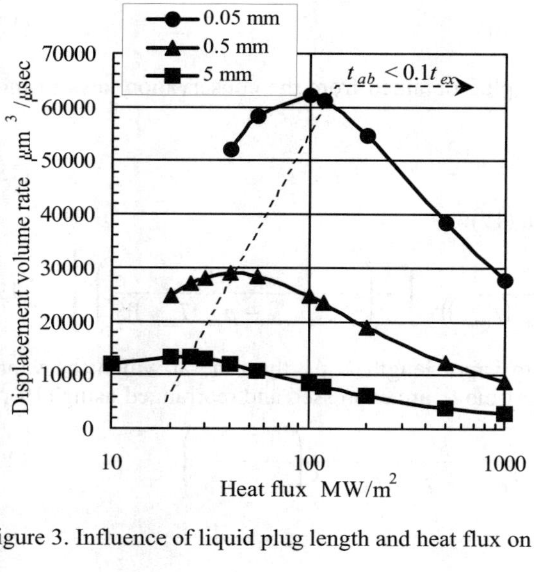

Figure 3. Influence of liquid plug length and heat flux on displacement volume rate

$$t_{max} = \frac{t_{ex}}{2} < t_{ab} \tag{24}$$

(24) means that rapid vaporization and liquid plug motion progress simultaneously but the dynamics are governed by inertia, so that t_{ab} may be included in t_{ex}.

3.1 *Influence of heat flux and length of liquid plug on displacement rate*

Figure 3 shows displacement volume rate D for different heat flux and length of liquid plug in the case of water. The dashed line shows the limit of (2). Although it is very interesting that D has

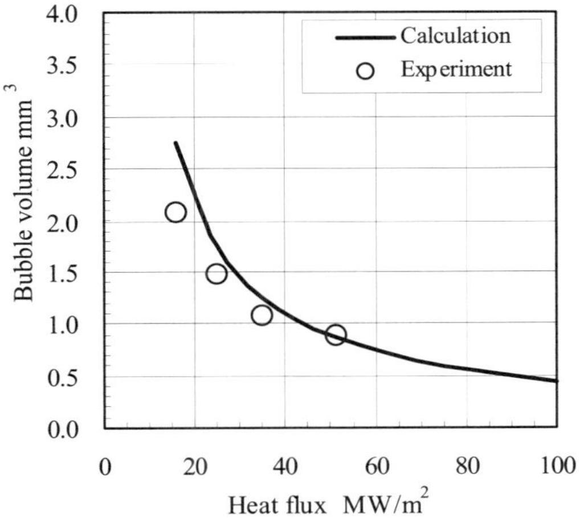

Figure 4. Comparison of model predictions for maximum bubble volume with experimental results (Asai (1991))

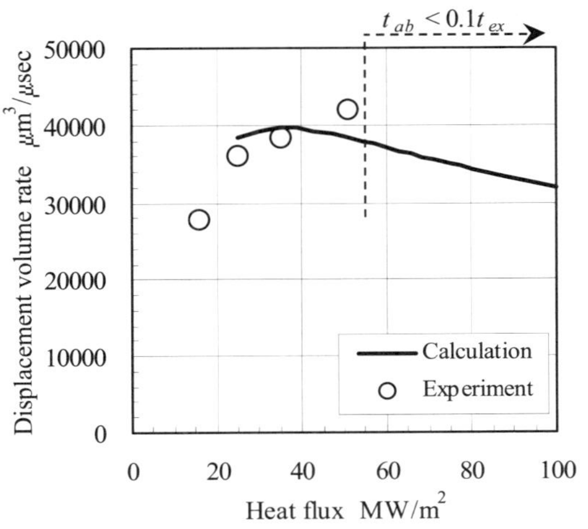

Figure 5. Comparison of predictions for displacement volume rate with experimental results (Asai (1991))

a maximum for each plug length, this value is outside the validity of relation (2). Additional investigation is needed to confirm its existence. These results also suggest that the condition of surrounding liquid affects the displacement volume rate.

3.2 *Comparison with experimental results*

Here we will compare the prediction with the experimental results reported by Asai (1991). Although there is no spatial constraint for bubble generation in Asai's experimental apparatus, the bubble dynamics are regarded as approximately one-dimensional. We set $T_{shl} = T_{cr}$ so that the heating period time t_h of the prediction agrees with the experimental conditions. We also set L_l 0.05 mm for the prediction so that the bubble existence time t_{ex} from the prediction agree with the experimental results reasonably. Figure 4 shows maximum bubble volume V_{max} under different heat fluxes. There is good agreement between the prediction and experiment. Figure 5 shows the comparison of the displacement volume rate D. Although there seems be good agreement in these comparisons, we need further investigation due to the reason mentioned above.

4 CONCLUDING REMARKS

The proposed model is significant in that it can be used to evaluate the basic performance of a liquid plug as an actuator without any empirical factor or fitting parameter in the descriptions. Although the authors believe that this paper addresses the fundamentals of fluid propulsion powered by vapor explosion, additional work is needed to refine the present model. The influences of neglecting surface tension at the bubble formation and neglecting thermal aspects after the bubble formation should also be taken into consideration. Additional experimental data for comparison with the model is also needed.

ACKNOWLEDGEMENTS

The authors acknowledge the support of this work by Hitachi Ltd.

REFERENCES

Asai A. 1991. Bubble Dynamics in Boiling Under High Heat Flux Pulse Heating. *ASME Journal of Heat Transfer*. 113: 973-979.

Carey V. P. 1992. *Liquid-Vapor Phase-Change Phenomena.* Washington DC: Hemisphere Pub. Corp.

Myers, G.E. 1998. *Analytical Methods in Conduction Heat Transfer, 2nd Edition.* Madison: AMCHT Publications.

Incropera, F. P. and Dewitt, D. P. 1990. *Fundamentals of Heat and Mass Transfer, 3rd Edition.* New York: John Wiley & Sons.

Koch M., Evans A. and Brunnschweiler A. 2000. *Microfluidic Technology and Applications*. Hertfordshire: Research Studies Press Ltd.

Mikic, B. B. and Rohsenow, W. M. 1969. Bubble Growth Rates in Non-Uniform Temperature Field. *Progress in Heat and Mass Transfer*. 2: 283-292.

Olsson A., Enoksson P., et al. 1997. Micromachined Flat-Walled Valveless Diffuser Pumps. *J. microelectromechanical systems*. 6: 161-166.

Ory E., Yuan H., Prosperetti A., Popinet S. and Zaleski S. 2000. Growth and Collapse of a Vapor Bubble in a Narrow tube. *Physics of Fluids*. 12: 1268-1277.

Tsai J. H. and Lin L. 2001. A thermal Bubble Actuated Micro Nozzle-Diffuser Pump. *Proc. Int. Conf. MEMS 2001., Interlaken, Switzerland, 21-25 January 2001*: 409-412.

Yuan H., Oguz H. N. and Prosperetti A., 1999, "Growth and collapse of a bubble in a small tube", Int. J. Heat Mass Transfer, Vol. 42, pp. 3643-3657.

Yuan H. and Prosperetti A. 1999. The Pumping Effect of Growing and Collapsing Bubbles in Tube. *J.Micromech. Microeng.* 9: 402-413.

Thermal Challenges in Next Generation Electronic Systems, Joshi & Garimella (eds)
© 2002 Millpress, Rotterdam, ISBN 90-77017-03-8

Application of thermoacoustic engines to heat transfer in microcircuits

O. G. Symko, E. Abdel-Rahman, Y. S. Kwon, M. Emmi & R. Behunin
Dept. of Physics, University of Utah, Salt Lake City, Utah, U.S.A.

Keywords: heat management, thermoacoustic, miniaturization, acoustic heat pump, engine

ABSTRACT: Thermoacoustic heat engines provide a practical solution to the problem of heat transfer in microcircuits where they can be used to pump heat or produce spot cooling of specific circuit elements. There are basically two types of thermoacoustic engines, a prime mover where heat is converted to acoustic energy, and a heat pump or cooler where sound can pump heat up a temperature gradient. Such devices are relatively simple and they are readily adaptable to microcircuit interfacing. Since this type of engine is usually operated in a resonant mode, the operating frequency determines its size. The devices presented here are pumped at frequencies ranging from 4 kHz to 24 kHz. They have been developed for interfacing with microcircuits as heat pumps or spot coolers. Results of their performance will be presented.

1 INTRODUCTION

A thermoacoustic heat engine consists of a working gas, in the present case it is air at one atmosphere, a medium for thermal rectification called the stack, heat exchangers at each end of the stack, a resonator, and a loudspeaker when the engine is a heat pump or a temperature difference when it is used as a converter of heat to sound (known as prime mover). In many of the large scale applications of thermoacoustic engines, discussed by Swift (1995), a typical operating frequency has been around 500 Hz. Since the devices are usually operated in the resonant mode with mainly standing wave components, it is possible to scale down their dimensions by raising the acoustic frequency. The devices presented here operate at frequencies in the range of 4 kHz to 24 kHz thus making it possible to have devices small enough for interfacing with microcircuits.

Theoretical aspects of thermoacoustics have been covered in a series of papers by Rott (1975), and Rott and Zouzoulas (1976); they were related to experimental results by Swift (1988), following the first demonstration of substantial acoustic cooling by Wheatley, Hofler, Swift, and Migliori (1983). Based on the above analysis of thermoacoustic phenomena, the total heat flux $\dot{Q}_2$ along the stack is given by

$$\dot{Q}_2 = -\frac{1}{4}\Pi\delta_\kappa T_m\beta \; p_1 u_1(\Gamma - 1) \tag{1}$$

where $\Pi\delta_\kappa$ is the total effective area next to the stack material perpendicular to the heat flow direction, δ_κ the thermal penetration depth, $T_m\beta$ are working gas parameters with β being the thermal expansion coefficient of the gas and T_m its mean temperature, $p_1 u_1$ represents the acoustic power with p_1 being the pressure oscillations and u_1 the speed oscillation of the sound; Γ is the ratio of the temperature gradient across the stack ∇T_m to a critical temperature gradient ∇T_{crit}, which depends on geometry, gas properties and acoustic drive. When $\nabla T_m > \nabla T_{crit}$ heat flux is toward the nearest pressure node in the resonator (as in a prime mover) while for

$\nabla T_m < \nabla T_{crit}$ heat flux is away from the pressure node as in a refrigerator. The total acoustic power produced or absorbed is given by

$$\dot{W}_2 = \frac{1}{4}\Pi\,\delta_\kappa \Delta x \frac{T_m \beta^2 \omega}{\rho_m c_p} p_1^{\,2}(\Gamma - 1) \tag{2}$$

where Δx is the length of stack, ω the acoustic angular frequency, ρ_m the gas density and c_p its specific heat.

Here as well the magnitude of Γ determines whether the engine is a prime mover or a heat pump. The efficiency of the prime mover is

$$\eta = \left(\frac{\Delta T}{T_m}\right)\frac{1}{\Gamma} \tag{3}$$

while the C.O.P. of the heat pump is:

$$C.O.P. = \left(\frac{T_m}{\Delta T}\right)\Gamma \tag{4}$$

In both cases the efficiency performance is lower than in a Carnot cycle. Equations 1 and 2 are approximate since the viscous losses were omitted. Both equations show the intimate connection between heat and sound in thermoacoustics. The presence of a stack provides heat exchange with a sound field and the generation or absorption of acoustic power. With a suitable geometry substantial amounts of heat can be moved as demonstrated, for example, by Garrett and Hofler.

2 INTERFACE WITH MICROCIRCUITS

2.1 *Acoustic cooler*

As part of the development of thermoacoustic engines for microcircuit heat transfer we discuss thermoacoustic refrigerators operated at frequencies in the range of 4 kHz to 24 kHz. An acoustic cooler at 4.5 kHz using a half-wave resonator has a length of ~ 4.0 cm. It is driven by a piezoelectric driver, of the bimorph type loaded with a cone; the stack consists of random material such as cotton wool or glass wool and it is in thermal contact with mesh copper heat exchangers. The working gas is air at 1 atmosphere chosen because of the convenience. This type of device can be used as a spot cooler on a circuit. Figure 1 shows its interfacing to an element of the circuit, perhaps a transistor, which needs to be cooled below room temperature. The driver produces a standing wave in the resonator; heat is pumped acoustically from the cold heat exchanger toward the hot heat exchanger, away from the nearest pressure node. The hot exchanger dissipates the heat to the ambient by convection-cooling of fins attached to the hot heat exchanger unit.

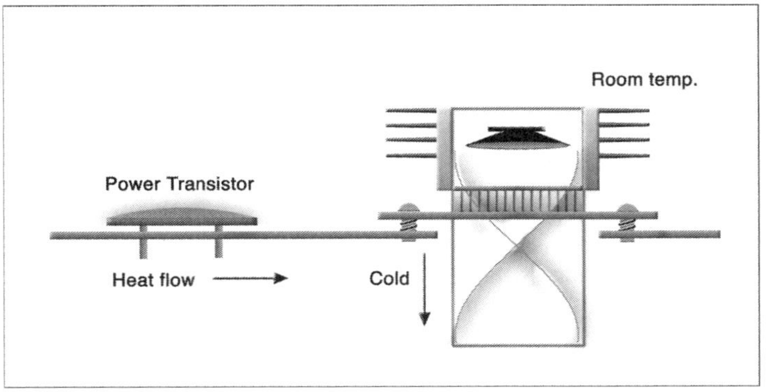

Figure 1. Acoustic spot-cooler interfaced with circuit.

A copper gasket is used for interfacing the cold heat exchanger of the cooler to the device which is to be cooled on the circuit. This provides simple access to the cold part of the cooler. One advantage of this type of interfacing is that the resonator and its gas do not need to be cooled, thus reducing the thermal load on the refrigerator. Tests on a 4.5 kHz refrigerator with both heat exchangers thermally floating are shown in Figure 2. The cold heat exchanger drops rapidly in temperature cooling by 4°C in 50 seconds. The hot heat exchanger with a substantially higher thermal load warms up by 7°C. This creates a temperature difference ΔT of 11°C between the two heat exchangers. By thermally anchoring the hot heat exchanger to ambient temperature lower temperatures can be reached by the cold exchanger. A cooling power of 1 watt can be achieved with a coefficient of performance of 2-3. The performance depends strongly on geometry and the performance of the driver. When driven beyond its specifications the driver produces excessive heating which needs to be dissipated by the hot exchanger. Further miniaturization of this type of refrigeration has led to one operating in the ultrasonic range at 24 kHz, its length at that frequency being 0.8 cm. Here as well the piezo is loaded resonantly by a cone.

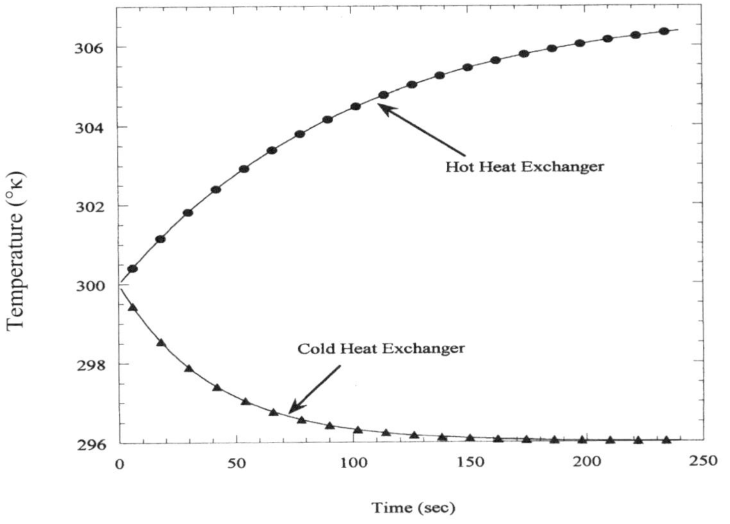

Figure 2. Performance of an acoustic cooler with both heat exchangers thermally floating.

2.2 *Prime mover*

A prime mover can also be used to remove heat from a circuit, a convenient application of the fact that heat produces sound. In this case we use essentially a ¼ wave resonator, typically 2 cm long. The hot heat exchanger is thermally anchored to the circuit while the cold exchanger is at room temperature. Due to heat flow from the circuit, a temperature difference ΔT across the stack, above the onset for oscillations, causes oscillations; heat is converted to sound. This is shown in Figure 3; with a heat input of 2 watts and a ΔT of 40°C across the stack, sound of ~ 2 kHz is produced at an intensity of 10 mW/cm^2. Such performance of this heat engine corresponds to ~ 10% of Carnot efficiency. On similar devices the efficiency is typically 10-16% of Carnot. Important features of this type of heat pump are the absence of moving parts, simple design, and small dimensions. (The possibly annoying feature of the sound can be overcome in various ways, such as operation in the ultrasonic range, or immediate power conversion to electricity). Further improvements of this type of engine can be made by lowering the threshold ΔT for oscillations and by incorporating the device into a parallel array.

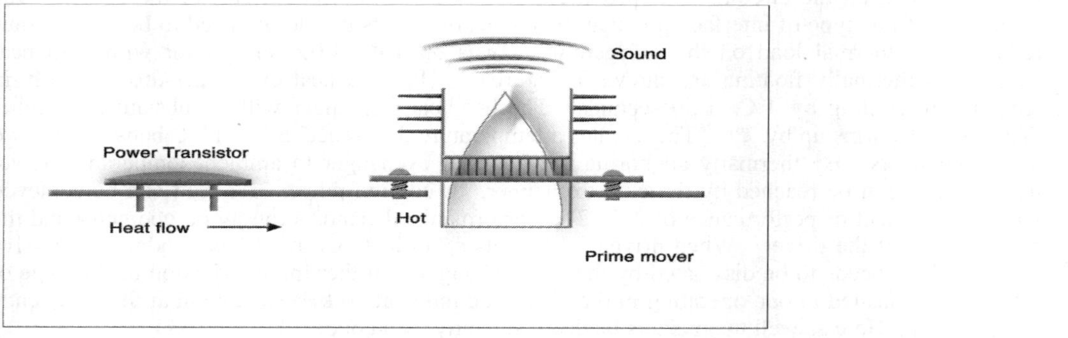

Figure 3. Prime mover interfaced with circuit.

Figure 4 shows the onset for oscillations of the prime mover at a temperature difference ΔT onset across the stack; at this point the random pressure fluctuations due to heat are converted to sound of essentially one frequency. Larger temperature differences lead to a non-linear regime of saturation of the level of acoustic oscillations. Figure 4 shows the sound level as a function of time as the hot exchanger is being continuously heated. Analysis below the onset has been discussed by Atchley. The approach taken here for heat transfer from a circuit can be compared to heat conduction along a solid conductor such as copper. In both cases a temperature gradient ∇T drives the heat flux, across an element of cross-sectional area A:

$$\dot{Q} = {}^{-}K\, A \nabla T \tag{5}$$

with K = thermal conductivity of solid conductor

or as in the acoustic case K = effective thermal conductivity of prime mover when producing sound

A prime mover similar to the one described here has been reduced in size to the point that it radiates sound as high as 20.4 kHz; this proves that this device can work in the ultrasonic range and that further miniaturization is possible.

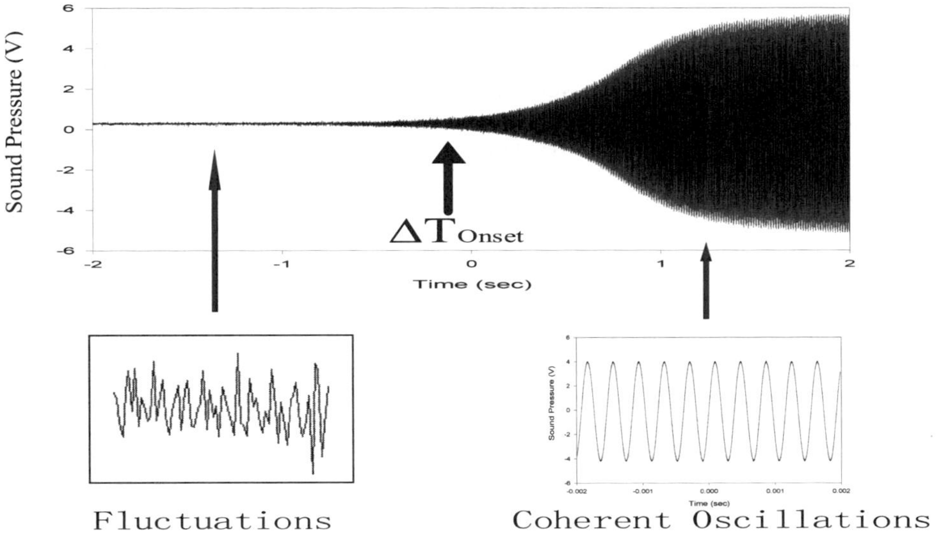

Figure 4. Acoustic oscillations in prime mover above ΔT_{onset} across stack.

3 CONCLUSIONS

We have presented two novel approaches using acoustics for removing heat from microcircuits. They have much merit for such applications, especially from the point of view of simplicity, performance, environmentally safe components, and ease of interfacing with microcircuits. The advantages of the acoustic approach compared to the standard passive copper/fins/convection approach shown in Figure 5 provide substantial improvements for microcircuit applications. Our active device, the acoustic cooler, can provide spot-cooling below ambient; this is important for some

Figure 5. Heat removal using copper/fins/convection approach.

applications. A cooling power of 1 watt with a ΔT of ~ 10°C or more can provide substantial improvements in the speed and electronic device performance. By raising the operating frequency, the device becomes smaller and its response time is reduced. Larger cooling power can be achieved by modifying the geometry, raising the gas pressure, and using gas mixtures of lower Prandtl number. Heat removal by acoustic radiation from a prime mover provides an approach which works especially well over short distances form the heat source. An onset temperature difference of 40°C presented here is a very large improvement over previous low frequency prime movers; with some development this onset can be lowered to the 10°C temperature difference range. Lowering this onset ΔT presents the circuit from getting too high a temperature rise. This method leads to substantial saving in the amount of heat conducting material that would have been needed in a passive approach. By raising our operating frequency, we have made the acoustic devices smaller so that they can be interfaced with microcircuits. The very high frequency range (the ultrasonic) provides an opportunity to develop arrays of microscale acoustic devices with higher power density. The results presented here point to further developments by using the acoustic devices in arrays. Although the demonstration of the performance is preliminary with room for further improvements, the potential of this new approach shows much promise for microcircuit and small scale electronic applications.

This research was supported by the Office of Naval Research and DARPA/SPAWAR.

REFERENCES.

Atchley, A. 1992. Standing Wave Analysis of a Thermoacoustic Prime Mover Below Onset of Self-Oscillation. *Journal of Acoustical Society of America* (92): 2907.

Garrett, S. and Hofler, T. 1991. Thermoacoustic Refrigeration. *TECHNOLOGY 2001. NASA Conference Publication* 3136, (2): 397.

Rott, N. 1975. Thermally Driven Acoustic Oscillations, Part III. Z. *Angew. Mathematical Physics* (26): 43.

Rott, N. and Zouzoulas, G. 1976. Thermally Driven Acoustic Oscillations, Part IV. Z. *Angew. Mathematical Physics* (27): 197.

Rott, N. 1980. Thermoacoustic. *Advances in Applied Mechanics* (20): 135

Swift, G. 1988. Thermoacoustic Engines. *J. Acoustical Society of America* (84): 1145.

Swift, G. July 1995. Thermoacoustic Engines and Refrigerators. *Physics Today*: 22-38.

Wheatley, J. C. Hofler, T. Swift, G. and Migliori, A. 1983. Experiments with an Intrinsically Irreversible Acoustic Heat Engine. *Physical Review Letters* (50): 499.

HIGH HEAT FLUX TRANSPORT CHALLENGES

Thermal Challenges in Next Generation Electronic Systems, Joshi & Garimella (eds)
 ISBN 90-77017-03-8

Pool boiling experiments for HFE-7100 dielectric liquid

Mohamed S. El-Genk and Huseyin Bostanci
Institute for Space and Nuclear Power Studies and Chemical and Nuclear Engineering Department.
University of New Mexico, Albuquerque, NM, USA

Keywords: dielectric liquids, nucleate boiling, critical heat flux, immersion cooling of electronics.

ABSTRACT: Pool boiling experiments are conducted to investigate the potential of the hydrofluoroether (HFE-7100) dielectric liquid for immersion cooling of high performance computer chips. In the experiments the inclination angle of the 1 cm^2 smooth copper surface, θ, varied from 0° (upward-facing) to 180° (downward facing) in 10°-30° increments and subcooling, ΔT_{sub} = 0 (saturation), 10, 20, and 30 K. For 90° ≤ θ ≤ 180° and ΔT_{sub} = 10-30 K, the nucleate boiling heat flux decreases as θ increases, consistent with earlier data for FC-72 dielectric liquid, but for θ < 90°, the nucleate boiling heat flux is highest at θ = 0° and increases very little or none with increased θ. The saturation nucleate boiling heat flux at surface superheats > 13 K is highest at 0° and decreases with increased θ to 180°, but at lower surface superheats it is lowest at 0° and increases with increased θ to 180°. The developed correlations for saturation nucleate boiling, as a function of θ, and for the Critical Heat Flux (CHF), as a function of θ and ΔT_{sub}, agree with the data to within ± 10%. Correlations for surface superheat and temperature at CHF are also developed. While CHF decreases slowly as θ increases from 0° to 120° then faster for θ > 120°, it increases linearly with ΔT_{sub}, consistent with earlier work for dielectric and non-dielectric liquids. Photographs of saturation nucleate boiling are presented for θ = 30° and 120°, which support present results. Results show that removal of heat dissipation from a smooth copper surface up to 25 and 35 W/cm^2 is possible in saturation and subcooling nucleate boiling of HFE-7100 and at surface superheat, ΔT_{sat} < 31.7 and 36.7 K, respectively.

1 INTRODUCTION

The dissipation heat flux from computer and microprocessor chips vary with their type and application. While heat dissipation for memory (DRAM) chips, at a junction temperatures ≤ 100 °C, is ~ 0.24 W/cm^2, the dissipation heat flux for high performance chips is ~ 28 W/cm^2 at same junction temperatures, and could exceed 30 W/cm^2 by 2006. For dissipation heat fluxes > 10 W/cm^2 immersion cooling by nucleate boiling of dielectric liquids is quite effective, offering the advantages of uniform surface temperature, use in "tight" spaces, environmentally friendly properties such as a zero Ozone depletion potential, excellent corrosion and chemical compatibility properties, safe handling, low boiling temperatures (e.g. 56 °C and 61 °C for HFE-7100 and FC-72 at 0.1 MPa, respectively), and a wide range of dissipation heat fluxes within a narrow range of surface temperatures. However, owing to the high wetting property of dielectric liquids (e.g., surface tension ~ 0.01 and 0.008 N/m for HFE-7100 ($C_4F_9OCH_3$) and FC-72 (C_6F_{14}), respectively), they fill surface crevices, delaying incipient nucleation until high surface superheat, which need to be measured and taken in consideration by the electronics-packaging engineers. Other concerns are the complexity of electronics packaging for removing the dissipated heat from the boiling liquid to an eventual heat sink, and the need to use a tightly sealed container to prevent leakage of the dielectric liquids.

The Fluorinert FC-72 liquid, and to some extent FC-86 and FC-87, has been used in electronics cooling applications and in a number of pool boiling experiments (Danielson et al., 1987, Carvalho and Bergles, 1990, Change and You 1996 and 1997, and Watwe et al., 1997, Rini et al. 2001). Little work, however, has been reported, on the recently introduced Novac liquids such as HFE-7100 and FC-7200 ($C_4F_9OC_2H_5$) (Liu et al 2001, and Arik and Bar-Cohen, 2001). These liquids have much lower Global warming potential and higher latent heat of vaporization than fluorinert liquids, thus capable of removing higher dissipation heat fluxes. The reported work on dielectric liquids has investigated only the effect ΔT_{sub} at inclinations of 0° and 90° (Hwang and Moran, 1981, Mudawar and Anderson, 1989, Carvaho and Bergles, 1990, McNeil, 1992, Lee et al., 1992,

Watwe et al., 1997, Arik and Bar-Cohen, 2001, and Liu et al., 2001), indicating a linear increase in CHF with increased subcooling. The effect of surface inclination, θ, has only been investigated for saturation boiling using FC-72. Chang and You (1996) indicated that for 10 x10 mm, smooth copper surface, the nucleate boiling heat flux at high surface superheat increased as θ increased from 0° to 90°, but for θ > 90°, there was a noticeable decrease in nucleate boiling heat transfer as θ increased. For a 20 x 20 mm surface, Chang and You (1997) reported that nucleate boiling heat flux increased as θ increased from 0° to 45°, but decreased with θ > 45°. Recently, Rainey (1999) has reported that for saturation boiling of FC-72 from 50 x 50 mm smooth copper surface at low surface superheat, the nucleate boiling heat flux increased slightly with θ from 0° to 45°, then decreased with a further increase in θ. Therefore, there is a need to systematically investigate the effects of θ and ΔT_{sub} on nucleate boiling heat flux and CHF for dielectric liquids, using the same experimental setup.

In this work, boiling experiments are performed using HFE-7100 to investigate the effects of θ and ΔT_{sub} on nucleate boiling and CHF from a smooth copper surface measuring 10 x 10 mm. In the experiments, $\Delta T_{sub} = 0$ (saturation), 10, 20, and 30 K and θ = 0°, 30°, 60°, 90°, 120°, 150°, 170°, and 180°. In addition, for saturation boiling the effect of θ on the surface superheat at incipient boiling and at CHF is investigated. The saturation nucleate boiling data are correlated in terms of θ, while the CHF data are correlated as a function of θ and ΔT_{sub}. Still photographs of the boiling surface are recorded using a high quality digital camera at different values of the nucleate boiling heat flux, θ, and ΔT_{sub}, however, only photographs of saturation nucleate boiling at 30° and 120° are presented in this paper.

2 EXPERIMENTAL FACILITY

The experimental facility shown in Figure 1a, consists of a heated water bath in an acrylic tank (318 x 318 x 514 mm), a smaller acrylic test vessel (Figure 1b) (149 x 130 x 343 mm), the copper test section (Figure 2) mounted to a rotation assembly (Figure 1b), a water bath temperature and electrical power control circuit, and a computer controlled data acquisition system. The water bath is heated using a 1500 W, AC immersion heater, having two rectangular coils (4.78 mm in diameter), which fit at the bottom of the water bath tank, separated ~ 20 mm from the side walls and raised 15 mm from the bottom using small copper-topped Bakelite cubes. The immersion heater is controlled using a temperature controller (Figure 1a) to maintain the bath temperature at desired values. A magnetic stirrer unit is used to speed up the outgassing of HFE-7100 and for ensuring a uniform pool temperature prior to conducting the experiments (Figs. 1a and 1b). The test vessel is sealed tight, using a rubber O-ring and eight stainless steel bolts, which fasten it to the vessel wall, to prevent leakage of HFE-7100 in the experiments (Figure 1b). To condense the vapor generated in experiments, a water cooled reflux condenser, made of a copper tube, 4.76 mm O.D, is installed on the inside of the cover plate of the test vessel. In addition, a small, water-cooled copper condenser coil is immersed into the HFE-7100 pool to adjust its temperature in subcooling experiments. The water bath temperature is measured using a K- type thermocouple and that of the HFE-7100 pool is taken as the average reading of two K-type thermocouples immersed in the pool, on opposite sides of the test section. The rotation assembly for adjusting the inclination angle is fabricated of acrylic material and can be rotated freely from 0° to 180° in small increments (Figure1b).

The 1.63 mm thick, copper surface block (Figure 2) simulates a computer chip and the underlying heating element is made of a Nichrome wire, 0.125 mm in diameter and a total resistance of 14 Ω, which is covered by a thin layer of high temperature and high thermally conductivity epoxy. Two, 0.5 mm diameter, 5 mm deep holes are drilled ~ 0.8 from the copper surface to install two, K-type thermocouples, whose average reading is taken as the copper surface temperature in the experiments. The cavity at the top of the test section (Figure 2) encased in a Lexan frame, is filled with a translucent, two-part epoxy adhesive, flush with the copper surface, to prevent bubbles nucleation outside the surface. Prior to the experiments, the copper surface is prepared using emery paper #1500. Then, the test section in mounted onto the rotation assembly and the electric power to the heating wire is supplied by a DC power supply (Figure 1a). A shunt resistor (~99.462 mΩ) is used to measure the current to the heating element while the voltage across the element is measured directly.

The dissipation heat flux from the copper surface is calculated from the measured current and voltage across the heating wire. The 3-D thermal analysis performed, using "Algor", a thermal analysis commercial software assuming a uniform heat transfer coefficient at the copper surface, indicated that heat losses from test section are negligible and surface temperature is uniform and only < 0.2 K below that indicated by the two thermocouples in the copper block (Figure 2b). In the experiments, the test section is immersed in the HFE-7100 such that the copper surface in the upward-facing position (θ = 0°) is at least 10 cm below the free surface of the pool. Owing to the high elevation of Albuquerque, NM (~ 1500 m above sea level) the saturation temperature of HFE-7100 is only 54 °C, versus 61 °C at 0.1 MPa. After reaching saturation temperature for the first time, it took up to two hours to outgas the HFE-7100 pool. Then, experiments begin by turning the magnetic stirrer off and adjusting the pool and water bath temperatures. The electric power to the heating wire in the test section is increased, via increasing the applied voltage in small increments (0.3 V in saturation and 0.5-1.0 V in subcooling experiments). The values of the surface heat flux and temperatures are recorded after reaching steady state, which typically took ~ 60 s. The data, however, are stored after 80 s to ensure steady state condition. To avoid burning the heating wire at CHF, when the difference in the steady state surface temperature following two successive incremental increases of the applied voltage, exceeds 10 °C, it is assumed that CHF is reached and the experiment is automatically terminated. The estimated uncertainty in measurements of temperature is $\pm$ 0.7 K, in surface heat flux is $\pm$ 3.7%, and in θ is $\pm$ 0.5°. The pool boiling curves obtained for saturation and ΔT_{sub} = 10, 20, and 30 K, and at θ = 0°, 30°, 60°, 90°, 120°, 150°, 170°, and 180°, are presented and discussed in the next section.

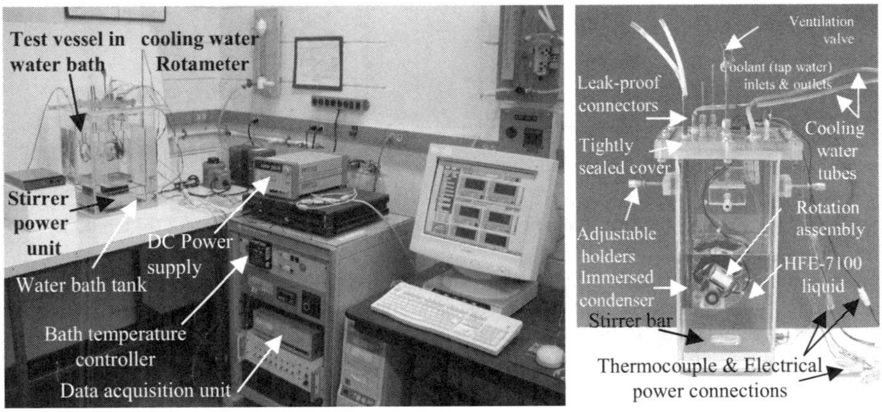

(a) Experimental setup (b) Pool boiling test vessel

Figure 1. Experimental facility and setup.

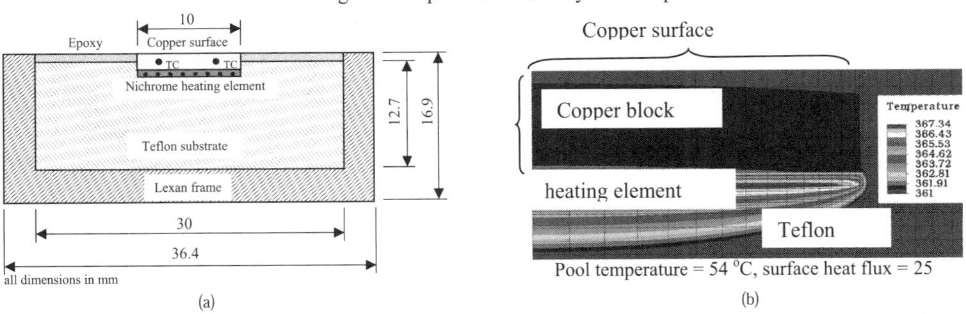

Figure 2. Test section: (a) A schematic with dimensions, (b) thermal analysis using Algor FEM commercial software, temperatures are in degrees Kelvin.

3 RESULTS AND DISCUSSION

Figure 3a presents two saturation boiling curves obtained at θ = 90°; in one, the heat flux is incrementally increased, while in the other heat flux is incrementally decreased and Figure 3b presents two saturation-boiling curves obtained at θ = 120° in two separate tests. Figure 3b demonstrates the good reproducibility of the data in the experiments. Figure 3a indicates that, except for the surface temperature overshoot in the boiling curve obtained with an incrementally increasing heat flux, the two pool boiling curves are almost identical. Similar results are obtained at other inclination angles, consistent with those reported by Rini et al. (2001), confirming that the method of changing the heat flux in the experiments does not affect the pool boiling curves.

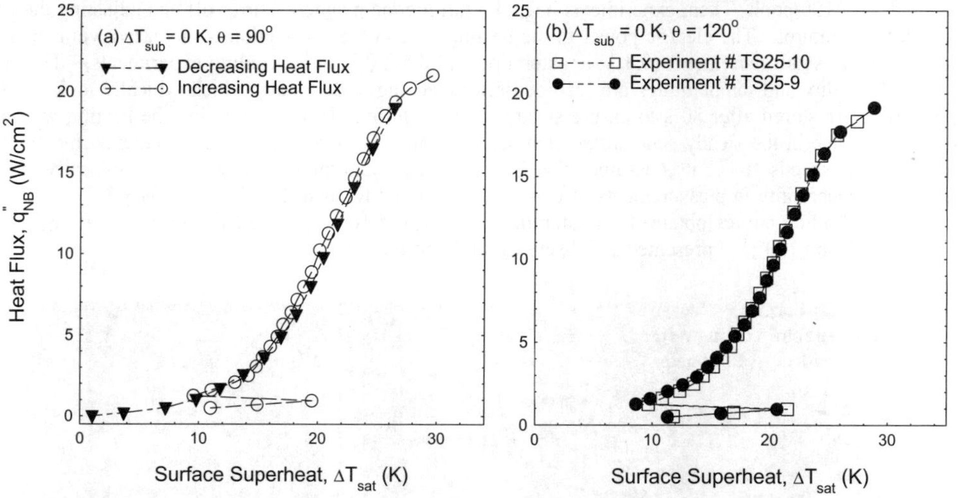

Figure 3. Nucleate boiling curves for saturated HFE-7100 showing: (a) hysteresis at θ = 90°, (b) reproducibility at θ = 120°.

Figure 3b also indicates that the surface superheat at incipient boiling varies from 21 to 22 K, versus only 8-10 K in nucleate boiling at same heat flux. The saturation boiling data do not show a clear trend of the dependence of surface superheat at incipient boiling on θ, averaging 22 ± 4 K (Figure 4a), which corresponds to a surface temperature overshoot of 11.6 ± 4K (Figure 4b) and a surface temperature of 349 ± 4K (or 76± 4°C). Previously reported superheats at incipient boiling of FC-72 are 25-40 K and 20-35 K for 10 x10 mm and 40 x 40 mm surfaces, respectively (Chang and You 1996 and 1997), and 16-17 K for a 50 x 50 mm surface (Rainey, 1999). The upper most data points in Figures 3a and 3b indicate the critical heat flux.

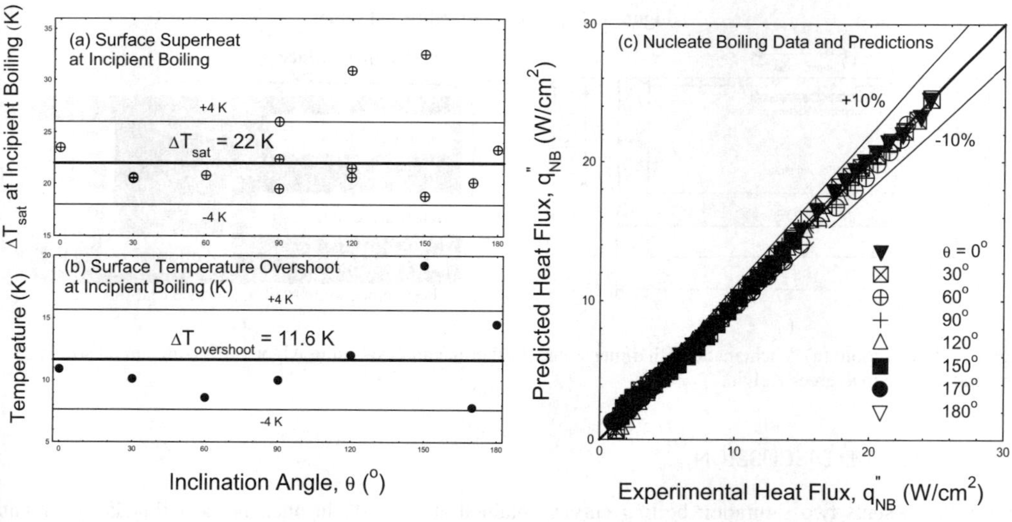

Figure 4. Saturation, nucleate boiling data for HFE-7100: (a) Surface superheats at incipient boiling, (b) comparison of saturation nucleate boiling correlation (Equation 1) and data.

3.1 *Saturation pool boiling curves and correlation*

The saturation nucleate boiling data presented in Figure 5a are compared with the developed correlation in Figure 4b. The boiling curves in Figure 5a indicate that nucleate boiling heat flux at high surface superheat (> 14 K) is highest at 0° and decreases as θ increases to 180°, in agreement with previously reported data for water (El-Genk and Guo, 1993) and FC-72 (Chang and You, 1996, and Rainey, 1999). Conversely, at lower surface superheats, nucleate boiling heat flux increases slightly as θ increases, consistent with previously reported data for water, liquid helium and liquid nitrogen (Nishikawa et al., 1983, and El-Genk and Guo, 1993). The saturation nucleate boiling heat flux data (Figure 5a) are correlated as follows:

$$q''_{NB} = \left[\left(a\, \Delta T_{sat}^{\,n} \right)^{M} + \left(b\, \Delta T_{sat}^{\,m} \right)^{M} \right]^{-(1/M)} \tag{1}$$

The coefficients and exponents in this correlation depend on θ and are listed in Table 1. The first term on the right hand side of equation (1) is obtained from correlating nucleate boiling data at low and intermediate surface superheat, while the second term is obtained from correlating nucleate boiling heat flux data at high surface superheat. The exact values of the surface superheats for these two regimes vary with θ. Equation (1), agrees with the present saturation nucleate boiling heat flux data to within ± 10% (Figure 4b). The subcooling nucleate boiling curves shown in Figures 5b-5d indicate that increasing ΔT_{sub} generally increases nucleate boiling heat flux, although the boiling curves are close for θ < 90°. For higher θ , however, nucleate boiling heat flux decreases as θ increases to 180°. Unlike saturation boiling, the boiling curves at 10 K subcooling show that at high surface superheats, nucleate boiling heat flux increases as θ increases from 0° to 60°, then decreases with increased θ > 60°. For ΔT_{sub} = 20 K, the boiling curves for θ = 0°-90° overlap, indicating a negligible effect of θ on the nucleate boiling heat flux. For θ > 90°, however, nucleate boiling heat flux decreases as θ increases to 180° (Figure 5c). The boiling curves at ΔT_{sub} = 20 K (Figure 5d) are similar to those in Figure 5b, except for the larger decrease in the nucleate boiling heat flux values with increased inclination.

Table 1 Coefficients and Exponents for Nucleate boiling correlation, Equation (1).

θ (°)	0	30	60	90	120	150	170	180
a	3.10×10^{-5}	4.59×10^{-5}	9.32×10^{-5}	2.48×10^{-5}	4.40×10^{-4}	7.53×10^{-4}	2.49×10^{-4}	7.84×10^{-4}
b	3.60	3.80	7.30	1.90	0.94	0.34	0.0295	0.0
n	4.25	4.09	3.85	3.53	3.35	3.10	3.42	2.13
m	0.56	0.54	0.33	0.71	0.90	1.11	1.81	2.13
M	(-) 5.0	(-) 5.0	(-) 5.0	(-) 8.0	(-) 10.0	(-) 10.0	(-) 10.0	(-) 10.0

3.2 *Effect of inclination angle on saturation critical heat flux*

The CHF data for saturation and subcooling boiling of HFE-7100 at ΔT_{sub} = 10, 20, and 30 K, are compared in Figure 6a, as a function of θ. As indicated later, CHF increases linearly with increased subcooling, but decreases with increased inclination, reaching its lowest values at 180°. The decrease in CHF with increased inclination is small for θ < 120°, but much faster for θ >120°. The saturation boiling CHF data are correlated using the form suggested by Kutatelatze (1961) and El-Genk and Guo (1993) as:

$$CHF_{sat} = C_{CHF,sat}(\theta)\, \rho_v^{0.5}\, h_{fg} \left[\sigma\, g \left(\rho_\ell - \rho_v \right) \right]^{0.25}, \tag{2}$$

$$C_{CHF,sat}(\theta) = \left[\left(0.229 - 4.27x10^{-4}\theta \right)^{-6} + \left(0.577 - 2.98x10^{-3}\theta \right)^{-6} \right]^{(-1/6)} \tag{3}$$

Equation (3), obtained from the best least square fit of the saturation CHF data indicated in Figure 6a by the solid circles, agrees with the data to within + 3%. The lowest and the highest values of the coefficient in equation (3) of 0.041 and 0.229 are for 180o (downward-facing) and 0o, respectively, and the values in between, in an increasing magnitude, are for • = 170o, 150 o, 120 o, 90 o, 60 o, and 30 o, respectively. In equation (2), •v, •l, hfg, and • are the liquid density, vapor density, latent heat of vaporization, and liquid surface tension at saturation temperature, respectively, and g is the acceleration of gravity. The coefficient in equation (3) has been shown to varying, not only with •, but also with the properties of the boiling liquid (El-Genk and Guo, 1993). For saturation boiling of water at • = 0o, El-Genk and Guo (1993) reported a value of

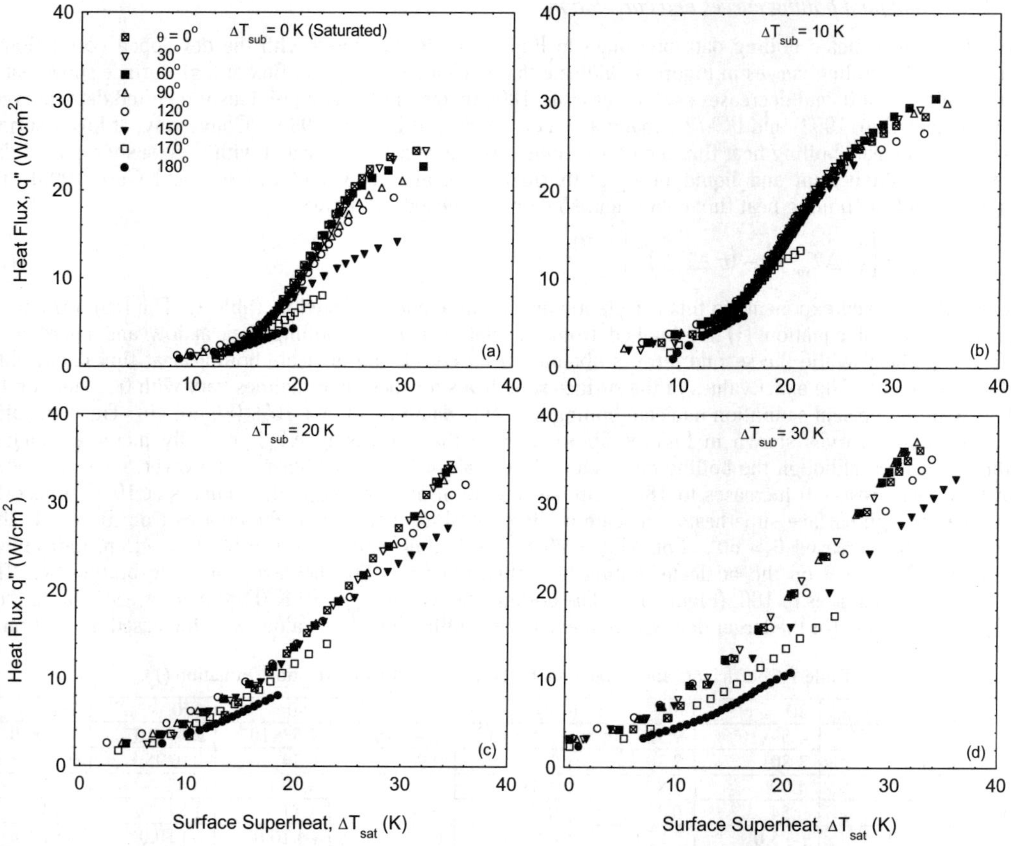

Figure 5. Nucleate boiling curves for HFE-7100 liquid at different inclination angles and subcooling.

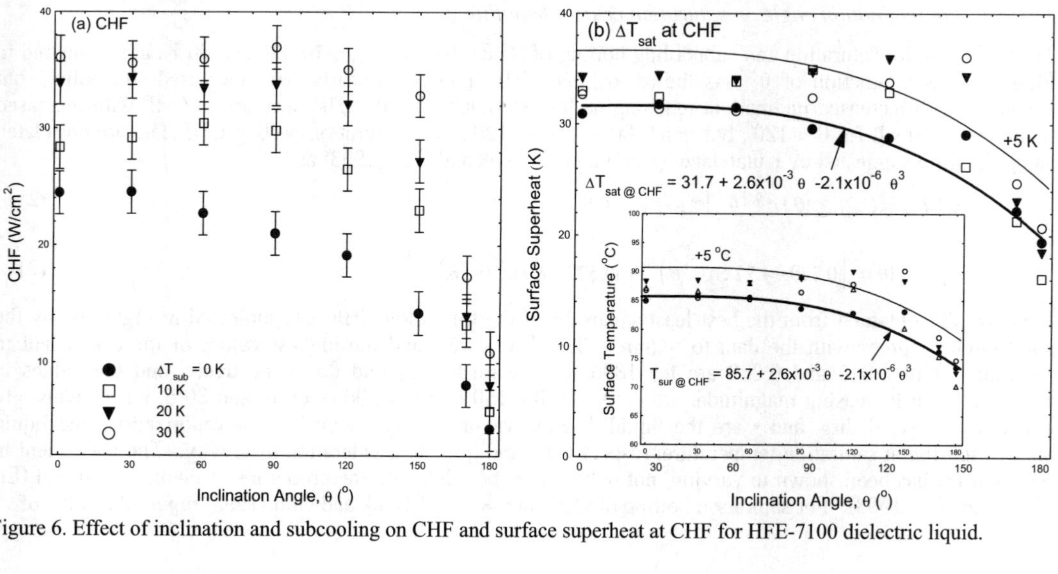

Figure 6. Effect of inclination and subcooling on CHF and surface superheat at CHF for HFE-7100 dielectric liquid.

0.145, which is about 11% higher than that reported by Kutatelatze (1962). They have also reported values of 0.1485 and 0.1385, for saturation boiling of liquid helium and liquid nitrogen, respectively, at • = 0o.

For saturation boiling, the data delineated in Figure 6b demonstrate the dependence of surface superheat at CHF on θ. This figure show that the effect of subcooling is non-conclusive, with only a 5 K increase in the surface superheat at CHF with increasing subcooling by 30K. For saturation boiling, the surface superheat at CHF is correlated as a function θ as follows:

$$\Delta T_{sat@CHF} = 31.7 + 2.6x10^{-3}\,\theta - 2.1x10^{-6}\,\theta^2. \tag{4}$$

The data in Figure 6b and in the insert figure show that the average surface temperature at CHF in saturation boiling is 85.7 °C, less than the 100 °C limit for the junctions temperatures of most high performance computer chips. Note that the measured saturation temperature of HFE-7100 in Albuquerque, NM, atmospheric pressure ~ 0.085 MPa, is only 54 °C. Figure 6b also shows that the decrease in the surface superheat at CHF with increasing inclination angle is less steep than for CHF at θ > 120° (Figure 6a).

3.3 *Effect of subcooling on critical heat flux*

Figure 6a shows that at • = 0°, CHF for HFE-7100 increases from ~ 24.45 W/cm^2 at saturation and to ~ 28.35, 33.9, and 36.06 W/cm^2 at 10, 20, and 30 K subcooling, respectively. At • = 180° however, CHF is as low as ~4.30 W/cm^2 at saturation, but increases to ~ 5.85, 8.07, and 10.85 W/cm^2 at 10, 20, and 30 K subcooling, respectively. The entire CHF data for saturation and sucbooled boiling in Figure 6a are correlated as a function of θ and ΔT_{sub}, using the same general form of equation (2) for saturation CHF as:

$$CHF(\theta, \Delta T_{sub}) = C_{CHF}(\theta, \Delta T_{sub})\,\rho_v^{0.5}\,h_{fg}\,[\sigma\, g(\rho_\ell - \rho_v)]^{0.25}, \tag{5}$$

$$C_{CHF}(\theta, \Delta T_{sub}) = C_{CHF,sat}(\theta).(1 + C^*(\theta)\Delta T_{sub}) \text{ , and} \tag{6a}$$

$$C^*(\theta) = 0.016 + 1.33x10^{-4}\,\theta + 2.469x10^{-7}\,\theta^2. \tag{6b}$$

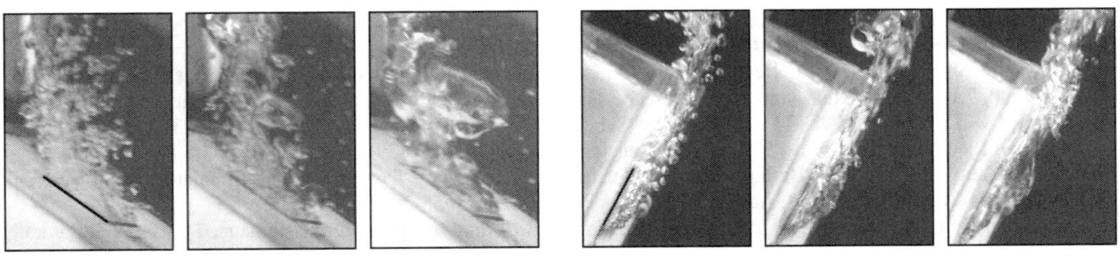

10.04 W/cm^2 ΔT_{sat}=20.7 | 16.53 W/cm^2, ΔT_{sat}=25.1 K | 23.41 W/cm^2, ΔT_{sat}=30.4 K | 5.04 W/cm^2 ΔT_{sat}=11.6 K | 12.73 W/cm^2 ΔT_{sat}=14.7 K | 18.87 W/cm^2 ΔT_{sat}=21.0 K

(a) θ = 30° (b) θ = 120°

Figure 7. Photographs of saturation nucleate boiling of HFE-7100 liquid in the experiments at θ = 0° and 120°.

Equation (5) agrees with the CHF data to within $\pm$ 10% (Bostanci and El-Genk, 2002). The linear dependence of CHF on ΔT_{sub} (Equation 6a) is consistent with that reported by previous investigators for dielectric and non-dielectric liquids (Carvalho and Bergles, 1990, Lee at al., 1992, and Ivey and Morris, 1962). The values of the coefficient C*(θ) (equation 6b) are 0.016, 0.019, 0.0275, 0.03, 0.0343, 0.0407, 0.0454, and 0.48 at θ = 0°, 30°, 60°, 90°, 120°, 150°, 170°, and 180°, respectively. The values of C* at 0° (0.016) and 90° (0.03) are the same as those reported or deduced from the reported data by other investigators for HFE-7100 and HFE-7200 (Arik and Bar-Cohen, 2001, Liu et al., 2001) and for FC-72 (Watwe and Bar-Cohen, 1997, Caralho and Bergeles, 1990, Mudawar and Anderson, 1989).

Figure 7a and 7b shows photographs of saturation nucleate boiling of HFE-7100 in the present experiments at θ = 30° and 120°, respectively, and at low, intermediate, and high nucleate boiling heat flux values. These figures indicate that the accumulation of vapor in the boundary layer increases with increased surface inclination, decreasing the nucleate boiling heat flux, consistent with the pool boiling curves in Figure 5a. In the discrete bubble region, the average departure bubble diameter from the copper surface, determined from the photographs taken for $\theta < 90^o$ in saturation and subcooling experiments is ~ 0.55 $\pm$0.07 mm.

4 SUMMARY AND CONCLUSIONS

The present experiments investigated the effects of θ and ΔT_{sub} on nucleate boiling heat flux and CHF for HFE-7100 dielectric liquid from 1cm^2 smooth, copper surface. The copper surface simulates a high performance computer chip with a measured dissipation heat flux in nucleate boiling of HFE-7100 of up to 25 W/cm^2 and 36 W/cm^2 in saturation and subcooled boiling, respectively. Results showed that for $\theta = 90^o$-180°, and $\Delta T_{sub} = 10$, 20, and 30 K, the nucleate boiling heat flux decreases as θ increases, but at lower inclinations ($\theta < 90^o$), however, the nucleate boiling heat flux changes very little with θ. In saturation experiments, the dependence of the nucleate boiling heat flux on θ at $\Delta T_{sat} > 14$ K, is similar to that in subcooled boiling at $\theta > 90^o$, decreasing with increased θ from 0° to 180°. Conversely, at lower ΔT_{sat}, the saturation nucleate boiling heat flux is highest at 0° and decreases with increased inclination. The saturation and subcooled boiling CHF decreases almost linearly with θ from 0° to 120°, then rapidly to its lowest value at 180°. The surface superheat at CHF also decreases as θ increases, increasing only by 5K in subcooled boiling experiments. The values of the surface superheat at CHF, in saturation boiling experiments, are correlated as a function of inclination angle of the boiling surface. Results showed that the decrease in the surface superheat at CHF with increasing inclination angle is less steep than for CHF at $\theta > 120^o$. The developed correlations for saturation nucleate boiling, as function of θ, and for CHF, as a function of θ and ΔT_{sub}, agree with the present data to within $\pm$ 10%. Results show that CHF increases linearly with ΔT_{sub}. The obtained slopes from the least square fit of the data of 0.016 and 0.03 for 0° and 90°, respectively, are the same as reported or deduced from most of the reported data by previous investigators for HFE-7100, HFE-7200, and FC-72 dielectric liquids.

REFERENCES

Arik, M. & Bar-Cohen, A. 2001. Ebullient Cooling of Integrated Circuits by Novec Fluids, Proc. Pacific Rim Int. Intersociety, Electronic Packag. Conference, Kauai, Hawaii, July 8-13, 2001.

Bastanci, H. & El-Genk, M. 2002, Effects of inclination angle and subcooling on nucleate boiling and critical heat flux of HFE-7100, Proc. 12th Int. Heat Transfer Conference, Grenoble, France, 18-23 August 2002 (in print).

Carvalho R. & Bergles, A. 1990. The Influence of Subcooling on the Pool Nucleate Boiling and Critical Heat Flux of Simulated Electronic Chips, Proc. 9th Int. Heat Transfer Conference, 2, 289-294.

Chang, Y. & You, S. 1997. Pool Boiling Heat Transfer from Inclined, Micro-Porous Surfaces Simulating Microelectronic Devices," ASME-EEP-Vol. 19-2, Adv. in Electronics Packaging, 2, 2055-2063.

Chang, Y. and You, S. 1996. Heater Orientation Effects on Pool Boiling of Micro-Porous-Enhanced Surfaces in Saturated FC-72, J. Heat Transfer, 118(11), 937-943.

R. Danielson, R., Tousignat, L., & Bar-Cohen, A. 1987. Saturated Pool Boiling Characteristics of Commercially Available Perfluorinated Inert liquids, Proc. ASME/JSME Thermal Engineering Joint Conference, 3, 419-430.

El-Genk, M. & Guo, Z. 1993. Transient Boiling from Inclined and Downward-facing Surfaces in a Saturated Pool, Int. J. Refrigeration, 6, 414-422.

Ivey, H. & Morris, J. 1962. On the Relevance of the Vapor-Liquid Exchange Mechanism for Subcooled Boiling Heat Transfer at High Pressure, Rept. *AEEW-R137*, UKAEA, Winfrith, England.

Kutateladze, S. 1961. Boiling Heat Transfer, Int. J. Heat and Mass Transfer,.4, 31-45.

Lee, T., Mahalingam, M. & Normington, P. 1992. Subcooled Pool Boiling Critical Heat Flux in Dielectric Liquid Mixtures, *ASME-HTD-* 206, 55-62.

Liu, Z., Lin, W., Lee, D. & Peng, X. 2001. Pool Boiling of FC-72 and HFE-7100, J. Heat Transfer, 123(4) 399-400.

Nishikawa, K., Fujita, Y., Uchida, S. & Ohta, H. 1983. Effect of Heating Surface Orientation on Nucleate Boiling Heat Transfer, ASME-JSME Thermal Engineering Joint Conference, 1, 129-136.

Rainey, K. 1999. Pool Boiling Heat Transfer from Plain and Microporous Finned Surfaces in Saturated FC-72, Master's Thesis, University of Taxes at Arlington, 1999,

Rini, D., Chen, R. and Chow, L. 2001. Bubble Behavior and Heat Transfer Mechanisms in FC-Pool Boiling, J. Experimental Heat Transfer, 14(1), 27 – 44.

Watwe, A., Bar-Cohen, A., & McNeil, A. 1997. Combined Pressure and Subcooling Effects on Pool Boiling From a PPGA Chip Package, " J. Electronics Packaging, 119, 95 – 105.

Thermal Challenges in Next Generation Electronic Systems, Joshi & Garimella (eds)

Single-phase and flow boiling cooling in multiple miniature curvilinear channels

R. Scott Downing
Hamilton-Sunstrand, Division of United Techn. Rockford, IL, USA
T: (825) 394-2811, F: (815) 226-2614, E: sdowning@snds.com

Gunol Kojasoy
Dept. of Mechanical Engineering, University of Wisconsin-Milwaukee, WI, USA
T: (414) 229-5639, F: (414) 229-6958, E: kojasoy@uwm.edu

Keywords: electronic cooling, two-phase flow, cold plate technology

ABSTRACT: This paper describes an experimental investigation of heat transfer in miniature helical flow passages under single- and boiling two-phase flow conditions. A series of test sections were designed and fabricated to cover a range of feature dimensions and hydraulic parameters of a typical miniature helical cold plate for electronic cooling. As a baseline for the curved channel experiments, additional linear channel test sections were also built and experimented by using R 134a as a working fluid. Results are discussed in terms of experimental data, and some experimentally based simple correlations are offered for the prediction of heat transfer coefficient in miniature helical channels under single- and two-phase flow boiling conditions.

1 INTRODUCTION

The miniaturization of electronic circuits, higher power levels per chip and increased packaging densities have driven the trend in electronics packaging toward higher heat fluxes. Bar-Cohen (1983) provided an excellent review of chip characteristics and heat dissipations of state of the art digital processors and concluded that heat dissipation of future digital processing chips could reach the 100 W/cm^2 level. Mackowski (1991) predicts that the most challenging problems associated with current and near term military electronic devices lie with power controllers, with steady-state heat fluxes in the 100 W/cm^2 and 200 W/cm^2 range. In fact, pulsed transient heat loads of short duration (approximately a second or less) may exceed 400 W/cm^2.

In order to accommodate such high heat fluxes, high performance cooling techniques are required to keep junction temperatures low for acceptable electronics reliability. High performance cooling must provide low junction coolant thermal resistance and the ability to absorb high heat fluxes. Additionally, for aerospace applications the cooling technology must be tolerant to adverse g-fields and be able to operate in all attitudes (gravitational orientations). The cooling system should have a minimum impact to the host aircraft, i.e., low coolant pumping power, low initial and operating costs, and be lightweight and compact.

The present study was motivated by the need for accurate experimental data in miniature curved flow passages in the design of small high performance heat exchangers devised at Hamilton-Sundstrand. This novel high-performance cooling approach is aimed at providing extremely low thermal resistance (approx. 0.08 K/W-cm^2), and operating up to very high burnout heat fluxes of greater than 200W/cm^2, anticipated in future aerospace power electronics. A unique feature of such a cold plate is the use of forced convective flow boiling in multiple miniature curvilinear flow passages providing enhanced heat transfer and higher critical heat fluxes as compared to flow in straight flow channels. A key design feature of this technology is that the centrifugal forces developed in curvilinear flow should be tolerant to external g-fields. This cold plate technology is acronymed the spiral flow compact high-intensity cooler (SFCHIC).

Among the several curved flow passages considered in the conceptual design stage, helically coiled channels were chosen since: 1) The effects of curvature can best be experimentally investigated with helical passages of constant curvature rather than with a spiral flow passage which has a continuously changing curvature, 2) Helically coiled channels provide an advantage in applications where the total liquid stream must be vaporized or nearly vaporized, 3) The centrifugal forces have been shown to maintain wet wall conditions to higher qualities than would persist in a straight tube, 4) The heat transfer coefficient is also improved due to the secondary flows and wet wall conditions, and finally, 5) Helical flow geometry would be a logical choice for two phase flow heat transfers in adverse g-field or in micro gravity.

An extensive literature search conducted by Downing (1999) covered the topics of single- and two-phase flow and heat transfer in helical and small channels. This literature provided an understanding of basic principles and predictive tools relevant to predicting flow phenomena, pressure drop and heat transfer in small curved channels needed to design a SFCHIC type cold plate. Specifically, boiling two-phase flow phenomena in small curved channels have not been studied and extrapolating correlation in the current literature base must be done cautiously. It is therefore, important to experimentally confirm that correlations developed outside the current published data range could be extrapolated, or modifications to those experimentally based predictive methods need to be established.

Since the fundamentals of single-phase flow in straight channels provide the basis for understanding and modeling two-phase flow phenomena, experimentation was done in miniature straight and curved channels in single-phase liquid flow as the basis for curved channels and two-phase flows. The philosophy of experimentation was to build a database and experimentally based correlation methodology starting with simple flow conditions and modifying models to account for the complicating effects. For example, the first set of tests was performed with single-phase, straight channel flow under adiabatic flow conditions. The effects of curvilinear and boiling two-phase flow were then experimentally evaluated and correlated as modifying factors to the straight channel single-phase results. This paper presents the results of the single- and two-phase pressure drop experiments in both straight and helical channels.

2 EXPERIMENTAL SETUP

2.1 *Flow loop*

The flow loop schematic is shown in Figure 1. The loop contains a pump, two accumulators--one bladder type and the other gravity type--two condensers, a pre-heater, a heater and a heat-flux amplifier. The entire loop is mounted on a two-level uni-strut frame structure except for the pump pallet. The loop is fully instrumented with absolute pressure and differential pressure transducers, thermocouples and high accuracy Coriolis type flow meters.

The bladder accumulator provided sufficient back pressure from a nitrogen tank for adequate NPSH to pump the Refrigerant 134a. The refrigerant passed through a throttling valve and a series of flow meters before entering into an electrical preheater. This automatic PID controlled electrical heater regulated the downstream temperature to user specified set point. For the experiments reported here the preheaters provided a subcooled liquid at the test section. The preheated refrigerant then passed through an expansion valve of appropriate size before being led to the test section. These expansion valves finely adjusted the flow and pressure to the test section.

Much of the experimentation was carried out under two-phase flow conditions. Because the single-phase portion before the onset of boiling must be analytically removed from two-phase flow data, it was preferred to minimize the inlet subcooling. The expansion valve, after the preheater was also used for fine tuning of the inlet subcooling to the test section. After regulating the degree of subcooling the refrigerant entered the test section. Heat is applied to the test section by a large pyramid-type copper heat flux amplifier.

2.2 *Test sections and instrumentation*

Eight test section patterns were designed to cover a wide range of geometric variables and be compatible with the heat flux amplifier. Two test patterns were combined into one test block, and four test blocks were fabricated. As a baseline to testing of curvilinear flow passages, two test patterns were designed to have straight flow channels. The linear patterns designated L1 and L2, were build into one test block. By design, the linear channels are similar triangles of different size. The flow passage area of the larger passage, L2, is 50% larger than the smaller pattern, L1. The number of parallel flow paths was selected so that the hydraulic resistance of each pattern was similar. As shown in Figure 2, the smaller passage pattern has 20 parallel paths and the larger passage pattern has 12 parallel flow channels. The spacing of the paths was selected to spread the selected number of channels over the width of the heat flux amplifier footprint.

Each of the other three test blocks contained two helical flow test patterns. The test pattern pairing was randomized to provide blocking of uncontrolled factors in both fabrication and experimentation stages. The three helical test blocks contained patterns B and F, patterns C and D, and patterns E and G. The nominal geometric characteristics of these helical patterns as well as the linear patterns, L1 and L2, are summarized in Table 1. As an example for the helical sections, a drawing of patterns C and D is illustrated in Figure 3.

The final operation in fabricating the test block was to attach thermocouples into both the heated and unheated surfaces of the block. High quality type-T thermocouples with 32 gauge wire attached with thermally conductive epoxy. Surface slots, 0.030 inches deep by 0.0040 inches wide were milled into the test blocks for the thermocouples.

Each test pattern was instrumented by eight thermocouples to evaluate the temperature distributions along each test section. Detailed thermocouple locations for linear and helical test patterns can be found in Downing (1999). Six thermocouples were located on the heated side whereas the other two on the unheated side of the test block.

In addition to the test section instrumentation, instrumentation in the flow loop that was directly tied into the data acquisition included the following:

- Six pressure transducers
- Eighteen T-type thermocouples -- 10 on heat flux amplifier and 8 on flow loop,
- Coriolis flow meter, and
- Turbine flow motor

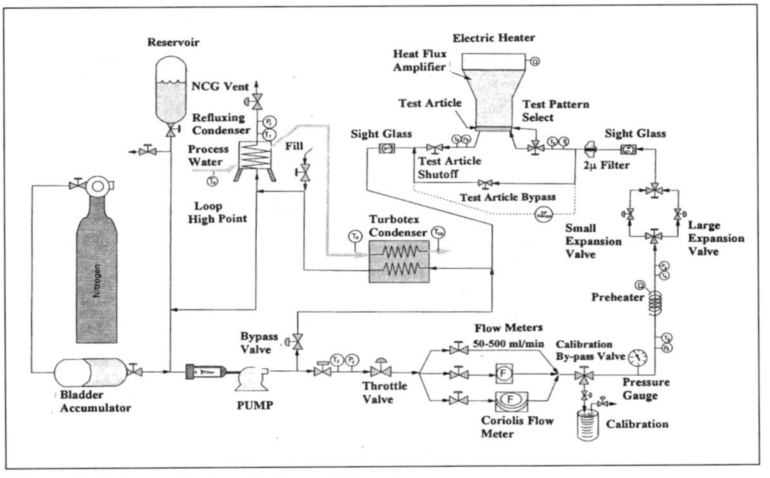

Figure 1. Schematic description of flow loop

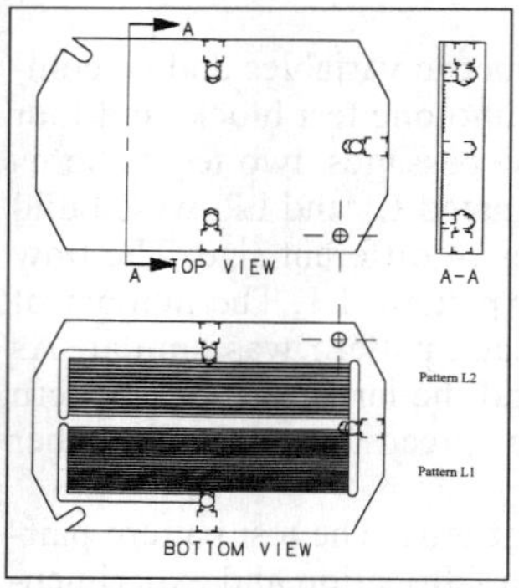

Figure 2. Linear test section. Patterns, L1 and L2

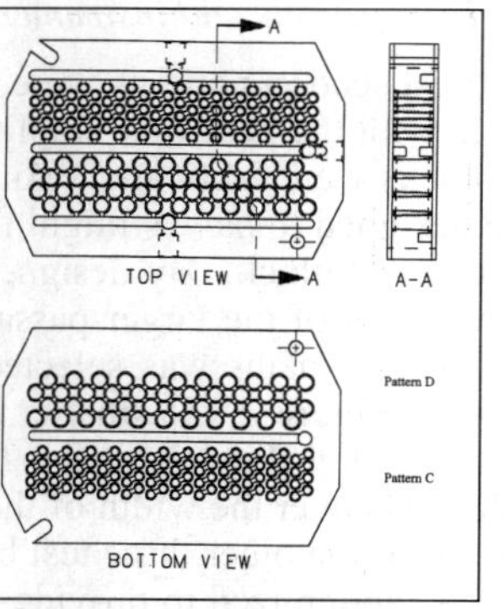

Figure 3. Hellical test section, C and D

Table 1. Test section channel geometries

Text Pattern	No. of Helices In Parallel	No. of Helices in Series	Helix Diameter [inch]	Hydraulic Diameter of Channel [inch]	Inverse of Curvature (D/d)
L1	20	1	----	0.0092	----
L2	12	1	----	0.0149	----
B	24	4	0.11	0.0204	5.5
C	12	8	0.11	0.0334	3.6
D	22	2	0.2055	0.0219	10.7
E	11	4	0.2055	0.0347	6.6
F	20	1	0.3125	0.0288	10.8
G	10	2	0.3125	0.0733	13.4

3 RESULT AND DISCUSSIONS

The purpose of the present study is to uniquely relate the single-phase flow in helical channels to flow behavior in linear channels providing a basis for a two-phase flow in helical channels. The experimentally based modeling process builds from the basic geometries (straight channels) to helical passages and from single- to two-phase flows. For example, curved tube heat transfer coefficients are modeled with a heat transfer enhancement factor on the established straight tube heat transfer coefficients. In a similar approach, two-phase multipliers are used to relate two-phase heat transfer performance to single-phase performance.

3.1 *Single-phase flow*

The single-phase flow data in straight channels, L1 and L2, provided insight into the heat transfer coefficient behavior and the transition from laminar to turbulent flow. A striking feature of the present single-phase flow data was the trend of a "blended" transition from laminar to turbulent flow. This is a unique characteristic of channel shapes of flow regions bounded by wall at acute angles This was observed by Eckert and Irvine (1969), in their extensive experimentation on isosceles-triangular passages of several apex angles.

Single-phase heat transfer results from the linear test sections compared with the Dittus-Boelter correlation. The data is reasonably well represented by the correlation in the turbulent flow region. Although the over-all data shows a blended transition from the laminar to turbulent flow, at the lower Reynolds number the measured heat transfer coefficient falls about 30% under the predicted value. The lower than predicted results at low Reynolds number is due to the fact that the Dittus-Boelter correlation was based on the fully-developed turbulent flow.

Following the single-phase flow in circular helical channel studies of Mori and Nakayama (1965 & 1967), Nag and Chakrabarty (1982), Clark (1974), Kalb and Seider (1972, 1974 & 1983),

Xin and Ebadian, (1997), and Kadombi et al. (1986), either of the following form of correlations was adapted for the present study of the single-phase flow in miniature non-circular flow channels.

$$Nu = f(\mathrm{Re}, d/D, \mathrm{Pr}) \quad \text{or} \quad Nu = f(De, d/D, \mathrm{Pr}) \tag{1}$$

Here De is the Dean number defined by

$$De = \mathrm{Re}(d/D)^{1/2} \tag{2}$$

In these equations Re is the Reynolds number based on the hydraulic diameter, d, of the flow channel, and D is the helical coil diameter. In fact, d/D represents the relative curvature of helix. It is to be noted here that for tubes of modest curvatures (d/D < 0.1), the Dean number fully characterizes dynamic similarities of fluid flow through a helix. However, for tubes of very tight curvature there is an additional dependence on the relative curvature d/D above that contained in the definition of the Dean number (Kalb and Seider, 1974).

The above referenced predictive correlations were compared with the experimental data of the present study covering the helical test patterns of B, C, D, E, F and G. Although there was a large scatter observed between the measurements of heat transfer coefficient and the predictions, a unique feature of this comparison was the fact that the majority of experimental data fell well above the predictions for the range of experimental conditions. In view of this discrepancy, a new correlation was sought in the form of Eq. (1). The single-phase heat transfer in the miniature helical channels of the present study is best correlated by

$$Nu = 0.386(De)^{0.417}(d/D)^{-0.5}\mathrm{Pr}^{0.4} \tag{3}$$

for range of 3.6 < D/d < 13.4; 800 < De < 4000 and Pr $\cong$ 3.6

Figure 4 shows the ratio of the measured to predicted Nusselt number at each data point. From this figure it is evident that the experimental data is predicted by Eq. (1) within $\mp$20% margin. Considering the wide variations in curvatures, the Dean number and the Reynolds number, the data scatter is quite reasonable. With high heat transfer coefficients and large surface area densities of the present test articles the overall conductance between the test block and the fluid is high. In heat exchanger terminology the helical test articles have large NTU's and will perform at effectiveness of near unity. The outlet temperature closely approaches the test block temperature, and consequently, significant changes in the heat transfer coefficient results in only minor changes in the fluid outlet temperature. Unfortunately, this situation made it difficult to precisely measure the heat transfer coefficient. Limitations on the experimental accuracy by the poor resolution due to the high efficiency of the test articles as heat exchangers are the primary contributor to the scatter observed on Figure 4. However, in spite of this situation, a general correlation for single-phase flow in helical coils predicts the data with a reasonable accuracy of $\mp$20% range.

3.2 *Two-phase flow*

Recently, Kandlikar (1990, 1998) has published a series of flow boiling correlations that have gained wide acceptance for straight channels. It was observed that this method provides a better representation of the present data, generally within $\mp$40%. However, the method actually over predicts most of the present data, primarily for the higher heat flux data range. The correlation has convective and boiling terms, providing a way to tune the correlation for the relative strength of both mechanisms. The following equation, closely following the method for Kandlikar, was found to predict 90% of the linear channel two-phase flow boiling data to within $\mp$30%.

$$h_{tp}/h_{csp} = \left[1.18C_o^{-0.2}Fr_{csp}^{0.3} + 850B_o^{0.7}\right] \tag{4}$$

where h_{csp} is defined as the convective heat transfer coefficient for the single-phase liquid flow as given by the Dittius-Boelter correlation, Fr_{csp} is the single-phase liquid Froude number, B_o is the boiling number and C_o is the convection number. These are defined as follows

$$Fr_{csp} \equiv G^2/\rho_f^2 gD; \quad B_o \equiv \dot{q}''/Gh_f; \text{ and } \quad C_o \equiv \left(1-x/x\right)^{0.8}\left|\rho_g/\rho_f\right|^{0.5} \tag{5}$$

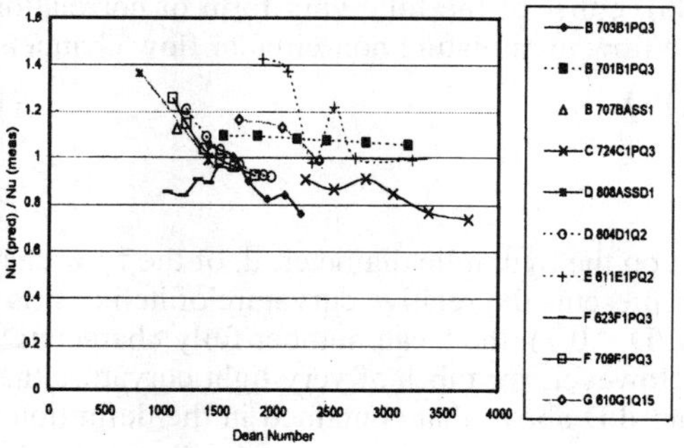

Figure 4. The ratio of predicted to measured Nusselt numbers for single-phase heat transfer helical channels

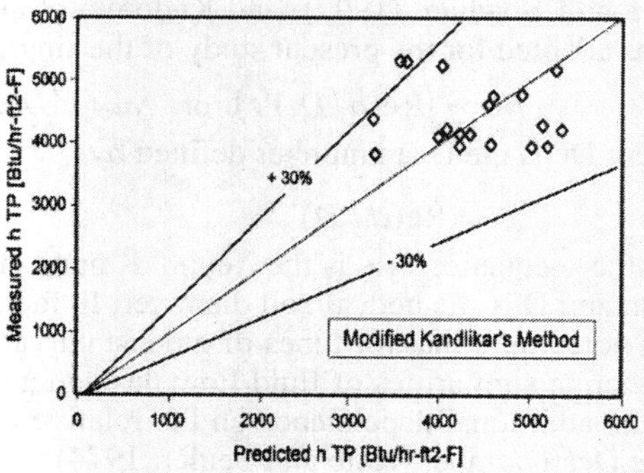

figure 5. two-phase flow in linear channels

The present experimental conditions that were used to arrive at the modified Kandlikar's correlation are: $1000 < Re < 6000$, $0.2 < \chi_{tt} < 5$, $5 < \dot{q}'' < 25$ and $50 < G < 500$.

The modified Kandlikar correlation is compared on Figure 5 to the measured heat transfer for linear test article, L2. The predictions are good within $\mp 30\%$ margin. Figure 6 shows the measured flow boiling heat transfer coefficient as a function of the mass flow rates for three data sets. Each data set corresponds to a constant heat input, 250 W, 350 W and 450 W, respectively. This figure illustrates a very interesting and important trend. At a given heat flux, the integrated heat transfer coefficient is nearly independent of mass flow and, therefore, quality. This would support Wambsganss' (1992) conclusion that at high heat flux levels the heat transfer performance is independent of mass flux.

The observed effective heat transfer coefficients in helical channels investigated here were nominally 50 to 100% higher than would be obtained in straight channels operating under the same mass fluxes and qualities. Most of this enhancement is expected to be attributed to the curvilinear flow effects. The controlling heat transfer mechanisms can be determined by inspection of the heat transfer performance as a function of various variables. Convective heat transfer component will increase at higher qualities, but at some point the nucleation degrades into film evaporation. A peak in heat transfer at some intermediate quality (and flow) was observed for most runs at constant heat input. It is assumed that this peak is at the location where the convective heat transfer enhancement is overtaken by the loss in nucleation. Figure 7 is representative of this trend. The peak in heat transfer coefficient occurs at outlet qualities between 35% to 55% range, depending on the heat flux level. This phenomena is directly linked to the optimum value of Martinelli parameter.

For almost all of the experimental data, including all helical coil test patterns, the highest measured heat transfer performance occurred at inverse Martinelli parameter of 0.5 to 1 range. The measured heat transfer coefficient against the inverse Martinelli parameter for the helical channel test pattern F is shown in Figure 8. The same trend was also observed in parametric studies of the heat transfer versus average liquid Dean number.

A phenomenologically based correlation that reflects the convective and nucleate boiling trends was found to be

$$h_{tp} = 0.15 h_{csp} \left(1/\chi_{tt}\right)^{-0.3} + 730 \dot{q}''^{0.45} \qquad (6)$$

where h_{csp} is the single-phase heat transfer coefficient of the liquid, flowing alone, in the coil. The curved channel single-phase algorithm which is presented in the preceding section is used to evaluate h_{csp}. A scatter plot showing the correlative error including all test patterns is shown in Figure 9. Considering the measurement difficulties and the uncertainties associated with the experimentation, an error margin of $\mp 15\%$ observed in this figure encouraging. However, it should be pointed out that although the approach taken in this study was phenomenologically based on the superposition principle, the results should be used with caution when applied to those cases out of the present experimental range. This is particularly true for the use of proposed correlations for other fluids since the fluid properties have not been scaled properly other than by the Prandl number for the single-phase flow and the Martinelli parameter for the two-phase flow.

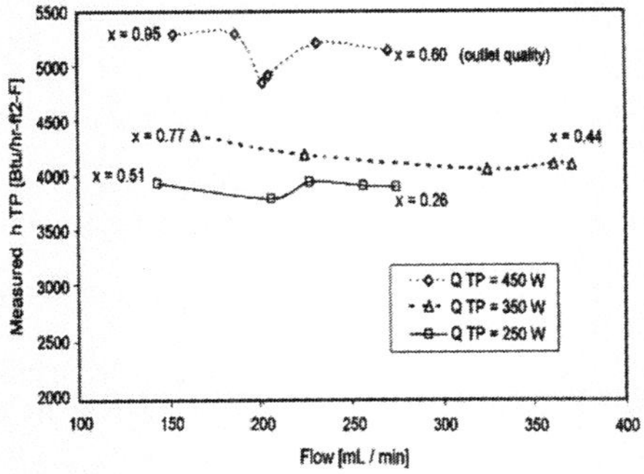

Figure 6. Measured heat transfer coefficient as a function of mass flow and heat input

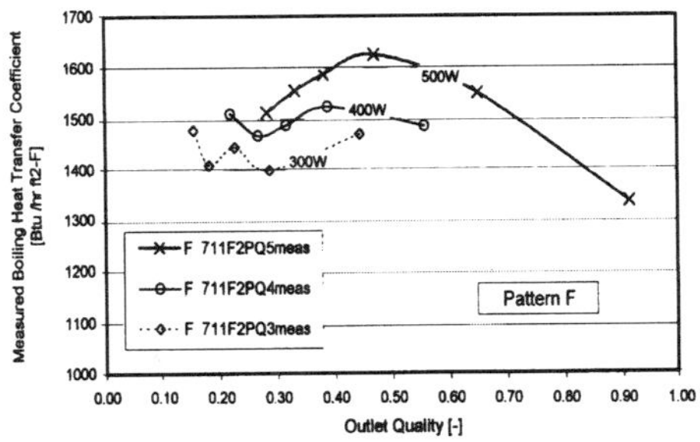

Figure 7. Measured two-phase flow heat transfer coefficient against outlet quality

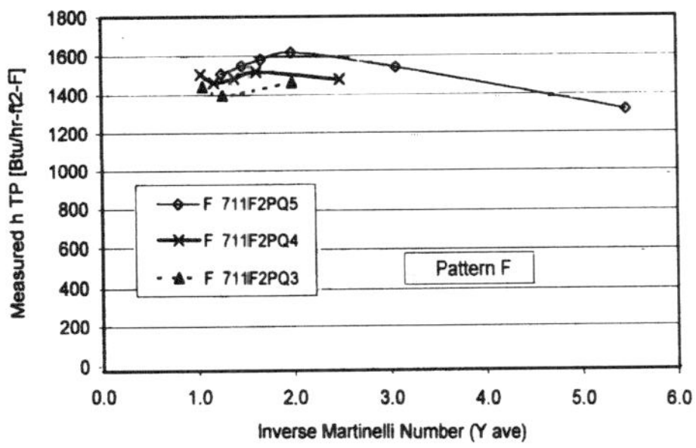

Figure 8. Measured two-phase flow heat transfer coefficient against inverse Martinelli parameter

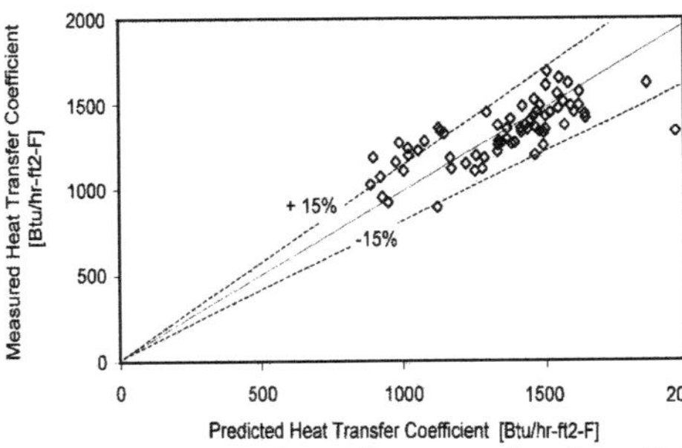

Figure 9. Comparison of measured and predicted heat transfer coefficient for helical channel

4 SUMMARY AND CONCLUSIONS

Experiments were conducted to develop heat transfer correlations valid for the geometrics and flow conditions expected in miniature helical cold plate technology. The heat transfer results for single- and two-phase flow in eight different test sections containing two linear and six curvilinear passages of different sizes and curvatures were correlated. Using both single-and boiling two-phase flow in these eight test sections, curvature effects on the heat transfer performance of helically coiled miniature flow passages were investigated in terms of heat transfer enhancements.

Experimental investigation showed that for the range of experimental conditions the Dittius-Boelter correlation represented the single-phase flow data of linear test sections reasonably well. However, a significant enhancement in heat transfer was observed in curvilinear test sections. A new experimentally based correlation was developed in terms of the Dean number.

Experiments under two-phase flow conditions in linear test sections revealed the fact that at a given heat flux, the integrated two-phase flow heat transfer coefficient is nearly independent of mass flux and therefore the quality. Although the higher heat fluxes have higher heat transfer coefficients, at high heat flux levels, the heat transfer performance is independent of the mixture mass flux. A slightly modified version of the Kandlikar's correlation represented the data in linear miniature channels. Influence of the heat flux and the Martinelli number were studied for the helical miniature channels, and a correlation based on the superposition principle was developed in terms of the nucleate boiling and the two-phase flow convective components.

REFERENCES

Bar-Cohen, A.. 1983. Thermal Design of Immersion Cooling Modules for Electronic Components, Journal of Heat Transfer Engineering, Hemisphere Press.

Clark, J. W. G. 1974. Heat Transfer in Coiled Tubes, ICI Report, CL-B, Research Note, JWG C/74/9.

Dean, W. R. 1927. Notes on the Motion of Fluid in a Curved Pipe, The London, Edinburgh and Dublin Philosophical Magazine and Journal of Science, Vol. 15, pp. 673-691.

Dean, W. R. 1928. The Steamline Motion of Fluid in a Curved Pipe, The London, Edinburgh and Dublin Philosophical Magazine and Journal of Science, Vol. 45, pp. 39-45.

Downing, R. S. 1999. Analytical and Experimental Study of Single and Two-phase Cooling in Miniature Straight and Helical Channels, Ph. D. Dissertation, Department of Mechanical Engineering, University of Wisconsin- Milwaukee.

Eckert, E. R. C. & Irvine, Jr. T. F. 1960. Pressure Drop and Heat Transfer in a Duct with Triangular Section, ASME Journal of Heat Transfer.

Kadombi, V. et al. 1986. Heat Transfer and Pressure Drop in a Helically Coiled Rectangular Duct, ASME, Journal of Heat Transfer, 83-WA/HT-1.

Kalb, C. E. & Seider, J. D. 1974. Fully-Developed Flow Heat Transfer in Curved Circular Tubes with Uniform Wall Temperature, AICHE J. Vol. 20, pp. 340-346.

Kalb, C. E. & Seider J. D. 1983. Entrance Region Heat Transfer in Uniform Wall-Temperature Helical Coil with Transition from Turbulent to Laminar Flow, Int. J. Heat and Mass Transfer, Vol. 26, pp. 23-32.

Kandlikar, S. G. 1990. A General Correlation for Saturated Two-Phase Flow Boiling Heat Transfer Inside Horizontal and Vertical Tubes, ASME Journal of Heat Transfer, Vol. 112, pp. 215-228.

Kandlikar, S. G. 1998. Heat Transfer Characteristics in Partial Boiling, Fully-Developed Boiling and Significant Void Flow Regions of Subcooled Flow Boiling, ASME Journal of Heat Transfer, Vol. 120, pp. 355-401.

Mackowski, M. J. 1991. Requirements for High Heat Flux Cooling for Future Avionics Systems, SAE Paper No. 912104.

Mori, Y. & Nakayama, W. 1965. Study on Forced Convection Heat Transfer in Curved Pipes--Part I. Laminar Region, Int. J. Heat and Mass Transfer, Vol. 8, pp. 67-82.

Mori, Y. & Nakayama, W. 1967 Study on Forced Convection Heat Transfer in Curved Pipes--Part II. Turbulent Region, Int. J. Heat and Mass Transfer, Vol. 10, pp. 37-59.

Nag, P. K et al. 1982. Turbulent Forced Convection Heat Transfer in Helically Coiled Tubes with Uniform Wall Temperature, J. Inst. Engrs. vol. 63, pp. 123-134.

Wambsganns, M. W et al. 1993. Boiling Heat Transfer in a Horizontal Small-Diameter Tube, Trans. ASME, Journal of Heat Transfer, Vol. 115, pp. 963-972.

Xin, R. C. & Ebadian, M. A. 1997. The Effects of Prandtl Numbers on Local and Average Convective Heat Transfer Characteristics in Helical Coils, ASME, Journal of Heat Transfer, Vol. 119, pp. 467-473.

Thermal Challenges in Next Generation Electronic Systems, Joshi & Garimella (eds)
© 2002 Millpress, Rotterdam, ISBN 90-77017-03-8

Microscale heat transfer to pressurized water at ultra-high power density

Robert H. Leyse
INZ, Inc., Sun Valley, ID, USA

Keywords: heat transfer, heat flux, phase change, thermal management

EXTENDED ABSTRACT: Recent advances in thermal management and characterization schemes have led to intense challenges to those who are charged with producing the forecasts and analysis of future trends in this area. Leyse, 2001, has reported phase change heat transfer phenomena that were discovered with a micron-sized heat transfer element operating in subcooled (20 degrees C) degassed, demineralized water over a wide pressure range (200 – 6000 PSIA) at ultra-high heat fluxes up to 3500 W/cm^2. The platinum heat transfer element (diameter 7.5 microns, length 1.14 mm) is installed within a one cm^3 stainless steel chamber. Sealed electrical terminals penetrate the chamber to effect direct current heating of the platinum element. Pressure is applied pneumatically. The adiabatic heating rate of the element is 6 degrees C per microsecond at 3700 W/cm^2; therefore, response is essentially instantaneous for the procedure described herein. The direct current voltage and current are measured from which the power and the resistance (temperature) are determined.

The following procedure applies: 1.) Pressurize the water-filled stainless steel chamber to 6000 PSIA 2.) Apply power at 3000 W/cm^2. 3.) Maintain constant heat flux as pressure is smoothly reduced from 6000 PSIA to 200 PSIA over a period of 20 seconds. Record voltage, amperage, and pressure at 0.1 second intervals.

Heat transfer phenomena thus discovered: See Fig. 1 for a plot of one set of data that serves as an introduction to the transitions in phase change heat transfer. Note the six distinct regions including the Transitions A, B, C, and D in phase change heat transfer at the ultra high heat flux, 2930 W/cm^2. For reference purposes, Fig.1 includes key thermodynamic properties of saturated water: Note the plot of saturation temperature of water as a function of pressure (the very lowest plot). Note also the vertical dashed line at the critical pressure of about 3200 PSIA. The critical point of water is at the intersection of these plots.

Recall that the procedure is to pressurize the 1 cm^3 chamber of degassed demineralized water to about 6000 PSIA and then apply full power. Power is maintained substantially constant as pressure is smoothly reduced over a period of about 20 seconds. The bulk water temperature is about 20 degrees C at the start of the run, and this increases by less than 5 degrees C during the course of the run (for example at 3360 W/cm^2 the power is 0.9 W or about 0.22 calories/sec and in 20 seconds the bulk water temperature increase for 1 gram of water is 20 X 0.22 = 4.4 degrees C).

Moving to Fig. 2, note the plots of four runs. In this discussion, we trace the curves from high pressure to low pressure as the pressure is reduced. In Run 1, the lowest heat flux, 2630 W/cm^2, the curve is smooth from the starting point of 6350 PSIA and 348 degrees C until power is turned off at 193 PSIA and 230 degrees C. Initially, the temperature of the platinum heat transfer element increases gradually as pressure is reduced and at the critical pressure, the temperature peaks at 358 degrees C. At this point the temperature is less than the saturation temperature and begins to smoothly decrease. At about 2430 PSIA the temperature is equal to the saturation temperature and then increasingly exceeds the saturation temperature as pressure is reduced until power is turned off at 193 PSIA and an element temperature of 230 degrees C.

In Run 2, 2930 W/cm^2, note the gradual increase in the element temperature as pressure is reduced. However, at 3700 PSIA there is a surprising upward step from 368 degrees C to 594 degrees C (arbitrarily called *Transition A* by the PI). The temperature then smoothly and relatively steeply increases to 695 degrees C at 3240 PSIA (*Transition B*). At this point, the temperature "turns around" and smoothly decreases to 482 degrees C at 2440 PSIA. (*Transition C*). Next there is a downward step to 352 degrees C at 2425 PSIA (*Transition D*). Then, the temperature smoothly decreases along a path that is narrowly above the path for the run at 2630 W/cm^2. This run ends at 272 PSIA and 253 degrees C.

At the next highest power, Run 3, 3010 W/cm^2, note the positions of the Transitions that are defined above. *In comparison with 2930 W/cm^2*, note: 1.) The higher starting temperature, 2.) The higher pressure for the start of Transition A, 3.) The lower temperature for the termination of Transition A, 4.) The higher temperature of the turnaround at the end of Transition B, 5.) The turnaround in the vicinity of the critical pressure, 6.) The termination of Transition C at a somewhat higher temperature and lower pressure, 7.) The termination of Transition D at a somewhat lower pressure, 8.) The smooth decrease in temperature is narrowly above the first two tracks (for most of the track) and terminates at 195 PSIA and 239 degrees C.

At the highest power, Run 4, 3360 W/cm^2, Transition A begins somewhat greater than the starting pressure. The turnaround at 3435 PSIA is well above the critical pressure. Transition D begins at a substantially higher temperature, but the pressure is in the range of the prior run. Transition D terminates at a slightly lower pressure than the prior runs, but the temperature is in the same range. The smooth decrease in temperature is narrowly above the first two tracks (for most of the track) and terminates at 261 PSIA and 245 degrees C.

Experts in the field will recognize that this collection of data discloses new territory in the field of phase change heat transfer. The ultra high heat fluxes that have been sustained over the wide pressure range are amazing in themselves. The Transitions A, B, C and D are truly new discoveries. Recently Schubert et al. (2001) have reported, "Applications in thermal and chemical process engineering may benefit from microstructure devices such as micro heat exchangers, micromixers, and microstructure reactors when high heat and/or mass transfer rates are mandatory." Herein we have addressed *ultra* high heat and mass transfer rates!

REFERENCES

Leyse, R., 2001, "Microscale Heat Transfer to Subcooled Water," *Proceedings of the Microgravity Transport Processes in Fluid, Thermal, Biological and Material Sciences II*, United Engineering Foundation, Inc.

Schubert, K., Brandner, J., Fichtner, M., Linder, G., Schygulla, U., and Wenka., A., 2001, "Microstructure Devices for Applications in Thermal and Chemical Process," *Microscale Thermophysical Engineering*, Vol. 5, pp. 17-39.

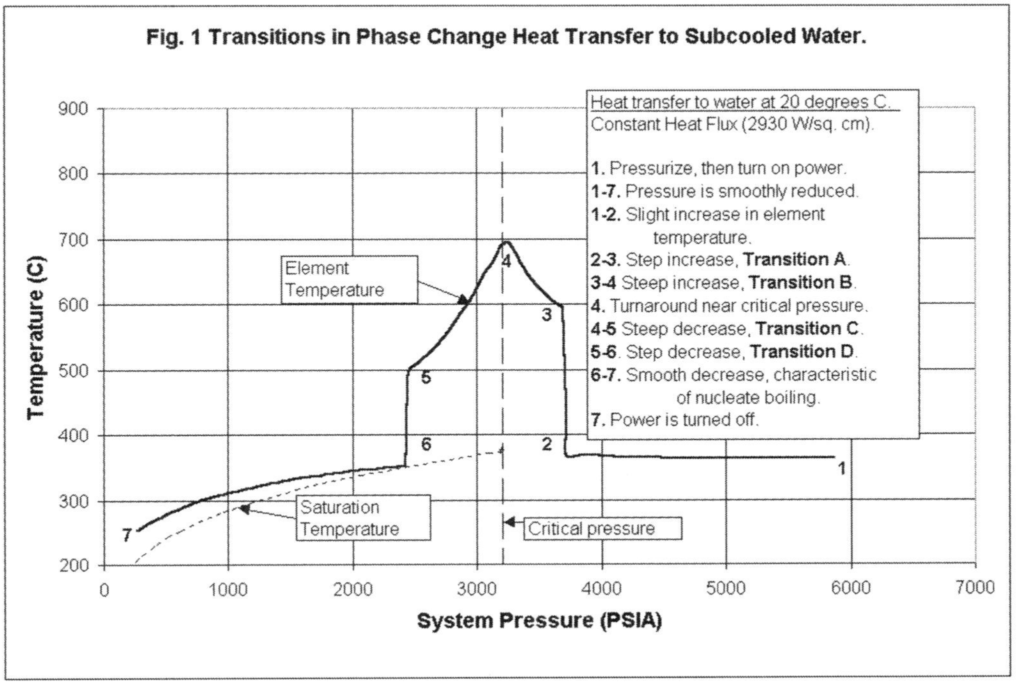
Fig. 1 Transitions in Phase Change Heat Transfer to Subcooled Water.
Heat transfer to water at 20 degrees C.
Constant Heat Flux (2930 W/sq. cm).
1. Pressurize, then turn on power.
1-7. Pressure is smoothly reduced.
1-2. Slight increase in element temperature.
2-3. Step increase, Transition A.
3-4 Steep increase, Transition B.
4. Turnaround near critical pressure.
4-5 Steep decrease, Transition C.
5-6. Step decrease, Transition D.
6-7. Smooth decrease, characteristic of nucleate boiling.
7. Power is turned off.
Element Temperature
Saturation Temperature
Critical pressure
Temperature (C)
System Pressure (PSIA)

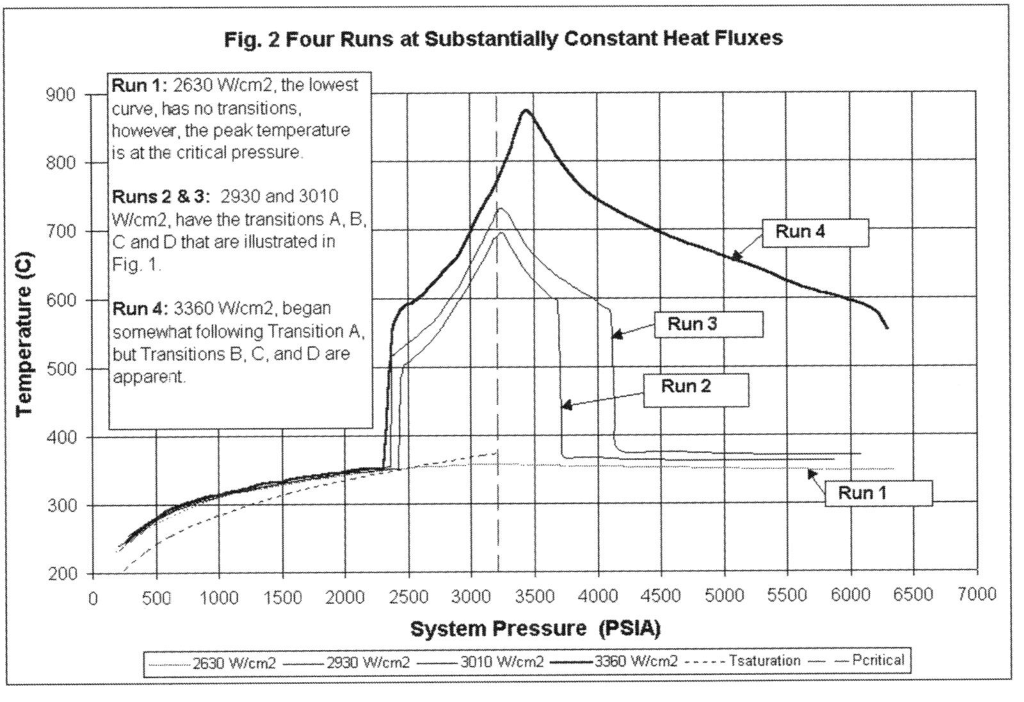
Fig. 2 Four Runs at Substantially Constant Heat Fluxes
Run 1: 2630 W/cm2, the lowest curve, has no transitions, however, the peak temperature is at the critical pressure.
Runs 2 & 3: 2930 and 3010 W/cm2, have the transitions A, B, C and D that are illustrated in Fig. 1.
Run 4: 3360 W/cm2, began somewhat following Transition A, but Transitions B, C, and D are apparent.
Run 4
Run 3
Run 2
Run 1
Temperature (C)
System Pressure (PSIA)
2630 W/cm2
2930 W/cm2
3010 W/cm2
3360 W/cm2
Tsaturation
Pcritical

Transport in thermal management materials and across interfaces

Thermal Challenges in Next Generation Electronic Systems, Joshi & Garimella (eds)
© 2002 Millpress, Rotterdam, ISBN 90-77017-03-8

Performance and testing of thermal interface materials

J.P. Gwinn* & R.L. Webb**
Dept. of Mechanical Engineering, The Pennsylvania State University, University Park, PA
*jpg119@psu.edu; **r5w@psu.edu

Keywords: thermal interface material, thermal interface resistance, contact resistance, electronics cooling

ABSTRACT: The increasing power and reduced die size of CPUs used in computers increases a need for significantly improved thermal interface materials (TIM). The TIM is used to reduce contact resistance at the CPU-heat sink interface. This work provides a state-of-the-art assessment on "thermal interface materials," including fundamentals, materials used, their performance, and how interface resistance is measured. The performance of new commercial interface materials is given, as well as discussion of the advantages and disadvantages of different materials. The report notes that the recommended interface test method does not necessarily duplicate the installation and operating conditions in an actual computer installation. Recommendations are presented on the design and operation of an apparatus intended to simulate actual computer installation conditions. The innovative Penn State "low melting point alloy" thermal interface material is described and compared to other commercial materials.

1 NOMENCLATURE

A: Surface area of the electronic component through which heat is emitted
k_{TIM}: Bulk thermal conductivity of the thermal interface material
Q": Power per unit area of the heat-emitting component
R_{int}: Thermal interface resistance (commonly presented with units of [K-in^2/W])
$R_{contact}$: Thermal contact resistance
$R_{contact1}$: Thermal contact resistance between the TIM and the CPU surfaces
$R_{contact2}$: Contact resistance between the TIM and the heat sink baseplate
R_{cond}: Conduction resistance across the thickness of the TIM
t: Thickness of the thermal interface material
T_{cpu}: CPU or other electronic package surface temperature
T_{hs}: Heat sink baseplate temperature
TIM Thermal interface material

2 INTRODUCTION

Thermal interface material (TIM) is used to minimize the contact thermal resistance between the central processor unit (CPU) and the heat sink. The plan area of central processor units CPUs has been steadily decreasing. Further, the power dissipation has been increasing. For example, die size of the Pentium 2 was 25.4 mm square and the power dissipation was about 33 W. However, the Pentium 4 has 12 x 12.5 mm die size and will dissipate up to 80 W. Intel has attached a 1.5 mm thick, 31 mm square copper heat spreader to the CPU die in attempt to increase the die physical strength and reduce the heat flux. However, our CFD calculations show that the heat flux is very non-uniform and the heat spreader has the same thermal resistance as for uniform heat flux on a 16 mm square area. Comparing the Pentium 4 (16 mm square effective die size) to the Pentium 2, we see that the heat flux has increased a factor of 6.1. For the Pentium 2, the TIM accounted for approximately 15% of the total allowable thermal resistance. If this same TIM were used on the Pentium 4, the TIM would use 80% of the allowed total thermal resistance. Further, increased CPU power is expected. By 2003, the CPU power should be approaching 120 W, which will further increase the interface heat flux. Because of this, significant work has been done to develop new higher performance thermal interface materials. In addition, the desire for increased yield and reliability requires lower operating temperatures (Viswanath et al., 2000).

The path of heat removal from an electronic package such as a CPU, involves conduction across the interface of the CPU case surface, through a thermal interface material (TIM), into a heat sink and then convection to the environment. There are many TIM products on the market today and it is difficult to perceive which are better than others. Data reported by a TIM manufacturer may not be reproduced by others, or validated in an actual computer installation. The first section of this paper presents a fundamental understanding of contact resistance, and the second section discusses options to reduce the interface resistance. The third section presents a classification of high performance thermal interface materials, with their advantages and disadvantages, and an assessment of recent advancements in interface material technology. Section four follows with a discussion of the measurement of thermal interface resistance and possible errors associated with the resistance measurement

3 BACKGROUND

Thermal interface resistance is a measure of how well heat is transferred across the interface of two mating rigid surfaces, such as a CPU and a heat sink baseplate (Figure 1a).

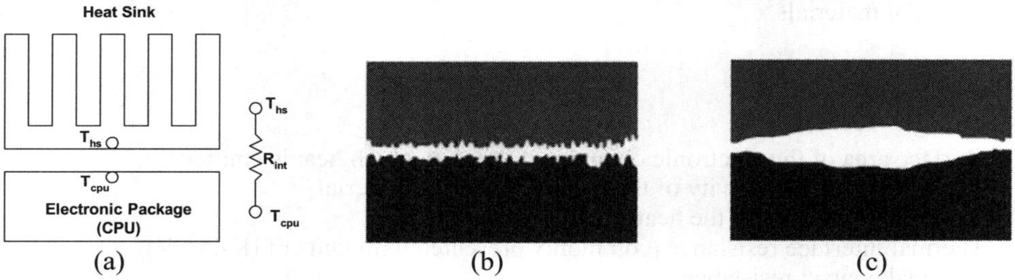

Figure 1: (a) Definition of thermal interface resistance, R_{int}, (b) heat sink and CPU surface roughness, and (c) poor heat sink and CPU surface flatness (deSorgo, 1996)

In the case of the CPU-heat sink interface (the focus of the present work), the interface resistance (per unit area) is defined as follows:

$$R_{\text{int}} = \frac{T_{cpu} - T_{hs}}{Q''} \tag{1}$$

where, T_{cpu} is the CPU (or other electronic package) surface temperature, T_{hs} is the heat sink baseplate temperature, Q" is the power per unit area of the heat-emitting component. The R_{int} is commonly presented with units of (K-in^2/W). As can be seen in the equation for R_{int}, the lowest possible interface resistance is reached when the heat sink temperature approaches that of the CPU.

The CPU and heatsink surfaces being joined contain a combination of surface roughness and surface non-flatness as shown in Figure 1b and 1c. The micro roughness is superimposed on a macroscopic non-planar in the form of a concave, convex or wavy surface, resulting in as much as 99% of the interface surface area being separated by air filled gaps (Figure 2a) unless a thermal interface material is used. Theoretically, the contacting surfaces will only contact one another at no more than three points. Because air is a very poor thermal conductor (k_{air} = 0.026 W/m-K at room temperature), it acts as a thermal barrier preventing efficient heat transfer across the interface. If no TIM is used to enhance heat transfer across the interface, the total interface resistance, R_{int} is due to contact resistance, $R_{contact}$ as shown in Figure 2a. Contact resistance, $R_{contact}$, is composed of two parallel resistances: (1) conductive resistance at the points where the mating surfaces contact and (2) conductive and/or radiative resistance across the gaps of non-contacting area. The existence of TIM material adds a thermal conduction resistance across the material of thickness "t."

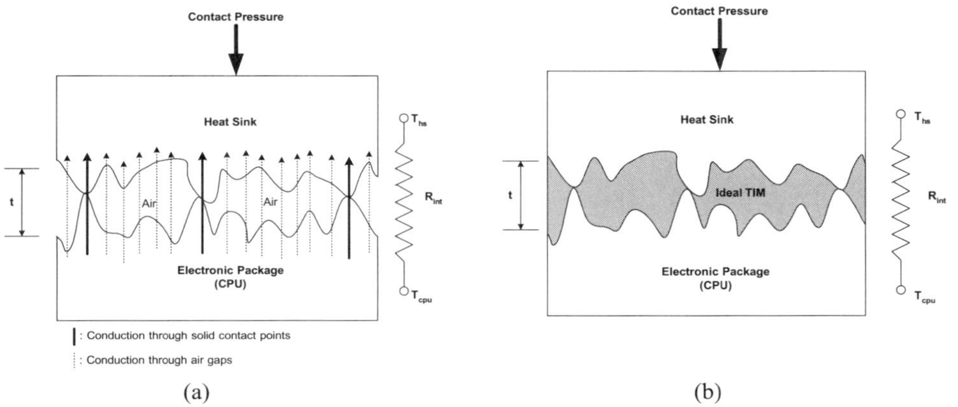

Figure 2: Exploded view of thermal interface when (a) no TIM is used. Note R_{int} is composed only of contact resistance, $R_{contact}$ and (b) when an ideal TIM is used.

4 METHODS TO REDUCE CONTACT RESISTANCE

For solids of high thermal conductivity, the contact resistance may be reduced by the following two methods: (1) increasing the area of contact spots, accomplished by (a) increasing contact pressure which will "flatten" the peaks of the micro roughness, and deflecting the mating surfaces to remove any non-flatness, or (b) reducing the roughness of the surfaces before the interface is formed by grinding the surfaces to remove non-flatness and buffing the surface to remove micro-roughness; and (2) using a TIM of high thermal conductivity that can conform to the imperfect surface features of the mating surfaces (Incropera & DeWitt, 1996).

Load constraints on electronic components and circuit boards make it unfeasible to use high contact pressure. Manufacturing highly finished surfaces is not practical due to cost constraints. Therefore, the practical alternative is to use a thermal interface material applied at a moderate contact pressure. Intel specifies 25 lbf. maximum heat sink clip force can be applied for the P4 processor in the 423-pin package, which results in 16.78 psi. on the 31mm^2 integrated heat spreader (Intel Corp., 2001), and their reference design heat sink clip applies 7 to 23 lbf or 5 to 16 psi. (Intel Corp., 2000). Any interstitial substance that fills the gap between contacting surfaces, and whose thermal conductivity exceeds that of air, will decrease the contact resistance (Incropera & DeWitt, 1996). The ideal thermal interface material (Figure 2b) would have the following characteristics:

1. High thermal conductivity
2. Easily deformed by small contact pressure to contact all uneven areas of both mating surfaces, including surface pores, eliminating $R_{contact}$.
3. Minimal thickness.
4. Would not leak out of the interface.
5. Would maintain performance indefinitely.
6. Non-toxic.
7. Manufacturing friendly (easy to apply and remove).

Ideal thermal interface material has not yet been discovered, but several options are available to provide reduced interface resistance. However, an actual thermal interface material will not be able to completely fill the cavities, and small air pockets will likely remain on both sides of the interface material (See Figure 3a).

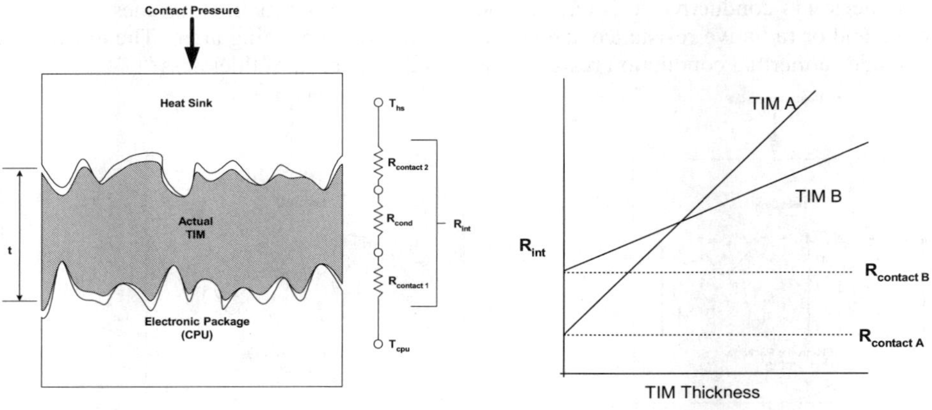

(a) (b)

Figure 3: (a) exploded view of thermal interface with actual TIM and (b) example plot of Rint vs. t for two TIMs (Chu et al., 1997)

As shown in Figure 3a, the interface resistance is actually composed of three resistances in series. The total interface resistance is expressed as follows:

$$R_{\text{int}} = R_{contact1} + R_{cond} + R_{contact2} \tag{2}$$

$$R_{cond} = \frac{t}{k_{TIM} A} \tag{3}$$

$R_{contact1}$ is the contact resistance between the TIM and the CPU surfaces, R_{cond} is the conduction resistance across the thickness of the TIM, $R_{contact2}$ is the contact resistance between the TIM and the heat sink baseplate, *A* is the surface area of the electronic component across which heat transfer occurs, k_{TIM} is the TIM bulk thermal conductivity and t is the thickness of the thermal interface material (Chu et al. 1997).

One can determine the total contact resistance ($R_{contact} = R_{contact1} + R_{contact2}$) by measuring thermal resistance data for a range of TIM thickness and plotting R_{int} vs. t. The R_{int} vs. t data exhibits the following linear relationship:

$$R_{\text{int}}(t) = R_{contact} + \frac{t}{k_{TIM} A} \tag{4}$$

The slope is equal to the inverse of the product of the TIM conductivity, k_{TIM} and the area A. The intercept of the R_{int} axis is equal to $R_{contact}$ (Chu et al. 1997). An example of the plot that will be obtained for two different interface materials is presented in Figure 3b. The $R_{contact1}$ and $R_{contact2}$ resistances may differ, because each of the two contact surfaces (the silicon die and the heat sink base) may differ in roughness or flatness.

Increasing the contact pressure will reduce R_{int}. Pressure increases the actual contact area and reduces the amount of air remaining at the interface, thus reducing contact resistances, $R_{contact1}$ and $R_{contact2}$, on either side of the interface material. Increasing contact pressure can also decrease the conduction resistance, R_{cond}, of the interface material itself, by reducing its thickness, t. The pressure required to reduce the thermal interface resistance of a given TIM is a function of its compressibility and surface properties as well as the surface properties of the mating surfaces (Latham, 1996).

Interface resistance is application specific, because of its dependence on mating surface properties. R_{int} data measured in a given test facility will not necessarily be applicable to different installations of a TIM, because the surface roughness and flatness of the mating surfaces will likely differ in each application. Thus, the TIM resistance values reported by TIM manufacturers may not be applicable to an actual computer installation. Further, it is not clear that the use of a heat sink clip will provide uniform pressure on the interface material, and estimation of interface resistance based only on the conductivity of the TIM will likely result in an under prediction of R_{int}.

5 POSSIBLE THERMAL INTERFACE MATERIALS

Incropera & DeWitt (1996) lists two classes of materials that are well suited for the reduction of R_{int} - thermal greases and soft metals. In addition to these, commercially available TIMs have been developed that make use of phase change materials (PCM), thermally conductive elastomers, and thermally conductive adhesive tapes. Most thermal interface materials, except pure metals, are composed of a polymer matrix filled with thermally conductive particles (i.e. metals, ceramics) (Viswanath et al. 2000). Whatever the TIM used, CPU manufacturers recommend that it be sized and positioned on the heat sink baseplate to insure the entire processor die area is covered to compensate for heat sink package attach mis-alignment (Intel Corp, 2000).

Many TIMs available are not suitable for use for CPU cooling. Soft metals require unacceptably high contact pressures to conform to the mating surfaces. Tests at Penn State showed that 350 psi contact pressure is required to produce a readable impression of a U.S. 25¢ coin in an ingot of pure indium. Thermally conductive elastomer pads also require unacceptably high contact pressures (300-500 psi) to fill interstitial voids (deSorgo, 1996). Thermally conductive compounds (elastomer paste filled with thermally conductive particles) are an improvement on elastomer pads that can be applied in paste form and cured in-situ to conform to mating surface, allowing metal to metal contact and minimal thickness. This, however, only yields moderate thermal performance (Chromerics, 2001). Thermally conductive adhesive tapes have poor thermal performance (test data show only slightly better performance than a dry joint) (deSorgo, 1996).

Feasible TIM candidates for CPU cooling are categorized below with advantages and disadvantages. Commercially available candidates are listed in Table 1 with manufacturer published interface resistance data and the test contact pressure. The reader should note that performance data are published at various contact pressures. The highest performance (most expensive) TIM listed in Table 1 is Arctic Silver thermal grease with suspended silver conductive particles (R_{int}= 0.0028 K-in^2/W at 12 psi). The next highest performance thermal grease (and more cost effective) is ShinEtsu G751 (R_{int}= 0.0157 K-in^2/W at unspecified contact pressure). The highest performance phase change material listed is Thermax HF-60110-BT; a low-melting-temperature-alloy PCM concept (R_{int} = 0.007 K-in^2/W @ 3-100 psi). However, Penn State tests have shown that the performance of the Thermax TIM significantly degraded under thermal cycling conditions. The performance of phase change materials varies by a factor of 4.5 (not including Thermax HF-60110-BT). The highest performance, phase change material reported is PowerStrate 60(AG) and 51 (AF) (R_{int}= 0.022 K-in^2/W at 10 psi.).

Table 1: CPU Cooling Commercial TIM Candidates

	Manufact. Spec.			Notes
	R_{int} (K-in^2/W)	Pressure (psi)	Conductivity (W/m-K)	
Thermal Grease				
Arctic Silver	0.0028	12	>7.5	Silver particle filler
ShinEtsu G751	0.0157	NP	4.5	Silicon fluid w/ alumina powder filler
ShinEtsu G765	0.06	NP	2.89	Silicon fluid w/ alumina powder filler
ShinEtsu G750	0.0257	NP	3.5	Silicon fluid w/ alumina powder filler
Berquist TIC-7500	0.035	50	7.5	- -
Phase Change Material (PCM)				
Thermax HF-60110-BT	0.007	3-100	20	LMTA Concept (failed in thermal cycling tests)
PowerStrate 60 (AG)	0.022	10	3	Al foil substrate
PowerStrate 51 (AF)	0.022	10	3	Al foil substrate
Chromerics T725	0.03	50	0.7	No Substrate
Orcus inc. FSF 52	0.03	4.5	NP	No Substrate
Chromerics T454	0.04	50	0.6	No Substrate
Thermagon T-pcm 905c	0.049	10	0.7	No Substrate
Bergquist HiFlow	0.05	40*	1	No substrate (printable)
Bergquist 200U	0.04	10	1.6	No Substrate
Bergquist HiFlow 225-U	0.08	10	1	No Substrate
Bergquist HiFlow 225-UT	0.09	10	0.7	No Substrate (natural tack)
Chomerics T443	0.1	50	1	Fiberglass substrate
Thermally Conductive Elastomer				
Chromerics Therm-A-Form T642	NP	NP	1.2	Thermally conductive compound (elastomer matrix w/)

*: Assumed since based on ASTM E1530
NP: Not Published

5.1 *High performance TIM categories.*

THERMAL GREASE is composed of a thermally conductive filler dispersed in silicone or a hydrocarbon oil to form a paste (de Sorgo, 1996). Recent improvements have been made in the application of thermal greases by developing "printable" greases, which can be screen-printed onto a heat sink baseplate at a specified thickness. Thermal interface resistance values for several commercially available thermal grease products are presented in Table 1.

- Advantages: High thermal performance at small contact pressures. Silicon-based thermal greases are attractive on the basis of their ability to fill the interstices and eliminate the interstitial air with a material whose thermal conductivity is much higher than that of air (Incropera & DeWitt, 1996). For example, the thermal conductivity of ShinEtsu G751 Thermal Grease, k = 4.5 W/m-K, which is 172 times greater than that of air at room temperature. As shown in Table 1, high performance thermal greases have higher effective thermal conductivity than other classes of TIMs (Viswanath et al., 2000). Interface pressure causes excess grease to flow from the interface and form a thin thermal joint and surfaces to contact. It is probable that all of the air pockets are not eliminated. They do not require curing (Viswanath et al., 2000). Intel recommends the use of thermal grease, and uses it in a thermal solution reference design. (Intel Corp, 2000).
- Disadvantages: Thermal greases are not manufacturing friendly in that they are messy and difficult to apply and remove. Excess grease that flows out of joint must be removed to prevent contamination and possible electrical shorts. Joint integrity must be maintained with mounting

hardware. Heat sink removal and re-installation requires cleaning the grease from the interface and a new application. Grease does not provide electrical insulation between contacting surfaces. Grease can be "pumped out" of the interface under thermal cycling because of the relative motion of the mating surfaces under thermal expansion and contractions. Grease joints can dry out with time, resulting in increased thermal resistance. Grease degradation rate is a strong function of operating temperature and number of thermal cycles (Viswanath et al., 2000).

PHASE CHANGE MATERIALS (PCMs) are made of a mixture of suspended particles of high thermal conductivity, such as fine particles of a metal oxide and a base material. Although phase change of the base material is implied, this is not the case. These materials do not actually change phase, but their viscosity diminishes so that they flow (Ollila, 1999). The base material can be a natural material such as fully refined paraffin, a polymer, a co-polymer, or a combination of all of these (Rauch, 2000). The base material is solid at low temperatures (i.e. room temperature), but behaves much like grease after reaching the "phase change" temperature (50-90°C for typical products available). Phase change materials can be supplied in the form of compound only, or in composite form in a specified thickness applied to a substrate. Thermal interface resistance values for several commercially available PCM products are given in Table 1.

- Advantages: High published thermal performance at moderate contact pressures. Material flows throughout the thermal joint to fill air gaps and provide minimum thickness by allowing the mating surfaces to come into contact. When the joint becomes thin, the viscosity prevents "pump out" from mechanical flexure of the interface surfaces (Viswanath et al., 2000). It handles easily (like a thermal pad) for installation.
- Disadvantages: Requires compressive force to bring materials together and cause the TIM to flow. The desired "phase change" property limits the choice of polymer and filler combinations; limiting the thermal performance of these materials (Viswanath et al., 2000). Published thermal performance is difficult to reproduce in independent test facilities, as found in work done at Penn State (Gwinn et al., 2002) and Showa Denko K.K. (Showa, 2000). Intel Corp. discourages the use of PCMs, because of low thermal performance (in contrast to Table 1) as determined by their thermal characterization tests. Intel also states that the strong adhesive bond formed by PCMs between the CPU package and heat sink may cause damage to the motherboard and CPU, because of high loads measured during their mechanical shock or drop tests (Intel Corp, 2000).

5.2 *New interface material technology*

This section assesses new concepts that have recently been advanced including the following technologies: corrugated copper substrate with thermal grease, gel TIM and low-melting-temperature-alloy (LMTA).

Corrugated copper substrate with thermal grease is described by Raiszadeh, 2001. This TIM construction has high lateral conductivity and distributes the heat in the lateral direction more evenly. This would be of value if the heat flux were non-uniform in the interface region. It is constructed of high performance thermal grease and a proprietary corrugated pure copper substrate. A thermal interface resistance of 0.021 K-in^2/W was found for this new TIM construction at 50 psi contact pressure. Work is being done to find a corrugation geometry that will reduce the required contact pressure. This construction achieved a 40% lower interface resistance than measured for the best thermal grease alone. This does not seem to be a desirable TIM because its reliance on thermal grease.

A gel-like TIM has been developed by Thermoset of Lord Chemical Products (Wells et al., 1998) that has thermal performance similar to high thermal conductivity greases, but does not leak out of the interface. Test data was obtained on a "custom built apparatus" and no error analysis is presented. This TIM can be dispensed or printed onto the surface of a heat sink baseplate. While this product may eliminate the problem of the TIM leaking out of the interface, the problem of messy assembly and disassembly remains.

Low-Melting-Point-Alloy (LMTA) TIM is a solder alloy that has a low melting temperature. It can reduce thermal interface resistance of a metal-to-metal joint by a factor of 10 (Cook & Token,

1984). The LMTA is selected to melt before reaching the maximum CPU operating temperature at which point, it flows to fill the air gaps and significantly reduces contact resistance. The composition of the LMTA is chosen for a desired melting point. LMTA interface materials make use of the high conductivity of metals while eliminating the problems mentioned above associated with implementing soft metals. Since the alloy is liquid at the critical operating condition, low contact pressure is required to force the liquid LMTA into the interstitial voids. Another advantage is the ease in disassembly of the joint after cool down since many LMTAs do not adhere to surfaces unless they are clean and free of oxidation (Cook & Token, 1984). This is a very promising thermal interface solution. Lehmann & Davidson (2001) have shown that the use of LMTA decreases the thermal resistance of a bare interface of two oxidized aluminum surfaces having 22 μm flatness deviation by 50% in the solid phase and 95% in the liquid phase. Their results show that the liquid phase performance is competitive with phase change materials of Table 1 (R_{int} = 0.03 in^2-K/W) with 3.4 psi contact pressure) (Penn State has developed a TIM consisting LMTA tinned on a 50μm thick copper substrate. Tests show that the performance is comparable to that of high performance grease. Further detail is provided in (Gwinn & Webb, 2002).

6 MEASUREMENT OF THERMAL INTERFACE RESISTANCE

ASTM D-5470 describes a method to measure R_{int}. The apparatus consists of two copper blocks, across which heat is transferred. The TIM is inserted at the interface between the two blocks. Electric heat is applied at the bottom of one block. Two or more temperature sensors (i.e. thermocouples) are installed in each copper block. The surface temperature of each block is measured from the temperature gradient in each block.

The analysis should be performed at steady state because transient methods are not suitable for analyzing materials with temperature-sensitive thermal and mechanical properties such as phase change materials (PCMs) (Tzeng et al., 2000). Also, transient systems such as the Laser flash method and the "Fitch" calorimetery method require additional material properties including density and heat capacity that introduce additional sources of error.

The heat flux must accurately be measured to resolve interface resistance. Typical constructions use electric cartridge heaters installed in bocks of high thermal conductivity material. A finite distance below the heaters is required for the heat flux to become uniform. It is extremely important that the temperature sensors are installed in a *uniform* heat flux location and that guard heaters or thermal insulation be used to limit heat losses (Tzeng et al., 2000). It is suggested that non-uniform heat flux may be a problem with some testers. Further, the heat flux measurement must be calibrated to ensure accuracy and allow estimation of the error range.

Mating surface temperatures must be accurately measured to determine R_{int}. Factors effecting surface temperature measurement accuracy are (1) temperature sensor location uncertainty, (2) resolution of temperature sensor, (3) distortion of temperature gradient by temperature sensors, (4) uniformity of heat flux at the temperature sensor (Gwinn et al., 2002). Contact pressure must be applied and measured to represent the end use of a TIM in a thermal solution. A consistent and high-quality compression mechanism is needed to insure the repeatability and sensitivity of test (Tzeng et al. 2000).

Surface flatness and roughness affects the performance of a TIM. The results obtained in an "interface test apparatus" will be specific to the surface roughness and flatness of the block faces used in the apparatus (Chu et al., 1997). Surface cleanliness may also affect the repeatability of results. Further commentary on these factors is given below.

The cleaning method and test sequence is important because thermal interface materials such as thermal grease will contaminate the test surfaces. An unavoidable residue of different performance TIM left in minute surface roughness pores contaminates the test results of another TIM by combining the performance of the second TIM with that of the first. Data are presented showing a calibrated CPU internal diode temperature is reduced by more than 10% after applying and "removing" thermal grease because of thermal compound remaining in the microscopic valleys of the interfacing surfaces (House, 2001). A cleaning method is recommended that includes a boiling bath to open the pores of the interface surfaces so that TIM residue can be removed (Arctic Silver LLC, 2001). One must take care that the temperature of the bath does not change the flatness of the heater block surfaces before the next test.

We emphasize that the results obtained in a "thermal interface tester" may not duplicate the installation and operating conditions of an actual CPU-heat sink installation (Eaton, 2001). For example, the surface roughness and flatness of the CPU baseplate and the heat sink may be different from that existing in the thermal tester. Further, the heat sink clip used in a computer application may not place uniform pressure on the interface. If the results are to be meaningfully interpreted for actual CPU-heat sink applications, tests should seek to duplicate the contact pressure, roughness and flatness of the CPU and heat sink interfaces. For this reason, the ability to swap out test blocks with different materials and surface finishes is desirable. However, this would require relatively expensive and time-consuming tests (Chu et al. 1997). There is currently no test method existing that can accurately define the thermal interface resistance of TIM in an application. Rank order comparisons are the most relevant tests that can be performed.

7 CONCLUSIONS

Current test methods do not reliably define the thermal resistance of a TIM in an actual CPU/heat sink application, because the thermal resistance is sensitive to flatness, roughness and contact pressure. Surface roughness, flatness, and contact pressure are critical factors in affecting the performance of a thermal interface material. There is a need to develop an acceptable cleaning method to remove TIM material from the surface pores of interfacing test surfaces. The highest performance interface materials available today are greases and phase change materials (metal and non-metal). The Penn State LMTA has performance competitive with the best thermal greases, but does not have their disadvantages.

8 REFERENCES

Arctic Silver LLC. 2001 Private correspondence with N. House, July 2001.

Chu, C.P., Solbrekken, G. L. & Chung, Y. D. 1997. Thermal Modeling of Grease-Type Interface Material in PPGA Application. *Proc. of 13th IEEE SEMI-THERM 1997,* Vol.1, pp. 57-63.

Cook, R.S., Token, K. H.& Calkins, R.L. 1984. A Novel Concept For Reducing Thermal Resistance. *J. Spacecraft*, Vol. 21, No. 1, pp. 122-124.

deSorgo, M. 1996. Thermal Interface Materials. *Electronics Cooling*, Vol.2, No.3.

Chromerics. 2001. Private correspondence with Dr. M. deSorgo, October 8, 2001.

Eaton, M. 2001. Thermal Resistance Testing. <http://www.orcusinc.com/test_therm.htm>

Gwinn, J.P., Saini, M. & Webb, R.L. 2002. Design of an Experimental Apparatus for Accurate Thermal Interface Material (TIM) Thermal Resistance Measurement. *Proc. of ITherm 2002.*

Gwinn, J.P. & Webb, R.L. 2002. Development of a Low Melting Temperature Alloy Thermal Interface Material (TIM). *Proc of ITherm 2002.*

House, N. 2001. Thermal Interface Basics. <http://www.arcticsilver.com/thermal_interface_basics.htm>

Incropera, F. P., DeWitt, D. P. (4th Ed.). 1996. *Fund of Heat and Mass Transfer.* John Wiley & Sons.

Intel Corp. 2000. Intel® Pentium® 4 Processor in the 423-pin Package Thermal Solution Functional Specification. # 249204-001 <http://developer.intel.com/design/pentium4/datashts/249198.htm>

Intel Corp.2001. Intel ® Pentium ® 4 Processor in the 423-pin Package at 1.30, 1.40, 1.50, 1.60, 1.70, 1.80, 1.90 and 2 GHz. # 249198-005, August 2001

Latham, C. 1996. Thermal Resistance of Interface Materials as a Function of Pressure. *Electronics Cooling Magazine*, Vol 2. No.3, p.35.

Lehman, Dr. G.L. & Davidson, D.A. 2001. Thermal Performance of Liquid Solder Joint Between Metal Faces. *Proc. of IPACK* 2001. Paper # IPACK2001-15890.

Ollila, T. 1999, Navigating the maze of thermal interface materials. *Electronic Products Fall 1999 Packaging Supp.*

Raiszadeh, F. 2001. Thermal-Mechanical Measurement and Analysis of an Advanced Thermal Interface Material Construction. *Proc. of 17th IEEE SEMI-THERM.* Number 01CH37189. pp 63-70.

Rauch, B. 2000. Understanding the Performance Characteristics of Phase-Change Thermal Interface Materials. *ITherm 2000 Proc.* Vol. 1, pp 42-47.

Showa Denko K.K. 2002. Private communication with Y. Furukawa, March 2000.

Tzeng, J. J-W., Webber, T. W. & Draqssowski, D. W. 2000. Technical Review on Thermal Conductivity Measurement Techniques for Thin Thermal Interfaces. *Proc. of 16th IEEE SEMI-THERM.* pp. 174-181.

Viswanath, R. Wakharkar, V., Watwe, A. and Lebonheur. V., 2000. Thermal Performance Challenges from Silicon to Systems. *Intel Technology J.*, Q3. *<http://developer.intel.com/technology/itj/q32000.htm>*

Wells, R., Sanders, Jason, Pecorari, Lisa & Hunadi, Dr. Ron. 1998. GELEASE^{TM}- A High Performance Alternative to Thermal Grease and Phase Change Materials. <http://www.thermoset.com/thermal_i.html>

Thermal Challenges in Next Generation Electronic Systems, Joshi & Garimella (eds)
 ISBN 90-77017-03-8

Spreading in the heat sink base: Phase change systems or solid metals??

Ioan Sauciuc, Greg Chrysler, Ravi Mahajan & Ravi Prasher
Intel Corporation, Chandler, AZ, USA

Keywords: spreading, phase change, boiling limit, heat pipe, electronics cooling

ABSTRACT : Presently, the micro-electronics industry needs thermal solutions that are able to dissipate high heat flux at low thermal resistance. The majority of OEMs (Original Equipment Manufacturers) within the microelectronics industry would like to achieve this by extending the application of air-cooling technologies since it implies minimal impact to the design of computer systems and is a cost effective solution space. Spreading resistance through the base of the heat sink is one major component of the total thermal resistance from the silicon junction to the local ambient, especially if larger volume heat sinks are to be used. Until now, most of the research has focused on using phase change systems (i.e. vapor chambers) for enhancing the spreading resistance of the heat sink base. Since no significant improvements have been achieved, there is a need to determine the envelope of the limitations for phase change-heat spreaders used in processor cooling, and to compare their performance against high thermal conductivity solid metals. Two simple models are presented to address the heat transfer limitations in phase change systems: the first one addresses wickless devices (i.e. thermo-siphons) and the second one deals with wick heat pipes. The first model predicts the heat transfer coefficients and thermal spreading resistance using Rohsenow's equation for nucleate pool boiling. The prediction results show that over the normal temperature ranges and heat fluxes encountered in cooling computer processors, the boiling heat transfer coefficient may achieve a maximum of 30000 W/m^{2}K for water. The second model proposed by Prasher et al. (2001) is a conduction-based model, which can predict the thermal resistance of the wick heat pipes/vapor chambers. In this conduction model, the wick is assumed to be completely saturated with the fluid and it is assumed that only conduction heat transfer takes through the wick. This model predicts performance limits based on the conduction resistance of the wick, whereas the first model predicts performance based on the boiling limits. Using these models, the ratio of phase change spreading resistance over solid metal spreading can be estimated. Contrary to common perceptions, this paper will show that there is a threshold envelope where a solid metal spreader (used as heat sink base) may have lower thermal spreading resistance than a copper/water-phase change system (i.e. vapor chambers or immersion boiling). It is also concluded that the phase change systems can be efficiently used only when they conduct heat to a remote heat exchanger.

NOMENCLATURE

A	Area [m^2]
$C_{s,f}$	roughness coefficient
P	input power [W]
Pr	Prandtl number
T	temperature [K]
cp	specific heat at [J/kgK] constant pressure
g	acceleration due to gravity [m/s^2]
h	heat transfer coefficient [W/ m^{2}K]

h_{fg} latent heat [J/kg]
k thermal conductivity [W/ m K]
n coefficient in equation 2
q heat flux[W/ m^2]

Greek symbols
μ dynamic viscosity [N s/ m^2]
ρ density [kg/ m^3]
σ surface tension [N/ m]

Subscripts
evap evaporator section
l liquid
v vapor

1 INTRODUCTION

The last two decades have seen a steady increase in microprocessor performance as silicon technology continued to scale in accordance with Moore's law. This increasing performance of the microprocessors is also associated with an increase in thermal design power and an increase in both average power density and local power density (commonly referred to as "hot spots"). Additionally as computer systems continue to become more compact and offer more functionality, other components adjacent to the processors (i.e. chipsets) have experienced an increase in power dissipation. The result of this is that the ambient temperatures local to the microprocessor heat sinks have increased and in some cases temperatures in excess of 45 ºC have been reported. Heat sink design for today's microprocessors must comprehend this increase in the local ambient temperature.

The thermal solutions must also ensure that the junction temperature of the processor (die temperature) does not exceed temperatures in the 90-110 ºC range, typically at the hot spots, to ensure device performance and reliability. These two temperature boundary constraints on the heat sink design, combined with an increasing trend in processor power have led to the realization that improvements are required in all aspects of the design of the cooling solution i.e. packaging, thermal interfaces and air cooled heat sinks. The focus of this paper will be on heat spreading in the heat sink base for air-cooled heat sinks.

The thermal performance of heat sinks can be improved in two areas: 1) by increasing the heat transfer coefficient and 2) by reducing the spreading resistance primarily in the heat sink base. The spreading resistance of conventional heat sink bases was not considered to be a major component of the total thermal budget in the past, especially when the heat source was centered on the heat sink base, and the heat sink base area was not appreciably larger than the heat source. However, the base spreading resistance needs to be further improved for air-cooling to be a viable option for cooling future micro-processors.

One of the approaches frequently adopted in the recent past has been the use of flat vapor chambers instead of conventional solid metal bases. Most of the emphasis in heat pipe and vapor chamber research was focused on the prediction of the maximum heat load and heat flux capacity (Faghri, 1995). In the case of vapor chambers, the general perception has been that phase change technologies have a higher effective thermal conductivity compared with solid metals. There has been little done to quantify the spreading resistance difference between phase change technologies and solid metals and in developing design envelopes to help choose between the two technologies.

In the early stages of thermal solution design and development a thermal engineer is more concerned with what-if scenarios such as whether it makes sense to use a heat pipe/vapor chamber instead of copper or other solid materials for better spreading of heat. At this stage, the exact design of the heat pipe/vapor chamber in terms of parameters such as heat carrying capacity and the burnout flux is not very important. Typically the time available to thermal designers to perform design sensitivity studies is very small and complete fluidic modeling for various configurations and scenarios is not possible. In addition, the commercially available thermal tools cannot perform the necessary complete fluidic modeling of heat pipes or vapor chambers. There is therefore a need for a tool that allows the thermal designers to perform these studies more quickly.

This paper will introduce two simplified modeling schemes for the prediction of "minimum" thermal resistance for current conventional vapor chambers. The first model assumes that the vapor chamber resistance is mainly due to evaporator resistance and all other resistances are neglected. The second model is a conduction model, which assumes equivalent thermal conductivities for the wick and vapor. Using these models and test data, it will be shown that there is threshold envelope where copper may have a lower spreading resistance than phase change technologies. The modeling of the solid metal conduction resistance can be accomplished through any number of commercially available software tools and so we will not discuss it in detail in this paper. In this paper, we will compare the thermal spreading performance of vapor chambers and solid metal for conventional rectangular heat sink base geometries.

2 SPREADING RESISTANCE: DEFINITION

Figure 1 shows the typical temperature distribution through a conventional heat sink base made of solid metal (aluminum or copper) with a uniform heat source. Due to the symmetry conditions with a centrally applied heat source, only half of the heat sink base is shown.

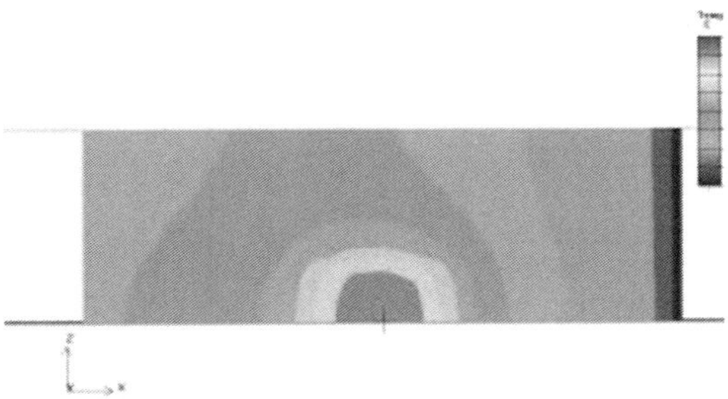

Figure 1. Typical temperature distribution over a conventional solid metal heat sink base

The spreading resistance of the base is defined as:

$$\theta_{spreading}\left[{}^{\circ}\mathrm{C}/\mathrm{W}\right]=\frac{T_{hot}-\overline{T}_{top}}{P} \qquad (1)$$

Where T_{hot} [°C] is the temperature of the hottest point on the base (usually into the center of the base) and T_{top}[°C] is the average temperature of the base top surface (see Figure 2). This definition can be used for both solid metal and conventional vapor chamber heat spreaders.

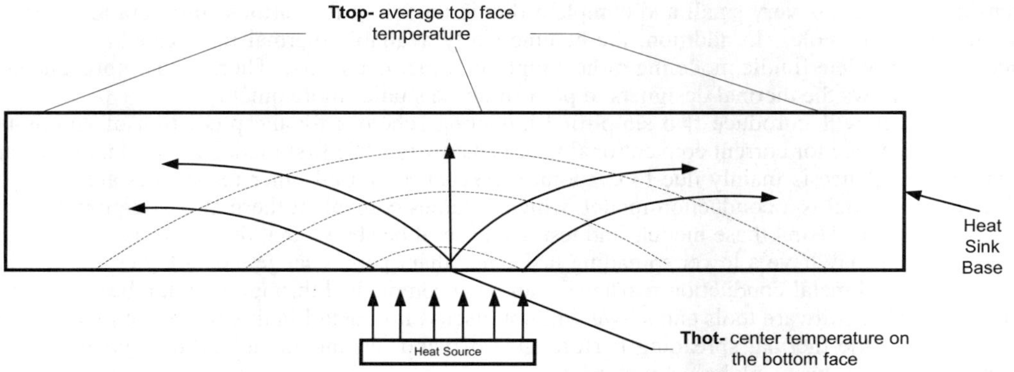

Figure 2. Temperature measurement location

3 HEAT TRANSFER LIMITATIONS FOR WICKLESS PHASE SYSTEMS (- BOILING LIMITATION-)

To predict the minimum thermal spreading resistance for a vapor chamber, a simplified model has been developed. This model assumes that the minimum vapor chamber spreading resistance($\theta_{spreading,vc}$) is approximately the evaporator (boiling) resistance $\theta_{evaporator}$ [Equation 2], where h_{evap} [W/m^{2}K] is the boiling heat transfer coefficient and A_{evap}[m^2] is the area of evaporator (heat input area)

$$\theta_{spreadingvc} \cong \theta_{\mathrm{evaporator}} \cong \frac{1}{h_{\mathrm{evap}} \times A_{evap}} \tag{2}$$

It is also assumed that the boiling regime inside the vapor chamber is nucleate pool boiling. This is a conservative assumption, since in reality the spreading resistance in a vapor chamber is larger than just the boiling resistance. If the spreading resistance calculated out of this simplified model is higher than that of a solid copper base, then a vapor chamber should not be used.

The boiling model is based on Rohsenow's (Incropera et al. 1990; Rohensaw et al. 1985) equation for nucleate pool boiling (equation 3) on a metal surface is given by:

$$q = \mu_l h_{fg} \left[\frac{g(\rho_l - \rho_g)}{\sigma} \right]^{1/2} \left[\frac{c_{p,l}}{C_{s,f} h_{fg} \Pr_l{}^n} \right]^3 \Delta T^3 \tag{3}$$

It can be seen that the heat flux is mainly a function of fluid properties {μ-dynamic viscosity [N s/ m^2]; h_{fg}-latent heat [J/kg]; g-acceleration due to gravity [m/s^2], ρ-density [kg/ m^3],σ-surface tension [N/ m], cp-specific heat at [J/kgK] and Prandtl Number}, surface properties and fluid/material combination { roughness Csf and n -material/fluid coefficients}, and the superheat required for boiling (ΔT). For the range of electronics cooling application, it is widely accepted that water/copper is the optimum combinations for vapor chamber fabrication. Applying this equation to such water/copper systems, the variation of the boiling heat transfer coefficient as a function of operating temperature is presented in Figure 3.

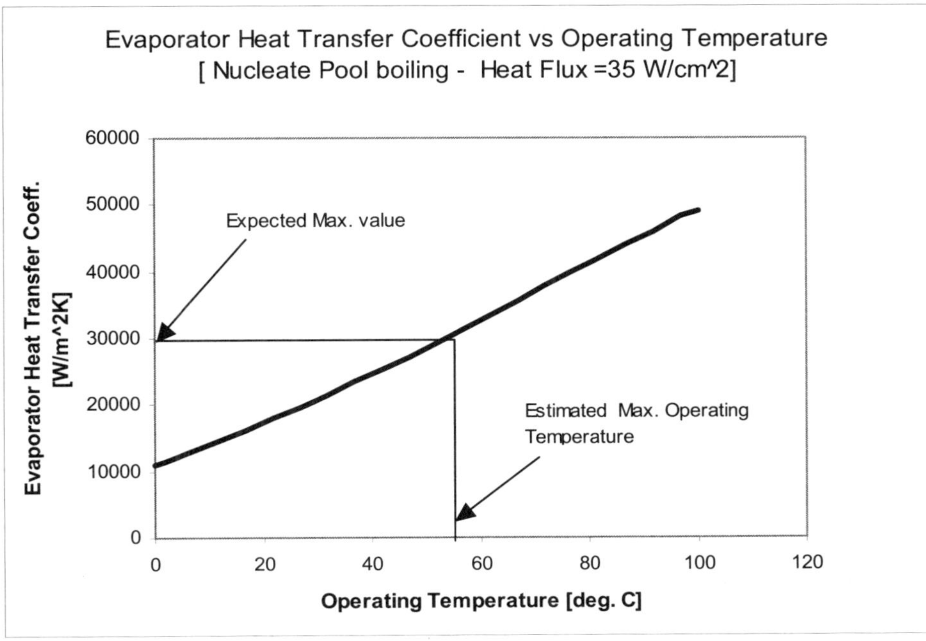

Figure 3 : Evaporator Heat Transfer Coefficient vs. Operating Temperature

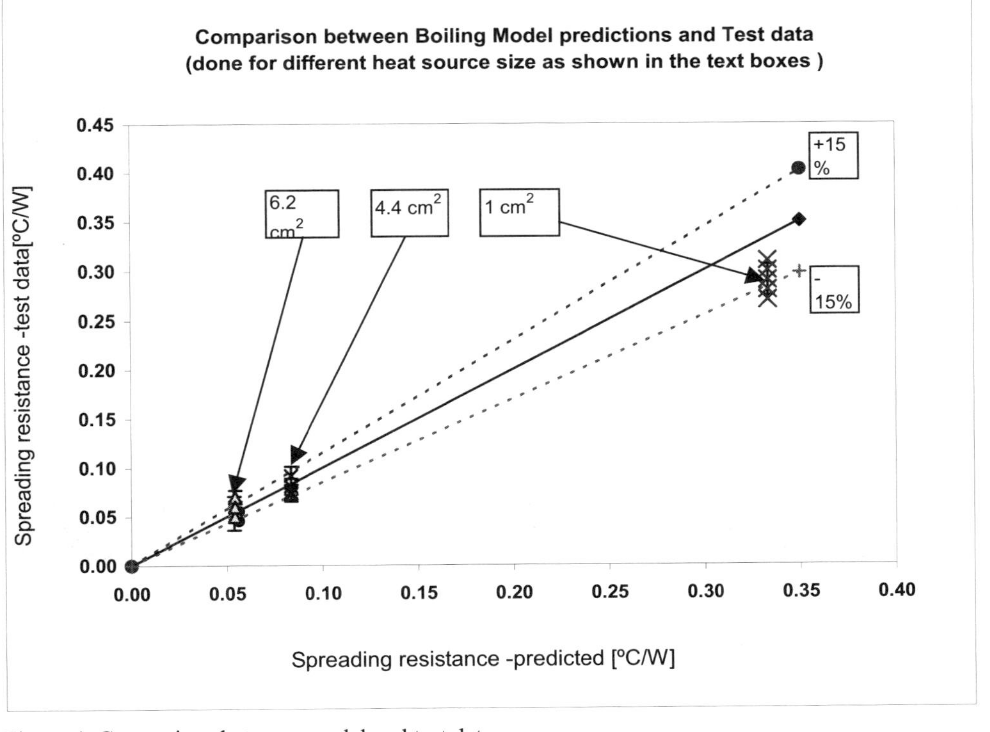

Figure 4: Comparison between model and test data

It can be seen that over the temperature ranges and heat flux ranges encountered in cooling computer processors, the water boiling heat transfer coefficient may achieve a maximum of 30000 W/m^2K. A few vapor chambers have been tested and the results have been compared with the above-simplified model. Figure 4 shows this comparison for different heat source sizes. It can be seen that the prediction model does a good job of fitting the test data, with less than 15% error.

4 HEAT TRANSFER LIMITATIONS IN WICKED HEAT PIPE

The conduction model proposed by Prasher et al. [2001] will be briefly mentioned here. The conduction model takes into consideration the following:

1) To determine the temperature drop or equivalently the thermal resistance of the various components one does not need to solve for the flow of liquid in the wick, as it is only needed for predicting the capillary limit of the heat pipe.
2) The wick is material of low effective thermal conductivity. This can be calculated by assuming that the wick is completely saturated with the fluid. There are various analytical models available in the literature to calculate the effective thermal conductivity of a fully saturated wick.
3) Very high thermal conductivity of heat pipes arises due to low temperature drop in the vapor. The temperature drop in the vapor can be calculated by modeling the pressure drop of the vapor along the length of the heat pipe and then applying the Clausius-Clapeyron equation. This in turn can be used to find the effective thermal conductivity of the vapor which for a vapor chamber is given by

$$k_{vapor_{VC}} = \frac{L^2 p_v \rho_v d_v^{\ 2}}{12 R \mu_v T^2}$$

where k_{vapor} is the effective thermal conductivity of the vapor, L is the latent heat, p_v is vapor pressure, ρ_v is the vapor density, d_v is the thickness of the vapor space, μ_v is the viscosity of the vapor, T temperature of the vapor and R is the gas constant. Prasher et al. [2001] conducted tests on water based vapor chamber in two configurations: remote cooling mode and active cooling mode. Figures (5) and (6) show the schematic of the remote cooling mode and the active cooling mode.

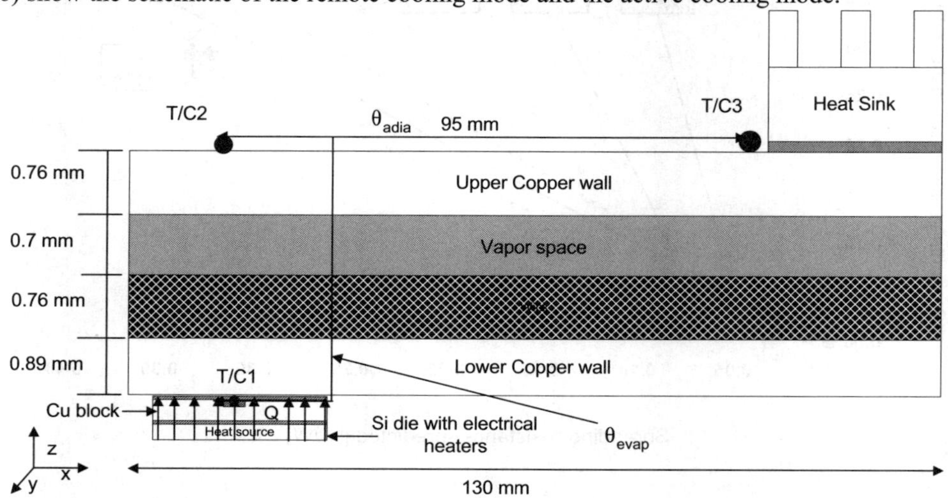

Figure 5 Experimental setup for the vapor chamber in remote cooling

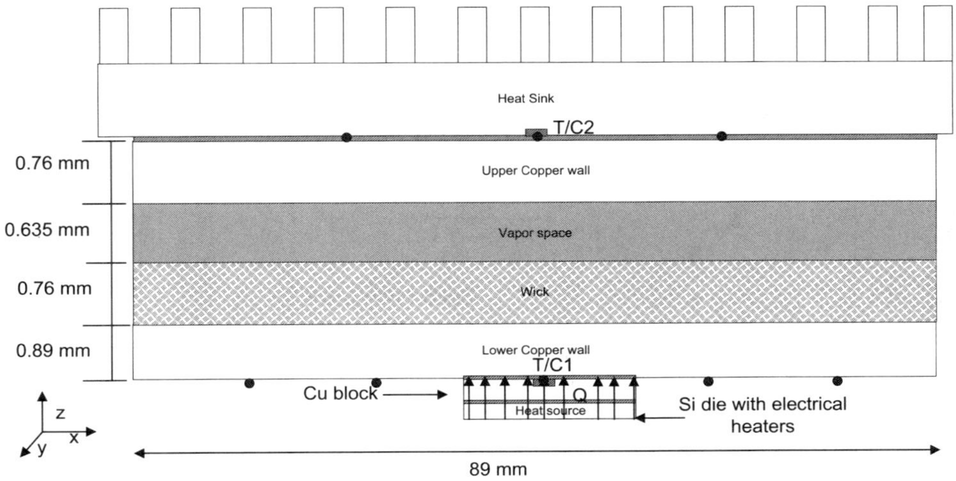

Figure 6 Experimental setup for the vapor chamber in active cooling

Figures (7) and (8) show the results of Prasher et al. [2001] for the remote cooling mode using the conduction model.

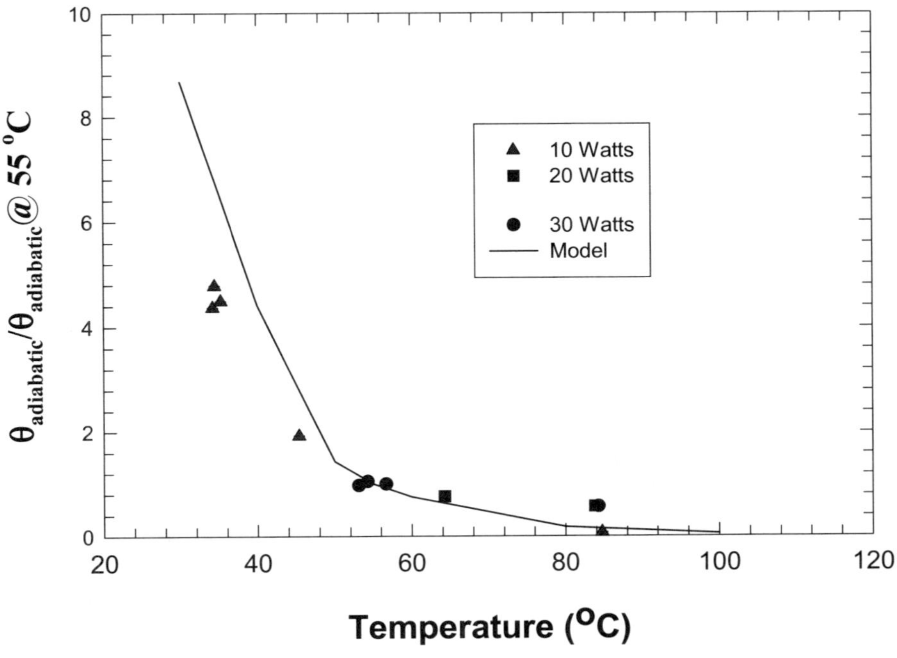

Figure 7 $\theta_{adiabatic}$ vs. temperature for different powers applied to the vapor chamber

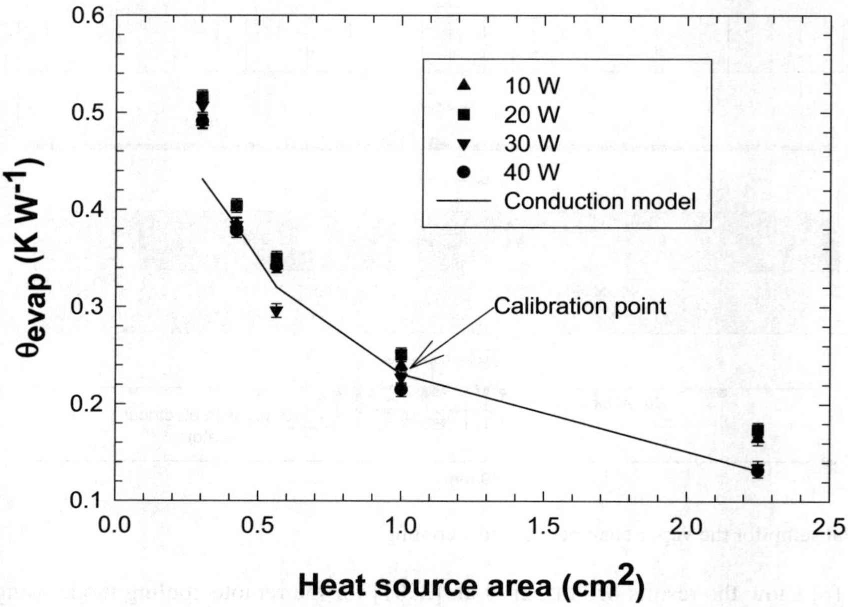

Figure 8 θ_{evap} vs. heat source area for different powers applied to the vapor chamber

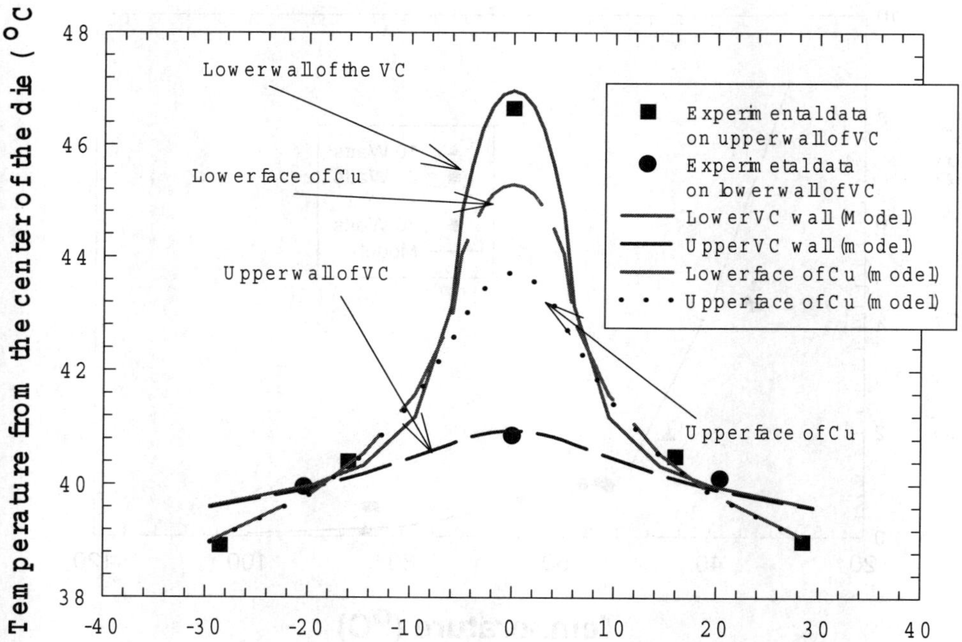

Figure 9 Comparison of thermal model with experimental data for active cooling mode

Figure (9) for active cooling mode shows that maximum temperature difference between the upper and lower face of vapor chamber is higher than that for copper, showing that copper for this configuration is better as heat spreader than the vapor chamber. Figure (9) shows the results for the active cooling mode. The conduction based model by Prasher et al. [2001] is in good agreement with the experimental data for both active and remote cooling mode. In their study Prasher et al. [2001] concluded that depending on the thermal conductivity of the wick, a vapor chamber technology could be worse than pure copper in the active cooling mode. The conclusions of the conduction model are:

- Wick thermal conductivity is a major source of thermal resistance
- Benefit of a vapor chamber over copper is dependant on the configuration in which this technology is being used.

5 SOLID METAL VS. PHASE CHANGE SPREADING RESISTANCE.

Using the boiling model, the ratio of phase change spreading over copper spreading can be estimated for conventional rectangular heat sink bases.

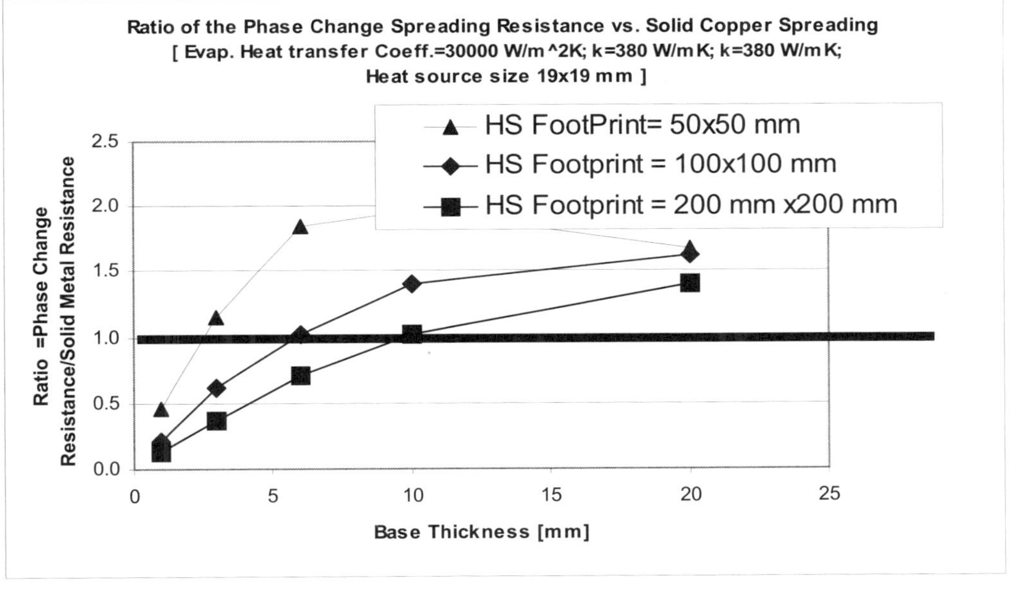

Figure 10: Ratio of the Phase Change Resistance vs. Solid Metal Resistance (boiling model)

Figure 10 shows this Ratio vs. Base thickness relationship for different footprint sizes. The heat input area is kept constant for this analysis. A number of concepts are evident, at least over the range of parameters studied, from the figure. For spreading resistance ratios greater than 1.0, the ratio decreases with increasing condenser size. This implies that the vapor chamber type base is better situated for larger condenser sizes. It can also be seen that the crossover point (ratio = 1.0) occurs at greater base thickness for larger condensers. For example, with a 200x200 mm footprint, a vapor chamber with a base thickness of 10 mm or less would out perform a corresponding copper base heat sink, however with a 50x50 mm footprint the base thickness would have to be less than about 2.5 mm in order for the vapor chamber to make the same claim. The figure also shows that there is a "worse case point" in the comparison of the thermal performance of a vapor chamber and solid copper base heat sink. This is identified by the maximum in the curve for the 50x50 mm footprint at a base thickness of 10 mm. At this point the spreading resistance ratio is at its largest value indicating the worst performance for the vapor chamber (when compared with the corresponding solid copper base). In general, there will be a maximum base thickness (dependent on heat source size and footprint) for consideration of a vapor chamber base. Unless weight is a major concern, with a base thickness above this maximum, a vapor chamber base should not be considered. Conversely, for a heat sink base thickness below this maximum a vapor chamber base should be considered.

6 CONCLUSIONS

Two different approaches for preliminary modeling of vapor chambers are presented in this paper, which substitute conventional heat sink bases with centered heat source. It can be concluded that:

- Two models presented can be used for preliminary design of vapor chambers. The prediction fits well with the test data.
- There is threshold envelope where copper has a lower spreading resistance than phase change technologies (water/copper).
- For spreading resistance ratios greater than 1.0, the ratio decreases with increasing condenser size.
- The cross-over point (ratio = 1.0) occurs at greater base thickness for larger condensers.
- There will be a maximum base thickness (dependent on heat source size and footprint) for consideration of a vapor chamber base.
- There is a "worse case point" in the comparison of the thermal performance of a vapor chamber and solid copper base heat sink where the vapor chamber to solid base resistance ratio is at a maximum.

REFERENCES

Faghri, A., 1995. Heat Pipe Science and Technology. Washington D.C: Taylor and Francis.

Incropera, D. P. De Witt, 1990. Fundamentals of Heat and Mass Transfer. Canada: John Wiley & Sons Inc.

Prasher, R.S., Shipley, J.C., and Devpura, A., 2001. A Simplified Modeling Scheme for Design Sensitivity Study of Thermal Solution Utilizing Heat Pipe and Vapor Chamber Technology. to be published in The Proc. Of International Mechanical Engineering Congress and Exposition, New York, 2001 .

Rohensaw, J., P., Hartent, J., P., and Ganic, M., 1985. Handbook of heat transfer. Second Edition. New York : McGraw-Hill Book.

Thermal Challenges in Next Generation Electronic Systems, Joshi & Garimella (eds)
© 2002 Millpress, Rotterdam, ISBN 90-77017-03-8

Spreading resistance with convective heat transfer from top and bottom surfaces

K.C. Toh, X.Y. Chen & J. C. Chai
School of Mechanical and Production Engineering, Nanyang Technological University, Singapore

Keywords: thermal resistance, spreading resistance

ABSTRACT: A nondimensional axisymmetric model is used to calculate the thermal resistance and maximum temperature with convective heat transfer from the top and bottom surfaces of a typical spreader plate. This configuration occurs commonly in advanced microelectronic packages. The conduction inside the substrate and the convective heat losses are predicted using the Finite-Volume Method. The effect of the ratio of the outer diameter to the heat source diameter, the ratio of the substrate thickness to the heat source diameter, and the Biot numbers of the top and bottom surfaces on the maximum temperature and the thermal resistance are examined. A total of six thicknesses, seven diameters, and five top surface and bottom surface Biot numbers are examined. The heat flow lines are presented to show the heat transfer paths and to explain the trends observed in the thermal resistance.

1 INTRODUCTION

In microelectronic packages, the heat generated from the die, first spreads within the package internals and is then dissipated to the surroundings via convective heat transfer from the top and bottom surfaces of the package directly or through the printed wiring board on which it is mounted. Effective cooling is crucial to the lowering of the peak temperatures but the linkage between the geometrical parameters and the external boundary conditions to the spreading resistances is still not well understood.

Package designers have, in recent years, recognized the importance of spreading resistance and the part it plays in determining the thermal performance of their packages (Lee et al. 1995). But in their quest to obtain a means for estimating the spreading resistance in the increasingly complex package structure, many have resorted to simplified formulae from handbooks without a proper appreciation of their limitations, while others have resorted to expensive, finely modeled, numerical solutions. Most of the currently available tools depend on configurations that enable a closed form analytical solution, where the heat flow is not split into multiple paths, and subjected to different external conditions (Yovanovich et al. 1979, Negus & Yovanovich 1984). In practice, the heat dissipates through all available paths in the package. At the very least, because of the current trend towards thin packages, heat dissipation via two parallel paths from the top and bottom surfaces should be considered.

The configuration for our current study is based on a comparison with the work of Nelson & Sayers (1992). They reported spreading resistances for various geometrical and flow conditions. In the actual physical system, the temperature field changes in all three coordinate directions. A possible configuration for semiconductor devices is shown in Figure 1. For clarity, the shaded region of Figure 1 is enlarged in Figure 2. However, an axisymmetric model will reduce computational time and the number of characteristic parameters. The axisymmetric model proposed by Nelson & Sayers (1992) is shown in Figure 3. A uniform heat input is placed at the *centre* of the top wall, while the remaining portion of the top wall is assumed to be insulated. The bottom wall exchanges heat with the surroundings. The effects of the ratio of the disk diameter to the heat source diameter, the ratio of the thickness to the disk diameter and the Biot number on the bottom wall were examined.

In our study, the top wall, outside the heat source area, is allowed to exchange heat with the surroundings with its own set of Biot number. This is a more realistic model which will cover almost many of the situations likely to be encountered in package design. The conduction inside the substrate and the convective heat losses will be predicted using the Finite-Volume Method (FVM) of Patankar (1980). With this model the heat flow within a substrate from a heat source and dissipating through convective heat transfer from the top and bottom surfaces can be studied. Effects of the various geometrical ratios similar to those of Nelson & Sayes (1992) on the maximum temperature and the spreading resistance are examined but with the addition of the Biot numbers for the top and bottom surfaces. The accompanying splitting of the heat flow will be explained in terms of the heat flow path lines presented for selected conditions.

The remainder of this article is divided into three sections. The mathematical formulation of the problem examined in this article is presented in the next section. This is followed by a section on the discussions of the results. Finally, some concluding remarks are made.

2 MATHEMATICAL FORMULATION

2.1 *Problem description*

Figure 4 shows the computational domain used in this article. The dimensionless radius and thickness (normalized by the diameter of the heat source) of the substrate are $B/2$ and L respectively. The bottom surface of the substrate is exposed to an environment with a heat transfer coefficient h_b and an ambient temperature of T_0. Part of the top surface ($R < 0.5$) is the heat source which is modeled as a region subjected to a uniform heat flux. The remainder of the top surface exchanges heat with the surroundings convectively. To minimize the number of parameters, the ambient temperature above the substrate is assumed to be the same as the temperature below the substrate. The heat transfer coefficient is h_t. The outer radius is assumed to be an adiabatic surface, while the inner surface is a line of symmetry.

2.2 *Governing equations*

The nondimensional energy equation in axisymmetric coordinates (Fig. 4) can be written as

$$\frac{\partial}{\partial X}\left(\frac{\partial \theta}{\partial X}\right)+\frac{\partial}{\partial R}\left(R\frac{\partial \theta}{\partial R}\right)=0 \tag{1}$$

2.3 *Boundary conditions*

The boundary conditions can be written as:

Top surface ($X = 0$):

$$\frac{\partial \theta}{\partial X}=\begin{cases} -1 & 0<R<0.5 \\ \dfrac{Bi_t}{L}\theta_t & 0.5<R<\dfrac{B}{2} \end{cases} \tag{2a}$$

Bottom surface ($X = L$):

$$\frac{\partial \theta}{\partial X}=\frac{Bi_b}{L}\theta_b \tag{2b}$$

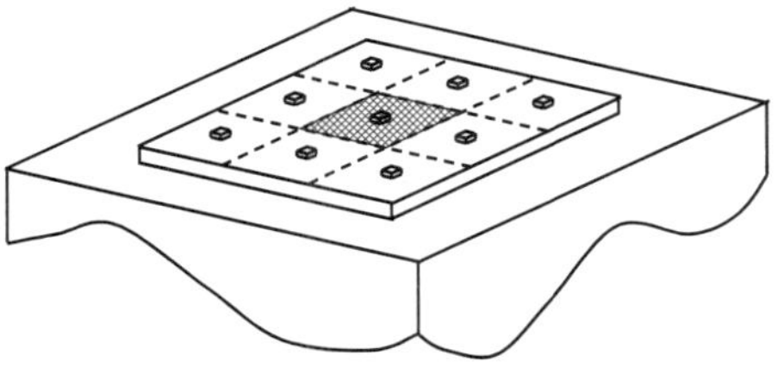

Figure 1. Symmetrically located heat sources

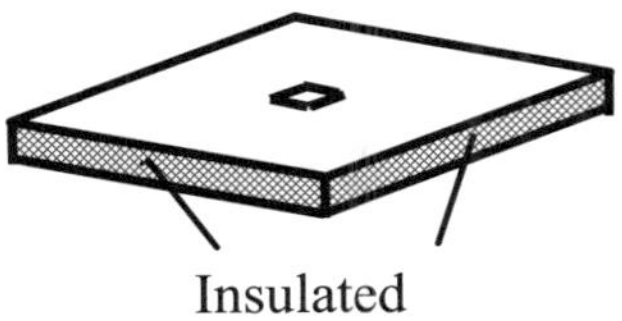

Figure 2. The problem modeled using a three dimensional model

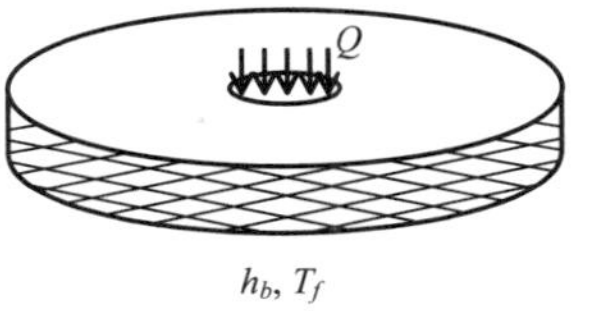

Figure 3. An axisymmetric model of the three-dimensional model shown in Figure 2

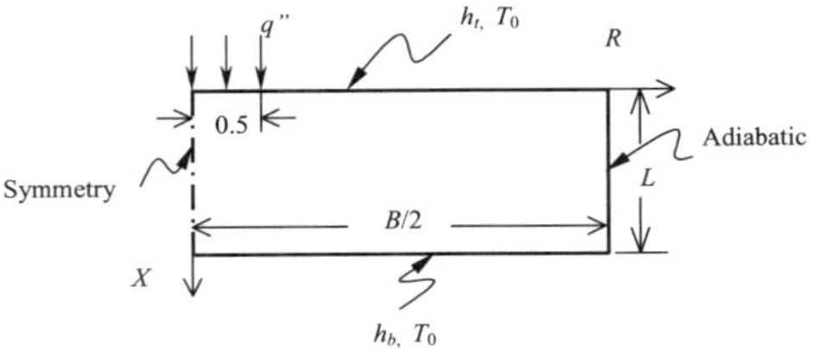

Figure 4. Computational domain: axisymmetric model

Outer surface ($R=B/2$):

$$\frac{\partial \theta}{\partial R} = 0 \tag{2c}$$

Inner surface ($R = 0$):

$$\frac{\partial \theta}{\partial R} = 0 \tag{2d}$$

2.4 *Solution procedure*

The energy equation (Eq. 1) is solved using the FVM of Patankar (1980). Discretization equations are formulated for every control volume in the solution domain. This is achieved by integrating Eq. (1) over a finite control volume. The resulting algebraic equation ensures perfect energy balance for each control volume and as a result, for the whole solution domain also. These algebraic equations are solved using the line-by-line TriDiagonal-Matrix Algorithm. A block-correction procedure (Settari & Aziz, 1973) is used to accelerate the convergence. For all the cases examined in this article, 100 control volumes were used in both the radial and axial directions respectively. Further grid refinement does not change the solutions indicating that the grid independent solutions have been obtained.

3 RESULTS AND DISCUSSIONS

The effects of the thickness L, diameter B. top surface Biot number Bi_t and bottom surface Biot number Bi_b on the maximum dimensionless temperature and the spreading resistance are examined in this article. Results are presented for five thicknesses namely, L = 0.1, 0.2, 0.5, 1, 2 and 10, seven diameters namely B = 1.2, 1.6, 2.4, 4, 6, 8, and 20, five bottom surface Biot numbers namely Bi_b = 0.01, 0.1, 1, 10 and five top surface Biot numbers namely, Bi_t = 0.01, 0.1, 1, 10. The maximum dimensionless temperatures are tabulated in Tables 1 through 5. Tables 6 through 10 show the spreading resistances for the same combinations of various parameters. The thermal resistance is defined as

$$R_t = \frac{T_{max} - T_s}{q} \tag{3}$$

where T_s is the average temperature of the top and bottom surfaces. This average was performed over the portion of the surfaces which exchange heat convectively with the surroundings. It should be pointed that the thermal resistance given in Eq. (3) contains both the spreading and material resistances. This will be explained in the next section on the resistances.

3.1 *Dimensionless maximum temperature*

It is also useful to compare the present results with the results obtained using an insulated top surface. The results of Nelson & Sayers (1992) are used for the purpose of this comparison. For convenience, they will be referred to as the *reference* values. Figures 5-7 show the maximum dimensionless temperature for various parameters.

3.1.1 *Effect of thickness*

Figures 5 shows the plot of the ratios of the maximum dimensionless temperatures with the reference temperatures as functions of the substrate thicknesses L, for five top surface Biot numbers. As expected, the ratios are less than 1. This is because the top surface provided an additional area for heat transfer which leads to a lower maximum temperature as compared to the reference condition. When the top wall is almost insulated (small Bi_t), the ratios approach the insulated top wall results of Nelson & Sayers (1992), leading to ratios of one for all L. For a given L, it can be seen from Figure 5 that as the Biot numbers increase, the ratios decrease. This is due to the increase in the top-wall heat transfer coefficient leading to more effective cooling of the substrate. The ratios also decrease with increasing L, indicating that the rate of decrease with non-adiabatic top wall is larger than the reference case with adiabatic top wall. Increasing the thickness, increases the material resistance to the bottom surface, making heat flow to the bottom less favored as compared to the top.

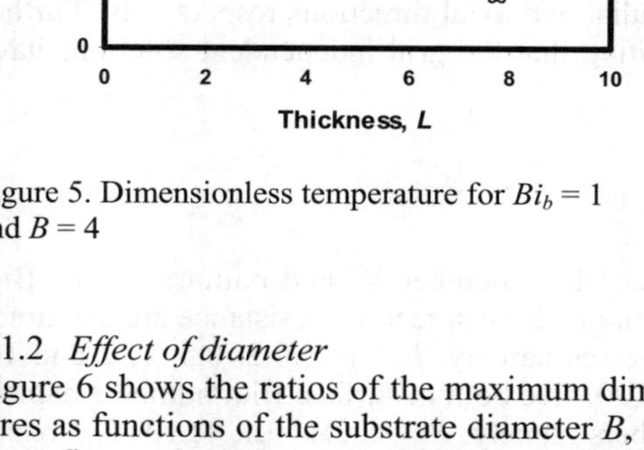

Figure 5. Dimensionless temperature for $Bi_b = 1$ and $B = 4$

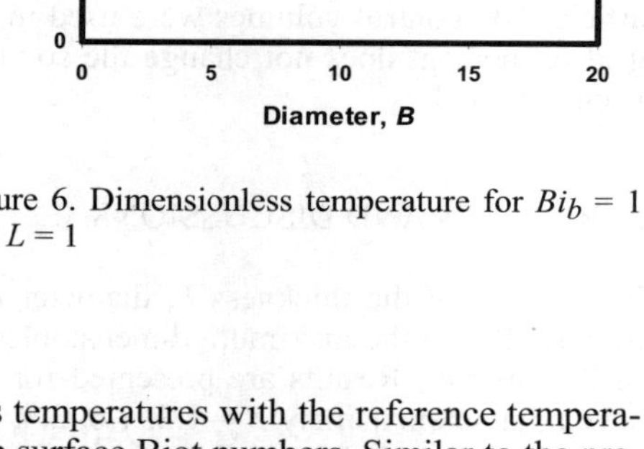

Figure 6. Dimensionless temperature for $Bi_b = 1$ and $L = 1$

3.1.2 *Effect of diameter*

Figure 6 shows the ratios of the maximum dimensionless temperatures with the reference temperatures as functions of the substrate diameter B, for five top surface Biot numbers. Similar to the previous figure, the ratios are less than one. This is due to the additional area for heat transfer from the top surface. The ratios approach one when the top surface Biot number decreases. Although less than one, the ratios increase with increasing B for most values of Bi_t. This indicates that as the diameter increases, the percentage improvement in cooling with the additional top wall heat transfer becomes less and asymptotes to a constant value. Spreading is a fairly localized mechanism, such that an increase in diameter beyond about 10 times the heat source diameter has no further effect. The reversals in the trends of the ratios for $Bi_t = 0.1$ and 1 do not mean that there are reversals in the trends of the dimensionless temperatures. This is shown in Figure 7 where the dimensionless temperatures decrease monotonically with B for three top surface Biot numbers. Under these limited conditions, very near to the heat source and with the presence of top wall cooling, the rate of decrease in the maximum temperature initially increases with an increase in diameter but subsequently decreases with further increases in diameter.

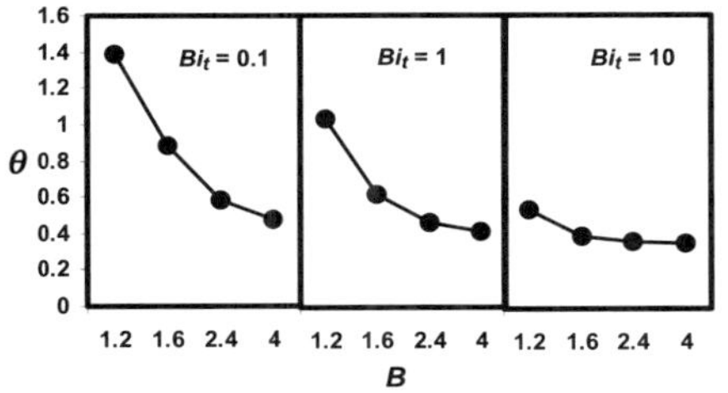

Figure 7. Dimensionless temperatures for $L = 1$ and $Bi_b = 1$. Note that the x-axis is not drawn to scale

3.2 *Dimensionless thermal resistance*

To examine the effects of various parameters further, two important trends are shown in Figures 8 and 11.

3.2.1 *Effect of Bi_t*

Figure 8 shows the dimensionless thermal resistances for $B = 1.2$, $Bi_b = 0.01$ and five thicknesses as functions of the top surface Biot numbers. It can be seen that the resistance increases for small thicknesses (for $L = 0.1$ and 0.2) and decreases for larger thickness. The same trends, namely, increasing thermal resistance for small thicknesses and the reverse trends for larger thicknesses can be observed for other values of B and Bi_b. To explain the reversal in trends, heat flow lines for two situations are shown in Figures 9 and 10. Figure 9 shows the heat flow lines for $B = 1.2$, $L = 0.1$, $Bi_b = 0.01$ and four top wall Biot numbers. Figure 10 shows the same information for a thicker substrate with $L = 1$.

3.2.2 *Behavior of thin substrates*

When the thickness is small, $L = 0.1$, the material resistance is small and the thermal resistance consists mostly of spreading resistance. This situation is shown in Figures 8 and 9. From Figure 8, when top surface Biot number Bi_t is small, the heat is forced to leave through the bottom surface. The heat flow lines in Figure 9a shows the paths of heat flow. Most heat travels from the top to the bottom surface. The heat flow lines spread as heat travels from the top to the bottom surface. As the top surface Biot number increases, it provides a path of lesser resistance to the heat flow. As a result, heat flows towards the top surface resulting in the heat flow lines bending towards the top surface. Due to the bending of the heat flow lines, the spreading resistance increases with the top wall Biot number as seen in Figure 8.

3.2.3 *Behavior of thick substrates*

Figure 10 shows the heat flow plots for the same conditions with a thicker substrate, namely $L = 1$. Similar to the previous situation, when the top wall Biot number is small, heat flows towards the bottom wall. Due to the large thickness and the small diameter B, the thermal resistance consists mostly of material resistance. As the top wall Biot number increases, more of the heat travels along the shorter path leaving the substrate through the top surface. This can be seen from the bending of the heat flow lines towards the top surface for higher top surface Biot numbers. However, since there is more room for the bending, it is not as severe as in Figure 9, so that spreading resistance is relatively low. With less of the heat passing through the material to the bottom surface, there is a corresponding reduction in material resistance. The thermal resistance, which is the combined material and spreading resistances, hence decreases in value with increasing top surface Biot number as seen in Figure 8 (for $L \geq 0.5$.)

3.2.4 *Effect of thickness*

Figure 11 shows the thermal resistance versus thickness plots for $B = 1.2$, $Bi_b = 0.01$ and four top surface Biot numbers. Generally, for a thicker plate, the spreading resistance tends to be lower because the heat flow lines are less distorted, but material resistance increases (Fig. 12). With a low top surface Biot number, when most of the heat exits through the bottom surface, it can be seen that as L increases, the thermal resistance first decreases (the reduction in spreading dominates), and then starts to increase (material effect dominates), However, this increase in thermal resistance at larger thicknesses is not manifested when the top surface Biot numbers is quite high (Fig. 11 for $Bi_t = 10$), because most of the heat exits from the top surface. There are almost no heat flow lines in the bottom half of the substrate (Fig 10 d) in this case so further increases in thickness has no effect.

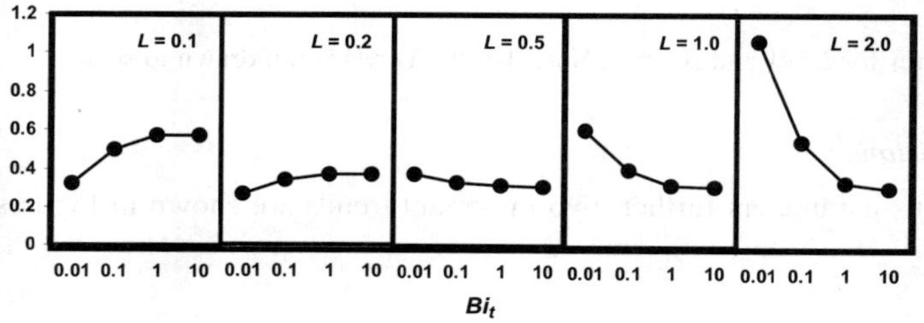

Figure 8. Dimensionless spreading resistance for $B = 1.2$ and $Bi_b = 0.01$. Note that the x-axis is not drawn to scale

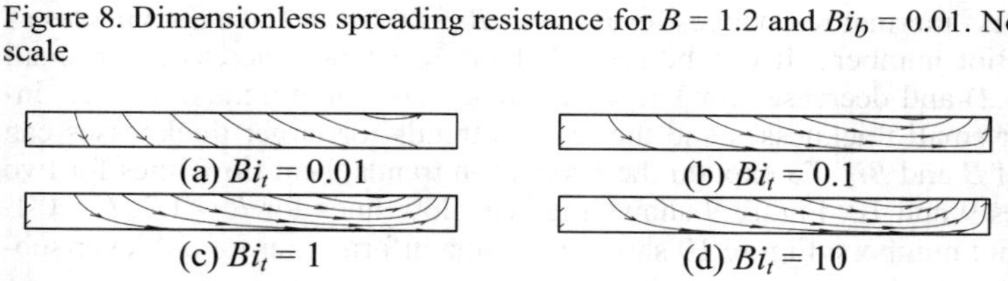

Figure 9. Heat flow lines for $B = 1.2$, $L = 0.1$, $Bi_b = 0.01$

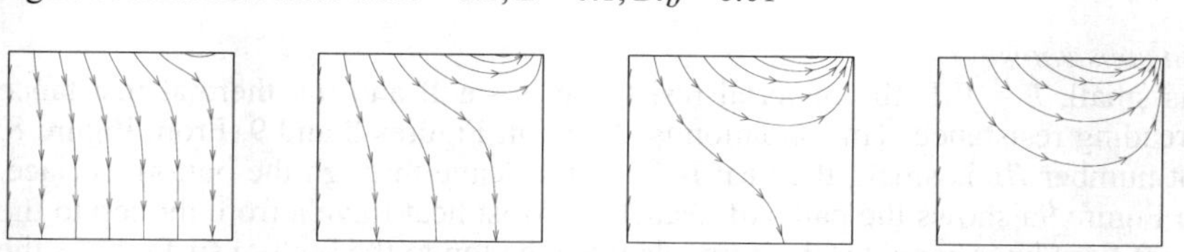

Figure 10. Heat flow lines for $B = 1.2$, $L = 1$, $Bi_b = 0.01$

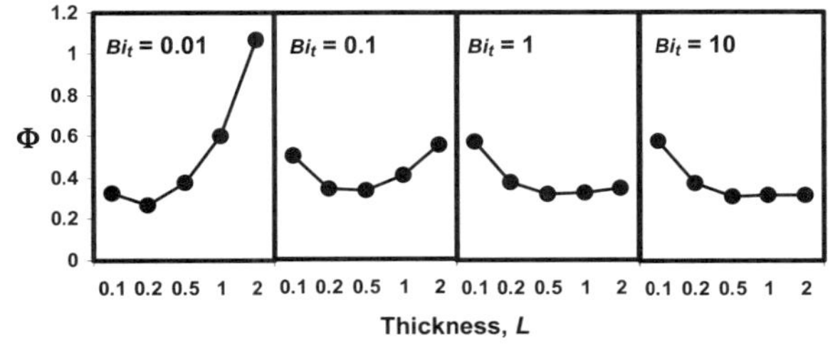

Figure 11. Dimensionless spreading resistance for $B = 1.2$ and $Bi_b = 0.01$. Note that the x-axis is not drawn to scale

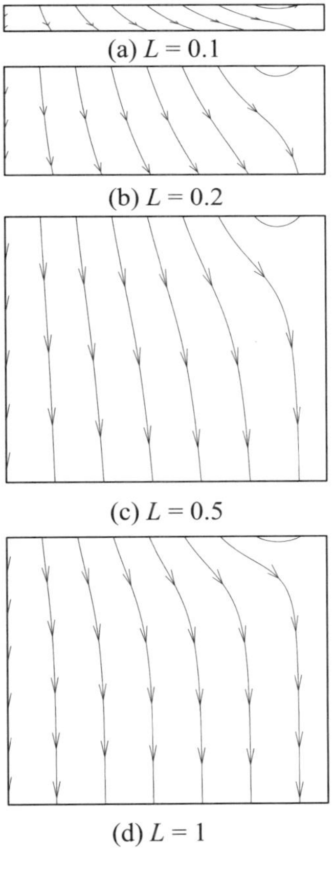

Figure 12. Heat flow lines for $B = 1.2$, $Bi_t = 0.01$, $Bi_b = 0.01$

4 CONCLUDING REMARKS

An axisymmetric model has been applied to predict the maximum temperature and spreading resistance of a heat source on a typical spreader plate with convective heat transfer from the bottom and top surfaces. A numerical scheme, based on FVM, was used to solve the energy equation. The dimensionless maximum temperature θ and spreading resistance ϕ were presented, and the effects of four parameters: thickness *L,* diameter *B*, top surface Biot number Bi_t and bottom surface Bi_b on their values have been analysed. A mechanistic explanation is attempted using plots of heat flow lines to show how the behavior deviates from models based on heat flow out from only one surface.

The results show that when there is an additional surface for heat transfer on the same side as the heat source, the effect of an increased thickness in the spreader plate is less detrimental on the maximum temperature of the heat source, because of the alternative path available. Since it function is as a spreader, increasing the diameter helps the heat to be distributed over a larger area, more so with two surfaces for convection, but past a certain effective diameter, further increases is of little significance. There are also significant differences in behavior for relatively thin and relatively thick plates. In thin plates, increasing the convection heat loss from the same side as the heat source, forces the heat flow to constrict through the thin layer increasing its spreading resistance. However for relatively thick plates, increasing the convection will always decrease the thermal resistance. But like the effect of diameter, it is unnecessary to increase the thickness past a certain value, as the additional material will not contribute to the heat flow path.

NOMENCLATURE

A_d	Area of the equivalent diameter of the heat source
b	equivalent diameter of disk
B	nondimensional diameter of disk = b/d
Bi	Biot number= hl/k
d	equivalent dimensional diameter of heat source
h	convective heat transfer coefficient
k	thermal conductivity of the disk
l	disk thickness
L	nondimensional thickness = l/d
q	heat flow rate
q''	heat flux
r	radial coordinate
R	nondimensional radial coordinate = r/d
R_t	thermal resistance
T	temperature
x	x coordinate
X	nondimensional x coordinate = x/d

Greek symbols

θ	nondimensional temperature = $(T-T_0)/(q''d/k)$
Φ	nondimensional resistance = $kR_t\sqrt{A_d}$

Subscripts

0	ambient
max	maximum
s	surface

REFERENCES

Lee, S., Song, S., Au, V. & Moran, K. P., 1995. Constriction/Spreading Resistance Model for Electronic Packaging, ASME/JSME Thermal Engineering Conference, Vol. 4, pp 199-206.

Yovanovich, M. M., Tien, C. H. & Schneider, G. E. 1979. General Solution of Constriction Resistance Within a Compound Disk, AIAA Paper 79-0179, The 17th Aerospace and Science Meeting, New Orleans, LA.

Negus, K. J. & Yovanovich, M. M. 1984. Constriction Resistance of Circular Flux Tubes With Mixed Boundary Conditions By Linear Superposition of Neumann Solutions, ASME-HT-84, NHTC, Niagara Falls, NY.

Nelson, D. J., & Sayer, W. A. 1992. A Comparison of Two-Dimensional Planar, Axisymmetric and Three-Dimensional Spreading Resistances, Eighth IEEE SEMI-THERM Symposium, pp. 62-68.

Patankar, S. V., 1980. *Numerical Heat Transfer and Fluid Flow*, Taylor and Francis, New York.

Settari, A. & Aziz, K. 1973. A Generalization of the Additive Correction Methods for the Iterative Solutions of Matrix Equations, SIAM Journal of Numerical Analysis, Vol. 10, pp. 506-521.

Thermal Challenges in Next Generation Electronic Systems, Joshi & Garimella (eds)
© 2002 Millpress, Rotterdam, ISBN 90-77017-03-8

In-plane effective thermal conductivity of plain-weave screen laminates

Jun Xu & R.A. Wirtz
Mechanical Engineering Department/MS 312, University of Nevada Reno, Reno, Nevada, USA

Keywords: thermal conductivity, plain-weave, porosity, screen-laminates.

ABSTRACT: A simple-to-fabricate woven mesh, consisting of bonded laminates of two-dimensional plain-weave conductive screens is described. Geometric equations show that these porous matrices can be fabricated to have a wide range of porosity and specific surface area, β. A heat transfer model is developed. It shows that the laminates can have a highly anisotropic thermal conductivity vector, with in-plane effective thermal conductivities ranging to about 78.5% of base material values. A technique to measure the laminate in-plane effective thermal diffusivity is described. Measurements of the in-plane effective thermal diffusivity of copper plain-weave laminates are used to benchmark the model.

1 NOMENCLATURE

Ay	effective unit cell section area for Y-direction conductivity
c	specific heat
cf	compression factor
dx, dy	diameters for the X and Y-direction filaments
$\overline{d}$	dx/dy
k_{ey}	effective thermal conductivity in Y-direction
k_s	thermal conductivity of base material
k_f	thermal conductivity of second phase material
$\overline{K}_{fs}$	k_f/k_s
$\overline{Key}$	key/k_s
Mx, My	mesh numbers along X and Y direction
$\overline{M}$	My/Mx
Sx, Sy	wire filament lengths
t_n	thickness of screen laminates with n layers thickness
R	effective thermal resistance
α	thermal diffusivity
β	surface area-volume ratio
ε	porosity

2 INTRODUCTION

A thermally conductive, open cell porous matrix such as an unconsolidated bed of small particles, or foamed metal such as foamed aluminum, make excellent heat exchanger surfaces due to their large surface area-to-volume property, β. Also, when used as a fill material, these matrices are effective solid composite conductivity enhancers. However, due to high porosity (ε) coupled with the tortuosity effect, the effective thermal conductivity of these materials, k_e is relatively small, so that much of the gain in performance obtained by having a large β is lost by having a relatively small k_e. Typical values of k_e in spherical particle packed beds are 10% - 15% of the particle thermal conductivity, k_s [Kaviany, 1995]. Commercially available metal foam such as aluminum foam, has an effective thermal conductivity that ranges from 2% to 6% of the base metal value [Ashby et al, 2000].

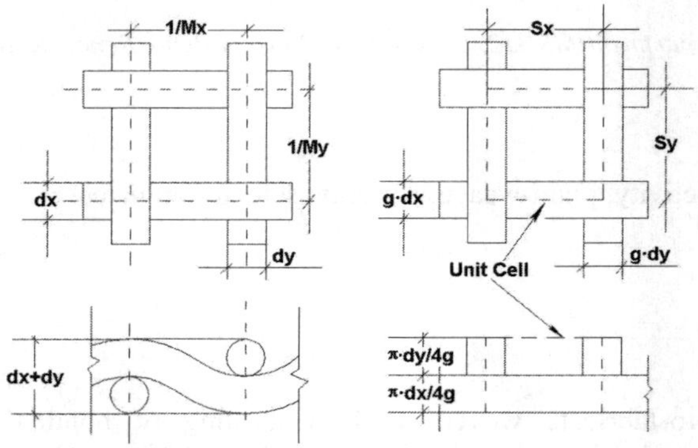

Fig. 1 Plain-weave unit cell.

An anisotropic porous matrix having a large surface area to volume ratio and high effective thermal conductivity in a particular direction will result in a very effective heat exchange device. In this paper, we show that stacked laminates of conductive screening can be configured to have these characteristics. Such a porous matrix can be incorporated into the design of a flow-through module or cold plate heat exchanger, resulting in a compact, high-flux device with reasonable pressure-drop characteristics [Park et al, 2002]. Furthermore, these laminates, when incorporated into a polymer-based composite sandwich structure, are promising conductivity enhancers [Zhao et al, 2002].

The earliest model for the effective thermal conductivity of a porous material is attributed to Rayleigh [1892, as referenced in Alexander, 1972]. The model assumes that the porous material is isotropic. More recently, Alexander [1972], Koh and Fortini [1973] and Chang [1990] have reported models and empirical correlations specifically directed at the cross-plane effective thermal conductivity component of screen material. We have found no reference in the technical literature for the in-plane component of the effective thermal conductivity.

3 PHYSICAL MODEL

Figure 1 summarizes the development of the thermal/physical model of a plain-weave screen. The left segment of the figure shows plan and edge views of a section of screen. Serpentine wire filaments have diameter dx and dy, and corresponding mesh numbers Mx and My. The wire filament pitch in the x- and y-directions are $Mx^{-1} = wx + dy$ and $My^{-1} = wy + dx$, respectively. In the absence of crimping, the screen has thickness $t_l = dx + dy$. Filament lengths are [Luo and Mitra, 1999]

$$Sx = \frac{1}{Mx}\left[1 + 9.6\left(\frac{dy \cdot Mx}{4}\right)^2 - 49.2\left(\frac{dy \cdot Mx}{4}\right)^4\right] \tag{1}$$

$$Sy = \frac{1}{My}\left[1 + 9.6\left(\frac{dx \cdot My}{4}\right)^2 - 49.2\left(\frac{dx \cdot My}{4}\right)^4\right] \tag{2}$$

Following Chang [1990], we transform the section of screen into the layered, rectangular cross section segment shown in the right segment of the figure. Each rectangular wire filament has thickness $\pi d / 4g$ and width gd so that the cross section area of each filament is $\pi d^2 / 4$. The geometric index, g is a measure of the contact between wire filaments at intersections.

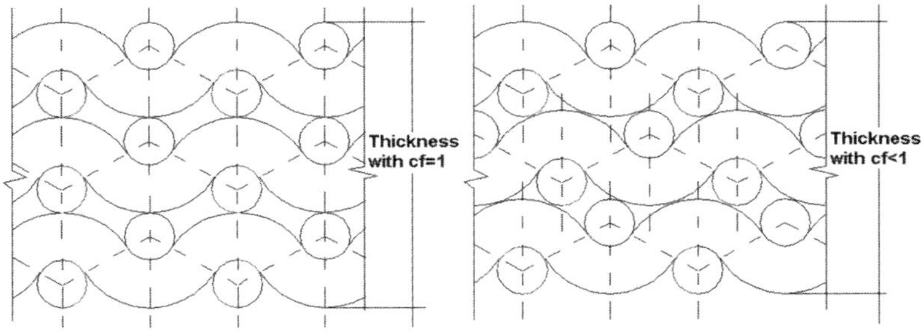

Fig. 2 Screen stacking configurations.

Figure 2 shows two possible arrangements of stacked plain-weave screens. The left hand view shows the situation where successive wire filaments are aligned so that the thickness of a laminate consisting of n screens is $t_n = n(dx + dy)$. The right element of the figure shows the situation where successive wire filaments are not necessarily in line so there is some interleaving of wire filaments when the screens are stacked. In this case, the thickness of the laminate is $t_n = n \cdot c_f (dx + dy)$ where c_f is the compression factor. The magnitude of the compression factor depends on the weave pattern, mesh numbers and wire filament diameters of the screen, and the stacking arrangement of the screen layers. If we restrict our attention to screens with $t_1 = dx + dy$, then $cf \leq 1$.

Consideration of Figs. 1 and 2 leads to expressions for the porosity, ε and specific surface area, β of plain-weave screen laminates.

$$cf \cdot (1 - \varepsilon) = \frac{\left(\pi \cdot dy^2 Sy + \pi \cdot dx^2 Sx\right) \cdot Mx \cdot My}{4(dx + dy)} \tag{3}$$

where $cf(1-\varepsilon)$ is the reduced "metal" fraction.

$$\beta = SF(1 - \varepsilon) \tag{4}$$

SF in Eq. (4) is the shape factor for the particular weave pattern and stacking arrangement

$$SF = \frac{4(dy \cdot Sy \cdot Mx + dx \cdot Sx \cdot My)}{dy^2 \cdot Sy \cdot Mx + dx^2 \cdot Sx \cdot My} \tag{5}$$

If we introduce mesh number and wire filament diameter ratios $\overline{M} = My / Mx$ and $\overline{d} = dx / dy$, then it can be shown [Xu, 2001] that the reduced metal fraction is a function of $\left(Mxdy, \overline{M}, \overline{d}\right)$ and β is a function of $\left(cf(1-\varepsilon), \overline{M}, \overline{d}\right)$.

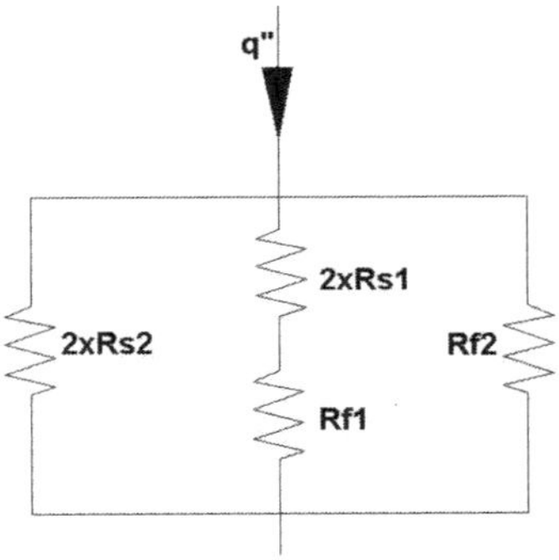

Fig. 3 Thermal circuit.

The in-plane effective thermal conductivity in the y-direction may be determined by considering the thermal circuit for conduction across the transformed unit cell, shown in Fig. 3. The figure shows three heat-flow paths: R_{s2} is the thermal resistance for conduction along the axis of the y-wire filaments; R_{s1} is the thermal resistance across x-wire filaments; R_{f1} is the thermal resistance to conduction in the y-direction, across the intervening "fluid" lying between x-wire filaments; and, R_{f2} in the thermal resistance of the fluid slab that lays between the y-wire filaments. These resistances are given as

$$R_{s1} = \frac{2g^2 \cdot dx}{\pi \cdot Ks \cdot dx \cdot Sx} \tag{6}$$

$$R_{s2} = \frac{8Sy}{\pi \cdot K_s \cdot dy^2} \tag{7}$$

$$R_{f1} = \frac{4g \cdot Sy(My, dx) - 4g^2 dx}{\pi \cdot K_f \cdot dx \cdot Sx} \tag{8}$$

$$R_{f2} = \frac{4g \cdot Sy}{\pi \cdot K_f \cdot (Sx - g \cdot dy) \cdot dy} \tag{9}$$

We further require that the volume of wire filaments be preserved across the geometric transformation shown in Fig. 1. Then $g = \pi / 4 \cdot cf$

If we define the effective thermal conductivity in the y-direction as $ke_y = qMy^{-1} / \Delta T$, then

$$cf \cdot \overline{K}_{ey} = \frac{1}{(\overline{d}+1)}\left[\frac{160\pi Mxdy \cdot (\overline{K}_{fs} - 1) + C \cdot \overline{K}_{fs} \cdot cf^{-1}}{D \cdot \overline{K}_{fs}} + \frac{C \cdot \overline{d} \cdot \overline{K}_{fs}}{160\pi \cdot \overline{d} \cdot \overline{M} \cdot Mxdy(1 - \overline{K}_{fs}) + C \cdot cf^{-1}}\right] \tag{10}$$

$$C = 123(Mx \cdot dy)^4 - 384(Mx \cdot dy)^2 - 640 \tag{11}$$

$$D = 123(My \cdot dx)^4 - 384(My \cdot dx)^2 - 640 \tag{12}$$

where $\overline{Ke}_y = ke_y / k_s$ and $\overline{K}_{fs} = k_f / k_s$ are the dimensionless effective conductivity and "fluid" conductivity, respectively.

Since most available commercial woven mesh products are isotropic structures, i.e *dx=dy=d and Mx=My=M*, the above general results can be simplified.

$$cf[1 - \varepsilon(iso)] = -3.906 \times 10^{-4} \pi \cdot C \cdot Md \tag{13}$$

$$\beta(iso) = 4(1-\varepsilon)/d \tag{14}$$

$$cf \cdot \overline{K}_e(iso) = \frac{1}{2}\left[\frac{160\pi \cdot Md \cdot (\overline{K}_{fs} - 1) + C \cdot \overline{K}_{fs} \cdot cf^{-1}}{C} + \frac{C \cdot \overline{K}_{fs}}{160\pi \cdot Md(1 - \overline{K}_{fs}) + C \cdot cf^{-1}}\right] \tag{15}$$

Equation 13 shows that the reduced "metal fraction", $cf(1-\varepsilon)$ is solely a function of the *Md*-product of the mesh. We note that $Md \to 1$ designates a "tightly" woven screen. In actuality, there is a physical limit on the magnitude of *Md.* For isotropic plain-weave screens where the thickness is limited to $t_l = 2d$, $Md_{2d}(\max) = 1/\sqrt{3} = 0.577$ so that the porosity is limited such that $0 < cf \cdot (1-\varepsilon) < 0.534$.

Equation (14) shows that the specific surface area, β follows the functional form of other porous media, with $SF = 4/d$. By contrast, $SF(spheres) = 6/d_s$ for an unconsolidated bed of spherical particles. Note, ε(spheres) is generally limited to a constant value of about 0.38.

Finally, Eq. (15) shows that the effective conductivity $\overline{Ke}_y(iso)$ is solely a function of $\overline{K}_{fs}$ and *Md.* It can be shown [Xu, 2001] that $\overline{Ke}_y(iso)$ is solely a function of $\overline{K}_{fs}$ and ε.

Figure 4 plots the reduced metal fraction as a function of *Mxdy*. The case for an isotropic screen laminate is shown as a solid line. This case always produces screen laminates having the highest

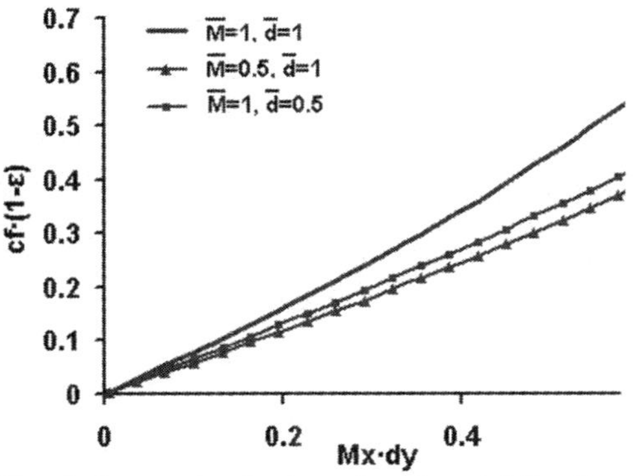

Fig. 4 Screen laminate reduced metal fraction.

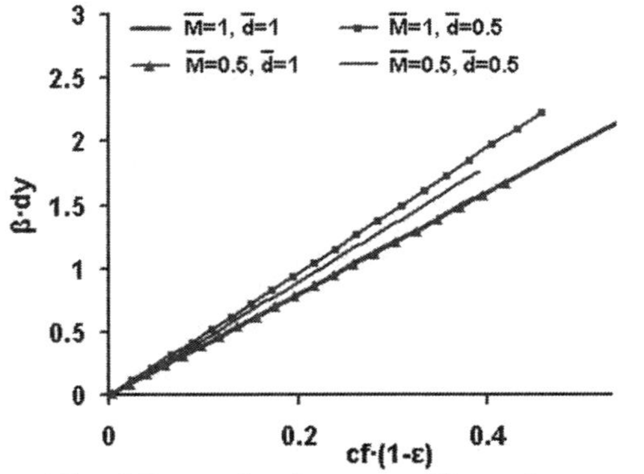

Fig. 5 Screen laminate specific surface area.

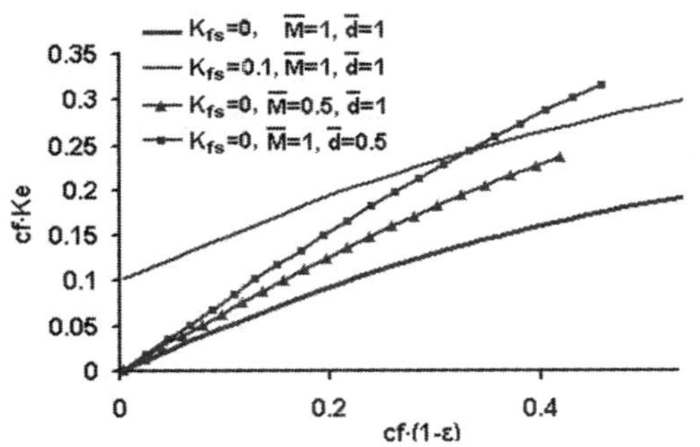

Fig. 6 Screen laminate conductivity.

metal fraction (lowest porosity). The reduced metal fraction for two anisotropic plain-weave laminate cases ($\overline{M} < 1$, $\overline{d} < 1$) are also shown in the figure.

Figure 5 plots βdy as a function of reduced metal fraction. The case for an isotropic screen laminate is shown as a bold solid line. The shape factor for this condition is SF = 4 (the slope of the plotted line). This is true for all conditions where $\overline{d} = 1$. The case $\overline{M} = 0.5$, $\overline{d} = 1$ is shown superimposed on the isotropic case. All conditions with $\overline{M} < 1$, $\overline{d} < 1$ lead to larger values of βdy at fixed $Mxdy$. Therefore, the condition $\overline{d} = 1$ always produces the smallest specific surface area.

Figure 6 plots the reduced effective conductivity as a function of reduced metal fraction. The case for an isotropic screen laminate with $\overline{K}_{fs} = 0$ is shown as a bold solid line. An additional case, isotropic screen laminate with $\overline{K}_{fs} = 0.1$ is also shown (solid line). Under these conditions, the effective conductivity is shifted upward approximately proportional to the value of $\overline{K}_{fs}$. In many applications where the second phase material is a hydrocarbon with the first phase a "good" conductor, $\overline{K}_{fs} < 0.005$.

Two additional, anisotropic screen laminate cases are shown in the figure. Cases with $\overline{M} < 1$ or $\overline{d} < 1$ always produce effective conductivities that are greater than obtained with isotropic plain-weaves.

It is easy to find the maximum possible value of $\overline{Key}$. When the $Mxdy \to 1$, $\overline{M} \to 0$ and $\overline{d} \to 0$, then, Eq. (10) gives,

$$Ke_y(\max) = \frac{(1-\pi)k_{fs}}{4} + \frac{\pi}{4} \tag{16}$$

Therefore, $Ke_y(\max)$ can be greater than 0.785.

4 MODEL VERIFICATION WITH EXPERIMENT

4.1 *Measurement of thermal diffusivity*

Consider a uniform cross section, long and slender test article with well-insulated periphery, as shown in Fig. 7. The test article is initially at a uniform temperature, T_i when a time-varying heat flux is applied to one end. We measure the temperature response at three locations, as shown in the figure. Assume constant thermophysical properties and one-dimensional transient conduction in the domain between points 1 and 3. The temperature response at point 2 will be given as

$$\Delta T_2(time) = fct(\Delta T_1(time), \Delta T_3(time); \alpha, \Delta y_1, \Delta y_2) \tag{17}$$

The temperature response at point 1 describes the heat input to the test domain; that at point 3 describes the heat outflow from the test domain. Then, the temperature response at point 2 can be used to determine the thermal diffusivity, α.

The test rig consists of a radiant heat source and shutter mechanism, and well-insulated screen laminate test articles that are approximately 200mm long. Temperatures at points 1, 2 and 3 are sensed with 0.25mm diameter Type T thermocouples that are soldered to the test article. Measurement points are located with an accuracy of ±0.1mm. Temperatures are recorded at a rate of 10 – 20 samples per second, with an accuracy of approximately ±0.2°C.

Three bar-shaped oxygen-free copper test articles (Alloy C10100, $\alpha = 11.1x10^{-5} m^2/s$) were tested. Each test was repeated six times. The mean measured thermal diffusivity for each of the test articles was $11.76x10^{-5} m^2/s$, $11.62x10^{-5} m^2/s$ and $11.72x10^{-5} m^2/s$, giving an overall average value of $11.7x10^{-5} m^2/s$. The 95% confidence interval for all 18 measurements (2σ about mean values) is $0.7x10^{-5} m^2/s$ and the implied offset error, based on the handbook value of thermal diffusivity of alloy C10100, is +5.2%.

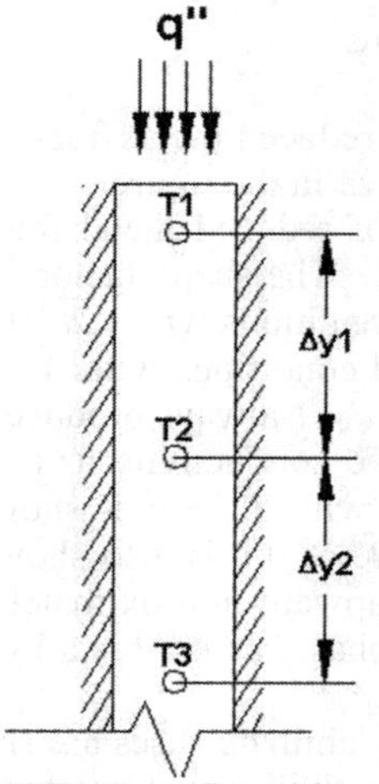

Fig. 7 Test configuration.

4.2 *Screen laminates test articles*

Isotropic, plain-weave copper screens are stacked and bonded together (using 95/5 Sn/Pb solder) to form the screen laminates. Various screen laminate samples were fabricated with this method: the mesh numbers range from $6.3cm^{-1}$ $(16inch^{-1})$ to $15.75cm^{-1}$ $(40inch^{-1})$, the bare copper wire filament diameter ranges from *0.28mm* to *0.46mm*, and the number of layers ranges from 3 to 10. Measured compression factors range from 0.70 to 1.00. Figure 8 shows a typical copper plain-weave screen-laminate. The wire filaments have an effective diameter of *0.48mm*, and the mesh number is $6.3cm^{-1}$ $(16inch^{-1})$. The sample has $\varepsilon = 0.689$, $\beta = 2592m^{-1}$.

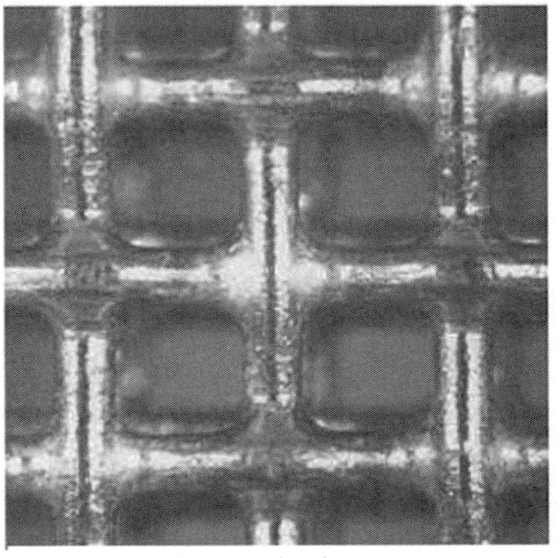

Fig. 8. Bonded screen laminate.

Twenty (20) test articles of the above mentioned fabricated isotropic screen laminates specimens are tested. In the calculation, the effective density/specific heat product, ρc of the screen laminate is calculated based on the measured weight ratios of copper and solder material in the composites. The copper wire properties are: $\rho = 8940$ kg/m^3, c = 393.5 J/(kg·K, $k_s = 400$ W/mK). The Sn-Pb solder material's properties are: $\rho = 7317$kg/m^3, c = 230.0J/kg·K, $k_s = 56$ W/mK. As

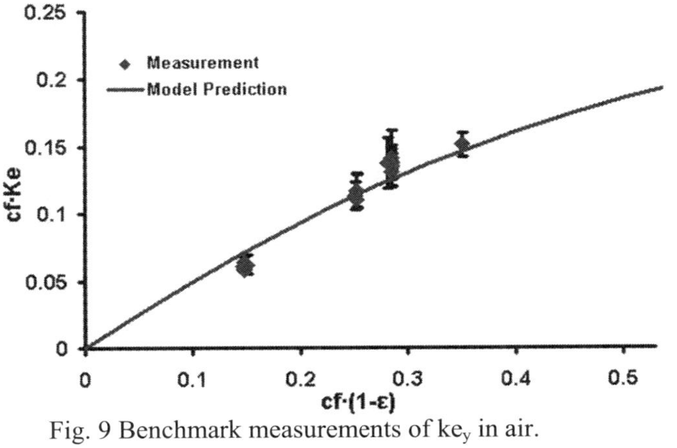

Fig. 9 Benchmark measurements of ke_y in air.

different specimens have different weight ratios of copper and solder material, so the calculated effective densities for the laminates range from 8694kg/m^3 to 8796kg/m^3, the specific heats range from 369.1J/kg·K to 379.0J/kg·K and, the thermal conductivity range from 349.7 W/mK to 371.1 W/mK. The thermal conductivity of air at temperature of 323K, 0.0278W/m·K is used.

4.3 *Test results.*

The average test result of reduced dimensionless in-plane equivalent conductivity for each specimen is shown with 2σ error bars in Figure 9. Each data point shown represents six or more test runs of the same sample. The model prediction for the isotropic plain-weave screen laminate saturated with air is also shown in the figure. It shows that the prediction of the present model is within 10% of all of the measurements. With consideration of the non-uniform solder layer's thickness on the filaments, and non-uniform properties of the high porosity screens, the prediction of the present model is quite accurate.

5 CONCLUSION

Simple porous media structures can be fabricated from conductive screening. The resulting structures can be configured to have a wide range of porosity, specific surface area and in-plane effective thermal conductivity which could be up to 23% and 78% of the filament material's conductivity for isotropic and anisotropic plain-weave screen laminates, respectively. The in-plane effective thermal conductivity of screen laminates, in particular for anisotropic screen laminates is much greater than that can be achieved with other porous media configurations. The material can be configured to have highly anisotropic thermal properties while at the same time high porosity is maintained.

The simple universal model for effective in-plane thermal conductivity and porosity has been verified with benchmark experiments. A technique for thermal conductivity (diffusivity) is set up and calibrated.

6 ACKNOWLEDGEMENT/DISCLAIMER

The Ballistic Missile Defense Organization through the Air Force Office of Scientific Research, USAF, sponsors this work under contract number F49620-99-0286. The views and conclusions contained herein are those of the authors and should not be interpreted as necessarily representing the official policies or endorsements, either expressed or implied, of the Ballistic Missile Defense Organization, the Air force Office of Scientific Research, or the U.S. Government..

REFERENCES

Ashby, M., Evans, A., Fleck, N., Gibson, L., Hutchinson, J. & Wadley, H. 2000. Metal Foams, A Design Guide. Butterworth Heinemann.

Alexander, E. G. Jr. 1972. Structure-Property Relationships in Heat Pipe Wicking Materials. *Ph.D. thesis, Department of Chemical Engineering, North Carolina State University*, Raleigh, NC

Chang, Woo Soon. 1990. Porosity and Effective Thermal Conductivity of Wire Screens. *Journal of Heat Transfer, Feb. 1990, Vol. 112.*

Kaviany, M. 1995. *Principals of Heat Transfer in Porous Media, 2-nd ed.*, Springer.

Koh, J. C. Y. & Fortni, A. 1973. Prediction of thermal conductivity and electrical resistivity of porous metallic materials. *Int. J. Heat Mass Transfer, Vol. 16, pp. 2013-2021*

Luo, S.Y. & Mitra, A. 1999. Finite Elastic Behavior of Flexible Fabric Composite Under Biaxial Loading. *Journal of Applied Mechanics, Vol. 66, pp 631-638.*

Park, J-Wook, Ruch, D. & Wirtz, R.A. 2002 Thermal/Fluid Characteristics of Plain-Weave Laminates as Heat Exchanger Surfaces. *Accepted, 40-th AIAA Aerospace Sciences Meeting and Exhibit*, Reno, NV

Xu, J. 2001. In-Plane Effective Thermal Conductivity of Plain Weave Screen Laminates. *M.S. Thesis, Mechanical Engineering Department, University of Nevada, Reno*, NV 89557, USA.

Zhao, T-W, Narla, V, Fuchs, A., Jiang, Y., & Wirtz, R. A. 2002. Thermal and Mechanical Properties of Woven Mesh Thermal Energy Storage Composites. *pending, AIAA/ASME Thermophysics and Heat Transfer Conf.*

Heat pipes and thermosyphons

Thermal Challenges in Next Generation Electronic Systems, Joshi & Garimella (eds)
 ISBN 90-77017-03-8

Metal foam heat sinks for laptop cooling application in a heat pipe-heat sink arrangement

A. Gupta
Flomerics, Inc., Southborough, MA, USA

N.F. Dean
STaR Center, Honeywell Electronic Materials, Spokane, WA, USA

Keywords: FLOTHERM, Computational Fluid Dynamics, CFD, metal foam, heat sink, heat pipe, laptop

ABSTRACT: The purpose of this study was to determine the viability of metal foam heat sinks for thermal management of laptop computers through numerical modeling, experiments and analysis. FLOTHERM was used as the Computational Fluid Dynamics (CFD) tool for numerical modeling. The focus was situations where heat is transferred away from the heat source by a heat pipe and dissipated remotely. Two cooling schemes, which differ in the type of fan-heat sink combination at the remote location, were employed. One is a duct-covered fan and heat sink and the other, a fan-sink, with a fan mounted directly on top of the heat sink. Experiments were conducted in a temperature-controlled wind tunnel to simulate a hot laptop enclosure. Results show that the proposed schemes are capable of dissipating 9-12W in forced convection, with significant advantage in weight over conventional cooling schemes. Tube bundle theory and FLOTHERM simulations were used to determine the pressure drop characteristic of heat sinks. The fan operating point was determined for each configuration. The study demonstrates that high performance, low-weight and reliable thermal management schemes can be developed using metal foam heat sinks.

1 INTRODUCTION

One common way of removing heat from processors in laptops and mobile computers is to transfer the heat to a remote location using a heat pipe and dissipate it using a forced convection air-cooled heat sink. Conventional heat sink designs like extruded sinks, bonded fin heat sinks and machined heat sinks have poor heat transfer efficiency and/or high cost associated with their production. The surface area available in a given volume often limits the heat transfer in these cases. This is mainly due to manufacturing aspects like fin aspect ratio (fin height to fin spacing ratio), which need to be within a certain maximum limit. Hence, as the microprocessor power output goes up, such systems become bulky and expensive. These problems can be alleviated with the use of metal foam heat sinks, which provide a high surface area to volume ratio, good heat transfer efficiency and considerable weight advantage over conventional heat sinks.

The heat dissipation capacity of a heat sink, Q, is a function of the heat transfer coefficient, h, the surface area of the heat sink exposed to airflow, A, and the temperature difference between the heat sink and air, ΔT. Mathematically,

$$Q = hA\Delta T \tag{1}$$

An irregular cell structure and a staggered arrangement of strands causes a high level of turbulence in metal foams, which results in a higher heat transfer coefficient, h, as compared to conventional heat sinks. Also, depending on the pore size, a very large surface area, A, is available for heat transfer within a given volume. Hence, metal foam heat sinks are shown to be very effective in heat dissipation (Bastawros 1997, Lee et al 1993). This study describes the use of metal foam heat sink assembly in ducted and fansink arrangements for laptop cooling application.

2 EXPERIMENTAL SET-UP

A heated aluminum block, 1.5 in (L) x 1.5 in (W) x 0.25 in (H), was used as a thermal dummy to simulate the heat generated by a computer microprocessor. This block served as a heat source as well as a heat pipe holder. The evaporator end of a heat pipe, 12 in long and 0.16 in diameter, was inserted into this aluminum block. The condenser end of the heat pipe was inserted into a central hole in the heat sink base plate, thus establishing a thermal path from the heat source to the remotely located heat sink.

The metal foam heat sink consisted of a piece of aluminum foam of the desired size bonded to an aluminum base plate. Aluminum base plates were machined to the required dimensions and a heat pipe hole was drilled at the center of one face of the plate for heat pipe insertion. The tolerance between the heat pipe hole and heat pipe diameter was tightly controlled to ensure good contact. Highly conductive thermal grease was used between the mating surfaces. The foam and the base plate were bonded after a thorough chemical cleaning. A high strength, highly conductive silver filled epoxy was used as the bonding agent. A 92-94% (6-8% metal density) porous foam was chosen for all the experimental runs. The number of pores per linear inch (PPI) ranged from 5 to 40. A plastic duct was placed over the metal foam heat sink assembly for heat sinks in ducted arrangement. A fan was placed next to the heat sink in the duct to draw air through the heat sink assembly. For heat sinks in fansink arrangement, a fan was placed directly on top of the heat sink. These arrangements are shown in Figure 1.

During operation of this set up, the heat pipe block was heated by powering a 1 in x 1 in, 10 mil thick resistive heater in contact with the block. The fan was connected to another power supply. 1 and 1.5-in diameter, 0.4-in thick, low profile fans (Comair Rotron data sheets) were employed. Data was collected for various heat dissipation levels at the heater block. Thermocouples were used for temperature measurement of the heater block and the heat sink base. The power supply system and the thermocouples were connected to a data acquisition system which recorded transient data and quit automatically after steady state was achieved.

3 SIMULATIONS

FLOTHERM was used to simulate the setup described above for validation of experimental and analytical results as well as for performing tests and obtaining results for the cases for which experiments were not conducted. Flotherm model showing the ducted and the fansink arrangement is shown in Figure 1. Please note that the duct has been hidden from the view to show the heat sink in the ducted configuration and the heat sink is represented in compact form in the fansink arrangement.

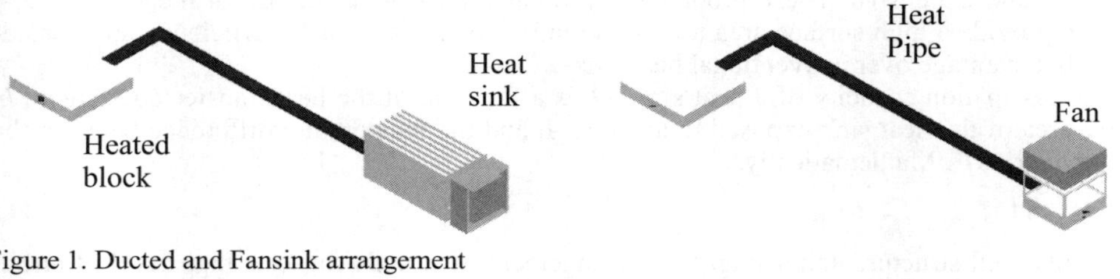

Figure 1. Ducted and Fansink arrangement

4 RESULTS

The entire heat pipe-heat sink setup was placed in the wind tunnel, which was maintained at 40° C, to simulate the actual conditions inside a computer enclosure. A small airflow of 10 FPM was maintained to homogenize the temperature across the cross-section of the wind tunnel. Experiments were done with heat sinks oriented cross wise as well as in line with the airflow to establish the effect of wind tunnel airflow on heat sink performance. The airflow rate in the wind tunnel was also varied in a narrow range. Such a small rate of airflow was found to have no effect on heat sink performance. The results of the experiments and simulations conducted to establish the heat dissipation limits of metal foam heat sinks are presented in the following sections. Data is mostly presented in the form of thermal resistance charts. While comparing these results with other applicable studies, it must be kept in mind that the present set up didn't utilize a heat spreader (commonly an aluminum plate mounted under the keyboard or on the chassis of the laptop to enhance heat transfer). Modeling work done in-house (Gupta 2000) has shown that heat spreaders can lower the overall thermal resistance by over 30% in such arrangements.

4.1 *Ducted heat sink tests*

Two heat sinks were tested in ducted arrangement – one high aspect ratio machined heat sink and a metal foam heat sink. The heat sinks were 2-in long, 1-in wide and 1-in tall, including a 0.25-in thick base plate for accommodating the heat pipe. A 94% porous, 10 PPI foam was used for all the experimental runs. Higher PPI aluminum foams could not be obtained in the desired size. The machined heat sink had eight fins, equally spaced at 0.067-in and each 0.067-in thick, resulting in an aspect ratio of about 12. The purpose of this heat sink was to serve as a benchmark for the performance of metal foam heat sinks. In this arrangement, both the fan and the heat sink were fully covered by a duct.

The overall thermal resistance of the system, from heat source to ambient air, is plotted for all the cases. A heat source to ambient temperature difference of 50° C is assumed acceptable throughout this study.

Figure 2 shows the results of fully ducted forced convection experiments. It is evident that the heat sinks are capable of dissipating 9-10W in this arrangement (50^0 C has been assumed to be the acceptable total temperature difference). High aspect ratio machined heat sink is thermally better than the foam heat sink by about 6%.

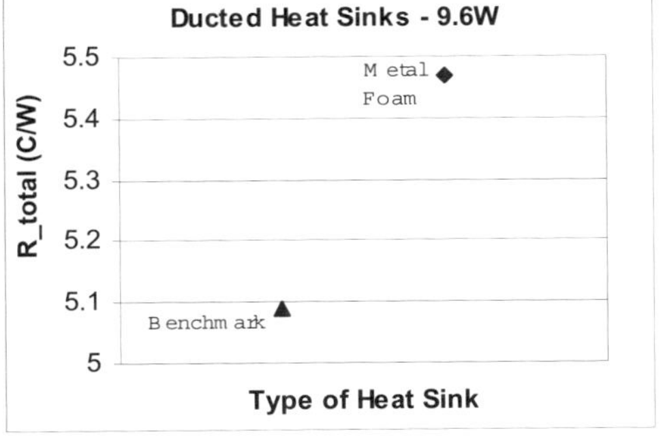

Figure 2. Total system thermal resistance for fully ducted heat sink arrangement

Even though the machined heat sink outperformed the foam heat sink thermally in all the experimental runs, the difference in the total thermal resistance was 6% or less in all cases. However, the metal foam heat sink (at 27 grams) was substantially lighter than the machined heat sink (58 grams) – offering over 50% reduction in weight. Good thermal performance and light weight make metal foam heat sinks a strong candidate for laptop/portable device cooling applications.

4.2 *Fansink tests*

Metal foam heat sinks were fabricated for testing with 1.5-in and 1-in diameter fans in fansink configuration. Four heat sinks, one each made out of 94% porous, 5, 10, 20 and 40 PPI aluminum foam respectively, were tested with a 1-in fan. These heat sinks were 1-in square and 0.75-in tall. They were compared to a machined (high aspect ratio) benchmark heat sink using FLOTHERM simulations. The benchmark heat sink had 64 equally spaced pins, each 0.065-in thick. Actual experiments were not conducted for this benchmarking test.

Three heat sinks, one each made out of 5 PPI, 10 PPI and 20 PPI aluminum foam, were tested with a 1.5-in fan. These were 1.5-in square and 0.75-in tall. A machined aluminum heat sink, having 144 equally spaced pins, each 0.065-in thick, was used as a benchmark for 1.5-in heat sink tests.

Figure 3 shows the results for 1-in fansink in forced convection for 9.6W heat dissipation at the heat source. As evident, about 10W is the maximum heat dissipation capacity of 1-in heat sinks. Benchmark heat sink resulted in the best performance. 10 PPI foam was thermally the best foam heat sink. This can be explained on the basis of pressure drop across these heat sinks, and will be discussed in greater detail in the discussion section.

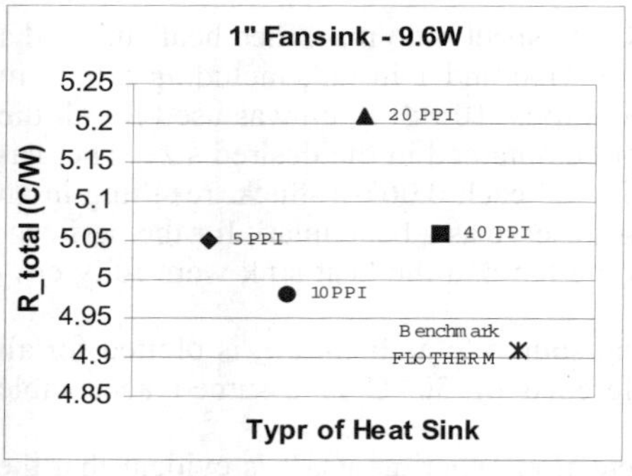

Figure 3. Total system thermal resistance for 1-in fansink arrangement

The results for 1.5-in heat sinks are shown in Figure 4. Over 12W could be dissipated in the best case.

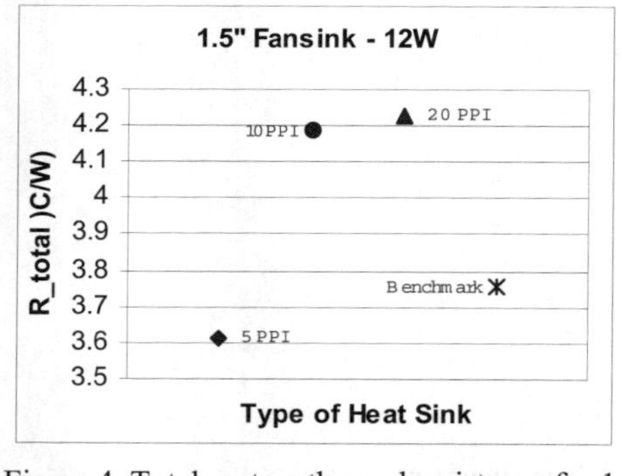

Figure 4. Total system thermal resistance for 1.5-in fansink arrangement

The 5 PPI foam heat sink performed better than the higher PPI foams as well as the machined heat sink. Metal foam heat sinks were about half the weight of the machined heat sinks. The performance of 1.5-in heat sinks was about 30% better than 1-in sinks on average, mainly due to a larger heat transfer area. Lower PPI foams performed better in both 1 and 1.5-in heat sink configurations. 5 PPI foam was the best for 1.5-in and 10 PPI for 1-in sink. This behavior will be discussed in further detail in the discussion section.

5 PRESSURE DROP PREDICTION FOR FLOW THROUGH METAL FOAM HEAT SINKS

Many interesting observations can be made based on the results presented above regarding the difference between ducted and fansink configuration and the effect of foam PPI on thermal performance. This section attempts to lay a theoretical foundation for the discussion of results to follow.

In general terms the pressure drop can be represented as (Geiger & Poirier 1980):

$$\frac{\Delta P}{L} = a\mu V + b\rho V^2 \tag{2}$$

where μ is the fluid viscosity, ρ, the fluid density, V, the average fluid velocity exiting the foam sink and a and b are constants for a particular foam cell structure. Numerous studies have been conducted to determine the constants a and b for pressure drop prediction. A few are briefly discussed here.

5.1 *Tube bundle theory and Ergun's equation*

This theory (Geiger & Poirier 1980) has been used to predict pressure drop for flow through packed bed of solids. It has been shown to yield reasonable results for metal foams, probably because it regards the bed of solids as a set of tangled tubes with weird cross-sections – a structure somewhat similar to metal foams. For turbulent flow, the correlation is given as:

$$\frac{\Delta P}{L} = \frac{4.2\mu V S^2}{\varepsilon^3} + \frac{0.292\rho V^2 S}{\varepsilon^3} \tag{3}$$

S is the total surface area per unit volume of bed (or foam, obtained from foam manufacturer's data) and ε is the porosity of the packed bed (or the metal foam).

5.2 *Plessis' correlation*

This theoretical model (Plessis 1994) is based on transport in a representative unit cell. A representative unit cell is imagined to be a cube with a rectangular arrangement of solid material (foam strands). In this manner, the foam is modeled as an isotropic porous material. The general correlation is:

$$-\frac{dP}{dx} = \frac{\mu V 36\chi(\chi-1)}{d^2\varepsilon^2} + \frac{\rho V^2 2.05\chi(\chi-1)}{d\varepsilon^2(3-\chi)} \tag{4}$$

where d is the length of the representative unit cell. χ is the tortuosity of the porous matrix and is given as:

$$\frac{1}{\chi} = \frac{3}{4\varepsilon} + \frac{\sqrt{9-8\varepsilon}}{2\varepsilon}\cos\left\{\frac{4\pi}{3} + \frac{1}{3}\cos^{-1}\left[\frac{8\varepsilon^2 - 36\varepsilon + 27}{(9-8\varepsilon)^{1.5}}\right]\right\} \tag{5}$$

5.3 *Volume averaged momentum equation*

Assuming a constant pressure gradient for flow through a porous matrix, the pressure drop can be derived from the volume averaged momentum equation as:

$$-\frac{dP}{dx} = \frac{\mu V}{K} + \frac{\rho f V^2}{\sqrt{K}} \tag{6}$$

K is the permeability and *f*, the inertia coefficient for fluid flow. Calmidi & Mahajan (1998) determined the permeability and the inertia coefficient experimentally for various foam densities and porosities.

These models were compared with manufacturer's data (ERG product data) for pressure drop across a 94% porous, 10 PPI foam for an accuracy check. This comparison is shown in Figure 5. Tube bundle theory and Plessis' model appear to match well with the experimental data while Calmidi's (1998) semi-empirical approach over predicts the pressure drop.

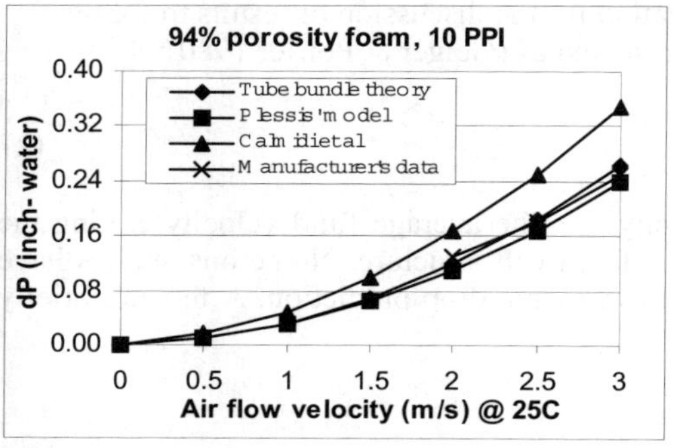

Figure 5. Pressure drop prediction for flow through a 94% porous, 10 PPI foam using various models

5.4 *Determination of the fan operating point*

The fan operating point can now be determined using the models outlined above for all the heat sink-fan configurations employed in this study. It is the point of intersection of the pressure drop characteristic of the heat sink and the fan curve (Comair Rotron data sheets). Tube bundle theory, because of its simplicity and accuracy, was chosen to derive the pressure drop-velocity curve for all the metal foam heat sinks. FLOTHERM simulations were used to obtain the pressure drop characteristics for the machined heat sinks. Figure 6 shows the fan operating point for all the cases.

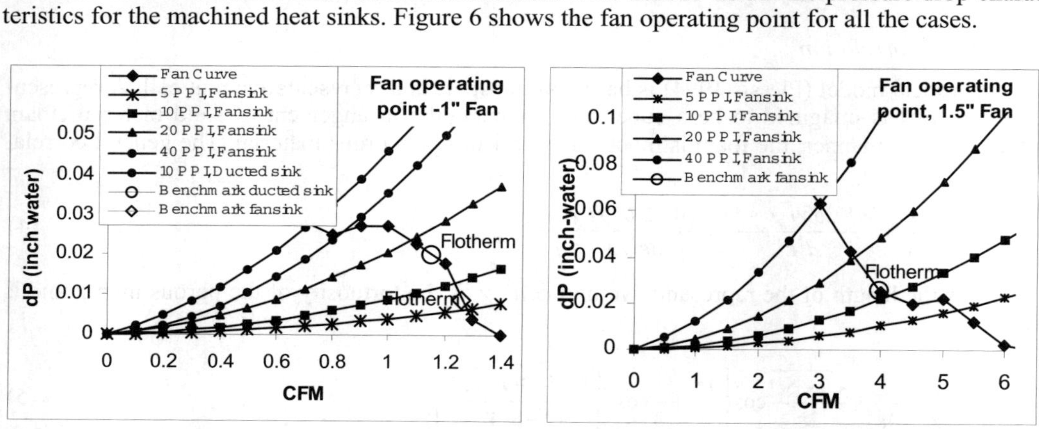

Figure 6. Fan operating points for 1-in and 1.5-in fans in various configurations

6 DISCUSSION OF RESULTS

The pressure drop across a metal foam structure is inherently high, which reduces the airflow through it as compared to conventional heat sinks. The advantage of larger surface area and greater turbulence can only be utilized if sufficient amount of air can be forced through them, either by using a powerful fan or a low PPI foam. The following discussion illustrates this.

Fansinks performed better than the ducted heat sinks. Figures 2 and 3 show that the performance of a 1-in fansink was about 5% better than a 1-in ducted heat sink, despite the "fin" area of the fansink sink being 50% smaller. Figure 6 shows the cause, a high pressure drop resulting in a linear speed of only 1.92 ft/sec (0.8 CFM) through the ducted heat sink as compared to 2.9 ft/sec (1.2 CFM) through the 10 PPI fansink – a 50% higher airflow. This confirms that airflow is the single most dominating factor in determining the heat transfer efficiency of a metal foam heat sink. An increase in surface area (such as with high PPI foams) doesn't proportionately compensate for a reduction in airflow.

1.5-in fansinks performed about 30% better than 1-in fansinks due to a much larger heat sink "fin" area and a bigger fan.

10 PPI foam performed better than the 5 PPI foam in 1-in fansink configuration (Fig. 3) because the 10 PPI foam has a much larger heat transfer area but a very similar air flow rate as the 5 PPI foam (Fig. 6). The effect of reduced airflow comes into play at 20 and 40 PPI, resulting in poorer performance. However, 5 PPI foam was the best in 1.5-in fansink arrangement because it had at least a 30% higher airflow as compared to other higher PPI foams (Fig. 6) in this arrangement.

The machined heat sink performed slightly better than all the foam heat sinks in the ducted arrangement (Fig. 2) while some foam heat sinks (5 PPI foam) performed better than the machined heat sink in 1.5-in fansink configuration (Fig. 4). This is again because of the high pressure drop through metal foam heat sinks in the ducted arrangement (Fig. 6).

It is clear that the knowledge of the fan operating point is critical in the determination of system performance. The methodology of analysis outlined above can help a designer identify optimum fan-foam heat sink combinations for a given application with only a limited number of experiments. Numerical simulations are particularly handy in filling up the gaps in experiments as in this study.

7 SUMMARY

Metal foam heat sinks were tested in ducted and fansink configurations in forced convection environment. Machined heat sinks were included in the tests to serve as benchmark for comparing the performance of metal foam heat sinks with the more conventional ones. Ducted heat sinks utilized a 1-in fan for cooling. The heat dissipation capacity in this arrangement was found to be over 9W in forced convection. Machined heat sink performed better than the foam heat sinks by about 6% in the ducted arrangement.

Fansink performance was found to be significantly better than the ducted heat sinks. A 1-in fansink was capable of handling 9-10W in forced convection – thermal performance very similar to the ducted heat sinks, but at less than half the weight and volume of ducted sinks (ducted foam sinks were bigger in size). 1.5-in fansinks were found to be capable of handling 12-13W. The 5 PPI metal foam heat sink outperformed the machined heat sink in 1.5-in fansink arrangement.

Foams with 5, 10, 20 and 40 PPI were used in fansink tests. In general, lower PPI foams showed better performance. 10 PPI foam was the most effective for 1-in and 5 PPI for 1.5-in fansinks. This was explained on the basis of larger pressure drop associated with higher PPI foams. The methodology of determining the fan operating point and its use for interpretation of results was outlined in the later sections of this report.

8 REFERENCES

Bastawros, A.F. & Evans, A.G. 1997. Characterisation of open-cell Aluminum alloy foams as heat sinks for high power electronic devices. *Proceedings of the symposium on the application of heat transfer in microelectronics packaging, IMECE, Dallas, 1997.*

Calmidi, V.V, & Mahajan, R.L. 1998. Convective transport in high porosity metal foams. *1998 IMECE Meeting, Nov. 15-19, Anaheim, CA.*

Comair Rotron, Product data sheets, 2675 Customhouse Court, San Diego, CA 92154.

ERG Materials and Aerospace Corporation, Product data sheets and technical literature, 900 Stanford Avenue, Oakland, CA 94608.

Geiger G.H., & Poirier, D.R. 1980. *Transport phenomena in metallurgy*. Addison Wesley.

Gupta. A. 2000. Unpublished modeling work done at Flomerics, 257 Turnpike Road, Southborough, MA, USA, 2000

Lee, Y.C., Zhang W., Xie H & Mahajan R.L. 1993. Cooling of a flipchip package with a 100W, 1 cm sq chip. *ASME International electronics packaging conference, New York, 1993.*

Plessis, P.D. 1994. Pressure drop prediction for flow through high porosity metallic foams. *Chemical Engineering Science*, Vol 49, No. 21, pp 3545-3553.

Thermal Challenges in Next Generation Electronic Systems, Joshi & Garimella (eds)
 ISBN 90-77017-03-8

Design and performance evaluation of a compact thermosyphon

Aniruddha Pal & Yogendra Joshi
Woodruff School of Mechanical Engineering, Georgia Institute of Technology, Atlanta, Georgia, USA

Monem H. Beitelmal & Chandrakant D. Patel
Hewlett-Packard Research Laboratories, Palo Alto, California, USA

Todd Wenger
Thermacore International Inc., Lancaster, Pennsylvania, USA

Keywords: thermosyphon design, electronics cooling, dielectric liquid cooling

ABSTRACT: Thermosyphons are a great prospect towards cooling of high heat dissipating electronics. The implementation of a compact thermosyphon for cooling of a Pentium 4 microprocessor in a Hewlett Packard Vectra PC is being presented in this paper. The thermosyphon involves four components in a loop: an evaporator with a boiling enhancement structure, a rising tube, a condenser and a falling tube. Experiments are done to assess the effects of working fluids and of system inclination on the performance of the thermosyphon loop. Numerical simulations are performed to reflect on the efficiency of the condenser in natural convection condition and the effect of the system fan on the performance of the condenser.

1 INTRODUCTION

Personal computers are currently air-cooled by either free or forced convection. Fans attached to the central processing unit (CPU) are the most popular cooling devices for low power dissipation systems. They are popular because of reliability, cost, efficiency and ease of implementation. However, for higher frequency chips (above 1000 MHz), the air-cooled heat sinks have some limitations in the form of bulky size, noise and insufficient cooling performance. As a result, current practice of dense packaging of electronics in compact spaces demands novel ways of heat dissipation, which will be able to dissipate as much as 100 W/cm^2 at chip levels while maintaining the device at acceptable temperatures, typically below 85°C.

One of the proposed solutions for dissipating high heat fluxes is the use of liquid cooling techniques. Two-phase heat transfer is a very attractive liquid cooling process, as high heat fluxes can be removed through vaporization of the fluid in an evaporator attached to or enclosing the heat source, but the added process of condensing the vapor and bringing it back to the evaporator adds complexity in the system.

A thermosyphon is such a device, which successfully implements two-phase liquid cooling by indirect contact with electronics. A two-phase thermosyphon basically consists of an evaporator and a condenser, which are connected through a passage, or a loop. Heat is transferred from the source through an interface to the evaporator, where the fluid vaporizes taking the latent heat of vaporization. The vapor then moves to the condenser, where it condenses. The released heat is dissipated to the ambient from the condenser and the condensed liquid is returned to the evaporator, thus completing a closed loop. The density difference between the liquid and vapor creates a pressure head, which drives the flow through the loop, and as such no driving force is needed. Thermal resistance at the interface between the heat source and the evaporator is a key design parameter. This must be minimized in order to get the benefit of low thermal resistance of the evaporator.

2 BACKGROUND

Tengblad and Palm (1996) tested several prototypes of thermosyphons with single and multiple evaporators. Ramaswamy et al. (2000) studied the combined effects of pressure and sub-cooling on the performance of two-chamber thermosyphons with gradually increasing heat fluxes. Lang et al. (2001) took the concept further and studied the effects of imposed circulation and location of condenser on the performance of a two-phase thermosyphon in a confined space. Garner et al. (2001) looked into the applications of loop thermosyphon (LTS) in high-density packaging. Recently, Webb et al. (2001) demonstrated a system capable of dissipating 100 W from a single CPU system.

The above-mentioned works established the importance and the prospect of thermosyphon as a compact thermal management device, both in terms of flexibility in size and cooling capacity range. While laboratory demonstrations of thermosyphons exist in the above literature, implementation of an actual device for a current microprocessor has not been presented. This work describes the design, construction and performance assessment of a two-phase compact thermosyphon for a Hewlett Packard Vectra VL800 PC with a peak microprocessor power dissipation of 80 W.

3 DESIGN PROCESS

For designing the thermosyphon from ground up, the following steps are taken:

- Verifying Cooling Requirements
- Developing Compact Loop Thermosyphon Design
- Fabrication of LTS unit

3.1 *Verifying cooling requirements*

The first step in designing the thermosyphon is to verify the cooling requirements and the available space for integrating the device with the CPU. The baseline specifications and parameters chosen are shown in Table 1.

Table 1. Design specifications and parameters for Vectra desktop

CPU	Pentium 4, 1400 MHz
Heat Load	80 W
Air Flow	0.00377 m^3/s (8 CFM)
Maximum CPU Case Temperature	70°C
Maximum Internal Ambient Temperature	45°C
Available Condenser Volume	75 (H) × 82 (W) × 26 (D) mm
Available Evaporator Volume	32 (L) × 32 (W) × 29 (D) mm
Allowed Temperature Rise Above Ambient	25°C
Target Thermal resistance	0.312 °C/W

3.2 *Developing compact Loop Thermosyphon (LTS) design*

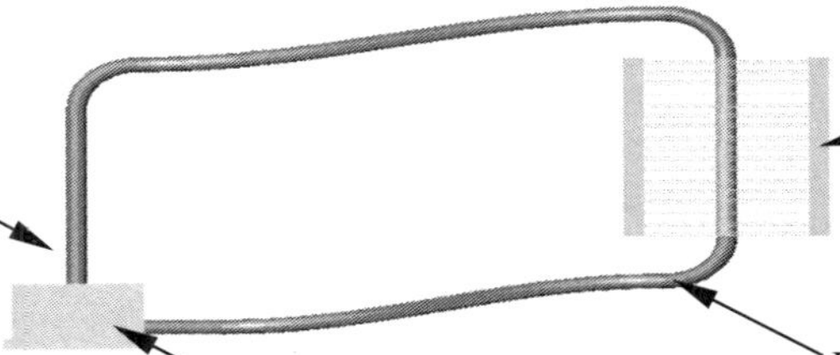

Figure 1. Conceptual design of the Loop Thermosyphon

Based on the design requirements, the conceptual design of the LTS is conceived as shown in Figure 1.

The design for the LTS is to have:

- An evaporator attached to the back of the chip with a boiling enhancement structure inside.
- A vapor outlet line from the top of the evaporator to the entrance of the condenser.
- An Aluminum finned condenser with horizontally oriented fins.
- A liquid line from the condenser outlet to the evaporator inlet.

3.3 *Fabrication of LTS unit*

Majority of the design efforts focused on the following issues:

- Sizing of liquid and vapor line.
- Easy manufacturability of the LTS unit.
- Suitable design of the condenser with appropriate number of fins.

Using Methanol as the working fluid at a power input of 240 watts, the liquid and vapor lines are sized. Except for Aluminum fins, Copper is used everywhere, because of its compatibility with the working fluid and the ease at which copper can be joined (brazed or soldered). The copper tubing also easily lends itself for custom bending of vapor and liquid lines in the compact space of a PC. The condenser is designed with 43 fins (38.1 × 82.5 mm), made of Aluminum 3003-H 14. The fins are pushed onto the condenser tube and attached using epoxy. Space restrictions inside the cabinet allowed only a horizontal orientation of the condenser with vertical fins.

The heat input block attached to the chip is approximately 12.7 × 12.7 mm. A spring-mounted thermocouple is placed in the center of the block. Using STYCAST TC-4 thermally conductive compound, the LTS evaporator is centered on the heat input block. The thermocouples for measuring temperature are placed as shown in Figure 2.

One critical factor in the design of a thermosyphon is incipience overshoot, which is the increase in temperature needed above the boiling point of the working fluid to initiate boiling. One solution is the use of boiling enhancement structure, which not only increases the surface area of the heat source in the evaporator but also increases the critical heat flux, as bubble generation rate is increased. A three-dimensional boiling enhancement structure (Ramaswamy, 1999) is fabricated and attached to the heat input block. The boiling enhancement structure (Figure 3) is made up of six layers of square (15 × 15 × 0.8 mm) copper pieces. Microchannels of approximately 0.31 mm width are machined on both sides of each copper strip in mutually perpendicular directions, and then the copper strips are soldered at 231°C (Pb-Sn liquidus temperature).

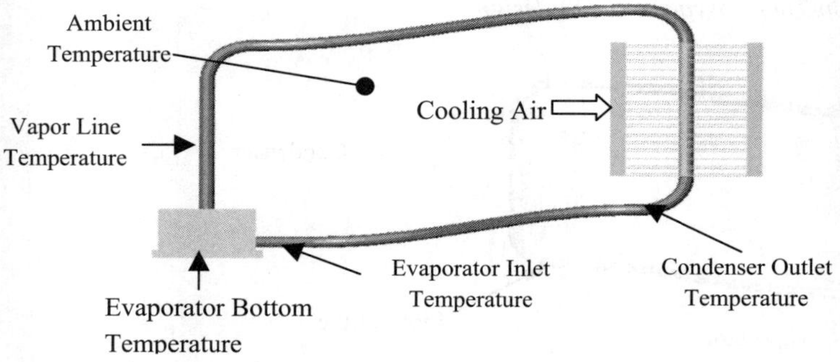

Figure 2. Thermocouple locations

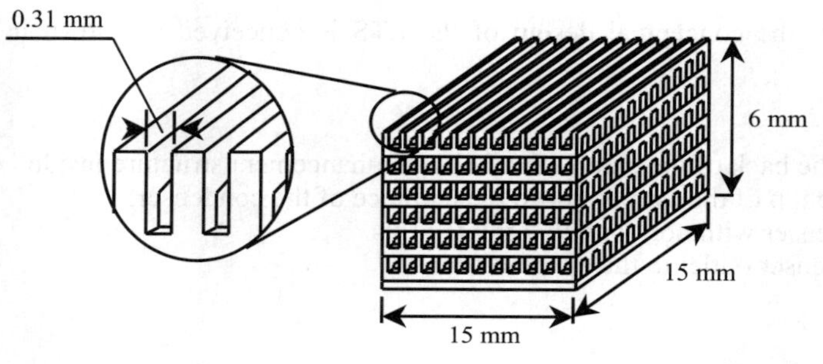

Figure 3. Magnified view of the boiling enhancement structure

4 STUDY OF THERMOSYPHON PERFORMANCE

The performance of the compact thermosyphon is studied based on the following topics:

- Effect of working fluids (water and dielectric liquid).
- Effect of inclination of the thermosyphon.
- Efficiency of the condenser in natural convection condition.

4.1 *Effect of working fluids*

The thermosyphon is tested with two fluids – deionized ultra filtered (DIUF) water and PF5060, a dielectric liquid. The LTS is mounted on a test stand and 0.0075 m^3/s (16 CFM) of cooling air is passed over the condenser fin stack. The parameters measured at various power inputs for two different working fluids are given in Table 2 and 3.

The water charged system clearly outperformed the dielectric liquid charged system as seen from the thermal resistance values. Water is a better working fluid, as the thermal properties of water are better than that of PF5060. The high thermal resistance in PF5060 may be attributed to the high amount of dissolved gases in it. It has been shown that the dissolved gases enhance heat transfer coefficient, however, they lead to a decrease in critical heat flux, thus increasing thermal resistance. The pressure of the dissolved gases may also have an adverse effect on the condenser and lead to overall performance degradation. However, with proper degassing procedure before charging the system, it may be possible to improve the performance of the PF5060 charged system.

Table 2. Thermosyphon charged with 8×10^{-6} m^3 (8 cc) DIUF, Test Data

Power (W)	Temperature (°C)						Thermal Resistance (°C/W)
	Evaporator Bottom	Vapor Line	Condenser Outlet	Evaporator Inlet	Ambient	ΔT (Evaporator to Ambient)	
25	34.6	32.7	31.4	32.0	23.2	11.4	0.46
41	40.5	39.0	38.0	38.3	23.3	17.2	0.42
55	45.3	43.6	42.5	42.3	23.2	22.1	0.40
71	52.3	50.3	49.6	48.8	23.3	29.0	0.41
82	54.5	53.9	52.7	52.2	23.1	31.4	0.38
86	57.4	56.0	55.2	54.5	22.6	34.8	0.41

Table 3. Thermosyphon charged with 8×10^{-6} m^3 (8 cc) PF5060, Test Data

Power (W)	Temperature (°C)						Thermal Resistance (°C/W)
	Evaporator Bottom	Vapor Line	Condenser Outlet	Evaporator Inlet	Ambient	ΔT (Evaporator to Ambient)	
25	45.6	37.5	36.5	37.1	22.1	23.5	0.94
41	61.6	50.3	48.4	48.7	22.2	39.4	0.96
51	71.5	58.8	56.3	56.3	22.3	49.2	0.97
65	83.3	68.9	65.2	65.1	22.4	60.9	0.94
81	101.0	83.7	79.0	78.7	22.6	78.4	0.97

4.2 *Effect of inclination on the performance of the thermosyphon*

Experiments are performed using a HP custom designed protractor assembly to tilt the loop thermosyphon under a controlled condition to assess the effects of inclination on the loop thermosyphon performance.

4.2.1 *Experimental method and apparatus*

A schematic of the experimental apparatus is shown in Figure 4.

The heating is provided by a copper heater block, attached to the bottom of the evaporator with thermal interface material (thermal conductivity = 0.39 W/m-K). Both the evaporator chamber and the heater block assembly are centered inside a bolted Plexiglas/Ultem enclosure to ensure a firm attachment between the evaporator and the surface of the heat source to minimize the interface resistance and to reduce heat loss to the environment. The evaporator dimension is 25 × 25 × 25 mm with a base of 30 × 30 × 10 mm, while the heat source dimension is 30 × 30 × 10 mm. A cartridge heater is inserted into the copper block to provide 60 W of total power. The heat is provided by a variable power supply monitored by a digital wattmeter. A fan powered by an HP E3636A DC power supply cools the condenser. The protractor assembly (Figure 5) allows the tilting of the thermosyphon loop in increments of 2° from the horizontal position in both the positive and negative direction with a ±0.2°. Temperatures of the evaporator, the condenser, the heat source and the inlet air to the condenser are measured using nine type T thermocouples. Steady-state temperature measurements are collected and recorded using an HP75000B data acquisition unit and an HP Vectra XU 6/200 utilizing an HP Visual Engineering Environment (HPVEE) software program.

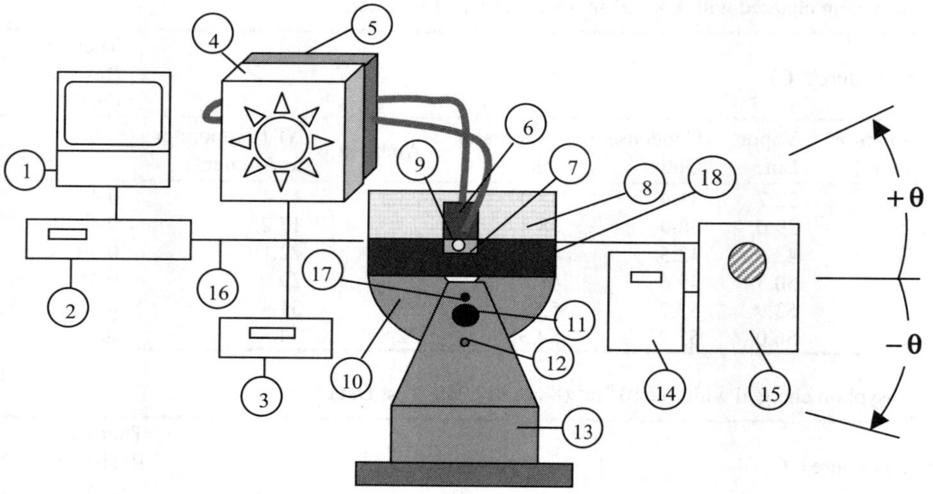

1. HP Vectra XU
2. HP 75000B Data Acquisition Unit
3. HP E3616 DC Power Supply
4. Condenser fan
5. Condenser
6. Evaporator chamber
7. Heater Block
8. Plexiglas enclosure
9. Cartridge heater
10. Protractor
11. Fine tuning pin
12. Locking pin
13. Assembly stand
14. Wattmeter
15. Variable Auto Transformer
16. Type T thermocouples
17. Shaft
18. Insulation

Figure 4. Schematic of the thermosyphon test assembly

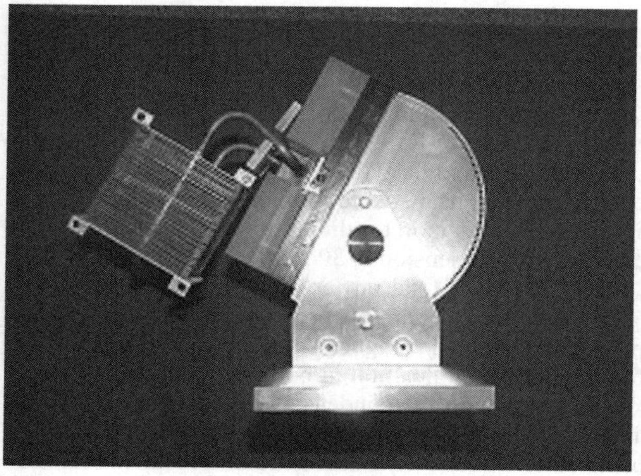

Figure 5. Protractor assembly with the thermosyphon

4.2.2 *Results and discussion*

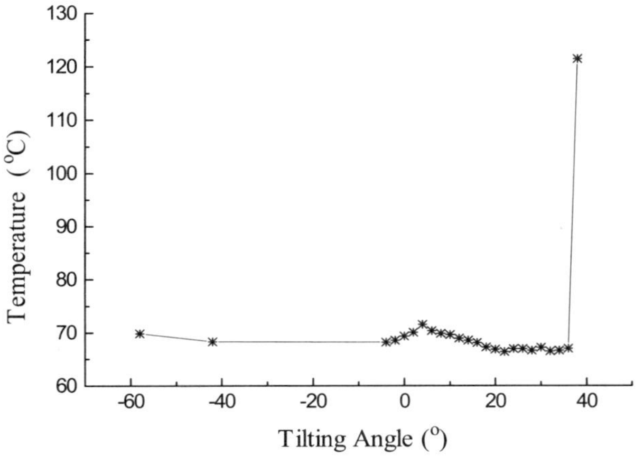

Figure 6. Evaporator temperature vs. tilting angle

Figure 6 shows that the evaporator maintains a uniform temperature for tilting angle up to -58°, which is the maximum tilting angle allowed by the protractor assembly fixture in the negative direction. In the positive direction the evaporator temperature is maintained between 66-71°C for tilting angles up to 36°, after which the evaporator temperature shoots up over 120°C. At this point, the severe inclination prevents the return of sufficient amount of liquid into the evaporator to replace the vapor leaving the chamber, leading to dry out. Thus the evaporator temperature increases suddenly due to the lack of cooling liquid present inside the evaporator chamber. The processes take place suddenly without any indication, and thus produce a discontinuity in the temperature readings. It is observed that the dry out angle (38°) is at the position when the condenser outlet level is at the same or close to the evaporator inlet level. This is expected since this loop is gravity aided and for liquid to return to the evaporator chamber, a positive gravity level has to be present between the condenser outlet and the evaporator chamber inlet. However, the length of the falling tube and the system charge may also be factors in determining the dry out angle.

An interesting point to observe in Figure 6, is that the evaporator temperature is much higher than 56°C (boiling point of PF5060 at atmospheric pressure), which suggests that the boiling might have occurred at a temperature higher than 56°C. This increase in boiling point may be attributed to higher pressure inside the evaporator, which can be caused by insufficient degassing or overcharging of working fluid in the thermosyphon.

Figure 7 shows temperature difference between the condenser and the evaporator versus the tilting angle. A key finding from this preliminary testing approach is that the thermal performance of the thermosyphon loop remained almost unaltered for a very large tilting angle.

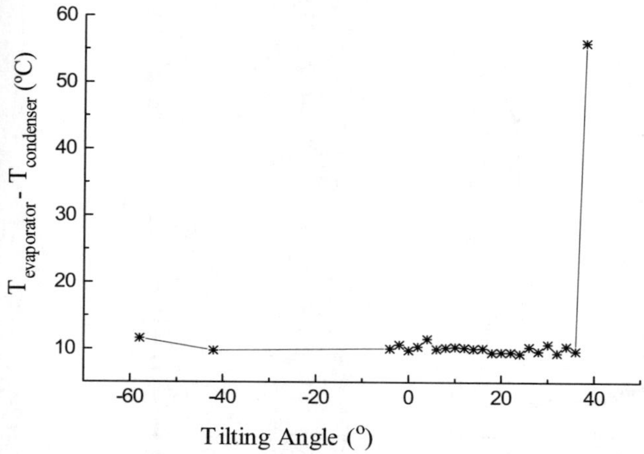

Figure 7. ($T_{evaporator} - T_{condenser}$) vs. tilting angle

4.3 *Efficiency of the condenser in natural convection condition*

4.3.1 *Numerical simulation*

To assess the performance of the condenser in the absence of a fan, a system based numerical approach is adopted to model the inside of the PC cabinet. The commercial software, ICEPAK is used for numerical modeling. The objective is to see how the condenser works without the aid of a system fan.

Two cases are considered: one with system fan and the other without the fan. For the first case, the system fan served as the major air outlet, while for the later case, the grill present next to the system fan in the chassis served as the air passage. The grill is represented in the numerical model as a vent with 80% opening. The openings at the front of the chassis are represented as vents (rectangular, 30% opening is assumed) in the model for both the cases. Since the power supply unit (PSU) is separated from the condenser by a perforated plate and has its own dedicated cooling fan, it is assumed not to heat up the condenser. Also, the PSU fan is expected to have very little effect through the perforated plate and moreover, since it is an exhaust fan, it can only increase the cooling performance. So the PSU and the PSU fan are not included in the model. The audio card, video card and the CD bays are considered to be the major obstructions to air flow, and as such they are included in the model. In the absence of actual measurement of airflow rate through the fan, the characteristic curve for the exhaust fan used in HP Vectra is given as input to ICEPAK for the forced convection case.

4.3.2 *Results and discussions*

For the case with fan, it is found that for an average condenser tube wall temperature of 80°C, the net heat dissipation from the condenser is about 66 W. The same boundary condition is used for the case where the system didn't have a fan. The results for both the cases are tabulated in Table 4. It is found that the condenser subjected to natural convection is able to dissipate 60% of the amount of heat dissipated by the condenser subjected to forced convection. So the presence of fan creates a big difference in heat dissipation. However, bigger sized fins may give better heat dissipation with natural convection. The isotherms for the 43rd fin (fin is vertically oriented) for both the cases are shown in Figure 8. In case 1, the forced convection has resulted in a symmetrical temperature profile in the fin (both about horizontal and vertical axes). In case 2, due to natural convection, temperature profile is symmetric only about the vertical axis. Due to heat conduction from the condenser tube, the isotherms are concentric at the vicinity of the tube, however at the upper and lower parts of the fin (away from the tube) due to natural convection, there is a difference in boundary layer thickness, which gives difference in temperature contours.

The average heat transfer coefficient from the condenser tube for the natural convection case is also much lower than the case with a system fan. These results show that the system fan improves the performance of the condenser a great deal, however, a better design of condenser may prove sufficient to dissipate heat through natural convection.

Table 4. Results from numerical simulation

Parameters	With Fan	Without Fan (Natural Convection)
Condenser Tube Temperature	80 °C	80 °C
Ambient Temperature	22.6 °C	22.6 °C
Heat Flow from Condenser Tube	66 Watts	40 Watts
Average Heat Transfer Coefficient from Condenser Tube	515 W/(K-m^2)	315 W/(K-m^2)

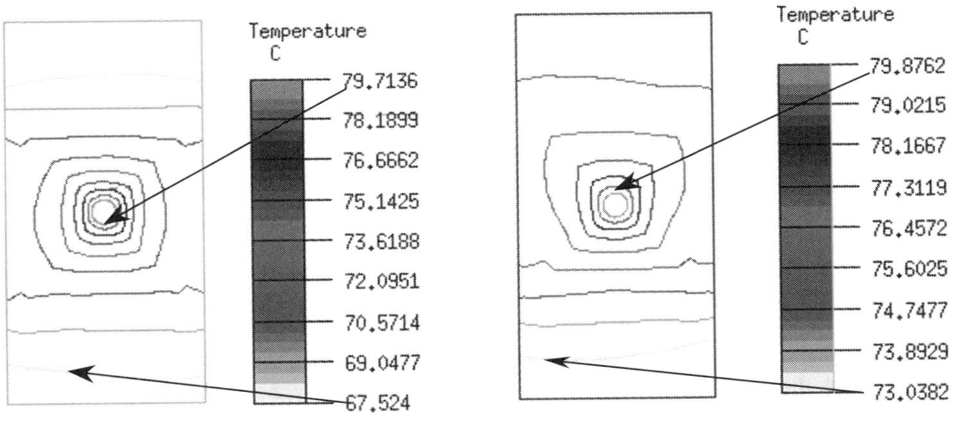

Figure 8. Temperature profiles in the 43rd fin for forced convection system (left) and natural convection system (right).

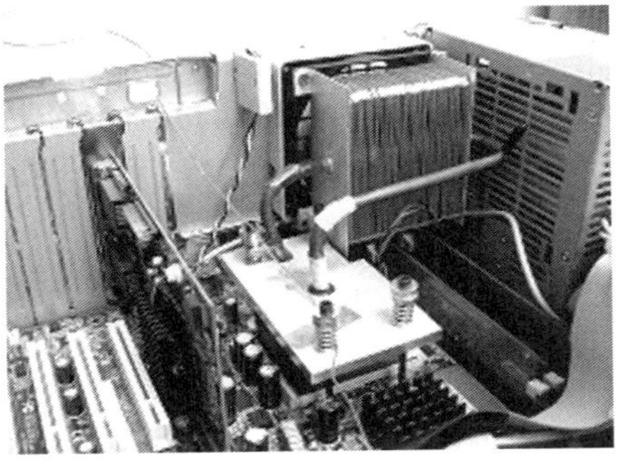

Figure 9. HP Vectra PC with the thermosyphon assembly

5 CONCLUSIONS

In this paper a design and performance evaluation approach for a two-phase compact thermosyphon (Figure 9) for cooling a Pentium 4 processor in a HP Vectra PC is presented. This study identifies the working fluid, the orientation and the system fan as important factors affecting the performance of the loop thermosyphon. Water is identified as the better working fluid with respect to dielectric liquid; however, optimum charging and proper degassing may improve the qualities of dielectric liquid as a working fluid. The thermal performance of the thermosyphon is also found to be unaffected for large inclination angles. Natural convection case is found to be inefficient, however a change in design may make the condenser efficient. The presented experimental and numerical results represent a preliminary study on the performance of the loop thermosyphon. Further detailed numerical and experimental studies are needed to accurately predict the behavior of the thermosyphon.

6 ACKNOWLEDGEMENT

The present research is sponsored through the DARPA HERETIC Program under contract N00164-99-C-0039, managed by Dr. Elias Towe and Dr. Dan Radack. Acknowledgement is also due to Mr. Lang Yuan for his suggestions during the design process.

7 REFERENCES

Garner, S.D. & Patel, C.D. 2001. Loop Thermosyphons and Their Applications to High density Electronics Cooling. *Proceedings of the International Electronic Packaging Technical Conference and Exhibition*, 2001.

Ramaswamy, C., Joshi, Y., Nakayama, W. & Johnson, W.B.1999. Thermal Performance of a Compact Two-Phase Thermosyphon: Response to Evaporator Confinement and Transient Loads. *Journal of Enhanced Heat Transfer*, 6, 279-288.

Ramaswamy, C., Joshi, Y. & Nakayama, W. 2000. Combined Effects of Sub-Cooling and Operating Pressure on the Performance of a Two-Chamber Thermosyphon. *IEEE Transactions on Components and Packaging Technologies*, March, 2000, 61-69.

Tengblad, N. & Palm, B. 1996. External Two-Phase Thermosyphons for Cooling of Electronic Components. *International Journal of Microcircuits and Electronic Packaging.*

Webb, R.L. & Yamauchi, S. 2001. Thermosyphon Concept to Cool Desktop Computers and Servers. *Proceedings of the International Electronic Packaging Technical Conference and Exhibition*, 2001.

Yuan, L., Joshi, Y. & Nakayama, W. 2001. Effect of Condenser Location and Tubing Length on the Performance of a Compact Two-Phase Thermosyphon. *Proceedings of 2001 International Mechanical Engineering Congress and Exposition, ASME*, November 11-16, 2001, New York, NY.

Thermal Challenges in Next Generation Electronic Systems, Joshi & Garimella (eds)
© 2002 Millpress, Rotterdam, ISBN 90-77017-03-8

Vibration actions on heat pipes as cooling element of electronic systems

K.V. Prisniakov, Yu.E. Nikolaenko & V.F. Prisniakov
Institute Geothecnical Mechanics of Nat. Acad. Science of Ukraine, Dnipropetrovsk, Ukraine

Keywords: optimal frequency, thermosiphon, evaporation, boiling, wetting angle

ABSTRACT: In the report the outcomes of researches of thermal modes of a water heat pipes in conditions of vibration effects of different frequencies and amplitudes are resulted. The boundary frequency bands and amplitude are determined, at which heat and mass transfer in heat pipes and thermosyphons is improved and is degraded. For an investigated heat pipes of boundary values of frequencies are equal 60 Hz, and amplitudes are equal approximately 3 or 5 mm.

1 INTRODUCTION

It is known, that oscillations and vibrations has influence on the heat and mass transfer process. The vibrations are for many equipment. But heat exchange intensification through superposition of oscillatory fields involves the complete resources of their artificial excitation. The investigation of impact of the vibrations on the processes with of phase in the working liquid of heat pipes and two-phase thermosyphons in very interesting. Available experimental researches display significant vibroeffect upon boiling system thermal modes. Therefore it is necessary to know the laws of process, on the one hand, and, on the other hand, to be able to actively influence the process in liquid.

The values of the internal characteristics of boiling process are imposed by vibration fields with their own frequency ω and amplitude A. The first stage should contain an idealised picture of determining the effect of certain vibration amplitude. We will investigate internal boiling parameters overheating surface $\Delta T = T_w - T_s$; the steam bubble growth $R(\tau)$; the bubbles departure R_d; the bubbling frequency f; density of steam generation centers n; and capillary pressure Δp; and the distribution of temperature on surface heat pipe T as a function of amplitudes and frequencies. It is well to bear in mind that at present the literatures have about influence of vibration on boiling processes, but information on there conditions for heat pipes is lacking.

2 ANALYSIS OF THE EXPERIMENTAL DATA OF BOILING

2.1 *The overheating*

The analysis of the experiments has show that at the frequencies close to the frequency of steam bubble departure there occurs an extremum ΔT depending on the value of heat flows. At very low values corresponding to transition from convection to bubble and film boiling there is a minimum and under developed boiling mode there is a maximum (Prisniakov et al. 1999). In case of heterogeneous nucleation the vibrations renders essential influencing on an overheating of a liquid. This overheating on value is less, than theoretical overheating in case of homogeneous nucleation (Fig. 1).

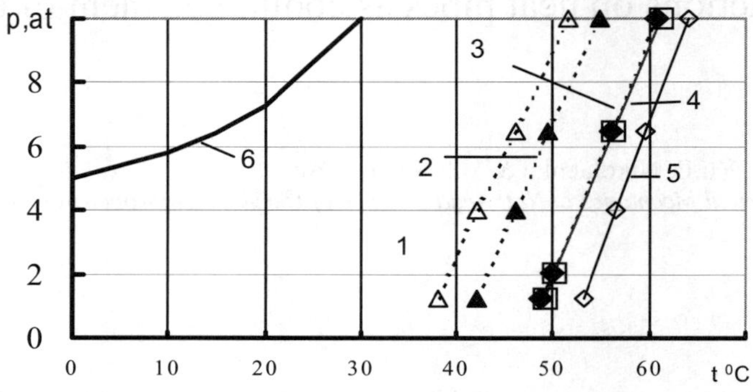

Figure. 1. A marginal overheating of a clean propane in conditions of an pressure: oscillation 1 -Δp = 3 at, 450 kHz; 2 - Δp = 2 at; 25 kHz; 3 - Δp = 4 at, 450 kHz; 4- Δp = 4 at, 25 kHz; 5 - without oscillation; 6 - line of saturation

2.2 *The steam bubble growth*

The analysis experiments of Chekhanow-Kulgina on the steam bubble growth with vibrations and without one has shown that rate bubble growth on a vibrating heater increases (Fig. 2).

2.3 *The bubble departure*

The size of the bubble departure is influenced first of all by vibration amplitude A (Fig. 3) (Prisniakov & Prisniakov. 1997).

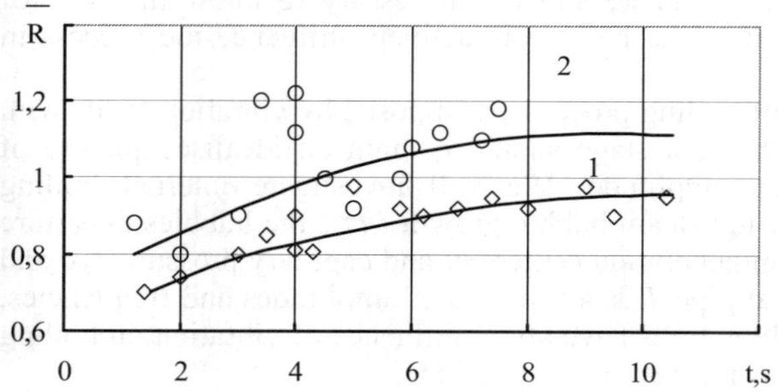

Figure. 2. Change of a function $\bar{R} = R/Ro$ during growth of steam bubbles: 1-A=0,05мм; 2- A=0,1 mm (ω= 120 Hz)

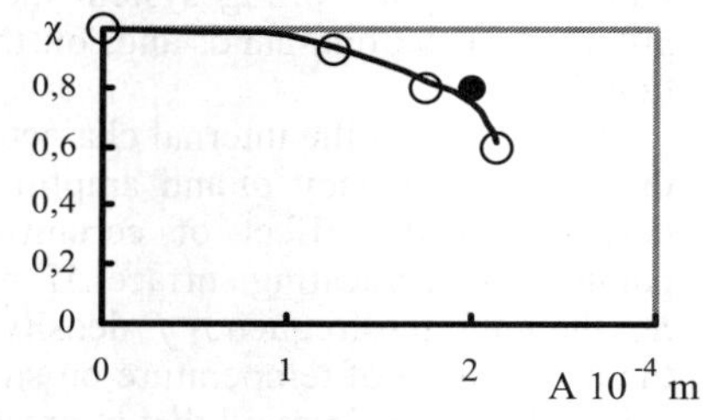

Figure. 3. Relation of a function $\chi = R_d/R_{do}$ from oscillation frequency A (for ω = 80 Hz)

2.4 *The bubbling frequency*

Vibration frequency influences bubbling frequency in the same way. At lower frequencies of heater vibrations the influences of amplitude increase. The experiments show that vibrations decrease steam bubbling frequency (Fig.4, 5, 6).

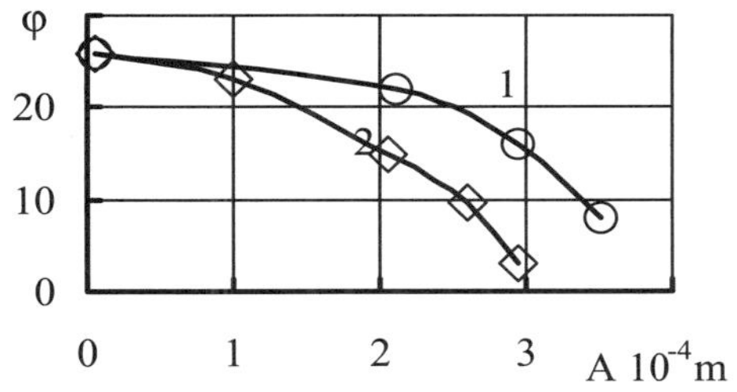

Figure. 4. Relation of a function $\varphi = f/f_0$ from oscillation frequency *A* at different frequencies: 1 - ω = 80 Hz, 2- ω =50 Hz

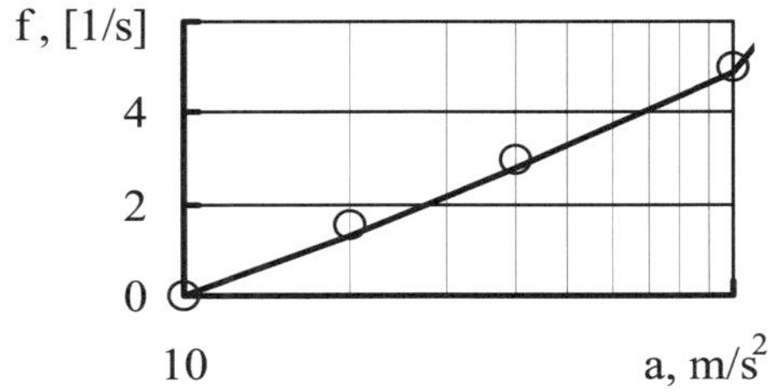

Figure. 6. Relation between minimum accessible frequency of generation of steam bubbles and acceleration vibration action

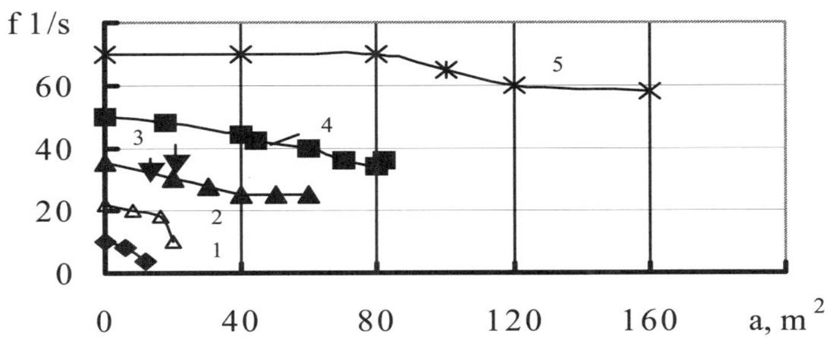

Figure. 5. Impact of acceleration vibration exciter on frequency of generation of single steam bubbles under conditions of coincidence of frequencies *F=f,* the arrows demonstrate a direction of activity DSGC ;1-q=380 W/^{m2}: 2- 430: 3- 500: 4 -710: 5- 970

2.5 *Heat change*

The experiments of Kulgina have shown, the ultrasonic oscillations under certain conditions can augment a heat-transfer coefficient on 5-30%. The heat-transfer coefficient was augmented on 15-63% at boiling of ethanol in experience Afgan & Zitko (1994, 1996). They also have received, that the effect of effect vibration on a heat-transfer coefficient on a horizontal surface is smaller after matching with a vertical surface.

3 THE EXPERIMENTAL DATE ABOUT VIBRATION IMPACT ON PROCESSES IN HEAT PIPES

We conducted experiments with a heat pipe arranged on a vibration table. The heat pipe manufactured of a stainless steel. A working fluid of heat pipe was water. The heat pipe was hardly consolidated on a shaker unit. The heating segment had the heater as a wire. On all length of the thermoelectric couple were attached to a surface of a tube by a spot welding. Thus 5 thermoelectric couples were fixed on surface condenser. The sixth thermoelectric couple was directly under an electrical spiral on surface evaporation. The purpose of experiments was analysis of influencing of vibrations on thermal characteristic ofl modes of thermal tubes. The outcomes of experiments are submitted in the tables 1 -6. The analysis of these outcomes has shown. That on frequency 40 Hz and at amplitude of 3.5 mm takes place full isothermal surface of a pipe. Is marked, that at low frequencies of oscillations (10-20 Hz) the vibrations do not render any of essential influencing on value of a surface temperature of a heat pipe. With increase of frequency stratification of temperature curves is watched. On frequencies 100 Hz this stratification reaches a maximum at amplitudes more than 6 mm.

Table 1 Value temperature on surface heat pipe °C under frequency 20 Hz (1 series).

a, mm/s²	***A, mm***	T_1	T_2	T_3	T_4	T_5	T_6
0	0	94	94	94	94	94	98
25	1.6	94	94	94	94	94	98
50	3.1	94	94	94	94	94	98
70	4.7	71	72	72	72	72	78

Table 2. Value temperature on surface heat pipe °C under frequency 10 Hz (2 series)

a, mm/s²	***A, mm***	T_1	T_2	T_3	T_4	T_5	T_6
0	0	117	118	118	118	118	119
25	1.6	99	99	99	100	100	101
30	1.9	86	88	88	89	89	92

Table 3 Value temperature on surface heat pipe °C under frequency 20 Hz (2 series)

a, mm/s²	***A, mm***	T_1	T_2	T_3	T_4	T_5	T_6
0	0	117	118	118	118	118	119
25	1.6	117	118	118	118	118	119
50	3.1	111	111	111	111	111	111
75	4.7	96	96	96	97	97	97

Table 4. Value temperature on surface heat pipe °C under frequency 40 Hz (2 series)

a, mm/s²	***A, mm***	T_1	T_2	T_3	T_4	T_5	T_6
0	0	117	118	11	118	118	119
25	1.6	117	118	11	118	118	119
50	3/1	117	117	117	117	117	117
75	4.7	116	116	116	116	116	118
100	6.3	108	108	109	109	110	111

Table 5. Value temperature on surface heat pipe °C under frequency 60 Hz (2 series)

a, mm/s²	***A, mm***	T_1	T_2	T_3	T_4	T_5	T_6
0	0	117	118	118	118	118	119
25	1.7	117	118	118	118	118	119
50	3.1	117	118	118	118	118	119
75	4.7	115	116	116	116	116	117
100	6.3	115	115	115	114	115	115

Table. 6 Value temperature on surface heat pipe °C under frequency 100 Hz (2 series)

a, mm/s²	**A, mm**	T_1	T_2	T_3	T_4	T_5	T_6
0	0	117	118	118	118	118	119
25	1.6	117	118	118	118	118	119
50	3.1	117	118	118	118	118	119
75	4.7	117	118	118	118	118	119
100	6.3	116	115	116	11	118	121

4 THEORETICAL PREREQUISITIES

4.1 *The main assumption*

To assess vibration effect upon the bubble growth rate let us make use of the simplest scheme of is growth on the heating surface suggested by D. Labuntsov. The basic hypotheses in an assumption of vibration effect on the course of all boiling processes with base angle θ_0 being changed. The motion of heating surface according to the periodic law will be designated after the formula (in zero phase)

$$X = A \sin \omega\tau \tag{1}$$

where A is vibration amplitude; ω is frequency; τ is time.
Let as assume that wetting angle θ changes according to the same law, i.e.

$$\theta = \theta_0 + A_\theta \sin \omega\tau \tag{2}$$

Let as determine A_θ, using the relation between nucleus radius R and mouth radius R_c:

$$R_c = R \cos(\alpha - \theta) \tag{3}$$

where α is semivertex angle.
Let us factorize R and θ into Taylor series leaving the 1st member, as it is usually done at linearization around point θ_0, then we obtain

$$\delta\theta = A_0 = \cos(\alpha-\theta) \csc(\alpha-\theta) A/R_0 = \cos^2(\alpha-\theta) \csc(\alpha-\theta) A/R_c \tag{4}$$

It is taken into account here that bubble radius deviation $\delta R = A$, i.e. it equals vibration amplitude. Formula (4) including, the law of changing the angle of bubble contact with heating surface will take the form

$$\theta = \theta_0 + \cos^2(\alpha-\theta) \csc(\alpha-\theta) \sin(\omega/f)\, A/R_c \tag{5}$$

4.2 *The overheating*

First let us view how liquid overheating occurs at boiling. For that we shall make use of *W.* Rohsenow model with our correction (Prisniakov 1988) $\Delta T = 2\sigma \cos\theta/(\Re\, \rho'' R_c)$ we find out the relation between liquid overheating at boiling without vibration ΔT_0 and with it ΔT:

$$\Delta T/\Delta T_0 = \Psi = \cos\theta \cos\theta_0 \tag{6}$$

Let us substitute θ in this equation with its value from expression (5) and after
$\Psi = \cos\Delta\theta - tg\theta_0 \sin\Delta\theta$, where

$$\Delta\theta = \cos^2(\alpha-\theta) \csc(\alpha-\theta) \sin(\omega/f)\, A/R_c \tag{7}$$

Let us investigate function Ψ for an extremum finding a vibration frequency at which function $\Psi(\omega)$ is of maximum (minimum) value. For that we determine derivative $\partial\Psi/\partial\omega$ and set it equal to zero

$$\partial\Psi/\partial\omega = -(\sin\Delta\theta + tg\theta_0 \cos\Delta\theta)\ [\cos^2(\alpha-\theta_0)\csc(\alpha-\theta)\cos(\omega/f)\ A/(fR_c)]=0$$

After making transformation we have

$$\omega_m = f\{k\pi - \arcsin[\sin(\alpha-\theta_0)\ \mathrm{sc}^2(\alpha-\theta_0)\ (\theta_0 R_c/f)]\} \quad (8)$$

Calculations after this formula produce a result for function Ψ extremum point equal to approximately 45 Hz which is close to experimental results.

4.3 *The density of steam generation centers*

The density of steam generation centers (DSGC) $\bar{n} = n/n_{max}$ is to be found according formula Prisniakov (1988)

$$\bar{n} = exp(-3.1\ \bar{h})/(b\ \ \bar{h}^v) \quad (9)$$

where $\bar{h} = h/h_m$ is the relationship of current and average quadratic height of surface microuneveuness; n_{max} is maximum DSGC (under crisis) (Prisniakov 1988)

$$n_{max} = \eta_1/\ \bar{R}_d^2 \quad (10)$$

Vibration field influence on DSGC in the following way:

$$n \sim \chi^{-2} \quad (11)$$

where $\chi = R_d/R_{d0}$.

5 HEAT PIPES

Motive force in heat pipes capillary pressure:

$$\Delta p = 2\sigma\{\cos\theta_e/r_e - \cos\theta_c/r_c\} \quad (12)$$

Rewriting Eqs. (12) in terms of the result (2) we obtain formula for analyses of impact the vibration on heat transfer in oscillating heat pipes:

$$\Delta p_c = 2\sigma\{\cos[\theta_{e0} + A_\theta \sin\omega\tau]/r_e - \cos[\theta_{c0} + A_\theta \sin\omega\tau]/r_c$$

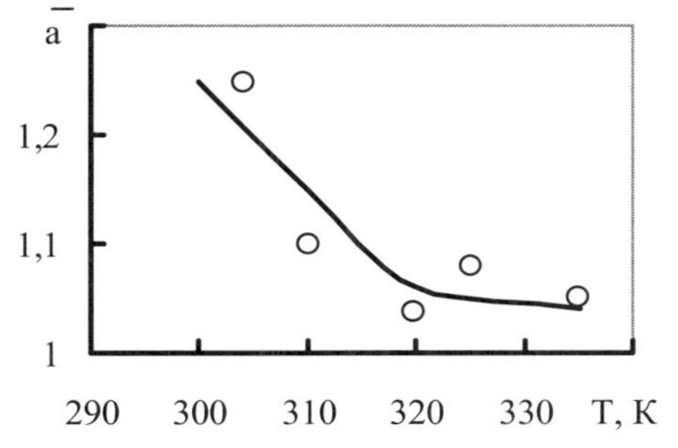

Figure. 7. Relative heat-transfer coeffic-ient as function from temperature of a liquid at ultrasound oscillations of the heater

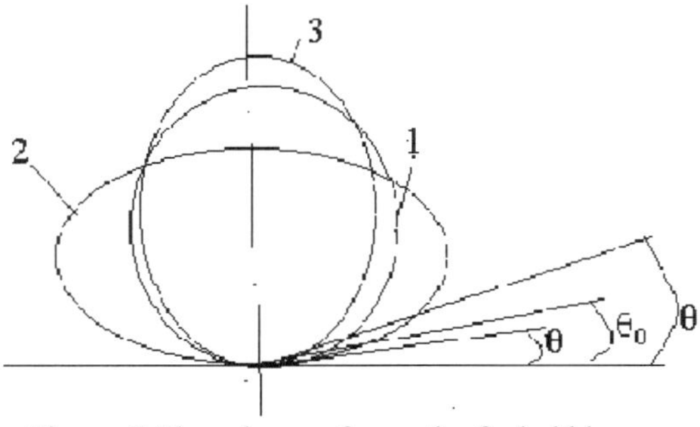

Figure. 8.The scheme of growth of a bubble on a heating surface in stage 1 at a phase of rise and in stage 2 at its lowering

6 CONCLUSION

Theoretical foundations of calculation of internal boiling characteristics and heat exchange under boiling condition have been elaborated assuming the mechanism of vibration influence through wetting angle. The experimental show that the processes in heat pipes are control by the vibrations.

SYMBOLS

a - vibration acceleration; A vibration amplitude; f departure bubble frequency; h height of surface microuneveuness; n –the density of steam generation centers; Δp- capillary pressure; r pore radius; radius; R radius; R_c –mouth radius; R_d – departure bubble radius; ΔT –surface overheating; θ - wetting angle; ρ - density; σ - surface tension; τ -time $(1/f)$; ω -frequency of vibration; $\Re$ - gas constant; $\Im$ - heat evaporation;
Index: e – evaporation; c – condenser.

REFERENCES

Prisniakov, K.V. & Prisnikov, V.F. 1999. The impact of the vibration on heat and mass transfer in the oscillating heat pipes and twophase thermosyphons. *Proc 11th intern. heat pipe conf., September 12-16 1999, Seikei University,* Tokyo.

Prisniakov, V.F. & Prisnikov, K.V. 1997. The physics of heat transfer in boiling and condensation. *Proc intern. symp. on the Physics of Heat Transfer in Boiling and Condensation, Moscow, Russia, May 21-24 1997*, Moscow.

Prisniakov, V. F. 1997. Density of vaporarization centers. *Industrial Heat Engineering.* 19 (2-5): 20-27; 33-39.

Prisniakov, V. F. 1988. *Kipenie (Boiling)*. Kiev: Naukova dumka.

Zitko V. & Afgan, N. 1996. Boiling Heat Transfer from horizontallyoscillating surface. *2nd European Thermal-Sciences and 14th UIT National Heat Transfer Conference 29-31 May 1996.Rome, Italy,* Pisa: Edizioni ETS.

Zitko ,V. & Afgan, N. 1994. Boiling Heat Transfer of binary mixtures from the oscillating heating surface. *Heat Transfer*. 5:201 –206.

Thermal Challenges in Next Generation Electronic Systems, Joshi & Garimella (eds)
© 2002 Millpress, Rotterdam, ISBN 90-77017-03-8

Innovative wick design for multi-source, flat plate heat pipes

M.J. Rightley, C.P. Tigges, R.C. Givler, C.V. Robino, J.J. Mulhall & P.M. Smith
Sandia National Laboratories, Albuquerque, New Mexico, USA

Keywords: heat pipe, micro-heat pipe, IC cooling technology, dryout, porous media, high heat flux devices

ABSTRACT: We present a novel micro-heat pipe wick design and fabrication technique to significantly boost the effective thermal conductivity of the heat pipe relative to the monolithic substrate material. Extensive porous flow modeling of the process has provided critical information on the key parameters and the resulting anisotropic wick designs have shown robust performance improvements. A methanol charged copper device reported in this paper showed a maximum thermal conductivity of 3240 W/m-K prior to dry out. This represents a better than 8-fold increase over the conductivity of solid copper.

1 INTRODUCTION

The development of the true heat pipe (in contrast to the gravity-assisted thermosyphon) was basically driven by the desire to effectively transfer heat across distances from the point of creation (the evaporator) to a location that facilitated its removal (the condenser) in the absence of any body forces. Several researchers in the early 1960's surmised that this process would be particularly useful in space applications when monolithic heat conduction devices were considered too heavy and too inefficient [Cotter, 1965, Grover et al., 1964, Grover, 1966]. The lighter weight of a heat pipe and the promise of significantly increased effective thermal conductivity more than offset the added expense associated with the design and fabrication research that was necessary to produce these devices.

In more recent times, the heat pipe has been identified as having potential in the increasingly more challenging field of thermal management of electronic devices [Dunn and Reay, 1994]. The primary challenges in the development of heat pipes for this application are related to constraints on geometry. Not only is it necessary to miniaturize the device to follow the trends in microelectronic devices, but there is also a need to have the heat pipe conform to the flat rectangular nature of their packaging. Additionally, to maximize the usefulness of a flat micro-heat pipe (as compared to a flat plate heat spreader), we are challenged to create a truly directional device capable of heat transfer from any one region to any other on the device.

Several researchers have investigated wick formation and performance at micro-scale geometries [Cao et al.,1997, Plesch et al., 1991, Wu et al., 1991] and have discovered that the nature of the grooves aligning the flow channels between condenser and evaporator is critical in achieving higher dryout fluxes. Plesch and co-workers created a water-based micro-heat pipe capable of dissipating 60 W/cm^2 prior to dryout using longitudinal grooves over the entire interior surface of the device. The superior performance of these wicks relative to other micro-heat pipes was attributed to the presence of the axially oriented grooves creating a high permeability path from the condenser to the evaporator. The creation of the miniature axial grooves, however, presents some difficulties that increase the basic costs associated with a device of this type. For example, the heat pipes tested in [Cao et al., 1997] were created using EDM (electro-discharge machining) processes which can be expensive and potentially time consuming.

Researchers at Sandia National Laboratories in New Mexico have developed a method for creating micro-scale heat pipe wicks based on photolithographic processing techniques borrowed from the semiconductor industry [Benson et al., 1998, Benson et al., 1999]. The wicks that were developed facilitated the significant reduction in characteristic dimension required to ensure fluid flow at the scale of a single CMOS chip. The development of these devices (materials and processing) was specifically geared toward mass production in an effort to eventually develop an economically viable device. Initial devices were fabricated with globally isotropic nickel wicks (i.e., locally anisotropic structures repeated numerous times over the interior surface of the heat pipe) on Kovar substrates and were tested using methanol as the working fluid. The results showed effective thermal conductivity improvements over a monolithic block of the substrate material of 63% [Smith, 2000] and provided wick processing information to simplify the photolithographic method.

Finally, computational modeling at Purdue [Vadakkan et al., 2000] produced interesting results suggesting a sensitivity of the liquid coolant flow from condenser to evaporator that is dependent on the porosity of the wick. Their data show that high porosity wicks tend to degrade the liquid flow back toward the evaporator.

The current work seeks to produce a micro-heat pipe that takes full advantage of the favorable permeability of the longitudinally grooved device while maintaining a photolithographic manufacturing process. Additionally, our design is intended to allow multiple heat sources to be cooled by a single heat pipe (so called "block heating") and, further, allows the restriction of the cooling boundary to be any surface or edge of the device. We have developed a sophisticated porous flow model that indicates that flow permeability, which, within a heat pipe, tends to vary inversely with the porosity, is one critical factor in the performance of a micro-scale wick. It also strongly suggests that wick anisotropy is a crucial parameter in the development of a multi-source micro-heat pipe.

2 MODELING

To facilitate the design process and identify the sensitive parameters for the block heated flat plate micro-heat pipe, we created a model of the heat pipe using a sophisticated porous flow computer code [Martinez et al., 1997].

The computer code, PorSalsa, was created originally to address subsurface multiphase flows and to utilize Sandia's massively parallel computer systems to solve them efficiently. Porous media included in the computational domain are assumed locally heterogeneous and phase transitions are allowed. Input parameters for any porous medium are flow permeability, porosity and the initial saturation level of the medium. For heat pipes, this later quantity is related to the initial coolant charge.

To cast the micro-heat pipe geometry into the appropriate style of the PorSalsa code, the domain was subdivided into a series of individual heterogeneous porous media, with separate input parameters used for each (Figure 1).

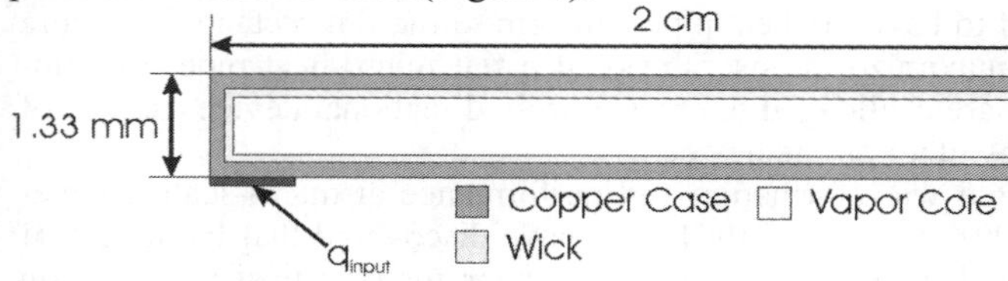

Figure 1. Heat Pipe Domain for PorSalsa Model

The solid nature of the copper casing was modeled as a porous material of extremely low fluid permeability (1E-14 m^2) and very low porosity (ε=5 %). In contrast, the wick itself was modeled by using the permeability and initial saturation (i.e., fluid charge) as parameters for a sensitivity study. A casing permeability of 1E-14 m^2 was sufficiently small relative to the wick permeability ranges studied to ensure all fluid flows were constrained to the interior of the heat pipe. The vapor core was set to a permeability of 1E-8 m^2 and initial saturation of 0.02%.

The original intent of the 2D modeling effort was a parametric-type study to provide guidance to the design and test activities as to reasonable wick permeability and initial fluid charging values. All model runs used water as the coolant so comparisons to the test data in which methanol was used must be considered qualitative. Later testing with water is planned to allow more quantitative evaluation of the predictive capability of the model.

The primary results obtained in the original 2D model are summarized in Figure 2.

Since the heat pipe is designed to offer a significant increase in the effective thermal conductivity of the device relative to a monolithic block of the casing material, all data are compared to the linear thermal response of solid copper. As is shown in Figure 2, flow permeability and, to a lesser extent, initial fluid charging of the device are key parameters affecting overall performance. In particular, the model indicates that maximization of the permeability must be a design goal while still maintaining sufficient capillary pressure (P_{cap}) to overcome expected orientation variations. This result is consistent with historical knowledge of heat pipes and has led to the development of composite wicks in which high permeability arteries or grooves are machined in the walls and coupled to a high porosity, high P_{cap} wick (such as a metal screen) [Dunn & Reay, 1994].

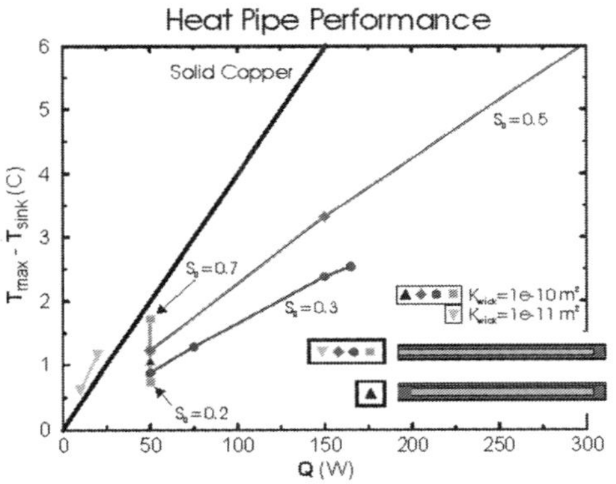

Figure 2. 2D PorSalsa Model Results

When the model was expanded to three dimensions, it became clear that to adequately cool multiple sources with a single edge involved in the cooling, an anisotropic wick was required. Figure 3 illustrates the liquid flow paths of the coolant in the 3D model with two equal sized power sources. Clearly, source shadowing will cause premature dryout of the source farthest from the cooling boundary.

3 WICK DESIGN AND FABRICATION

The concept with which we formulated our permeability controlled, multi-source wick design is based on the potential flow formulation of steady frictionless flow from a "source" (evaporator region near each heat source) to a "sink" (condenser surface). The solution of this problem, for the general case of multiple sources and cooling boundary geometry is obtained through the Laplace equation. Hence, we dubbed the device the "Laplace Wick".

After dividing the heat sources into equal strength point flow sources (the number based on the relative power of the source) and subdividing the condenser surface into an equal number of appropriately sized regions, Laplace's equation is solved to obtain the flow streamlines. These streamlines, since they are tangent to the fluid velocity vector at every point in the flow field, represent a natural return path for the coolant from the condenser to the evaporator. Consequently, each streamline was then considered the zeroth order, or primary, wick line and capillary pressure was maximized by adding higher order wick lines (i.e., lines that do not intersect either the source or sink) between the primaries. This densification was conducted by recursively controlling line spacing along the curvature of the primary wick lines to a set maximum. Finally, small gaps along the length of each wick line were included at regular intervals. They were included to prevent catastrophic blockage failures by allowing liquid to bypass around contaminates or defective wick regions.

Figure 4 illustrates the design results for a four source, rectangular heat pipe cooled along one short side. Note that the two larger sources near the center are each represented by a pair of Laplacian flow sources indicating relatively higher heat loads.

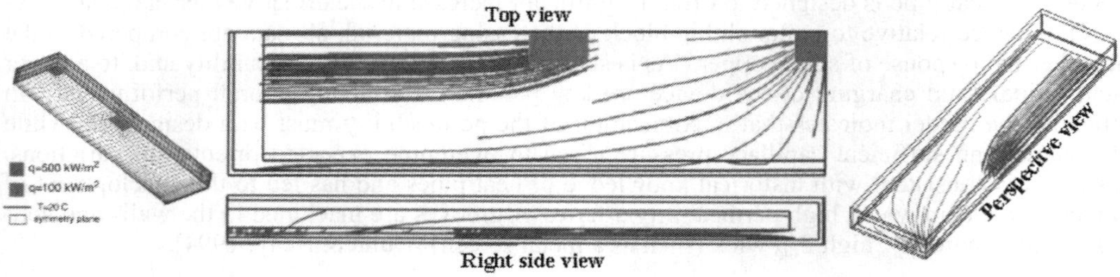

Figure 3. 3D PorSalsa Heat Pipe Results (Liquid Phase)

The highly anisotropic nature of the wick is expected to significantly reduce the source shadowing problem (coolant flow mismatches in favor of sources closest to the condenser) by greatly reducing the permeability in the direction transverse to the wick lines. Tangential permeabilities are controlled by varying the maximum pitch, wick height and wick width. To illustrate the wide range of permeability values that this design concept provides, we compared values for a commercially available sintered metal felt (*K* dependent on initial areal density and final sintered thickness) and our Laplace wick (*K* dependent on wick pitch, *P*, width, *W*, and depth, *H*). For the sintered felt, a modified version of the Blake-Kozeny equation was used as follows [Marcus, 1972]

$$K = \frac{d^2\varepsilon^2}{122(1-\varepsilon^2)}$$

where d is the wire diameter and ε is the wick porosity (calculated as a function of the areal density and final thickness). For the rectangular Laplace wick, K is provided in [Chi, 1976] as

$$K = \frac{2\varepsilon R^2}{(f \cdot \mathrm{Re})}$$

where

$$R = \frac{2(P-W)H}{(P-W+2H)}$$

and $(f \cdot \mathrm{Re})$ is a frictional coefficient given in [Kays, 1966]. Figure 5 illustrates the results of these calculations (W=40 µm, H=60 µm).

Notice that the tangential permeability of the Laplace wick is controllable through more than three orders of magnitude and, if one reviews the 2D results of the PorSalsa model (Figure 2), it is clear that this level of control will allow significant performance optimization through the design process.

After the wick design is completed, the drawings are converted to mask layout files and a photolithographic mask is created of the design. This mask is then used to control a proprietary electroplating process performed by Polar Group Inc. (Vernon Hills, IL) that plates copper wick material onto 150 mm copper substrate wafers (thicknesses of 0.51 and 0.76 mm were used).

The circular "wickless" areas in the interior of the heat pipe surface are the locations of copper support posts to maintain the rigidity of the device for charging and operation. Note that the top two wicks and the bottom two wicks are complementary but that variations in relative heat loading levels were used in the design for eventual characterization of the design process.

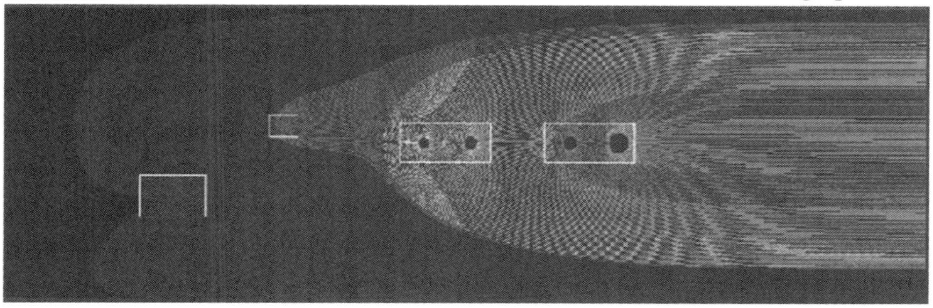

Figure 4. Four Source Rectangular Laplace Wick

Once the wicks have been completed, final assembly consists of EDM cutting of the wick parts including a spacer overlap region outboard of the wick. These spacers are used to provide the open central vapor core of the device. The wick plates, the spacer and a 1.6 mm fill tube are then assembled and hermetically sealed using a combination of electron beam welding (edge seals and fill tube) and resistance welding (support posts) as described in [Benson et al., 1998, Benson et al., 1999].

Figure 6 shows a finished heat pipe, with two mounted thermal test sources and a single source heat pipe mounted to the cooling frame and ready for test.

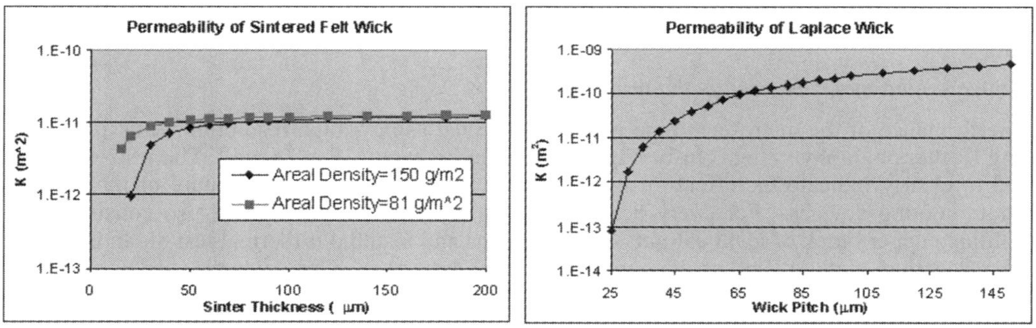

Figure 5. Permeability Comparison

4 EXPERIMENTAL PROCEDURE

4.1 *Heat pipe coolant charging*

Prior to preparation of the heat pipe for test, the device is charged with coolant at reduced pressure. To accomplish this, we first calculate the baseline initial charge level by subtracting the actual volume of wick material from the swept volume of the wick. The actual wick volume is determined from knowledge of the wick porosity and thickness (wick depth). This quantity represents the volume of liquid that would "fill" the voids around the wick up to the height of the wick and is considered the standard charge for a heat pipe. Although several manufacturers deliberately overcharge their devices a small amount (10%-20%, [Marcus, 1972]) in an effort to prevent premature dryout, the results of our modeling of the system indicates that, at the micro-scale of interest, overcharging may degrade performance unacceptably. As a result, for this report we charged the device to its standard charge amount (16% by volume for the device used in this study).

The charging process was completed by first evacuating the heat pipe and ensuring hermeticity against the vacuum pumping system. A valve to a methanol reservoir was then opened and the entire interior volume of the heat pipe was filled with the coolant liquid. Next, a chart recording of the pressure in the heat pipe fill tube was initiated and the valves reset to again evacuate the pipe. The pressure response of this process clearly indicates the moment when all of the liquid has been removed and only methanol vapor remains (t_0). The fill process is then repeated and another evacuation commenced. This time, the fill tube is crimped at a time corresponding to the product of 1-fill fraction (x_{fill}) and the overall pump out time:

$$t_{charged} = (1 - x_{fill}) t_0$$

The heat pipe was carefully weighed before and after the charging process, and the results used to confirm the timed charge procedure.

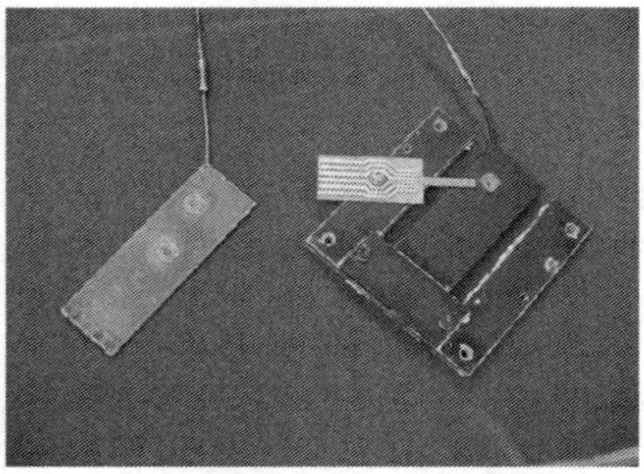

Figure 6. Assembled Laplace Wick Heat Pipes

4.2 *Infrared thermography setup and test procedure*

The performance of the micro-heat pipe was characterized using a high resolution, emissivity correcting IR thermography system (Infrascope I, Quantum Focus Inc, San Diego). The tests were designed to clearly indicate the effect on the maximum thermal test die temperature of the various substrate cooling devices. For every heat pipe geometry that was tested, we also constructed a monolithic copper block of identical outside dimensions and install our thermal test die in the same manner (location, method, etc.) as the one placed on the heat pipe. This device is used as a control to allow direct comparison of the measured temperature differences and to infer values of the effective thermal conductivity.

The Infrascope camera is capable of very high spatial resolution temperature measurement (less than 5 micrometers) if high power lenses are installed (with a commensurate limitation to the field of view). For our heat pipe testing, we were more interested in determining the temperature distribution from the evaporator to the condenser. We used a variable infrared lens (1-5X, roughly 1.5 mm per pixel) for our tests and focused the camera array to show the entire thermal test die and a small portion of the cooling frame metal.

The heat source for these tests was a thermal test chip produced in the Microelectronics Development Laboratory (MDL) at Sandia National Laboratories [Sweet et al., 1993]. The test chip is a 5.6 mm x 5.6 mm (0.31 cm^2 total area) and has four polysilicon resistance heaters arranged around the periphery of the chip. The chip also contains a temperature-calibrated diode that is useful in monitoring the average temperature of the die. To attach the die, we used JM7000 high thermal conductivity epoxy. And, since we were specifically interested in gathering comparitive data, we made an effort to ensure the bond line thickness and mounting procedures were identical for the heat pipe and solid monolithic blank. All die attach work on the two types of devices were performed at the same time by the same operator.

All copper and aluminum surfaces (the heat pipes and the cooling frames) that would be included in the IR field of view were painted black using a flat black spray paint to reduce the possibility of emissivity-induced errors in the anticipated low temperature measurements made of these cooler areas. The ATC04 die was, of course, not painted and we did not expect significant emissivity-induced errors since the die temperature was much higher than the background. Note that the Infrascope device also performs a two-temperature emissivity calibration for each use of the system. This calibration is performed at two different known, uniform temperatures and the ratio of the resulting radiance measurements provides an estimate of the emissivity of the various regions within the field of view (one per pixel) of the IR camera.

Wire bonding was then completed and the device was ready for testing. The basic test layout is shown photographically in Figure 7.

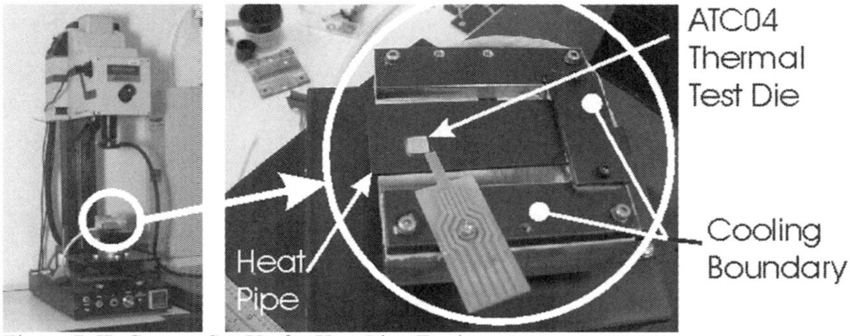

Figure 7. IR Camera Set-Up for Heat Pipe Testing

Due to the thermally massive nature of the cooling frame (see Figure 7), the test procedure assumed an invariant cooling boundary temperature over time. We performed the testing by forcing current through the polysilicon heaters of the ATC04 thermal die using an active power control system. The total power input to the die was monitored and the forced current was adjusted to maintain a constant power level. The Infrascope would then average the radiance images over 30 frames, starting at the instant power was turned on to the die (the frame rate of the camera in this situation is about 6 frames per second). Therefore, the image was the result of about 5 seconds of average IR emission.

To ensure repeatability and post-dryout rewetting of the wick, after we had ramped the device through all of the power settings, we soaked it at 5 W (16 W/cm^2) and 20 W (64 W/cm^2) and repeated the 20W single data point run.

To evaluate steady-state operation, the movie mode of the Infrascope was used to set frame rates and allow continuous frame recording after power on. Frame sampling rates were varied, based on the applied power, to ensure the devices (heat pipe and copper blank) reached steady state prior to power off.

5 RESULTS AND DISCUSSION

5.1 *5 Second integration results*

The initial data from the short time (5-second integrated) testing of the first device, called 'A-B Laplace,' are provided in Figure 8. The vertical axis is the measured difference between the maximum temperature (on a single pixel basis from the Infrascope image) and the average pixel temperature of the cooling frame. The field of view of the Infrascope was set to ensure that both the thermal test die and the cooling frame were visible to the detector. The actual IR images corresponding to the red blocked points in Figure 8 are provided in Figure 9.

To allow inference of the effective thermal conductivity from the IR data, we created a solid copper blank of the same outside dimensions and mounted a thermal test die in the same location. The heat pipe data are compared to the solid copper and the effective thermal conductivity is inferred as the ratio of the instantaneous slope of the two device performance curves. As expected, the copper blank response is linear (within the experimental error of the test system) indicating a uniform k_{eff} over the temperature range of the test.

The heat pipe response exhibits three regions of distinct performance characteristics as marked on Figure 9:

Region 1: < 48 W/cm^2
Region 2: 48 < q_{input} < 64 W/cm^2
Region 3: > 64 W/cm^2

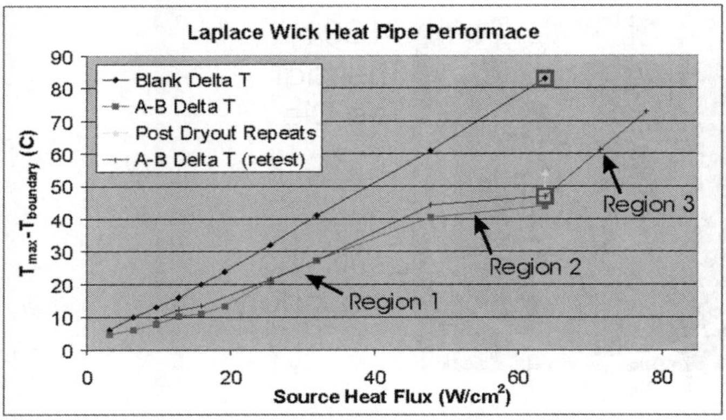

Figure 8. Laplace Wick Performance (5 Second IR Integration)

The drastic slope increase exhibited in region 3 appears indicative of liquid starvation (dryout). The dryout flux of 64 W/cm^2 is larger than previous versions of the flat plate, photolithographic wick design (isotropic) as discussed in [Benson et al., 1998, Benson et al., 1999, Smith, 2000] by roughly a factor of 2.5. This difference is attributed to the anisotropic flow protection afforded by the Laplace wick design.

We surmise that the "shelf" seen in region 2 represents the optimal operating condition for this device. In this region, the maximum source temperature is independent of the applied power indicating an equilibrium condition. This condition probably indicates that the coolant liquid layer beneath the heat source is sufficiently thin to support a high vaporization rate. In this condition, small increases in the heat load create variations in the vaporization rate but P_{cap} is still sufficient to cover the evaporator with a thin layer of liquid.

The ratio of the slope of this line to that of the monolithic blank yields an indicated effective conductivity improvement over copper of 8.1X, resulting in k_{eff} = 3240 W/m-K for the heat pipe (k_{copper}=400 W/m-K). The data suggest additional testing is needed to evaluate the nature of the region 2 performance.

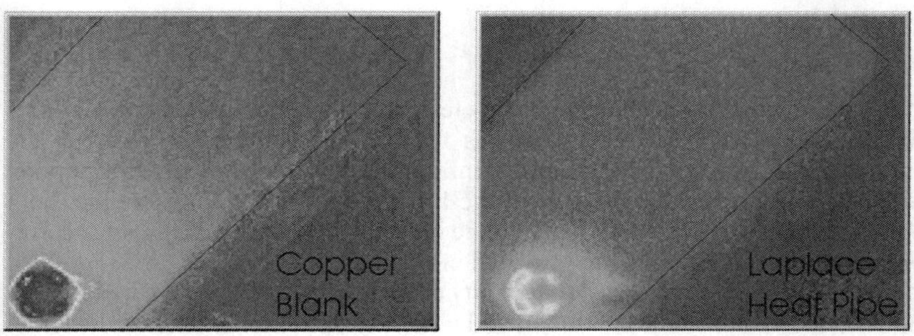

Figure 9. IR Images at 64 W/cm^2 (Equal Temperature Scales)

We are uncertain of the reason for the region 1 behavior in these data. It is probable that the true detail of the plot is not yet visible due to the lack of resolution at the higher heat fluxes. It may be that the liquid layer in the evaporator is too thick to support the rapid vaporization rate seen in region 2. Under these conditions, a relatively large temperature gradient would exist across the liquid layer in the evaporator that would significantly reduce the evaporation rate. This might indicate an excess of initial liquid in the device. The results of the PorSalsa model (Figure 2) encourage minimization of the initial saturation level and do show an inflection point in the direction of improved performance at high heat loads for all S_0 values tested for K_{wick} = 1e-10 m^2. Consequently, further testing to evaluate the effect of initial charge percentage is planned.

5.2 *Steady-state results*

Prior to testing of the A-B Laplace heat pipe, we had anticipated a long-term degradation of performance due to the lack of liquid communication flow paths from the bottom plate to the top. Although future versions of this device have had design improvements made to alleviate this problem, the A-B device did not benefit from those modifications. In essence, the orientation of the heat pipe in our test set up was particularly poor, as gravity would act to adversely affect performance. The effect would be observed as a transient degradation of the temperature difference with a clear indicator of the point of dryout.

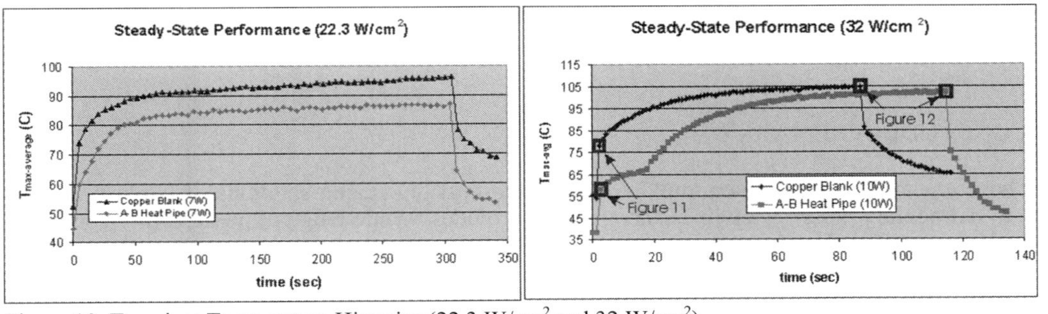

Figure 10. Transient Temperature Histories (22.3 W/cm^2 and 32 W/cm^2)

Consequently, our test matrix included gathering of data in "movie" mode (time based imaging) to evaluate this effect. Figure 10 illustrates the transient temperature history for the a four pixel spread around the location of the maximum temperature pixel at 22.3 W/cm^2 and 32 W/cm^2, left to right. The plot includes both the copper blank and the A-B device. As expected, transient dryout is observed after which the overall performance of the device approaches that of the monolithic material.

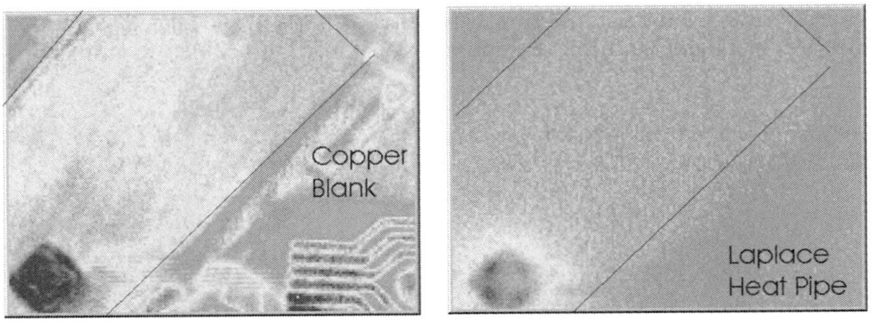

Figure 11. Early Time Images (32 W/cm^2, same temperature scale)

A comparison of the images near the beginning and end of the trajectories shown in Figure 10b is provided as Figures 11 and 12. Figure 11 represents the early-boxed points in Figure 10b while 12 is for the later-boxed points. Notice the clear temperature gradient for the heat pipe in 12. This clearly indicates that there is no liquid in the vicinity of the heat source. From the Figures, it is apparent that the orientation penalty for the tested device is severe. Future tests to evaluate the effectiveness of using sintered metal felt to provide the needed bottom-to-top flow paths are planned. Additionally, the use of annular sintered felt wraps to be installed around each of the support posts has been demonstrated in the assembly of newer versions of the Laplace heat pipe device.

Unfortunately, test data to evaluate the performance improvement, if any, are not yet available.

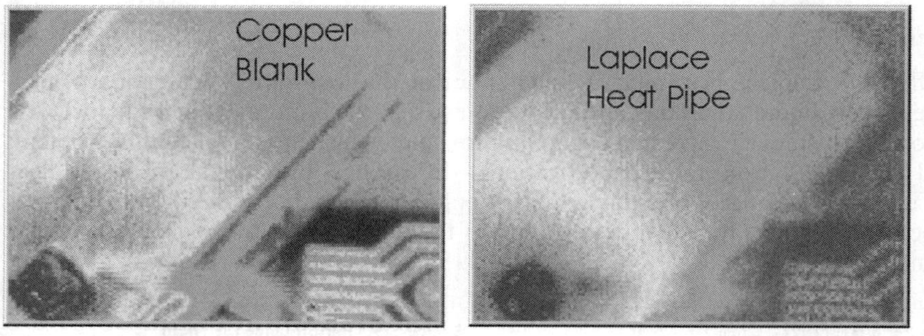

Figure 12. Steady State Images (32 W/cm^2, same temperature scale)

6 CONCLUDING REMARKS

We have created an edge-cooled, multi-source micro-heat pipe using solutions of Laplace's equation to define the primary wick lines. The process provides a wide range of permeability control in the anisotropic wick design stage.

A porous flow model, PorSalsa, was applied to guide the design process and evaluate key parameters, such as fluid charging and flow permeability. Two- and three- dimensional solutions provided strong evidence that the anisotropic wick was critical in reducing the effects of source shadowing.

A prototype device was assembled and tested (with a single heat source) and the results show significant improvement over more common wick materials (including sintered metal felt and other isotropic designs). The ratio of heat pipe conductivity to that of the monolithic material was measured to be 8.1X. In optimal operating conditions using copper, this corresponds to a conductivity value in excess of 3200 W/m-K. Dryout was observed at 64 W/cm^2. Transient dryout occurred at lower fluxes and was attributed to a lack of liquid flow between the unheated and heated surfaces.

Additional design refinements and performance optimization tasks are planned to further improve the device.

Sandia is a multiprogram laboratory operated by Sandia Corporation, a Lockheed Martin Company, for the United States Department of Energy under contract DE-AC04-94-AL85000.

REFERENCES

Benson, D.A. et al., 1998, Ultrahigh-Capacity Micromachined Heat Spreaders, *Microscale Thermophysical Engineering*, Vol. 2, pp. 21-30.

Benson, D.A. et al., 1999, Kovar Micro Heat Pipe Substrates for Microelectronic Cooling, SAND99-0714.

Cao, Y. et al., 1997, Experiments and Analyses of Flat Miniature Heat Pipes, *J. of Thermophysics and Heat Transfer, Vol.* 11, No. 2, pp. 158-64.

Chi, S.W., 1976, *Heat Pipe Theory and Practice*, MacGraw-Hill, New York.

Cotter, T.P., 1965, Theory of Heat Pipes, Los Alamos Scientific Laboratory Report No. LA-3246-MS.
Dunn, P.D., & D.A. Reay, *1994, Heat Pipes, Fourth Edition*, Pergamon.
Grover, G., Cotter, T., and G. Erickson, 1964, Structure of Very High Thermal Conductance, *J. Appl. Phys.*, Vol. 35, pp. 1990-91.
Grover, G., 1966, Evaporation-Condensation Heat Transfer Device, U.S. Patent Application # 3229759, Approved: 1/18/1966.
Kays, W.M., 1966, *Convective Heat and Mass Transfer*, MacGraw-Hill, New York.
Marcus, B.D., 1972, Theory and Design of Variable Conductance Heat Pipes, Report NO. NASA CR 2018, Washington D.C..
Martinez, M.J. et al., 1997, LDRD Final Report: Physical Simulation of Nonisothermal Multiphase Multicomponent Flow in Porous Media, SAND97-1766, Sandia National Laboratories.
Peterson, G.P., 1994, *An Introduction to Heat Pipes: Modeling, Testing, and Applications*, John Wiley and Sons.
Plesch. D. et al., 1991, Miniature heat Pipes for Heat Removal from Microelectronic Circuits, Proc. ASMI Annual Meeting, Atlanta, GA.
Smith, P.M., 2000, Sandia National Laboratories, personal communication, October.
Sweet, J.N. et al., 1993, Assembly Test Chip Version 04 (ATC04) Description and User's Guide, Sandia National Laboratories internal memorandum.
Vadakkan, U., Garimella, S., and C. Sobhan, 2000, Characterization of the Performance of Flat Heat Pipes for Electronics Cooling, Proceedings of the 2000 International Mechanical Engineering Congress and Exposition, Orlando, FL.
Wu, D., & G.P. Peterson, 1991, Investigation of the Transient Characteristics of a Micro Heat Pipe, *J. Thermophysics,* Vol. 5, No. 2, pp.129-34.

Compact models and system-level challenges

Thermal Challenges in Next Generation Electronic Systems, Joshi & Garimella (eds)
© 2002 Millpress, Rotterdam, ISBN 90-77017-03-8

The application of a genetic algorithm to develop a compact thermal model.

R.L. Clarksean and J. Torkelson
Leading Technology Designs, Inc., MN, USA

Keywords: genetic algorithm, compact, thermal model, electronics

ABSTRACT: Compact thermal models are used throughout the electronics industry to simplify complex analyses and to reduce the design time required for electronics cooling systems. A number of modeling approaches have been used to develop "compact" thermal models that accurately represent the temperature rise and heat transfer through a chip on a board. A technique is proposed to accurately model the "thermal resistors" of the compact model. The technique is based on the use of a genetic algorithm to optimize the resistors, based on numerical or experimentally determined temperatures and known heat generation. The circuit model consists of 11 resistors internal and external to the representative chip geometry. The model requires known heat generation and measured surface temperature data.

1 INTRODUCTION

Modeling the flow and heat transfer in electronic systems is a very complex problem. Analysis of a computer chassis may include heat sinks, power supplies, single or multiple boards installed within the system, heat sink fans, enclosure fans, and wire blockage of flow passages. Trying to model all of the length scales involved with the items discussed above can be overwhelming. The length scales will range from the board scale all the way down to the type of interconnect used for chip packages installed on the electronic boards.

In theory one could model all of the different length scales, but from a practical standpoint, attempting to model 3 or 4 orders of magnitude change is difficult. Depending on the numerical techniques used, the amount of mesh easily exceeds 106 or 107 nodal points. This ultimately results in large computer memory requirements and significant solution times. Clearly, this is unacceptable for the design engineer who is trying to analyze multiple designs as quickly as possible. The goal is to always decrease the time to market for a product from concept, to preliminary design, to the final design and ultimately production.

To decrease the computing requirements and to speed up development times, numerous researchers have worked on the development of compact thermal models (CTMs) to be used in conjunction with detailed flow and heat transfer models. The goal of any compact thermal model is to reduce the range of length scales in a problem to a manageable level. Chip-level details are replaced by a simpler models that estimate chip temperatures in conjunction with a system level flow and heat transfer analysis.

Research reported in recent electronics packaging conferences clearly indicates this is an active area of research and the need for more robust models. Aranyosi et al. (1999) and Aranyosi et al. (2000) analyzed a number of circuit configurations in an attempt to develop CTMs that were independent of external boundary conditions. Extensive experimental and numerical models were used in these works. Yu and Przekwas (1999) worked on the development of an automated system for defining resistance network type models. Numerous researchers have examined resistance type

CTMs, including Bar-Cohen and Krueger (1989) and Pashaie-Rad et al. (1998) as just two recent examples of such work. The extensive work of the DELPHI Project was geared towards a methodology for developing compact models (Rosten et al. 1997).

There are a number of factors that must be considered in the development of a CTM. These factors include:

- Topology of the network,
- Location and application of external boundary conditions,
- Optimization and verification of the proposed model,
- Location of the heat source in the model, and the
- Technique used to incorporate the CTM into the detailed model.

The present research focuses on the use of a genetic algorithm to assist in defining the resistors within the CTM. If experimental test data exists, this data could be used in conjunction with a genetic algorithm to accurately define the resistors. The paper will discuss a basic circuit topology, give a brief overview of the genetic algorithm approach, outline the numerical approach, and then discuss preliminary modeling results.

2 TEST MODEL CIRCUIT TOPOLOGY

A common and simple circuit topology was selected to test the use of a genetic algorithm. The goal was to use a genetic algorithm approach to define the resistors in the circuit topology. It was assumed that the topology was known in advance. Future work will consider the use of a genetic algorithm to assist in defining the proper topology.

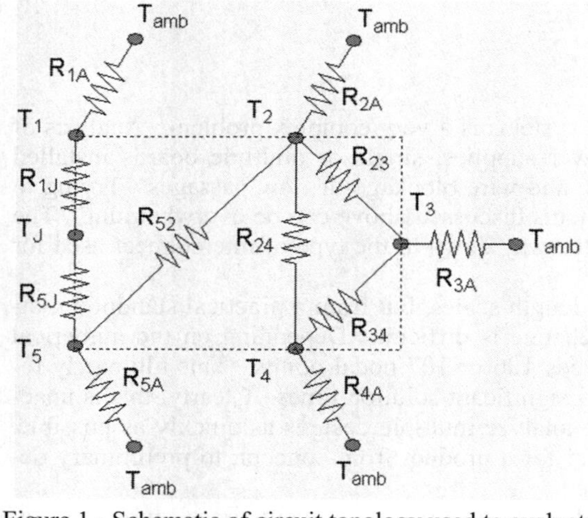

Figure 1 - Schematic of circuit topology used to evaluate genetic algorithm.

The circuit topology is shown in Figure 1. The topology considers 5 external temperature points (T1 to T5) and one internal temperature, the junction temperature (TJ). The heat generated within the model is lost to the surroundings at a temperature of Tamb. This topology is common and has been discussed in the work of Aranyosi et al. (1999) and Aranyosi et al. (2000). The heat generation is assumed to occur in the region of point J. The dashed line in the model represents the "chip" region of the model. The external resistors represent the heat loss to the surroundings.

3 GENETIC ALGORITHM OVERVIEW

The texts by Gen and Cheng (1997) and Michalewicz (1996) were used for the development of the genetic algorithm scheme. Only a general overview will be given here. As the genetic algorithm approach is used here, it is an optimization technique.

The terminology associated with genetic algorithms comes from the field of genetics, hence the name. As with any optimization scheme, the goal is to search for the best solutions in an efficient manner. The genetic algorithm approach strikes a balance between exploration and exploitation of the search space.

The process starts with the selection of an initial population of potential solutions. These potential solutions are then evaluated as to their "fitness" to the potential problem. As discussed in Section 4, this test requires the definition of a fitness function for the problem. Once the "best" solutions out of this initial population have been selected, they will be modified to create a new population to be evaluated for their fitness. The process of creating a new population involves crossovers and mutations.

The process continues until a specified number of "generations" have been created. The "fitness" value keeps improving and the number of generations increases. The process of creating the new populations keeps moving the solution closer to the desired outcome.

This general process was used for the worked conducted within this paper.

4 OUTLINE OF NUMERICAL APPROACH

Rather than search for experimental data upon which to test this modeling approach, it was decided to use numerically obtained data for the model. Two general models were developed, the first considered the situation where all of the resistors were of the same magnitude and the second considered resistors that have a magnitude typical of those found in CTMs.

Simple energy balance equations were written for each of the nodal points in the model. For example, the energy balance equation for Node 1 is shown below.

$$\frac{(T_J - T_1)}{R_{1J}} = \frac{(T_1 - T_{amb})}{R_{1a}} \tag{1}$$

where the variables are shown in Figure 1. T represents the temperatures, while R represents the resistors in the network. The units for temperature is °C and W/°C for the resistors.

Writing a simple energy balance for each node and then rearranging to solve that equation for the nodal temperature results Equations (2) to (7).

$$T_1 = \frac{\left\{T_J\left(R_{1a}/R_{1J}\right) + T_{amb}\right\}}{\left(1 + R_{1a}/R_{1J}\right)} \tag{2}$$

$$T_2 = \frac{\left\{T_5/R_{52} + T_4/R_{24} + T_3/R_{23} + T_{amb}/R_{2a}\right\}}{\left(1/R_{2a} + 1/R_{52} + 1/R_{24} + 1/R_{23}\right)} \tag{3}$$

$$T_3 = \frac{\left\{ T_4 / R_{34} + T_2 / R_{23} + T_{amb} / R_{3a} \right\}}{\left(1/R_{34} + 1/R_{23} + 1/R_{3a} \right)} \tag{4}$$

$$T_4 = \frac{\left\{ T_2 / R_{24} + T_3 / R_{34} + T_{amb} / R_{4a} \right\}}{\left(1/R_{24} + 1/R_{34} + 1/R_{4a} \right)} \tag{5}$$

$$T_5 = \frac{\left\{ T_J / R_{5J} + T_2 / R_{52} + T_{amb} / R_{5a} \right\}}{\left(1/R_{5J} + 1/R_{52} + 1/R_{5a} \right)} \tag{6}$$

$$T_J = \frac{\left\{ T_1 / R_{1J} + T_5 / R_{5J} + Q \right\}}{\left(1/R_{1J} + 1/R_{5J} \right)} \tag{7}$$

where Q = the heat generation rate.

The basic approach was to define a set of resistors (magnitude), solve for the resultant temperatures for a given heat source, and to then use this data as input to the genetic algorithm approach. Equations (2) to (7) were solved with a simple iterative scheme to obtain the temperatures.

The predicted temperatures for the "known" resistor values were then used as "measured" data in the genetic algorithm. These "measured" values, along with the known heat fluxes, were used to see how accurately the genetic algorithm could predict the resistors. The error was defined as the difference between the assumed values (used to obtain the measured temperatures) and the genetic algorithm resistor values. All 11 resistors were solved for through the use of the genetic algorithm. The use of simple convective flow relationships can be used to reduce the number of unknown external values from 5 in this case to 1. This possibility will be examined in future research efforts.

Within the genetic algorithm a fitness function was defined. This is the relationship that is to be maximized. For the present work, the fitness function is shown in Equation (8) and is similar to that used by Kuzuno et al. (1999).

$$Fitness = \frac{1}{1 + \sqrt{\sum_{i=1}^{6} (T_{i\_meas} - T_{i\_calc})^2 + \sum_{j=1}^{5} (q_{j\_meas} - q_{j\_calc})^2}} \tag{8}$$

where q_j = external heat fluxes for the measured (known) and calculated (genetic algorithm) conditions.

As the error between the calculated and measured values decreases, the fitness function approaches 1. For larger errors, the value of the function is less than one. Other fitness functions could be used and will be tested in the future.

A simple process flow for the genetic algorithm approach includes
1. Initial estimate for resistors.
2. Determination of temperatures at nodal points with the assumed resistors (iteratively).
3. Calculation of the fitness parameter.
4. Modification of the resistors through the use of the genetic algorithm.
5. Return to step 2 until all iterations have been completed.

5 PRELIMINARY MODELING RESULTS

Two sets of values for the circuit topology were tested. Test Case 1 is shown in Table 1. Test Case 1 considers equal values for all of the resistors. The genetic algorithm was effective in accurately determining the magnitude of the resistors for the circuit topology. The error between the estimated resistors and the known values were all below one percent, except for R_{34}. The error for this resistor is 15 percent. Note that this error has little impact on the predicted temperature field, nor does it impact the overall accuracy of the temperatures that are estimated for the genetic algorithm resistors. This error is insignificant because the temperatures of nodes 3 and 4 are equal, implying there is no heat transfer between these nodes. Hence, the magnitude of R_{34} will not impact the overall solution.

The second test case uses data estimated from the work of Aranyosi et al. (2000). The magnitude of the resistors were estimated based on testing they conducted on this particular circuit topology. The magnitude of the resistors range from 2 to 100, nearly two orders of magnitude change.

For the genetic algorithm approach to work, upper and lower bounds must be specified for each resistor to bracket the search region. The results shown in Table 1 are for the fitness function shown in Equation 8 and for upper and lower limits of 1 and 105 for each of the resistors. A total of 9,000,000 fitness tests were conducted for the results shown in Table 2.

For Test Case 2 (See Table 2), the errors were much larger this time for the same solution procedure. The errors associated with the resistors averaged 9.35 percent. The temperature field solution shows only an error of roughly 0.80 percent. Because this model has only been tested on one set of temperature data, one would be concerned how well this model would adapt to slightly different operating conditions.

Table 1 – Comparison of known and genetic algorithm estimated temperatures and resistors for a common circuit topology with equal resistors.

Test Case 1							
Comparison of Resistor Values				Comparison of Temperature Values			
Resistor	Known (°C/W)	GA Est. (°C/W)	Percent Error*	Node	Known (°C)	GA Est. (°C)	Percent Error
R_{1J}	5	5.019	0.38	T_1	47.22	47.23	0.02
R_{5J}	5	4.980	0.40	T_2	30.56	30.56	0.00
R_{23}	5	4.983	0.34	T_3	27.78	27.78	0.00
R_{24}	5	4.956	0.88	T_4	27.78	27.78	0.00
R_{52}	5	4.980	0.40	T_5	41.67	41.68	0.02
R_{34}	5	4.245	15.1	T_J	69.44	69.44	0
R_{1A}	5	5.025	0.50	T_{amb}	25.0	25.0	- na -
R_{2A}	5	4.989	0.22				
R_{3A}	5	4.989	0.22	Q	10.0	10.00	0
R_{4A}	5	4.971	0.58				
R_{5A}	5	4.989	0.22				
		Avg.	*1.75*				
Notes: * - Percent Error = Absolute value ((known-GA Est)/known)							

In an attempt to further reduce the error associated with the magnitude of the resistors, information from the first run (Table 2) was used to refine the search region. Rather than have the genetic algorithm look for possible solutions over a range from 1 to 105, these upper and lower bounds were readjusted based on the results shown in Table 2. These changes were made and additional tests were conducted.

Changes were also made to the fitness function as shown in Equation 8. The square root of the errors was eliminated as shown in Equation 9 and this relationship was used for the fitness function.

$$Fitness = \frac{1}{1+\sum_{i=1}^{6}(T_{i\_meas} - T_{i\_calc})^2 + \sum_{j=1}^{5}(q_{j\_meas} - q_{j\_calc})^2} \tag{9}$$

Using both the redefined fitness function and the knowledge of the parameters defined in Table 2 an additional test was conducted to see if the errors associated with the resistors could be decreased. The result of this effort is shown in Table 3. The rate of convergence to a better fitness number for the modified fitness equation (Equation 9) was more rapid than the function proposed in Equation 8. Because of the increased convergence speed only 3,000,000 tests were conducted.

The overall predictions for the resistor values improved to an average error of 2.81 percent, versus the previous error of 9.35 percent. The errors associated with the temperature field remained approximately the same. The error associated with the heat flux in these two situations actually increased slightly.

Table 2 – Comparison of known and genetic algorithm estimated temperatures and resistors for a common circuit topology with widely varying resistors resistors.

Test Case 2 – Wide Range 9,000,000 Tests							
Comparison of Resistor Values				Comparison of Temperature Values			
Resistor	Known (°C/W)	GA Est. (°C/W)	Percent Error*	Node	Known (°C)	GA Est. (°C)	Percent Error
R_{1J}	2	2.14	7.00	T_1	101.4	100.8	0.78
R_{5J}	5	4.95	1.00	T_2	82.89	82.48	0.71
R_{23}	15	15.14	0.93	T_3	73.68	73.26	0.86
R_{24}	43	39.06	9.16	T_4	69.13	68.76	0.84
R_{52}	6	5.68	5.33	T_5	91.96	91.39	0.85
R_{34}	37	59.2	60.0	T_J	103.0	102.4	0.77
R_{1A}	100	100.4	0.40	T_{amb}	25.0	25.0	- na -
R_{2A}	100	94.18	5.82				
R_{3A}	100	90.96	9.04	Q	3.0	2.976	0.80
R_{4A}	100	102.6	2.60				
R_{5A}	100	101.6	1.60				
		Avg.	*9.35*				
Notes: * - Percent Error = Absolute value ((known-GA Est)/known)							

Table 3 – Comparison of refined solution values for the known and genetic algorithm estimated temperatures and resistors. Only 3,000,000 tests were conducted in this situation.

Test Case 2 – Narrowed Range – Modified Fitness 3,000,000 Tests							
Comparison of Resistor Values				Comparison of Temperature Values			
Resistor	Known (°C/W)	GA Est. (°C/W)	Percent Error*	Node	Known (°C)	GA Est. (°C)	Percent Error
R_{1J}	2	1.93	3.50	T_1	101.4	100.6	0.79
R_{5J}	5	4.98	0.40	T_2	82.89	82.19	0.84
R_{23}	15	14.44	3.73	T_3	73.68	73.10	0.79
R_{24}	43	44.98	4.60	T_4	69.13	68.60	0.77
R_{52}	6	5.92	1.33	T_5	91.96	91.22	0.80
R_{34}	37	36.43	1.54	T_J	103.0	102.2	0.78
R_{1A}	100	96.71	3.29	T_{amb}	25.0	25.0	- na -
R_{2A}	100	96.38	3.62				
R_{3A}	100	95.61	4.39	Q	3.0	2.953	1.56
R_{4A}	100	102.6	2.60				
R_{5A}	100	101.9	1.90				
		Avg.	*2.81*				
Notes: * - Percent Error = Absolute value ((known-GA Est)/known)							

The results indicate that a genetic algorithm technique can easily be developed to assist in the design of CTM in conjunction with experimental test data.

6 CONCLUSIONS

A simple genetic algorithm procedure has been developed to assist in the definition of resistor values for compact thermal models. Two test cases were evaluated on a common circuit topology. The first test case considers equal values for all of the resistors. All but one of the resistors was estimated within one percent or less error. The resultant temperature field for these estimate resistor values found the predicted and baseline temperatures to be nearly identical.

The second test case considers nearly two orders of magnitude variation in the resistor values. A two-tiered approach was used here. First the resistors were estimated from very broad upper and lower limits. Then, the field of search was narrowed through the redefinition of the upper and lower bounds. In conjunction with a modified fitness function, the resultant errors were less than the first set of simulations (9.35 vs. 2.81).

These preliminary test results indicate that the genetic algorithm can be used successfully. Future work will work on refining the fitness function definition and on automating the circuit definition.

REFERENCES:

Aranyosi, A. Ortega, A., Griffin, R., West, S., and Edwards, D., 1999, ASME, *Advances in Electronic Packaging*, EEP-Vol. 26-1: 83-96.

Aranyosi, A., Ortega, A., Evans, J., Tarter, T., Pursel, J., and Radhakrishnan, J., 2000, Development of compact thermal models for advanced electronic packaging: Methodology and experimental validation for a single-chip CPGA package, *2000 Inter Society Conference on Thermal Phenomena*, IEEE, Las Vegas, Nevada, USA, vol. 1: 225-232.

Bar-Cohen, A. and Krueger, W.B., 1997, "Thermal characterization of chip packages – evolutionary development of compact models, IEEE CMPT, vol. 20, no.4: 399-409.

Gen, M. and Cheng, R., 1997, *Genetic Algorithms & Engineering Design*, New York: John Wiley and Sons.

Kuzuno, M., Nishio, T., Yamada, Y., and Koyamada, K., 1999, ASME, *Advances in Electronic Packaging*, EEP-Vol. 26-1: 65-73.

Michalewicz, Z., 1996, *Genetic Algorithms + Data Structures = Evolution Programs*, 3rd Edition, Berlin:Springer-Verlag.

Pashaie-Rad, A., Hermanson, J., and Rencis, J., 1998, Development of component-level thermal models of a ceramic pin grid array, IEEE, *1998 Intersociety Conference on Thermal Phenomena*, Seattle, Washington, USA: 512-519.

Rosten, H., Parry, J., Lasance, C., Vinke, H., Temmerman, W., Nelmans, W., Assouad, Y., Gautier, T., Slattery, O., Cahill, C., O'Flattery, M., Lacaze, C., and Zemlianoy, P., 1997, Final report to SEMITHERM XIII on the European-funded project DELPHI – the development of libraries and physical models for an integrated design environment, Thirteenth IEEE SEMI-THERM Symposium Proceedings: 73-91.

Shidore, S., Adams, V., Lee, Tien-Yu (Tom), 2000, A study of compact thermal model topologies in CFD for a flip chip plastic ball grid array package, *2000 Inter Society Conference on Thermal Phenomena*, IEEE, Las Vegas, Nevada, USA, vol. 1: 322-329.

Yu, E. and Przekwas, A., 1999, Reduced thermal models of electronic packages for system level design applications, ASME, *Advances in Electronic Packaging*, EEP-Vol. 26-1: 75-82.

Thermal Challenges in Next Generation Electronic Systems, Joshi & Garimella (eds)
 ISBN 90-77017-03-8

Thermal modeling and experimental evaluation of outdoor cabinets

K. Nevelsteen & M. Baelmans
K.U.Leuven, Leuven, Belgium

W. Nelemans, G. Swaelen & R. De Smedt
Alcatell-Bell, Antwerpen, Belgium

Keywords: harsh environments, thermal modeling, system level

ABSTRACT: In this paper the experimental evaluation of an outdoor telecommunication cabinet with double-walls under solar load is described. The separate influence of different system and environmental conditions such as internal dissipation, solar load, internal fans and external fans on the component temperature rise is evaluated. A simplified thermal model for the double-wall door is established. This model is implemented in a CFD code and validated against experimental results. The CFD model predicts heat transfer rates within 20 % leading to predicted temperatures with an accuracy of approximately 2 °C.

1 INTRODUCTION

Special cabinets with double-wall doors are used for outdoor telecommunication systems in order to limit the component temperature rise under solar load. Typically, the internal air flow inside these cabinets is completely sealed from the external induced convection cooling. Thus dust penetration is prevented. Figure 1 shows a picture and a schematic view of the investigated cabinet.

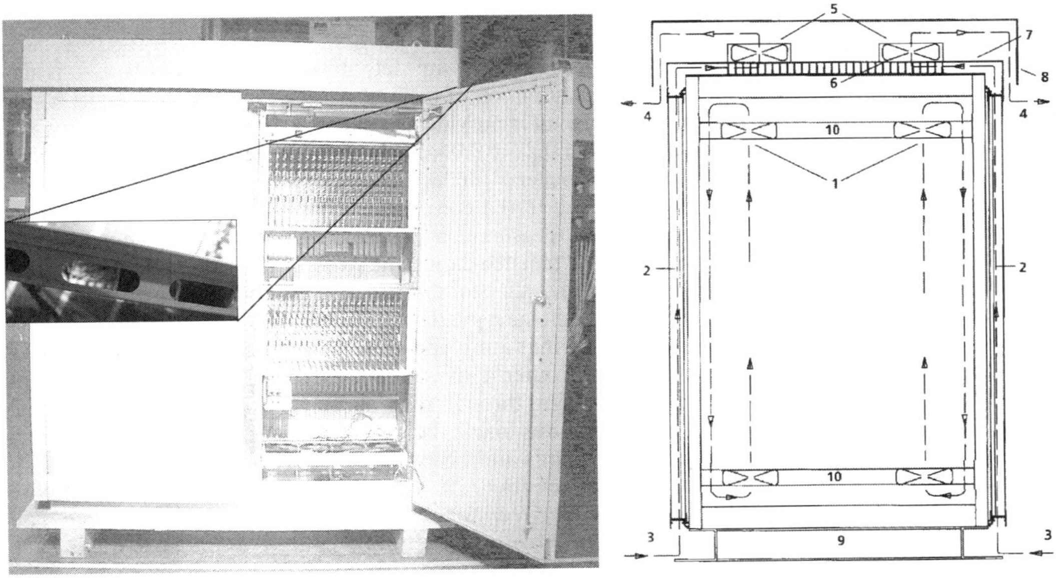

Figure 1. Photo and schematic view of the outdoor cabinet

Outside air is sucked through the double walls (2) of the cabinet by means of the external fans (5). The cool air flows in at the bottom of the cabinet (3), flows through the double walls (2), over the heat sinks (6) on the inner roof of the cabinet and leaves the system at all sides (4) of the external roof (8). Two internal fan units (10) provide air circulation (1) between the dissipating boards inside the sealed cabinet. Each fan unit consists of four constant speed fans. Inner and outer door walls are connected to each other by thermal vias which form channels in the double-wall door. The insert on figure 1 shows a top view of the double-wall door. The dimensions and complexity of the enclosure structure in general and of the doors in particular are too demanding for detailed CFD computations at system level. Therefore a simplified thermal model for the double-wall door is elaborated.

To this end the thermal behavior of an outdoor cabinet is experimentally evaluated for different solar loads and operating conditions. The solar load is provided by a solar simulator. Subsequently, these experiments are processed in order to derive a model for the heat transfer in the double-wall door of the cabinet and a model for use in a CFD package is deduced. The computational results are then validated against the experimental results for two sets of operating conditions.

2 EXPERIMENTAL INVESTIGATION OF THE ENCLOSURE UNDER SOLAR LOAD

In section 2.1 the conducted experiments for material characterization and system evaluation are described. Next a summary of the results and a thermal model are presented.

2.1 *Experimental set up and measurement campaign*

Prior to the experimental evaluation of the enclosure, additional parameters as solar load and radiation characteristics of the wall are experimentally determined. As both solar and thermal radiation should be taken into account, the corresponding wavelengths will significantly differ. The cabinet wall will not act as a black ($\varepsilon = 1$) nor as a grey body ($\varepsilon \neq f(\lambda)$). Therefore, radiation theory justifies the use of four wavelength dependent variables to characterize the radiation behavior of the wall: solar and far infrared absorptivity (α_{solar} and α_{FIR} respectively) and solar and far infrared emissivity (ε_{solar} and ε_{FIR} respectively). The solar (radiation with wavelengths smaller than 3 μm) and infrared (radiation with wavelengths larger than 3 μm) radiation intensity from the solar simulator are determined by a pyrano- (Kipp & Zonen CM11) and a pyrgeometer (Kipp & Zonen CG1) respectively. The average solar radiation intensity on the cabinet door amounts to 1063 W/m^2 and the average infrared radiation intensity amounts to 86 W/m^2.

Next, the radiation characteristics of the cabinet wall are determined. The emissivity ε_{FIR} is deduced from infrared measurements, with the assumption of thermal equilibrium such that $\varepsilon_{FIR} = \alpha_{FIR} = 0.81$. For the emissivity ε_{solar}, the paint manufacturer provides a value of 0.45. A separate experiment determines the absorptivity α_{solar} to be equal to 0.28 ± 0.04. In this experiment, a steady-state energy balance is elaborated for a piece of the wall material under the solar simulator. The inaccuracy on α_{solar} is due to the measurement errors and due to the correlations for the convection heat transfer. Measurement errors arise from the temperature measurements and the measurement of the velocity to estimate the natural and forced convection heat transfer during the experiment. The same correlations as in section 2.3 are used, with appropriate length scales for this separate experiment.

For the thermal characterization of the enclosure, wall and air temperatures as well as air velocities in the double-wall door and in the cabinet are measured under different conditions. Three thermocouples measure the temperature of the outer door wall and three the temperature of the inner door wall. The temperature of the air flowing through the double-wall door is measured at the top and at the bottom of the door. Air temperatures inside the cabinet are measured at the upper and the lower fan units, between the dissipating boards and under the inner roof of the cabinet. Both inner and outer roof wall temperatures are also measured. Air velocities are measured with a hot-wire anemometer at 5 positions in the air channel between the door and the boards. In the double-wall door, air velocity is measured in 4 different air channels, each time in 2 points situated above each other in the same air channel.

During the experimental campaign the voltage applied to the internal (0, 36 or 48 V) and external fans (0 or 36V), solar load (on or off) as well as the internal power dissipation (200, 300, 450, 600 or 670 W) are varied.

2.2 *Summary of the experimental results*

Twenty two experiments are carried out for which energy balances are elaborated for the thermal characterization of the door. Figures 2 and 3 summarize the separate influence of each boundary condition on the temperature rise between the boards above ambient temperature. For the cases in figures 2 and 3, the external fans are off and on respectively. Similar results for inner and outer door wall temperatures and for the air temperature at the fans are measured. The different experiments are indicated as function of the internal dissipation. Different symbols represent the different possibilities for the voltage of the internal fans (0, 36 or 48 V) and for the solar load (on, off). The two measurements without the roof on the cabinet are indicated as 'No roof'. Internal fan voltage during these experiments is 36 V.

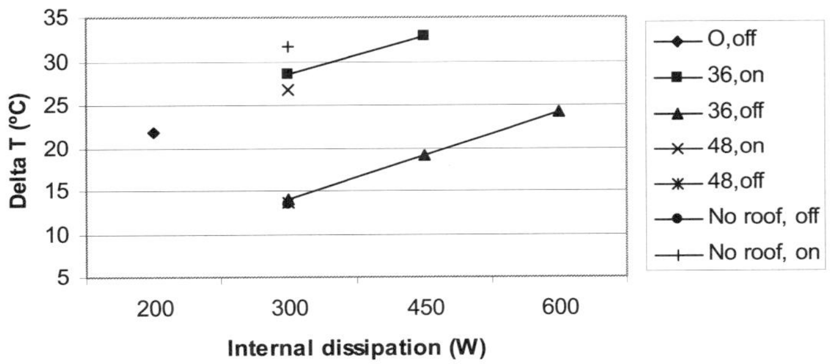

Figure 2. T increase between boards relative to environmental temperature, external fans off

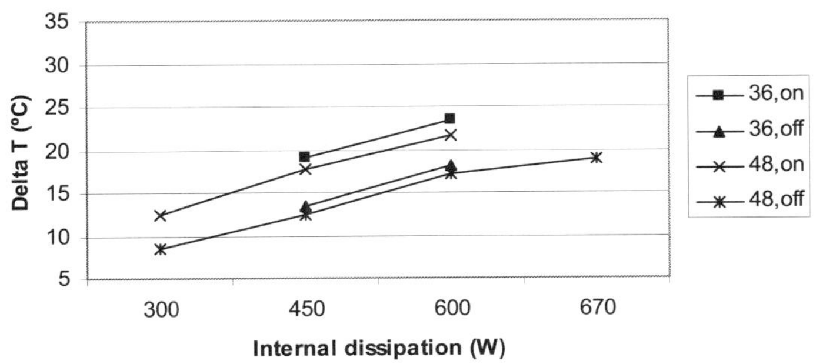

Figure 3. T increase between boards relative to environmental temperature, external fans on

From the experimental measurements, it can be concluded that when the solar load is on, all temperatures for the cabinet with the roof are approximately 4 °C lower than for the cabinet without the roof. When the solar load is off, there is no significant difference. Thus, it can be concluded that the cabinet roof acts well as a radiation shield.

Switching the external fans on when solar load is applied causes a temperature decrease of approximately 15 °C, independent of the internal fan voltage. When the external fans are off, solar load causes a temperature rise of approximately 23 °C and 21 °C for both door wall temperatures when the internal fan voltage is 36 or 48 V respectively. The temperature rise between the boards in this case is 14 and 13 °C respectively. For each extra 150 W of internal dissipation, the internal temperature rises approximately 4.5 °C.

2.3 *Heat transfer model*

Based on the experiments, energy balances for the double-wall door are evaluated. The external conditions consist of convection heat transfer as well as of radiation heat transfer. Because the ratio Gr/Re^2 (Gr = Grashof number) is between 0.1 and 10, both natural and forced convection should be taken into account and are based on correlations (Çengel 1997, Shah & London 1978). The global heat transfer coefficient h_{total} is deduced from a global Nusselt number

$$Nu_{total} = h_{total} . \delta / k, \tag{1}$$

where δ = characteristic length = cabinet height; k = thermal conductivity of the air.

Nu_{total} combines natural and forced convection Nusselt numbers:

$$Nu_{total} = (Nu_{nat}^3 + Nu_{forced}^3)^{1/3} \tag{2}$$

The Nusselt number Nu_{nat} for natural convection heat transfer is based on the Rayleigh number and is

$$Nu_{nat} = \left(0.825 + \frac{0.387 \cdot Ra^{1/6}}{\left[1 + (0.492/Pr)^{9/16}\right]^{8/27}} \right)^2 \tag{3}$$

where Ra = Rayleigh number = Gr.Pr (-); Pr = Prandtl number (-). The ratio of the area on the perimeter of the door surface is chosen as the characteristic length for the Grashof number (Çengel 1997). The forced convection Nusselt number depends on the Reynolds number, which is of the order of magnitude of 20 000 (based on the width of the door (L = 0.625 m)). For external flows, the critical Reynolds number above which the flow is turbulent is 500 000. Therefore, correlations for laminar flow can be used. The forced convection Nusselt number Nu_{forced} equals then to

$$Nu_{forced} = 0.664 \cdot Re_L^{0.5} \cdot Pr^{1/3} \tag{4}$$

where Re_L = Reynolds number (-). The width of the door is taken as a characteristic length.

The net radiation heat transfer intensity $q_{net,in}$ (W/m^2) to the cabinet surface is deduced from

$$Q_{net,in} = [\alpha_{solar} . q_{in,\,solar} + \varepsilon_{FIR} . q_{in,FIR} - \varepsilon_{solar} . 0.0003 .\ \sigma . (T^4 - T_{amb}^4)$$

$$- \varepsilon_{FIR} .\ 0.9997\ . \sigma . (T^4 - T_{amb}^4)] . A \tag{5}$$

using the experimental data. Hereby incoming radiation intensities for solar and infrared wavelengths are denoted with $q_{in,solar}$ (W/m^2) and $q_{in,FIR}$ (W/m^2) respectively. T and T_{amb} are the surface and ambient temperature respectively. The factors 0.0003 and 0.9997 take into account the distribution of the radiation energy for the wall across the wavelength spectrum.

Heat transfer between the air convected inside the cabinet and the double-wall door is also investigated. The velocity V in the double-wall door is measured. Based on this velocity a mass flow rate $\dot{m}$ (kg/s) is calculated. The total heat flow Q_{door} (W) extracted from the door by the airflow through it is given by:

$$Q_{door} = \dot{m}\ .\ C_p\ .\ \Delta T \tag{6}$$

where $\dot{m}$ = air mass flow rate (kg/s); C_p = specific heat of the air (J/kgK); ΔT = temperature difference of the air between inlet and outlet (°C). For the heat transfer $Q_{channel}$ from the air inside the cabinet, equal formulas are used. Furthermore radiation between the equipment front panel and the door is taken into account.

Based on the above mentioned assumptions, the different experiments are evaluated. Figure 4 (right side) shows the energy balance for the double-wall door for configuration 2 (see section 3.1). In this case, the heat input $Q_{channel}$ to the door is 38.3 W from the cabinet inner and 276.5 W radiation heat transfer from the solar load. Q_{door} = 267.9 W is extracted by the air flowing through the door, 49.0 W by external convection and 0.2 W by radiation to the equipment front panel.

Based on the evaluation of the energy balances on the door, the solar absorptivity can be refined. The value α_{solar} is calculated for each of the cabinet experiments with solar load, so that the heat balances for the cabinet door do fit. The average value α_{solar} = 0.31 is finally used in the evaluation of the energy balances. The average deviation on these balances amounts to 6.1 % relative to the amount rejected by the convective air flow through the double-wall.

3 COMPACT NUMERICAL SIMULATION OF THE DOUBLE-WALL DOOR

3.1 *CFD-model*

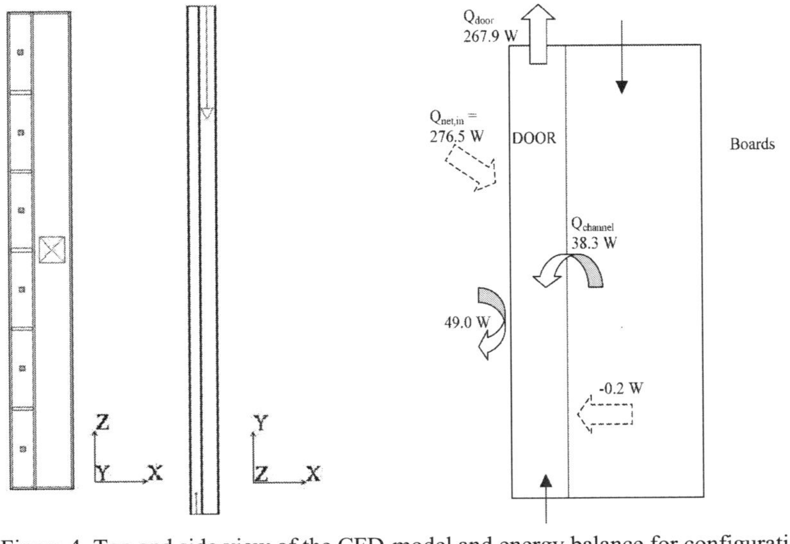

Figure 4. Top and side view of the CFD-model and energy balance for configuration 2

Starting from the experimental results it seems to be appropriate to compose a simple CFD model for the double-wall door of the cabinet based on a counterflow configuration (see fig. 4, left side). The left and right channels model the double-wall door and the channel in front of the printed circuit boards respectively. Surface area factors account for the enhanced heat transfer surface of the crenellated door walls (see insert on fig. 1). These area factors are equal to 1.194 and 1.138 for the air channels in the door and between the inner door wall and the boards respectively. Thermal vias in the door are modeled by a smaller set of equivalent thermal resistance. The area factor for the thermal vias is 2, because there are only 5 thermal vias in the model and 10 in reality. Further decreasing the number of thermal vias in the simulation is not recommended because in that case the temperatures of the inner and outer wall are much less homogeneous over the width of the door in comparison with the experimental data. Only three grid cells are needed in the thickness direction of the channels. To adequately model the conduction in the walls, 2 cells in the thickness direction are taken. All above mentioned geometrical simplifications significantly reduce the grid size and hence the computational effort.

Input parameters for this CFD model are the inlet air temperature and velocity in both channels corresponding to the experimental data set. Radiation and convection at the outer wall are simulated by a heat source in the outer wall. Heat transfer as well as air and wall temperatures are validated for 2 sets of boundary conditions:

1: Internal and external fans: 48 and 36 V, power dissipation 300 W, solar load off
2: Internal and external fans: 48 and 36 V, power dissipation 300 W, solar load on

In a simulation of the complete enclosure including PCB racks, the air velocity in the double-wall door is depending on the external fan voltage only. In contrast, air velocity and temperature in the channel in front of the printed circuit boards should then be calculated in the simulation.

3.2 *Numerical results*

The numerical (left) and experimental (right) results are summarized in table 1 and 2.

Table 1. Numerical and experimental results for the case 1 (without solar load)

	Numerical results		Experimental results	
	Q =Heat transferred	Air outlet	Q = Heat transferred	Air outlet
	W	°C	W	°C
Channel	59.3	27.6	76.3	27.1
Door	53.5	22.4	66.2	22.7
	Outer wall	Inner wall	Outer wall	Inner wall
	°C	°C	°C	°C
Top	23.5	23.6	24.7	26.0
Middle	22.8	23.0	23.4	23.8
Bottom	22.2	22.4	22.2	22.6

Table 2. Numerical and experimental results for the case 2 (with solar load)

	Numerical results		Experimental results	
	Q = Heat transferred	Air outlet	Q = Heat transferred	Air outlet
	W	°C	W	°C
Channel	23.0	41.6	38.3	40.8
Door	237	38.6	267.9	39.9
	Outer wall	Inner wall	Outer wall	Inner wall
	°C	°C	°C	°C
Top	44.8	44.0	46.2	43.5
Middle	42.2	41.4	/	44.7
Bottom	38.8	38.2	42.5	39.5

'Channel' denotes the air channel between the boards and the inner door wall, 'Door' denotes the channels in the double-wall door. The wall temperatures are shown for the inner and the outer wall of the double-wall door, at the top, the middle and at the bottom of the door.

For the case without solar load, the deviation between the numerically calculated heat flow and the experimental one is 19 % and 22 % for the door and for the channel respectively. The average deviation for the inner wall temperatures is 0.7 °C. Given the grid restrictions, it is concluded that the model adequately represents the heat transfer through the double-wall door.

Next, for the case with solar load, the deviation between the numerically calculated heat flow and the experimental one is 11 % and 40 % for the door and for the channel respectively. The deviation seems high, but is relatively small because the 23.0 W heat transfer from the channel should be interpreted with regard to the 237 W heat transfer in the door and gives rise to maximum 6.1 %

error on the energy balance over the door panel. The average deviation for the inner wall temperatures is 1.4 °C. The deviation between the numerical and the experimental results are caused by the fact that in reality the direction of heat transfer is not the same over the whole door. Indeed, due to the high solar load, heat is transferred in the upper part of the enclosure from the door to the cabinet inner. In contrast, in the lower part of the door, the direction of the heat transfer is from the cabinet inner to the door. Because of a simplified description of the flow inside the enclosure and because of homogeneous solar load, the deduced thermal model does not simulate this feature. Given the large temperature rise above ambient, the numerical results are still fairly accurate.

4 CONCLUSIONS

An outdoor cabinet with double-walls is experimentally evaluated under solar load. The thermal characterization of the door has led to a compact model with a reduced number of grid cells and a simplified geometrical structure. The model fits the experimental results well: the average deviation on the calculated energy balances for the double-wall door is 6.1 %. The model in the CFD package adequately predicts the thermal behavior of the outdoor cabinet within 20 % accuracy for the heat transfer. Temperatures are predicted with an accuracy of approximately 2 °C. From this, it can be concluded that heat transfer in the double-wall door is adequately modeled.

REFERENCES

Bramson, A.M. 1968. *Infrared radiation. A handbook for applications.* New York: Plenum Press.

Brewster, M. Q. 1992. *Thermal radiative transfer & properties.* New York: John Wiley.

Çengel, Y. A. 1997. *Introduction to thermodynamics and heat transfer.* New York: McGraw-Hill.

Hottel, H. C., & Sarofim, A.F. 1967. *Radiative transfer.* New York: McGraw-Hill.

Sayigh, A.A.M. 1977. *Solar energy engineering.* New York: Academic Press.

Shah, R. K., & London, A. L. 1978. Laminar flow forced convection in ducts. In Irvine, T. F., & Hartnett, J. P. (eds), *Advances in heat transfer. Supplement 1.* London: Academic Press.

Thermal Challenges in Next Generation Electronic Systems, Joshi & Garimella (eds)

Creating reduced-order models for electronic systems: An overview and suggested use of existing model reduction and experimental system identification tools

B. Shapiro
Dept. of Aerospace Engineering, University of Maryland at College Park, MD, USA[1]

Keywords: electronic systems, low-order models, model reduction, experimental system identification, conduction, convection, radiation, fluid dynamics, turbulence, analysis, design.

ABSTRACT: The micro-packaging community has a great need to reduce the size of its heat transfer simulations so that it can: simulate and analyze more complex systems, include additional physical phenomena, and improve its ability to search the micro-packaging design space. Aside from further improvements in machine speed and numerical algorithm efficiency, this is basically a question of model reduction and experimental identification: one would like to know how to dramatically reduce the size of heat transfer simulations when they are available, and one would further like to identify models directly from experiment when accurate, computationally feasible, numerical simulations are not available (as in the case of turbulent flows through complex geometries).

Fortunately, the topic of low-order modeling for design has been widely studied and successfully applied in other fields (mostly in control engineering and fluid dynamics, although also in structural mechanics, chemical deposition, heat transfer and combustion). Specifically, there are thousands of papers on the mathematical techniques of model reduction and experimental system identification. This paper gives a rough overview of these techniques, it suggests how these tools might be effectively used for electronic systems including cases that involve unsteady fluid dynamics, and it summarizes some of the reduced-order modeling lessons learned in other fields.

1 MOTIVATION

This paper is aimed at low-order modeling for design of electronic systems. Our basic point is this: for system design, small not-so-accurate models are far more useful than large accurate models. This sweeping statement is of course overly simplistic, but it is nevertheless correct in the following ways:

In the first place, design (especially preliminary design) is all about choosing system parameters (such as chip geometries and positions, cooling channel shapes) and operating ranges (such as fan speed and power inputs). Systems models are the only way that we can do this. Hence, at the very least, our models should provide a capability to combinatorialy search the parameter space: we should be able to set the parameters, run our model, and observe the results. This is not possible with big models. For example, if we have a simulation that takes 1 day to run, we have 10 parameters that we are interested in, each parameter can take on 5 values, and we want to search just 1% of the design space: that would take $(5^{10}\times 1)/100$ days or 267 years. If our model runs in one second, this same search would take 1.4 days. Not great, but at least it is reasonable.

Second, as our models get smaller, we can apply far more analysis and design tools (see Figure 1). These tools serve to guide us through the parameter design space by identifying critical parameters, and by providing an understanding of the parameter inter-dependences. If our model has 10 internal states and 5 parameters: we can examine the equilibrium behavior, check stability, analyze

[1] The author would like to thank Dr. Yogendra Joshi for his advice and answers on electronic systems.

the forced response, find limit cycles, compute chaotic attractors if they exist, and perform global optimizations such as branch-and-bound. If, instead, our system has 100,000 internal states, we cannot do any of these things---even if the model can be simulated in seconds on a super-computer. This is because the computation time associated with finding a bifurcation diagram, or with computing a branch-and-bound optimum, scales exponentially with model size and the required computation is not feasible for larger models.

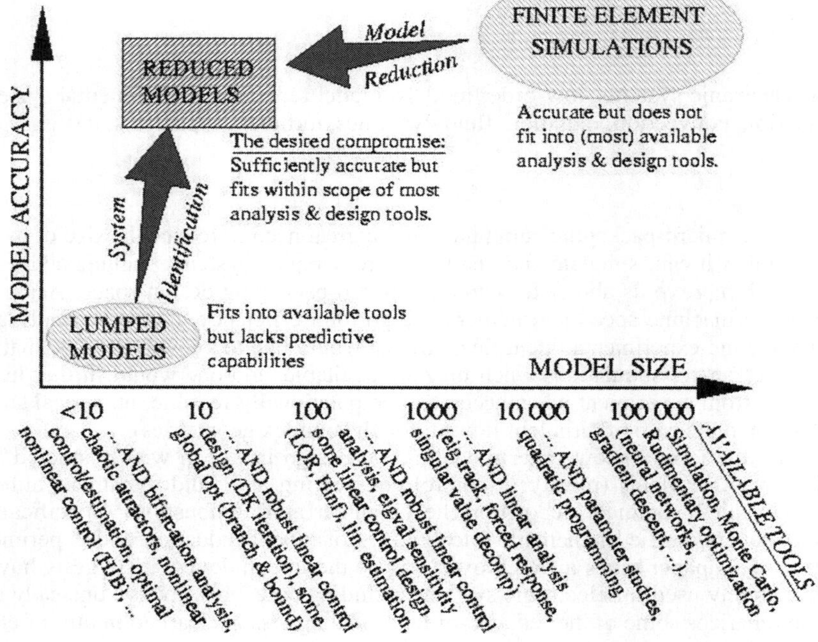

Figure 1: Model size versus model accuracy tradeoff: we desire small, sufficiently accurate models.

Finally, the process of creating low-order models forces the researcher to isolate and quantify the dominant physical mechanisms. This invariably leads to a better understanding of the system behavior, and it usually reveals effective design decisions that would not have been identified through numerical simulation, experimentation, or "black box" optimization methods such as neural networks and genetic algorithms. This increased understanding is the most valuable benefit derived from low-order modeling efforts, but, it is also the benefit that is hardest to convey to skeptical researchers (primarily because it cannot be quantified in any way). Here we will just point out two examples: insights from the low-order Moore-Greitzer 3-state model [1] have enabled design avoidance and practical nonlinear control (the system is linearly uncontrollable) of surge and stall in jet-engine experiments [2]; the understanding captured by the low-order combustion model [3] has enabled limit cycle control in practical combustion rigs [4].

Hence our goal in this paper is two-fold. First, we give a broad (and incomplete) overview of the available model reduction and experimental system identification techniques. Second, we provide an indication of how these methods can be used to create reduced-order models of electronic systems, and we summarize some of the lessons learned from model reduction efforts in other fields.

2 THE INPUT/OUTPUT COMPONENT MODEL REDUCTION APPROACH

Control engineers spend a great deal of time hooking up components and analyzing/designing the resulting system behavior. It turns out that the most natural way to approach this problem is to view each component as a dynamic input/output map. Here each component receives time-varying multi-dimensional information from all the surrounding components, this input along with the internal dynamics causes a continual change in the internal states, the resulting multi-dimensional time-varying output is then a function of the dynamic internal states. (For a solid chip: the inputs could correspond to the temperature or heat fluxes from the surrounding environment; the dynamic internal states would correspond to a representation of the internal temperature field; and the outputs would be space and time dependent heat fluxes out of the chip.)

The are three key points here: 1) Any systems can be broken up in this fashion. 2) There is no error or approximation associated with this split: it is simply a convenient point of view. 3) There exists a wide array of tools aimed at analyzing and designing the performance of such inter-connected input/output maps.

The component inter-connection structure is usually represented by `block diagrams' as shown in Figure 2, and the resulting system performance can be analyzed completely for linear input/output blocks, and partially for nonlinear input/output maps.

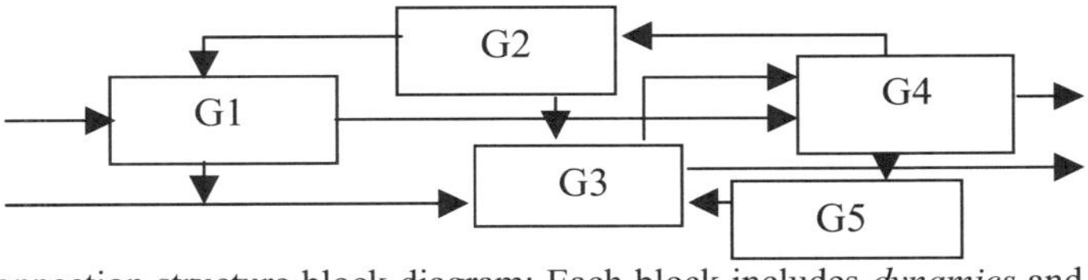

Figure 2: Inter-connection structure block diagram: Each block includes *dynamics* and can be nonlinear, the connection arrows can represent a tremendous amount of information (e.g. the discretized time-varying temperature field T(x,t) on a component face). The behavior of such inter-connections can be analyzed and designed using control engineering type tools [5].

Hence our first recommendation is that the electronic systems low-order modeling problem be couched in this input/output fashion. Accordingly, we discuss how to derive component input/output models from finite element simulations in Section 2.1 and from experiment in Section 2.2.

2.1 *Creating low-order input/output component models from FEM simulations*

For the case where numerical FEM simulations accurately predict component behavior, we can proceed by partitioning any FEM simulation into component blocks. This is illustrated in Figure 3.

The partitioning results in a large, but finite, dynamic input/output map for each solid component or chosen fluid region:

$$y(t) \Leftarrow \begin{bmatrix} \dot{x} = f(x,u) \\ y = h(x,u) \end{bmatrix} \Leftarrow u(t) \qquad (1)$$

Here, (f,h) corresponds to a discretization of the underlying heat transfer PDEs as written in [6], x is the internal state in the region (the value of all variables at all nodes), u lists all the inputs to the region (variables at immediately adjoining nodes if we follow the convention shown in Figure 3, or we can also think of this as incoming fluxes, plus any inputs due to power rates into the chip), and y lists all the system outputs (boundary nodes or outgoing fluxes, plus any additional variables that we may wish to observe).

It is the dimension of the internal state variable x that we are trying to reduce in equation (1). So for each component we desire a reduced-order map $(\tilde{f},\tilde{h})$ that has the same input u and output y,

$$y(t) \Leftarrow \begin{bmatrix} \dot{x} = f(x,u) \\ y = h(x,u) \end{bmatrix} \Leftarrow u(t) \quad \rightarrow \quad y(t) \Leftarrow \begin{bmatrix} \dot{\tilde{x}} = \tilde{f}(\tilde{x},u) \\ y = \tilde{h}(\tilde{x},u) \end{bmatrix} \Leftarrow u(t) \tag{2}$$

but with a dramatically reduced number of internal states ($\dim(\tilde{x}) << \dim(x)$), that provides essentially the same dynamic input-to-output behavior. (The analogy here is this: consider a drum where the input is the hand striking the surface [so a complex input], and the output is the instantaneous displacement of the drum at each point in space [a complex output]. A low-order input/output model would represent the internal states of the drum by the amplitude and phase of the first few fundamental drum modes (first radially symmetric mode, first traveling wave mode, … ,3rd traveling mode). Even the first few modes will adequately recreate the input/output behavior. This is true because these modes are the natural (i.e. dominant) modes of the drum. The key model reduction step, therefore, is to find a judicious set of modes onto which the dynamics may be projected. This is exactly what model reduction techniques endeavor to do.)

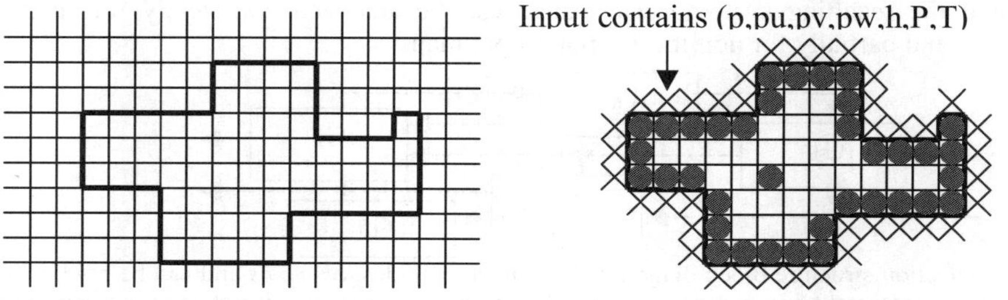

Figure 3: Partitioning a FEM simulation into arbitrary regions. This is just a matter of keeping track of all the region boundary node values: all states stored at the internal boundaries of a region act as outputs to other regions (o), all node just outside the region act as inputs (x). If desired, selected temperatures inside the region can also be listed as outputs (so as to monitor hot-spots). Power source rates to a chip will be listed as additional inputs. Thus each region has a large number of internal states, a *vector* input, and a *vector* output. We are going to pick regions corresponding to component geometries. If we pick a solid 'chip' region (no fluid), the internal nodes in that region will only include solid material properties. In the context of FEM, this is all we mean by a `dynamic input/output map': *there is no error, approximation, or assumptions associated with this partitioning.*

Below, we outline two techniques that can achieve dramatic reductions in model size.

Balanced Truncation Model Reduction [5]: Balanced truncation is aimed at large linear input/output models. Hence it applies to components whose *internal* behavior is linear or almost linear (e.g. solid components where conduction is dominant), but it can also be applied successfully to nonlinear maps *if the underlying modes do not change appreciably as the states deviate away from their nominal settings.* (It is, in fact, quite surprising how often nonlinear systems can be effectively reduced using linear model reduction techniques.)

Balanced truncation model reduction has the following properties: 1) The method is globally optimal. A 5th order balanced truncation model is the best possible 5 dimensional reduced-order model in the sense that it globally minimizes the model reduction error transfer function in the infinity norm. 2) Balanced truncation includes a rigorous model-reduction error bound: this means that the output generated by the reduced-order model is guaranteed to track the output predicted by the full-order model for *any choice of (time-varying) inputs.* 3) We can aim the reduced order model at any desired frequency range. So if we are interested in transients that have a 5-30Hz content, we can generate a reduced-order model that is focused on this frequency range.

Balanced truncation reduced-order models can be computed for models with up to thousands of states [7], hence it is possible to reduced some FEM input/output component models directly using balanced truncation. The method is most suited to solid components or empty regions that have only weak nonlinear terms, or those components that can be accurately linearized about an operating point (so cases with radiation are appropriate); the method is not suited to cases that include highly nonlinear fluid dynamic effects.

Proper Orthogonal Decomposition (POD) [8]: The advantage of POD model-reduction is two-fold: first, it is aimed specifically at nonlinear simulations; second, it has been demonstrated to work superbly in complex fluid dynamic systems [8, 9]. For example, POD methods have been shown to reduce RANS or LES turbulent simulations from millions of internal states down to 5 or 10 modes with only a 5 or 10% loss in accuracy [9]. The method works directly from simulation data: one runs the desired FEM simulation enough times to generate a statistically representative sample of data, POD then finds the low-order subspace that most closely approximates this data in the 2-norm sense, the full dynamics are then Galerkin projected onto this subspace to create the low-order input/output model. POD retains most of the desirable properties of balanced truncation: it is optimal with a reduction-error bounds (although now in a statistical sense), and since it can be applied in the frequency domain [10] it can be frequency weighted to focus on the desired frequency range.

POD is the preferred method of choice for reducing complex fluid simulations, especially those involving turbulence.

2.2 *Creating low-order input/output component models from experiment*

Experimental system identification methods will be necessary when accurate FEM simulations are not feasible. In these cases, we must identify the missing model elements experimentally.

Like model reduction, system identification is an intensely studied field in its own right [3, 11-16]. In the past, system identification has enabled low-order modeling of complex systems when it has been directed at specific, dominant, but unknown physical effects. For example, in the case of pre-mixed combustion [3], experimental identification was used to identify the time-delayed nonlinear pressure-to-heat-release input/output map. When this map was combined with a low-order model of the chamber acoustics derived from FEM, the resulting system model allowed the design and implementation of a limit cycle controller in practical combustion rigs [4].

Below we outline 3 system identification approaches.

Linear 'Black Box' System Identification: Linear input/output models are described equivalently by their transfer function in the frequency domain, or by their impulse response in the time domain. So if one performs a frequency sweep or impulse response on the experiment, then a linear model can be fitted to this data by a least squares or correlation method [16]. This identification method is well suited to estimating the dynamics and time-constants of the unknown physical effects: How fast are the relaxation times? Can the phenomena be represented effectively by a first-order differential equation? What are the natural frequencies? It also gives a good initial indication of the phenomena complexity: if the measured data has only a few well defined natural frequencies, chances are that the effect can be captured accurately with a low-order input/output model. The method does not work if the component or effect in question is unstable, and it does not work for effects that are strongly nonlinear.

Linear 'Grey Box' Parameter Estimation: Given a linear system with a defined model structure, but an unknown set of parameters *(p1,p2,...,pn)*, these methods identify the missing parameters from experimental data. The vector *p* is typically found by the least squares method, or by the maximum likelihood method [16]. The key point here is that there must be sufficient data compared to the number of free variables: one cannot identify 37 free parameters with 52 data points. Instead, one would need at least *37*(fairly big factor)* number of data points. This method takes into account the structure of the linear system, i.e. it takes into account known information, and is therefore a 'gray box' method. This kind of method is natural if the structure of the linear model is defined by the physics, but the rate coefficients (e.g. the coefficient of heat transfer in some unfamiliar substance) are badly known and must be identified from experiment.

Nonlinear parameter identification: The most general system identification problem can be phrased in this way: Given any nonlinear model that depends on a finite or infinite number of unknown parameters *(p1,p2,p3,...)*, plus some set of experimental data, find the vector p that minimizes the error between the predicted and observed data. The difficulty with such an all-inclusive formulation is that most problems so formulated cannot be solved, and for those that can be there is no standard method of solution. Hence the researcher must assess his practical needs versus the available nonlinear system identification schemes; he must then formulate a useful system identification problem that can be solved using known nonlinear optimization techniques. This is not easy.

Our advice here is the following: Phrase the smallest system identification problem possible (least number of parameters to be identified). The way to do this is to conceptually isolate the dominant physical effect, and to aim your identification at this effect alone. We suggest this because the complexity of almost any nonlinear solution scheme will scale exponentially with the number of parameters to be identified. Nonlinear system identification tasks with many parameters cannot be solved.

2.3 *An example of the type of results possible within the input/output model reduction framework*

In this section we want to give a small indication of the benefit of the input/output approach. The result outlined below is not possible without the input/output machinery advocated above.

We answer the following two questions: *How do component errors build up to give system errors? And how can we create a-priori component models that will minimize system wide errors?*

These questions can be answered by considering just two inter-connected linear components as shown (the results extends directly to n components, and we suspect that the basic idea remains valid for nonlinear systems although this is admittedly hard to show).

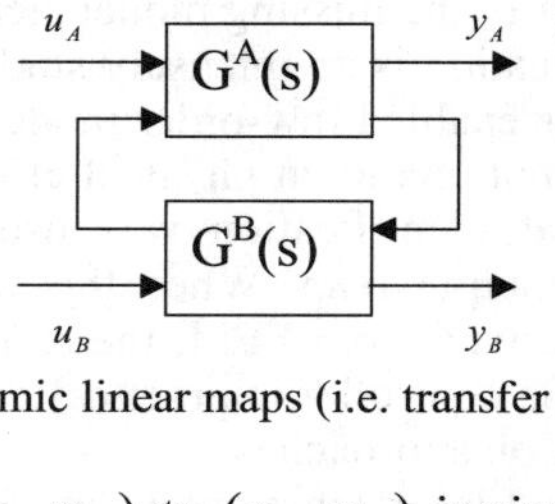

Figure 4: Two inter-connected dynamic linear maps (i.e. transfer functions).

The transfer function from (u_A, u_B) to (y_A, y_B) is given by

$$\begin{bmatrix} y_A \\ y_B \end{bmatrix} = \begin{bmatrix} G_{11}^A + G_{12}^A G_{22}^B (I - G_{22}^A G_{22}^B)^{-1} G_{21}^A & G_{12}^A G_{21}^B + G_{12}^A G_{22}^B (I - G_{22}^A G_{22}^B)^{-1} G_{22}^A G_{21}^B \\ G_{12}^B G_{21}^A + G_{12}^B G_{22}^A (I - G_{22}^B G_{22}^A)^{-1} G_{22}^B G_{21}^A & G_{11}^B + G_{12}^B G_{22}^A (I - G_{22}^B G_{22}^A)^{-1} G_{21}^B \end{bmatrix} \begin{bmatrix} u_A \\ u_B \end{bmatrix}$$

The inverse transfer function $(I - G_{22}^A G_{22}^B)^{-1}$, or equivalently $(I - G_{22}^B G_{22}^A)^{-1}$, which appears repeatedly in the equation above reflects the dynamics of the closed-loop system. It is large at resonant frequencies of the inter-connected system.

If we model reduce $G^A(s)$ and $G^B(s)$ individually, and express the system model-reduction error ε_{SYS} in terms of the component model reduction error ε_A and ε_B, we find that the component error is amplified by exactly this closed-loop dynamics transfer function:

$$\varepsilon_{SYS}(s) \approx \left[(I - G_{22}^A G_{22}^B)^{-1} \quad (I - G_{22}^B G_{22}^A)^{-1} \right] \begin{bmatrix} \varepsilon_A(s) \\ \varepsilon_B(s) \end{bmatrix} \tag{3}$$

So the conclusion is two-fold: One, we must resolve our model accurately at the resonant frequencies of the closed-loop system. Two, we can do exactly this by weighting our component-model-reduction-problem by an estimate of the closed-loop frequency response. Hence the input/output framework can tell us how to design reduced-order component models that minimize systems-level model reduction errors.

3 CONCLUSION: PUTTING IT ALL TOGETHER

In this final section, we give an indication of how the techniques above can be combined to form a complete low-order model of a complex electronic system. Here we focus on an electronic system with fan cooling (such as a personal computer or laptop).

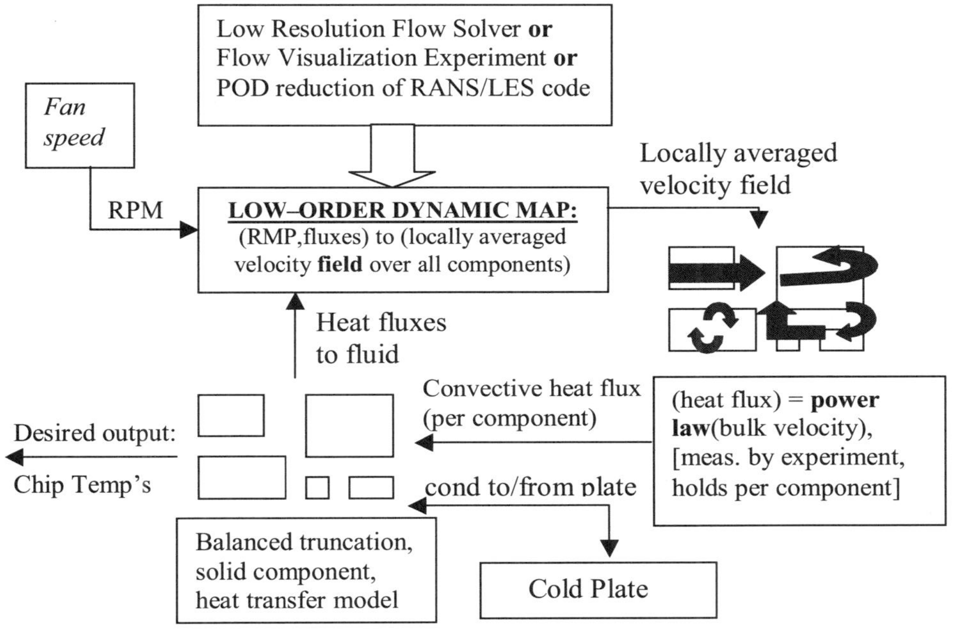

Figure 5: A conceptual flow diagram for low-order model of an electronic system with convective cooling. (Starting at the top left.) The fan speed largely determines the fluid velocity field in the system, but this velocity field can also depend on heat fluxes generated by the chips and breadboards (especially when the fan is off), hence the velocity map receives these two inputs. The dynamic velocity map is a spatially rough representation of the fluid dynamics, it resolves the fluid dynamics down to cm, or maybe mm, length-scales. This part of the model outputs a locally averaged velocity field that captures the path of the fluid over the chips (at time t=3, fluid first goes over chip 7 before it turns left to pass over chip 4) but it makes no attempt to resolve the fine scale turbulence structure. The heat carried away by convection at each chip is then given by the experimentally identified power law [17]: at chip 13, averaged velocity is V_{13}, so heat convected away, is $q=(V_{13})^P$. This is done for each chip. Now that the convective fluxes are known, it only remains to solve the conduction/ radiation heat transfer in the solid. This is done to low-order by the balanced truncation model. (If the nonlinear radiation effects are significant, we can replace this by a POD model that captures nonlinear radiation effects.) If necessary, this model can also be coupled to a cold-plate balanced truncation or POD model. To complete the loop, convective heat fluxes go to the reduced-order fluid model. One can pull out any desired system output: here we have decided to observe chip temperatures.

Figure 5, and the associated caption, illustrates how it is possible to split the electronic system behavior into a set of dynamic input/output models. The caption also outlines the flow of information from component model to component model (keep in mind that some simple looking arrows actually transmit a great deal of information, like the arrow that transmits the locally averaged velocity field).

The solid chip and cold plate input/output models will be created by balanced truncation: each chip will have incoming fluxes as inputs, and outgoing fluxes as outputs; selected chips will also output their temperatures for observation purposes. The turbulent flow field reduced-order model will be created in one of 3 ways: by reducing a full scale turbulent simulation through POD, by creating a low-resolution (mm length-scale) flow simulation, or by a flow visualization experiment. This model will *not* attempt to resolve the fine scale turbulence, but will only give a broad indication of the path and velocity of the flow over each chip. The convective heat transfer caused by the turbulent flow field will then be captured by the experimentally identified 'local averaged velocity' to 'local heat flux' map. It is this experimentally identified map that will capture the 'details' of the turbulent heat transfer mechanisms. Clearly, these last two aspects of the problem are the critical pieces, and the model reduction and system identification efforts should be focused here.

Our assessment is that this type of model will capture the dominant system dynamics, and it will reveal the tradeoffs due to parameters such as fan speed and chip placement.

4 REFERENCES

[1] E. M. Greitzer, "Surge and rotating stall in axial flow compressors: Part 1-Theoretical Compression System Model, Part 2-Experimental results and comparison with theory," *ASME Journal of Engineering for Power*, vol. 98, pp. 190-98, 1976.

[2] G. Gu, S. Banda, and A. Sparks, "An overview of rotating stall and surge control for axial flow compressors," presented at Proceedings of 35th IEEE Conference on Decision and Control, 1996.

[3] C. A. J. R. M. Murray, R. Casas, A. I. Khibnik, C. R. Johnson Jr., R. Bitmead, A. A. Peracchio, W. M. Proscia . "System Identification for Limit Cycling Systems: A Case Study for Combustion Instabilities," presented at American Control Conference, 1998.

[4] A. Banaszuk, M. K. K. Ariyur, and C. A. Jacobson., "An Adaptive Algorithm for Control of Combustion Instability," *Automatica*, 2000.

[5] K. Zhou, Doyle, J.C., Glover, K., *Robust and Optimal Control*: Prentice Hall, New Jersey, 1996.

[6] V. H. Adams, Y. Joshi, and D. L. Blackburn, "Three-Dimensional Study of Combined Conduction, Radiation, and Natural Convection from Discrete Heat Sources in a Horizontal Narrow-Aspec-Ratio Enclosure," *Transactions of the ASME*, vol. 121, pp. 992-1001, 1999.

[7] D. C. Sorensen, "Projection methods for balanced model reduction," Applied Math, Rice University March 19 2001.

[8] L. Sirovich, "Turbulences and the dynamics of coherent structures, part I & II," *Quarterly of Applied Mathematics*, vol. 45, pp. 561-571, 1987.

[9] C. W. Rowley, T. Colonius, and R. M. Murray, "Dynamical Models for Control of Cavity Oscillations," presented at 7th AIAA/CEAS Aeroacoustics Conference, 2001.

[10] B. a. W. Shapiro, K., "Analyzing Mistuning of Bladed Disks by Symmetry and Reduced-Order Aerodynamic Modeling," *Journal of Propulsion and Power*, 2000.

[11] A. C. Antoulas, Sorensen, D.C., Gugercin, S., "A survey of model reduction methods for large-scale systems," Applied Math, Rice University Dec 5 2000.

[12] K. J. Astrom, Eykhoff, P., "System identification-a survey," *Automatica*, vol. 7, pp. 123-162, 1971.

[13] S. A. Billings, "Identification of nonlinear systems-a survey," *IEEE Proceedings D*, vol. 127, pp. 272-285, 1980.

[14] J. A. M. Fellipe De Souza, "Control, state estimation and parameter identification of nonlinear distributed parameter systems using fixed point techniques: a survey," presented at Proceedings of the 4th IFAC Symposium, Control of Distributed Parameter Series, 1987.

[15] R. Isermann, "Parameter estimation of dynamic systems.," *Automatisierungstechnik*, vol. 36, pp. 199-207, 1988.

[16] L. Ljung, *System Identification Theory for the User.*: Prentice-Hall, 1999.

[17] M. Molki, M. Faghri, and O. Ozbay, "A Correlation for Heat Transfer and Wake Effect in the Entrance Region of an In-Line Array of Rectangular Blocks Simulating Electronic Components," *ASME Journal of Heat Transfer*, vol. 117, pp. 40-46, 1995.

Thermal Challenges in Next Generation Electronic Systems, Joshi & Garimella (eds)
© 2002 Millpress, Rotterdam, ISBN 90-77017-03-8

Remote heat sink concepts for high power heat rejection

R.L. Webb
Dept. of Mechanical Engineering, Penn State University PA, USA

S. Yamauchi
Research Engineer, Showa Denko, K.K., Tochigi, Japan

Keywords: thermo-syphon, thermal resistance, cooling device, heat exchanger, heat sink

ABSTRACT: This paper addresses heat rejection concepts intended to reject 200 W (or more) from the CPU of a desktop computer or server. A high-performance, air-cooled thermo-syphon heat rejection system is described that meets the goal. This concept is believed to represent the highest performance that has been reported for an air-cooled heat rejection system. A working fluid is boiled in a small chamber mounted on the CPU, and the vapor is condensed in the tubes of the remotely located air-cooled condenser. The air-cooled heat exchanger provides much higher performance than is obtainable with conventional "round-tube" technology. The boiling design is compared with other systems that have been proposed – e.g., liquid or two-phase cooling in micro-channels, and is shown to provide better performance. The heat exchange technology may also be applied to refrigerated cooling.

Nomenclature

A	Total heat transfer surface area, m^2
A_{fr}	Frontal heat exchanger area, m^2
G	Conductance, W/K
G_{avg}	Conductance based on the average surface temperature, W/K
G_{max}	Conductance based on the maximum surface temperature, W/K
h	Air-side heat transfer coefficient, W/m^2-K
h_{nb}	Pool boiling heat transfer coefficient, W/m^2-K
M	Molecular weight
P	Pressure, Pa
P_{cr}	Critical pressure, Pa
P_r	P/P_{cr}
q	Heat flux, W/m^2
Q	Heat dissipation, W
R_{AHS}	Air-cooled heat exchanger thermal resistance, K/W
R_{bo}	Boiling thermal resistance, K/W
R_{CPU}	CPU heat sink thermal resistance, K/W
R_{CV}	Convection thermal resistance, K/W
R_{int}	Interface thermal resistance, K/W
R_p	Surface roughness parameter, μm
R_{sp}	Spreading thermal resistance, K/W
R_{tot}	Total thermal resistance from CPU to air, K/W
$R_{tot,TS}$	Thermo-syphon thermal resistance, K/W
T_{amb}	Ambient temperature, °C
T_{hot}	Maximum CPU case or copper spreader temperature, °C
T_{sat}	Saturation temperature, °C
T_w	Wall temperature, °C
η	Fin efficiency
ΔP	Pressure drop, mm-H^2O

1 INTRODUCTION

Recently, the processor power density has been increasing, and it may reach 100 W/cm^2 in the near future. Currently, CPUs are typically cooled by a CPU mounted fan/heat sink device. Although a copper heat spreader is used on the CPU, it does not provide a uniform heat flux on the spreader surface area. For example, the allowable CPU case temperature of the Intel Pentium 4 is 76°C and the heat dissipation is 75.3 W. The Pentium 4 CPU case temperature specification applies the 1.5 mm thick × 31 mm square copper heat slug attached to the CPU. Our CFD analysis shows that the thermal resistance of the Pentium 4 heat spreader is equivalent to that for a uniform heat flux on a 16 mm square interface. Thus, the current chip heat flux is 30 W/cm^2 based on the 16 mm square equivalent heat source size. It is very unlikely that use of CPU mounted fan/heat sink devices will support cooling requirements of future CPUs used in computers. Saini and Webb (2002) identify anticipated cooling limits of mounted fan/heat sink devices. They estimate that the limit for a 60 × 80 mm heat sink size is about 100 W.

Penn State University has worked to develop a high performance heat rejection system (see Figure 1) that is intended to provide high CPU power heat rejection with heat rejection to ambient air. A working fluid is boiled in a small chamber mounted on the CPU, and the vapor is condensed in the tubes of the remotely located air-cooled condenser. This cooling concept is described and is compared to concepts that have been proposed by others. The key components of any heat rejection system are the heat sink at the CPU, and the heat rejection device - in this case an air cooled heat exchanger. It is expected that the air-cooled heat exchanger will have the controlling thermal resistance. However, the CPU heat sink must support a high flux and provide small thermal resistance.

Figure 1. Prototype thermo-syphon device

2 REQUIRED THERMAL RESISTANCES FOR DESKTOP COMPUTERS AND SEVERS

In this paper, the design specification is set at 100W CPU heat dissipation (Q), and the maximum CPU case or copper spreader temperature (T_{hot}) is 75°C. Assuming that the air entering (ambient) temperature to the heat exchanger is 40°C, the target value of the total thermal resistance (R_{tot}) is as follow:

$$R_{tot} = \frac{T_{hot} - T_{amb}}{Q} = \frac{70-40}{100} = 0.3 \text{ K/W} \tag{1}$$

2.1 *CPU mounted heat sinks*

If the heat sink is directly mounted on the CPU, the component thermal resistances are the interface resistance (R_{int}), the spreading resistance (R_{sp}) and the heat sink convection resistance (R_{cv}). Commercial manufacturers have developed high performance thermal greases and phase change materials to minimize the R_{int}. Based on 16 mm square effective interface size, the R_{int} provided by high-performance interface materials will be approximately 0.066 K/W. Note that this is 22% of the R_{tot}.

The R_{sp} is estimated by the spreading resistance equation of Lee (1995). For presently used CPU fan/heat sinks, the heat sink base is approximately 60 × 80 mm. Using a 6 mm thick copper heat spreader with this would result in R_{sp} = 0.1 K/W. So, the allowable convection resistance is 0.3 – (0.066 + 0.1) = 0.134 K/W, or only 44.6% of the total thermal resistance.

2.2 *Remote heat sinks*

The Figure 1 "remote" heat sink does not use a large base area in contact with a finned surface. Hence, the spreading thermal resistance is greatly reduced. The Figure 1 device has two key thermal resistances - the CPU heat sink (R_{CPU}) and the air-cooled heat exchanger (R_{AHX}).

$$R_{tot} = R_{CPU} + R_{AHS} \tag{2}$$

The R_{CPU} consists of three components – the interface (R_{int}) between the CPU and the heat spreader, the heat spreading (R_{sp}), and the boiling (R_{bo}) resistances. Therefore, the R_{CPU} is expressed as:

$$R_{CPU} = R_{\text{int}} + R_{sp} + R_{bo} \tag{3}$$

The R_{int} will be the same value as above (0.066 K/W). Boiling of the working fluid occurs on a very small surface area. The boiling performance of water and refrigerants are so high that a big spreader is not required. The CPU heat sink device used here is 22 mm square (or smaller). Assuming a 3.0 mm thick copper base, 22 × 22 mm size, the R_{sp} = 0.032 K/W. Hence, the available thermal resistance for $R_{bo} + R_{AHS}$ is 0.3 – (0.066 + 0.032) = 0.202 K/W, which is 67.3% of the total. The present remote heat sink concept and related technology are discussed below.

3 REMOTE HEAT SINK TECHNOLOGY

In "remote heat sink" technology, the air-cooled heat exchanger will be located in the back wall of the computer case (or in a duct for servers). Heat is added to a "working fluid" via a heat sink attached to the CPU. The working fluid is then transported to the remote heat sink for heat rejection to air. Possible concepts for the CPU mounted heat sink are:

- Liquid cooling. This will involve pumping of a liquid between the CPU attached heat sink and the remote heat sink.
- Thermo-syphon. This will involve evaporation of a working fluid in the CPU attached heat sink. The liquid condensed in the remote heat sink is gravity drained back to the CPU heat sink.
- Conventional heat pipe. This is similar in concept to the thermo-syphon, although an internal wicking may be able to provide some capillary pumping, as opposed to depending on gravity.

3.1 *Penn State thermo-syphon concept*

Webb and Yamauchi (2001) describe the Figure 1 "thermo-syphon" concept. The thermo-syphon concept uses a high-performance automotive type heat exchanger and gravity drainage of a boiling/condensing fluid. The working fluid is boiled on a high-performance porous boiling surface, in a small chamber attached to the CPU, and condensed in the tubes of the air-cooled heat exchanger. The internal tube passages are specially configured to yield high condensation performance.

Using aluminum construction, candidate working fluids are R-134A, propane, or iso-butane. A new low "Global Warming Potential" (GWP) fluorine based fluid has been developed by 3M Corp. for application to cooling electronic equipment. This is HFE-7200 and it has a significantly lower GWP than does R-134A. The GWP of HFE-7200 is 90, as compared to 1300 for R-134A. Bar-Cohen and Arik (2001) have measured the boiling performance of HFE-7200 on a plain surface. They took the critical heat flux data and subcooled performance on the plain surface. We have estimated the pool boiling heat transfer coefficient (h_{nb}) by Cooper correlation (1984).

Table 1. Properties and operating conditions of the refrigerants

Refrigerant	M	P_{cr} (MPa)	P (MPa) (Temp. = 70°C)	P_r
R-134a	102	4.06	2.162	0.533
HFE-7100	250	2.23	0.127	0.0569
HFE-7200	264	2.07	0.1	0.0483
Water	18.015	22.12	0.031162	0.00141

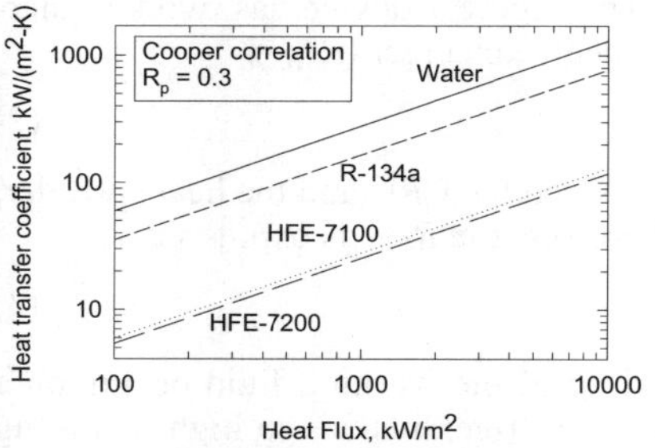

Figure 2. The heat transfer coefficient for the refrigerants

$$h_{nb} = 90(q)^{0.67} M^{-0.5} P_r^{\,m} (-\log P_r)^{-0.55} \tag{5}$$

where M is the molecular weight. P_r expresses P/P_{cr}, and P_{cr} is the critical pressure. m is as follow:

$$m = 0.12 - 0.2 \log R_p \tag{6}$$

where R_p is the surface roughness, and 0.3 is selected in the prediction. Table 1 shows properties of working fluids, and Figure 2 shows the predicted boiling performance. Figure 2 shows that the boiling coefficient of HFE-7200 is much less than that of R-134a.

If water is used as the working fluid, copper components are required. Water is the most environmentally acceptable working fluid. However, pure water cannot be used as the working fluid for refrigerated CPU applications, because it would freeze.

Using aluminum materials with R-134A and the prediction methodology of Webb and Yamauchi (2001), we have identified that the Figure 1 device having 75 × 90 mm frontal area heat exchanger (16 mm deep) will cool a single 100 W CPU. This device has been tested using electric heat input from a 20 mm square heater, and the results are shown in Figure 3. If the device is used with an 80 mm diameter case fan, the air-flow rate at the balance point is 17 cfm (1.18 m/s air frontal velocity). Figure 3 shows that the test results are approximately 4-15% better than the predictions, which were based on a 20 mm square heat source. The measured thermo-syphon total resistance ($R_{tot,TS}$), excluding R_{int} (corrected to 16 mm square heat source) yield $R_{tot,TS}$ = 0.207 K/W, at the balance point is 0.19 K/W. This is 12% lower than the design value 0.234 K/W (= 0.3 – 0.066).

A limitation of the Figure 1 concept is that the boiler must be located at a lower elevation than the air-cooled condenser so that liquid is "gravity drained" to the boiler. Special chassis layout consideration is required to insure this. If space is not available to allow gravity drainage, it may be possible to use a capillary structure in the vapor tube to use capillary force to pull liquid up to the boiler. However, our predictions show that it will not be possible to lift the liquid more than 80-100 mm for 100 W heat rejection.

Garner and Patel (2001) describe a thermo-syphon concept that is similar to the Penn State concept. Their concept is specifically directed at rack servers. Garner and Patel propose use of a "loop" thermo-syphon, which has separate vapor and liquid supply tubes. This avoids possible liquid entrainment in a single-tube device. However, entrainment was not a problem in tests of the

Figure 1 device, which uses a 13 mm inside vapor transport tube. The Garner and Patel paper describes a mechanical design to support multiple CPU racks, which allow hot swapping of the chassis racks. The Garner and Patel concept thus provides a practical implementation of the Penn State thermo-syphon concept for a server application.

3.2 *CPU heat sink concepts*

The Figure 1 concept uses nucleate boiling on a high performance sintered boiling surface. At 100 W, our test data show that the boiling coefficient will be approximately 8.8 W/cm^2-K for a 22 mm square porous aluminum surface with R-134A, or 13.4 W/cm^2-K for water boiling on a 16 mm square sintered copper surface. For these surfaces, the $(T_w - T_{sat})$ on the boiling surface will be below 6.0 K.

Heffington et al. (2001a) describe a thin film evaporation concept called "vibration-induced droplet atomization" (VIDA). It uses a piezoelectric transducer to vibrate a plate at about 2.5 kHz and produce a shower of small diameter drops. A multi-hole orifice plate (1.6 mm diameter holes) just above the driver assists in making the small drops. These drops impinge on the hot surface and evaporate as a thin film. Tests were performed using water at 100°C, and at lower temperatures. The data of Heffington (2001b) were taken at higher heat flux than the Penn State data, hence one cannot draw conclusions about its performance in the 40 W/cm^2 range. Their measured CHF was 109 W/cm^2 for 100°C heater temperature.

In the Heffington et al. (2001a) VIDA device, the excess spray is gravity drained to the vibrating driver. An apparent key limitation is that it must be operated in the horizontal orientation with the hot surface above the driver. This would require that the CPU die is on the bottom side of the mother board, as compared the typical installation on the top side of the board. The principal possible advantage of this concept is that it may be able to support higher heat flux than can exist for pool boiling on an enhanced surface. However, Penn State has tested water boiling on sintered copper surfaces at heat fluxes up to 50 W/cm^2 which is corresponds to 128 W on a 16 mm square surface. Simply by using a heat input area greater than 16 mm square, larger heat inputs can be supported. For example, use of a 22 mm square boiling surface at 50 W/cm^2 will support 220 W heat dissipation.

Amon and Murthy (2001) describe a similar "spray impingement, thin film evaporation device, which has the acronym EDIFICE. This concept uses coolant jets that are atomized and sprayed on the hot surface. The jets are formed by an air pressurized liquid reservoir. They have experimented with different nozzle orifice shapes and "textured" ceramic hot surfaces. The working fluid used was HFE-7200. Data taken at the highest heat flux (30 W/cm^2) show a heat transfer coefficient of 1.2 W/cm^2-K, which is quite low compared to the boiling coefficient for water measured by Penn State on the sintered surface (8.5 W/cm^2-K). They believe that this will support heat fluxes in the 50-100 W/cm^2 range. This system would require a high-pressure pump to form the liquid jets.

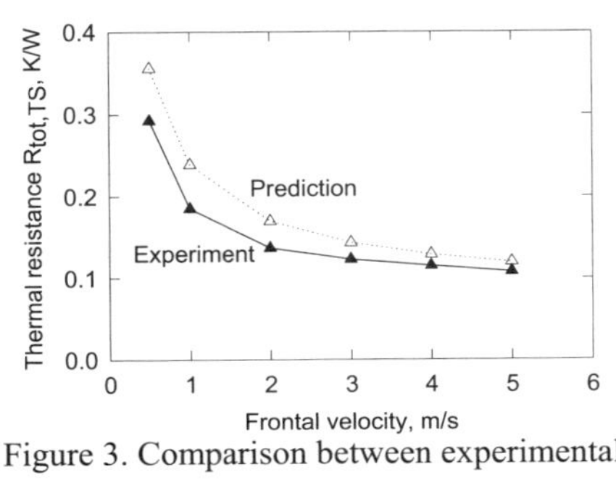

Figure 3. Comparison between experimental and predicted total thermo-syphon thermal resistance

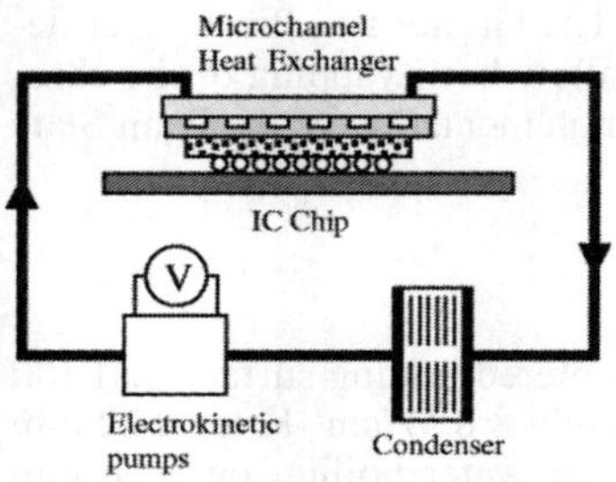

Figure 4. Schematic of an electkinetically-pumped microchannel cooler for VLDI circuits (Taken from Koo et al. 2000)

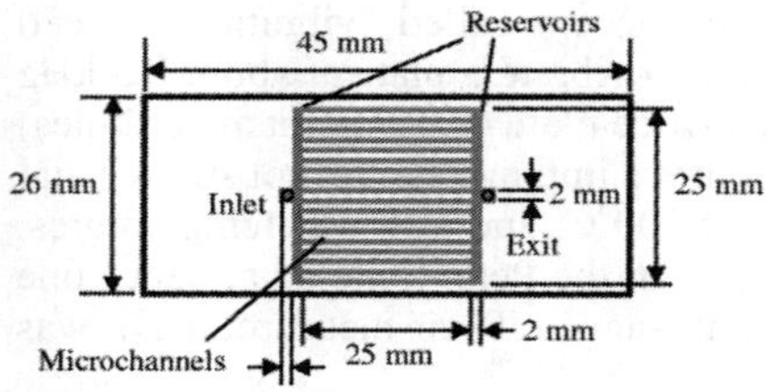

Figure 5. Schematic of a microchannel heat sink (Taken from Koo et al. 2000)

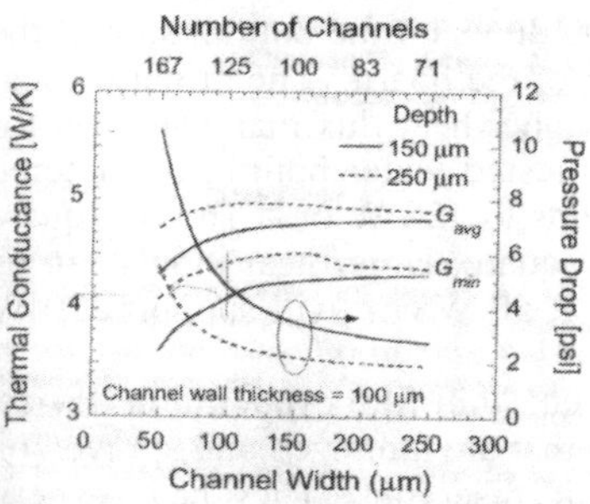

Figure 6. Thermal conductance of the microchannel heat sink (Taken from Koo et al. 2000)

3.3 *Water cooled remote heat sinks*

Others have proposed transferring the heat from the CPU to a two-phase or single-phase liquid that flows through multiple, parallel "micro-channels" that are etched in a silicon wafer. Koo et al. (2001) have proposed use of two-phase cooling using multi-micro-channels machined in a silicon substrate. Their concept is illustrated in Figure 4. They propose to use an electro-kinetic pump to force subcooled water through the micro-channels, where vaporization occurs. Koo et al. have provided predicted results (supported by single-channel tests) for a 25 mm square heat sink (see Figure 5) having either 150 or 200 µm channel depth. Figure 6 shows the thermal conductance (G in W/K) as a function of channel width (or number of channels). The conductance (G) is defined either in terms of the average surface temperature (G_{avg}) or the maximum surface temperature (G_{max}). Although Figure 6 shows that the best performance is provided by the smallest channel size, large pressure drop exists for channels below 100 µm wide. For 150 µm deep and 100 µm wide channels (125 channels in the heat sink), the thermal conductance is G_{avg} = 4.1 W/K. Converted to thermal resistance, the result is 0.039 K-cm^2/W. This thermal resistance is approximately half that of the present water boiling surface, which yields 0.074 K-cm^2/W. Note that the stated performance data are based on predictions and single-channel data. Koo et al. (2001) performed single-channel boiling tests in a 50 μm × 70 μm × 20 mm long and compared their predictions with the single-channel data. The data and predictions agreed reasonably well.

Jiang et al. (2001) constructed and tested a multi-micro-channel heat sink similar to the concept of Koo et al. (2001). They used 40 parallel channels and 100 μm square etched in a 20 × 29 × 0.5 mm thick silicon wafer with water as the working fluid. This device was able to reject only 30 W before flow instabilities occurred, which resulted in large temperature excursions. The R-value of the heat sink was 0.25 K/W (or 1.5 K-cm^2/W). Compared to the predictions of Koo et al. (2001) for 100 μm × 150 μm size channels, the experimental performance is far below that based on the Koo et al. analysis. The present authors suggest that flow mal-distribution in the 40 channel heat sink may be responsible for the poor performance. The concerns related to flow mal-distribution are addressed by Rao and Webb (2000).

A second etched silicon wafer water-cooled heat sink is described by Perret et al. (2000). Water flowed in a 20 mm square wafer having 230 μm wide rectangular micro-channels. For 0.5 l/min water flow rate and 100 W heat input, the measured thermal resistance was approximately 0.85 K-cm^2/W. The performance of the Perret et al. (2000) micro-channel heat sink is considerably better than that of Jiang et al. (2001). However, the R-value of the Perret et al. (2000) heat sink is approximately 11 times that of the present water boiling surface, which yields 0.074 K-cm^2/W. So, based on the work done to date, micro-channel cooling appears to offer significantly lower boiling performance that is provided by the present thermo-syphon boiling device.

Note that the micro-channel concept of Jiang et al. and Perret et al. requires use of a high-pressure micro-pump, which is not required with the present thermo-syphon concept. The pump adds considerably to the cooling system cost. However, the air-cooled condenser does not need to be located above the CPU/boiler. If the micro-channel concept were used in a refrigerated server application, use of a high-pressure micro-pump would not be required. Also note that the boiling resistance in the present air-cooled thermo-syphon concept is only 10% (R_{bo} = 0.029 K/W) of the total thermal resistance (R_{tot} = 0.3 K/W). If water cooling were used, the boiling resistance would be more important.

3.4 *Preferred CPU heat sink device*

The Penn State porous enhanced boiling surface will meet the performance requirements needed for the Figure 1 thermo-syphon device. This device will have no problem in supporting CPU heat loads as high as 220 W. Further work needs to be done on the VIDA and EDIFICE devices to establish their relevance and practicality. Existing micro-channel cooling concepts do not offer performance with existing boiling concepts. Liquid cooled micro-channel concepts that require a liquid pump introduce added cost and complexity, as well as reduced reliability.

3.5 *Refrigerated cooling*

Refrigerated cooling allows the CPU temperature to be maintained at approximately –40°F. This will allow faster CPU clock speed. Refrigeration considerably adds to the cost and complexity of the heat rejection system. These concepts are essentially applicable to server applications, which value holding the CPU temperature to a lower value than the presently used 75°C. For lower CPU temperature, the CPU speed can be increased. KryoTech, inc. makes commercial R-134A the vapor compression system, which is described in Peeples (2001a).

An acceptable working fluid must be identified for application to refrigerated servers. Vapor compression systems would typically use an HFC fluid, such as R-134A, or R-404C. However, these fluids have relatively high GWP.

The Stirling cycle has received serious consideration for refrigerated servers. It uses Helium working fluid, which is fully environmentally acceptable. One option is to use Helium in a micro-channel heat sink at the CPU. No information has been reported on this. However, it will probably not meet performance requirements.

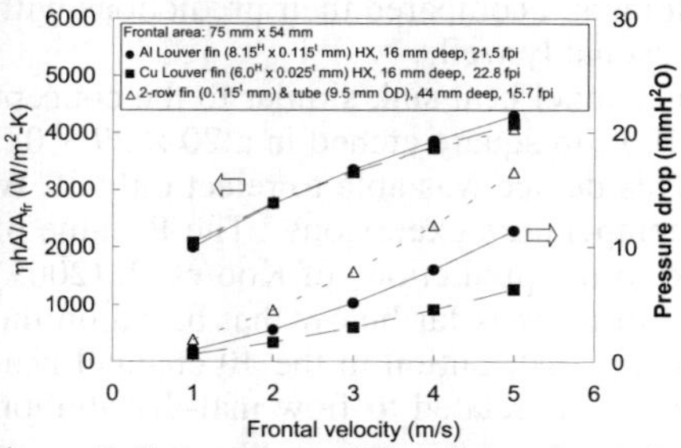

Figure 7. Comparison of flat tube and round tube heat exchanger performance

4 AIR-COOLED HEAT-EXCHANGER

The air-cooled heat exchanger used in the Figure 1 design has very high performance, relative to that of finned, round tube heat exchangers used in the refrigeration industry, or in the refrigerated cooler design of Peeples (2001b). The Figure 1 device is an automotive type, brazed aluminum heat exchanger that is typically used in automotive refrigerant condensers. The use of flat tubes yields air pressure drop considerably below that of typical 9.5 mm diameter round tubes used in residential air-cooled refrigerant condensers. Further, the flat tube design uses high performance louver fins. Figure 7 compares the heat transfer and pressure drop performance of the Figure 1 aluminum flat tube design (21.5 fins/in, 16 mm deep) with that of a round tube heat exchanger (9.5 mm tube diameter and plain fins) designed to give the same air-side thermal performance. The equal thermal performance round tube design has 2-row, is 44.0 mm deep, and has 15.7 fins/in. The flat tube/louver fin heat exchanger performance was predicted using the correlation of Webb (1995), and the round tube design was predicted by the Wang (2000) correlation. Both heat exchangers have the same finned frontal area (75 mm x 54 mm). The air-side heat transfer performance is defined by $\eta hA/A_{fr}$, where η is the fin efficiency, and A_{fr} is the heat exchanger frontal area. For the same $\eta hA/A_{fr}$ as the flat tube design the round tube design has 100% higher air pressure drop than that of the flat tube design. Also, the air flow depth of the round tube design is 1.75 times greater.

As previously stated, the Figure 1 heat exchanger may also be made of copper construction for use with water working fluid. The thickness of copper fins is only 25 µm vs. 115 µm for aluminum fins. For the same $\eta hA/A_{fr}$, Figure 7 shows that the flat tube pressure drop of the copper heat exchanger is 36% lower than that of the aluminum flat tube heat exchanger.

5 CONCLUSIONS

This paper has identified a high-performance remote thermo-syphon heat rejection device that can support CPU heat rejection as large as 220 W. Prototypes of the concept have been made of aluminum (R-134A working fluid) or of copper (water working fluid). Use of water in an all copper heat exchange device will meet possible future requirements for a "green" working fluid.

Because the air-side thermal resistance is 70% of the total, it is imperative that a high-performance air-cooled heat exchanger is used. The Figure 1 heat exchanger meets this requirement. The 16 mm deep Figure 1 brazed aluminum flat tube/louver fin heat exchanger provides the same thermal performance and 52% lower air pressure drop than a 2-row (44.0 mm deep) 9.5 mm tube diameter round tube heat exchanger. An all copper flat tube/louver fin design will provide 36% lower air pressure drop than the brazed aluminum design, because the fins are only 25% as thick.

The Figure 1 device uses a sintered porous boiling surface to achieve high boiling performance. Further work needs to be done on the VIDA and EDIFICE boiling devices to establish their relevance and practicality. Existing micro-channel cooling concepts do not offer performance with existing boiling concepts.

Liquid cooled micro-channel concepts do not appear to offer performance competitive with that of boiling heat exchange concepts. Further, they require a liquid pump, which introduces added cost and complexity, as well as reduced reliability.

REFERENCES

Arik, M., and Bar-Cohen, A., 2001. Ebullient Cooling of Integrated Circuits by NOVEC Fluids, *Proc. of IPACK'01,* July 8-13, 2001, Kauai, Hawaii, Paper IPACK2001-15515.

Cooper, M. G., 1984. Saturation nucleate, pool boiling - a simple correlation, *Int. J. Heat Mass Transfer*, Vol. 23, pp. 73-87.

Garner, S. D. and Patel, C. D., 2001. Loop Thermosyphons and Their Applications to High Density Electronics Cooling, *Proc. of IPACK'01,* July 8-13, 2001, Kauai, Hawaii, IPACK2001-15782.

Heffington, S. N., 2001a. Vibration-Induced Droplet Atomization Heat Transfer Cell for Cooling of Microelectronic Components, *Proc. of IPACK'01,* July 8-13, 2001, Kauai, Hawaii, Paper IPACK2001-15596.

Heffington, S. N., 2001b. Private communication, Sept. 26, 2001.

Jiang, L., Koo, J. M., Zeng, S., Mikkelsen, J. C., Zhang, L., Zhou, P., Santiago, J. G., Kenny, T. W., Goodson, K. E., Maveety, J. G., and Tran Q. A., 2001. Two-Phase Microchannel Heat Sinks for an Electrokinetic VLSI Chip Cooling System, *Proc. of the 17th IEEE SEMI-THERM Symposium*, March 20-22, 2001, San Jose, CA, 153-157.

Koo, J. M., Jiang, L., Zhang, L., Zhou, P., Banerjee, S. S., Kenny, T. W., Santiago, J. G., and Goodson, K. E., 2000. Modeling of Two-Phase Microchannel Heat Sinks for VLSI Chips, *Proc. of the 14th Annual IEEE International MEMS-01 conference,* Interlaken, Switzerland, pp. 422-426.

Lee, S., Song, S., Au, V., and Moran, K.P., 1995. Constriction/Spreading Resistance Model for Electronics Packaging, *ASME/JSME Thermal Engineering Conference*, Vol. 4, pp. 199-206.

Murthy, J. Y., Amon, C. H., Gabriel, K., Kumta, P., and Yao, S.C., 2001. MEMS-Based Thermal Management of Electronics Using Spray Impingement, *Proc. of IPACK'01*, July 8-13, 2001, Kauai, Hawaii, Paper IPACK2001-15567.

Peeples, J. W., 2001a. Vapor Compression Cooling for High Performance Applications, *Electronics Cooling*, Vol. 7, No. 3, pp. 16-24.

Peeples, J. W., 2001b. Mechanically Assisted Cooling for High Performance Applications, *Proc. of IPACK'01*, July 8-13, 2001, Kauai, Hawaii, Paper IPACK2001-15715.

Perret, C., Boussey, J., Schaeffer, C., and Coyaud, M., 2000. Analytic Modeling, Optimization, and Realization of Cooling Devices in Silicon Technology, *IEEE Transactions on Components and Packaging Technologies*, Vol. 23, No. 4, pp. 665-672.

Rao, P., and Webb, R. L., 2000. Single Phase Flow in Micro-channels - a Critical Review, *Proc. 2000 National Heat Transfer Conf.*, Pittsburgh, PA, Paper NHTC2000-12102.

Saini, M., and Webb, R. L., 2002. Thermal Performance Limits of Forced Convection Air Cooled Heat Sinks for Computer Cooling, accepted by *ITherm 2002.*

Wang, C. C., 2000. Recent Progress on the Air-side Performance of Fin-and-Tube Heat Exchangers, *International Journal of Heat Exchanger*, Vol. 1, pp. 49-76.

Webb, R. L., and Yamauchi, S., 2001. Thermo-Syphon Concept to Cool Desktop Computers and Servers, *Proc. of IPACK'01*, July 8-13, 2001, Kauai, Hawaii, Paper IPACK2001-15773.

Webb, R. L., Chang, Y., and Wang, C., 1995. Heat Transfer and Friction Correlations for the Louver Fin Geometry, *1995 Vehicle Thermal Management Systems Conference Proceedings, Soc. of Automotive Engineers*, Warrendale, PA, pp. 533-541.

Thermal Challenges in Next Generation Electronic Systems, Joshi & Garimella (eds)
 ISBN 90-77017-03-8

An integrated thermal architecture for thermal management of high power electronics

Z. J. Zuo, L. R. Hoover & A. L. Phillips
Thermacore International, Inc., Lancaster, PA, USA

Keywords: thermal architecture, electronics cooling, chip to system level

ABSTRACT: This paper presents a thermal architecture concept for analysis of thermal problems and solutions existing in electronics systems. The thermal problems are categorized into a total of seven levels from chip to system. Advanced thermal technologies for addressing the thermal problems at all seven levels are discussed. Integrating the thermal architecture with the electronic architecture can significantly improve the effectiveness of the thermal management.

1 INTRODUCTION

In following Moore's Law during the past decade, the electronics industry seemed able to defy the laws of economics by providing ever increasing computing power at less cost. However, the industry has not been able to suspend the laws of physics in that higher computing performance is accompanied by more heat generation. Many market sectors are reporting board level heat dissipation that several years ago was seen only at the system level. As a result, thermal management is becoming increasingly critical to the electronics industry. Figures 1 and 2 show the trends of increasing heat dissipation in microprocessors and telecommunication cabinets [Corcoran 2001, Thermacore 2001b]. The time urgency of the thermal problem is obvious: microprocessor power will exceed 1kW and cabinet power will exceed 25kW within five years. Clearly, finding effective thermal solutions will become a major constraint for the reduction of cost and time-to-market, two governing factors between success and failure in commercial electronics sales.

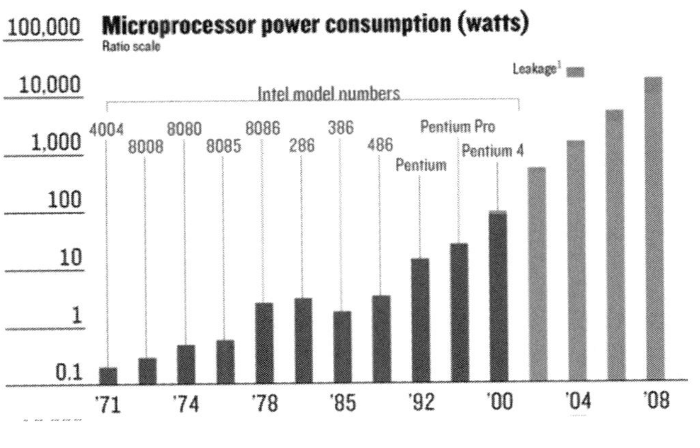

Figure 1. Increasing heat dissipation in microprocessors

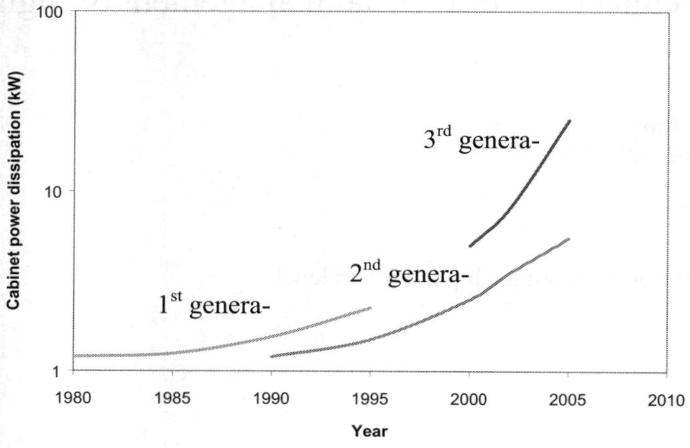

Figure 2. Increasing heat dissipation in telecommunication cabinets

The thermal problem caused by the increasing heat dissipation is further compounded by the trend of miniaturization, one of the main methodologies of the electronics industry to satisfy the increasing market demand for faster, smaller, lighter and cheaper electronic devices. According to Hewlett Packard, the number of devices per chip will approach 200×10^6 in 2005, performing 40 million logic operations per second [Thermacore 2001a]. The result of this miniaturization is increasing heat fluxes in electronics. Figure 3 shows the trend of increasing heat fluxes in mobile and performance PC's and high-end servers [Thermacore 1998].

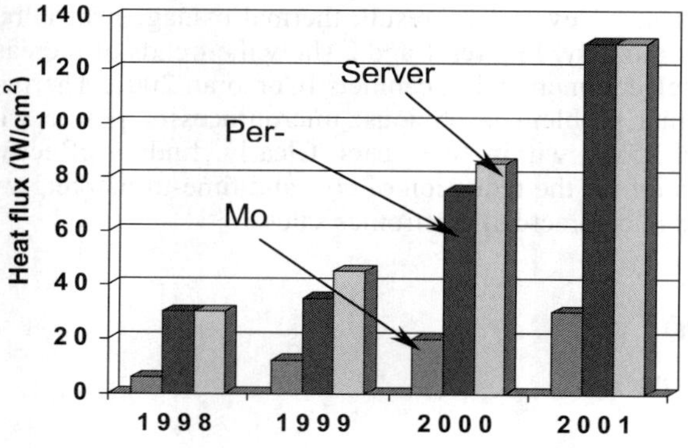

Figure 3. Increasing heat fluxes in chips

High heat fluxes are also found in power and opto-electronic equipment. For example, metal oxide semiconductor-controlled thyristors generate heat fluxes from 100 to 200W/cm^2 [Ponnapan et al. 1995]. Five years ago, the Department of Defense was looking into high voltage power electronics for applications such as the USAF More-Electric Aircraft Initiative that may generate heat fluxes of 300W/cm^2 [Cao et al. 1996]. Heat fluxes in some optoelectronics applications exceed 1,000W/cm^2 [North & Rosenfeld 1995].

A further challenging aspect is the non-uniform heat flux distribution in electronics. Figure 4 shows the heat flux distribution at the junction plane of a high performance server chip. To protect confidential information, all heat flux values in Figure 4 have been normalized (divided by the peak heat flux that is on the order of $150W/cm^2$). As shown, the peak heat flux is over 5 times the average heat flux over the entire chip surface ($\sim 30W/cm^2$). Under such conditions, integrating advanced heat spreading mechanisms onto the back of the chip becomes an attractive option due to the elimination of undesirable thermal interfaces.

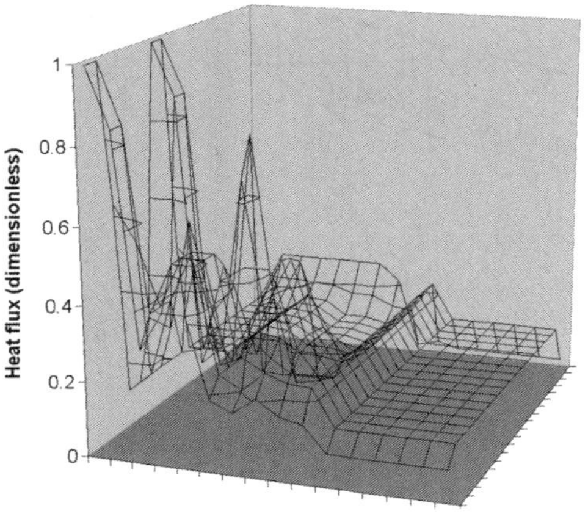

Figure 4. Non-uniform heat fluxes in a server chip

Figure 5 summarizes the thermal problems and challenges facing the industry [Thermacore 2001c]. The red line shows the trend of decreasing case temperature that is demanded by the increasing clock speed and heat dissipation. The blue line shows the trend of rising enclosure temperature that is caused by the increasing heat dissipation in the enclosure. The difference between the two lines indicates the "thermal budget" available for the design of the cooling devices/systems. As the two lines converge, the thermal budget shrinks, presenting increasing challenges to the thermal system design. As the thermal budget approaches zero, refrigeration systems will have to be used to provide the necessary cooling.

There are several negatives associated with the use of refrigeration systems. The additional electrical power required by the refrigeration system not only increases the cost to operate the electronic equipment but also causes adverse environmental impacts in the forms of pollution from power generation processes and noise from the equipment itself. Reliability issues are among the other negatives associated with refrigeration systems. Therefore, one challenge facing the industry is how to avoid or delay the use of refrigeration systems for electronics cooling. This article presents one possible approach to this challenging problem. We start by analyzing the different levels of thermal problems existing in typical electronic enclosures. Using the analogy to the commonly used interconnection architecture concept, we propose a thermal architecture concept as a baseline for such analysis. An advanced thermal architecture that utilizes effective and passive components is discussed. Finally, the benefits and challenges in implementing such architecture are examined.

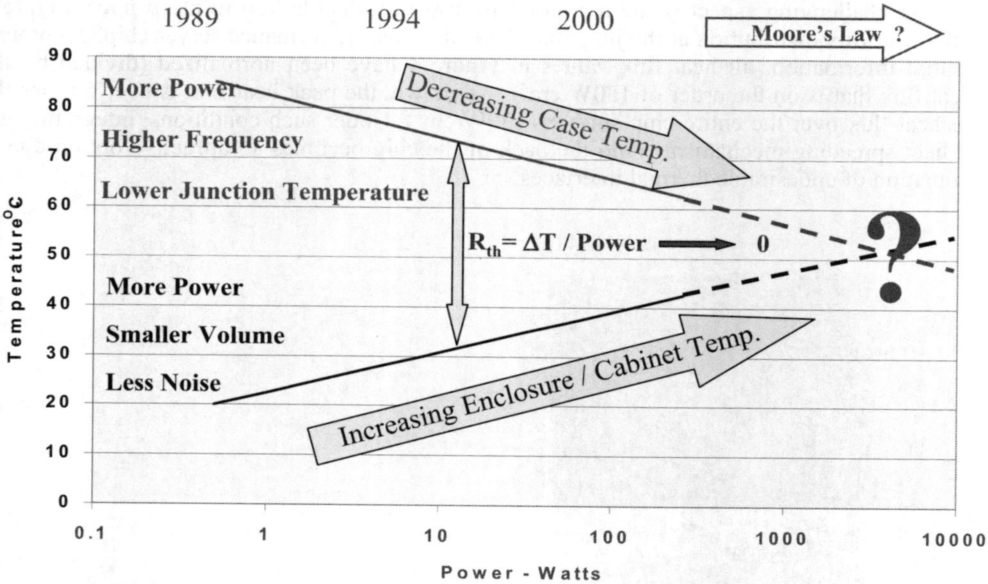

Figure 5. Trend of shrinking thermal budget for thermal management systems

2 LEVELS OF THERMAL PROBLEMS

Successful cooling applications must deal with thermal issues at the device, device cluster, PCB (printed circuit board), subassembly, rack and cabinet. In many situations, the problem is further complicated by the fact that the thermal solution is an "after thought". While a new OEM equipment design often utilizes the latest software and implements the fastest new semiconductor technology, the thermal issues are generally relegated to the "later phases" of the new product design.

This is not a new problem. The issues of electronic system packaging, power distribution, and interconnection are similarly "later phase" design issues. In fact there are many parallels between the solution process for thermal and for interconnection/electronic packaging. For example, both the thermal system solution and the electronic interconnection system solution must interface with semiconductors, chassis and all levels in between, and must accommodate the logic/functional partitioning, field repair/expansion and system maintenance/upgrade strategies for the OEM equipment.

There have been many years of research and development devoted to creating the products and innovations which have allowed the interconnection industry to "keep up" with the rate of change being driven by the chip speed, power, and bandwidth requirements. Some of these innovations came from within the large OEM's or device manufactures, but most came from corporate giants in the interconnection industry. AMP/Tyco, Molex, FCI, Amphenol and others are all billion-dollar companies, any one of which is larger that the sum of the thermal revenues from the top 5 thermal solution providers. These companies are driving the interconnection industry that is over $30 billion dollars worldwide (not including all the PCB or backplane business).

The interconnection industry has identified a consistent "levels of packaging" concept that accommodates the OEM equipment design, installation and field support requirements. Product families have been created to deal with these different levels of interconnections. Table 1 summarizes the 6 levels of interconnections and the representative products [Fleck Research 1999].

Table 1. Six interconnection levels in electronic systems

Level	Definition	Examples	Annual sales
1	Interconnections between wafer, die, substrate and multichip module.	Wire bonding, peripheral and area array interconnections.	$1.4B
2	Interconnections between chip and substrate or board.	IC sockets and solder bumps.	$1.2B
3	Interconnections between boards.	Cardedge, headers and board mount.	$4.1B
4	Interconnections between wire and board.	Receptacles, wire wrap and sockets.	$3.6B
5	Interconnections between wires.	Circular and coax connectors.	$8.9B
6	Others.	Backplanes, cable assemblies, etc.	

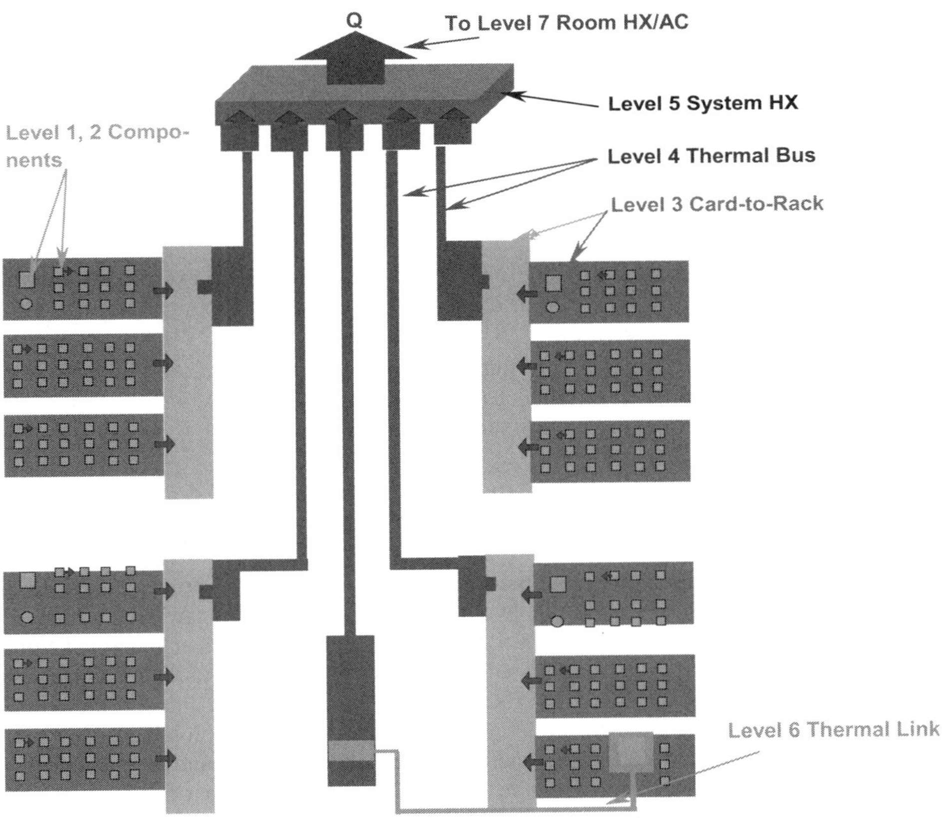

Figure 6. Thermal architecture diagram

What is the impact of all this on the thermal management industry? Perhaps we can learn from the experience and the structure above and construct thermal solutions to effectively address the different levels of thermal problems. With the help from the industry experts, Thermacore's Technology Center has developed a thermal architecture concept to describe the different levels of thermal problems in electronic systems. This thermal architecture concept is illustrated in Figure 6 which shows a schematic diagram of heat flows in a typical electronic enclosure cooling system [Phillips 2000]. Table 2 summarizes the definitions of thermal problems at each level. Also included in Table 2 is an assessment of the techniques that are currently used at each level. Thermacore has initiated patent applications on the thermal architecture concept, including the definitions of the seven levels.

Table 2. Definitions of thermal architecture levels

Level	Definition	Current techniques
1	Heat spreading within component (chip package) as characterized by junction-to-case thermal resistance.	Heat conduction through thin layers of diamond and other interface materials.
2	Heat spreading through interface between component and circuit card as characterized by chip-to-card interface thermal resistance.	Copper spreader, aluminum spreader with embedded heat pipes and vapor chamber heat spreaders.
3	Heat transfer from circuit card to rack/chassis as characterized by thermal resistances across circuit card and card-rack interface.	Aluminum based chassis (some with embedded heat pipes) and vapor chamber based chassis.
4	Heat transport (Thermal Bus) from card-rack interface to system heat exchanger as characterized by convective thermal resistance of coolant flow over rack/chassis.	Forced airflow over extrusion heat sinks attached to card/rack, and pumped liquid flow through cold plates attached to card/rack.
5	Heat exchange between cabinet inside and outside as characterized by system heat exchanger performance characteristics.	Air/air, liquid/air heat exchangers and refrigeration/air conditioning systems.
6	Direct thermal link between certain high power components and Thermal Bus, bypassing Levels 2 and 3 to avoid extra interface thermal resistances.	Air or liquid flow directed to high power components.
7	Thermal control of equipment room (not applicable for outdoor equipment).	Air/air, liquid/air heat exchangers and refrigeration/air conditioning systems.

Table 2 describes the thermal architecture in the most general sense. There can be numerous variations in practical systems. Figure 7 examines the layout of a server box to illustrate the thermal architecture [Thermacore 2001a]. At Level 1 (within the processor package), the heat generated at the junctions spreads across three layers of materials before reaching the outer surface of the package lid: the silicon material of the die, the epoxy interface between the die and the lid, and the AlSiC lid. Figure 8 [Thermacore 2000a] shows the CFD analysis results of the junction temperature profile of the chip under the power conditions described in Figure 4. The hot spots indicate the ineffectiveness of the current spreading mechanisms under the highly non-uniform heat fluxes. More effective approaches are thus required. Technical challenges to heat spreading at Level 1 include the lack of understanding of miniature or micro scale heat transfer, potential CTE mismatch problems between spreader and semiconductor materials, and potential adverse impact on chip packaging processes.

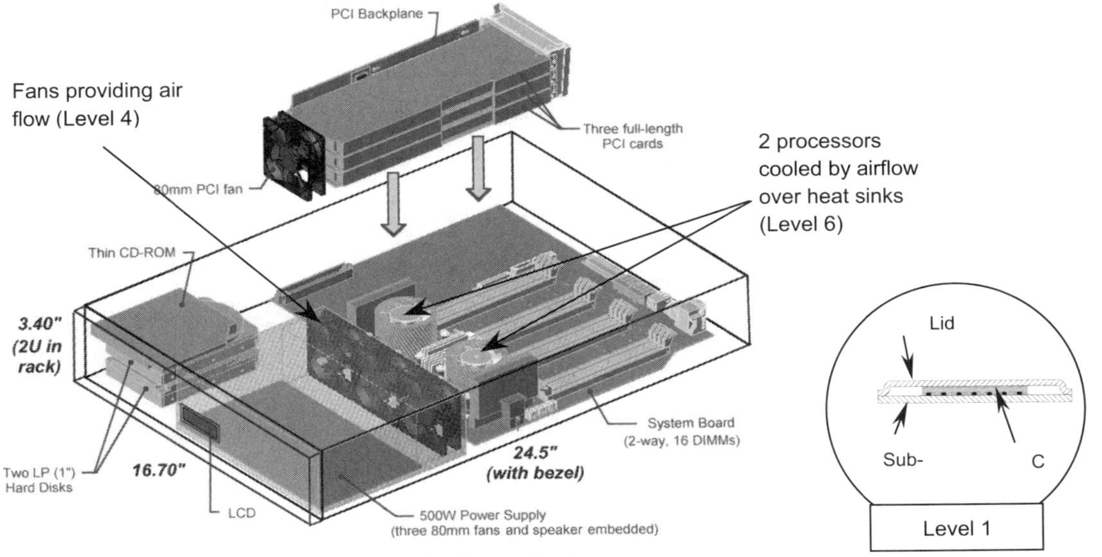

Figure 7. Thermal architecture case study: Server Cooling

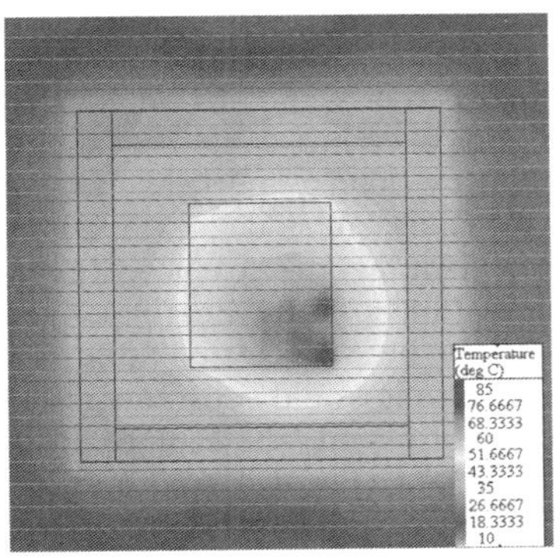

Figure 8. Junction temperature profiles

The circuit board in this case plays a minor role in cooling the processors, since the majority of the heat goes upward into the package lid. Therefore, Levels 2 and 3 are skipped in this architecture. The heat sinks on top of the processors provide the interface between the components (Level 1) and the airflow (Level 4), functioning as the direct thermal link (Level 6). The lower power components on the board also dissipate their heat into the airflow. Level 5 (system heat exchanger) is skipped since the airflow is discharged directly into the room air. At Level 7 in this case is the air conditioning system that provides thermal control for the room where the server situates.

The above case is probably the simplest form of the thermal architecture. As the number and variety of the components increase, the thermal architecture can become complicated, and all seven levels may exist. At this time, we cannot show the full details of such a case due to confidentiality of proprietary information. But we can show in Figure 9 several pieces of the system to illustrate the thermal architecture [Thermacore 2001c].

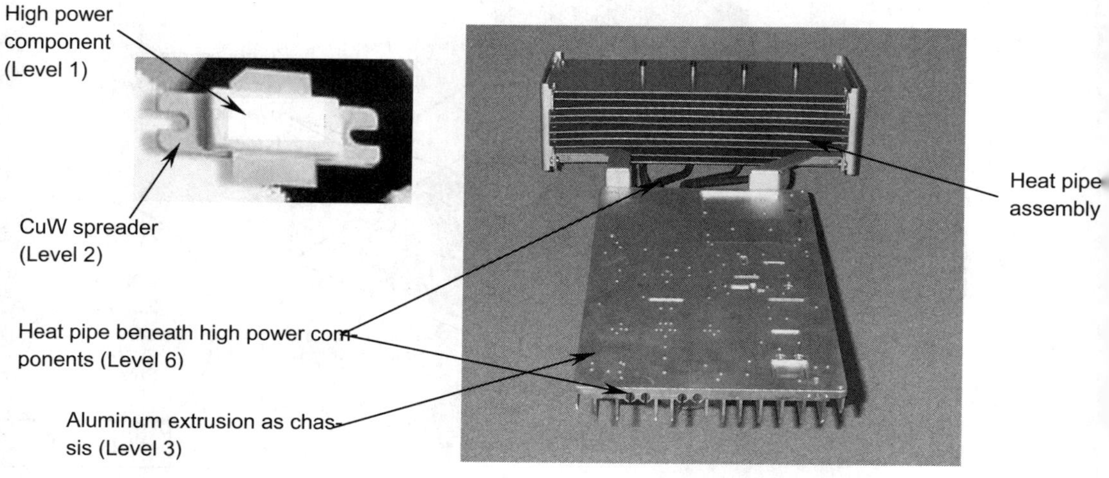

Figure 9. Thermal architecture case study – telecommunication cabinet cooling

The high power component in Figure 9 consists of a 120W amplifier (Level 1) mounted on a copper-tungsten (CuW) heat spreader (Level 2). Level 3 in Figure 9 is represented by an extruded aluminum heat sink which serves as the chassis; components and/or circuit cards are directly mounted to it. The aluminum heat sinks provides Level 3 cooling by conducting heat from the component side to the extrusion side for dissipation to air. Heat pipes are embedded in the aluminum so they pass directly beneath high power components. To the extent that the heat pipe directly cools these components they can be considered to be Level 6 thermal links. These heat pipes efficiently transport heat from the (aluminum) chassis to a fin stack where heat is rejected to air. The assembly in Figure 9 is actually a standalone unit, but in the generic case it would constitute a shelf in a cabinet and the fin stack would be located in a separate airway or outside the cabinet. In the generic case the cabinet inside airflow would constitute the thermal bus (Level 4) and a system air/air heat exchanger (Level 5) would be required to dissipate heat from cabinet inside to outside. Level 7 is not required since this is an outdoor application.

The main advantage of this cabinet thermal architecture is the maturity of the component technologies. A relatively well-established infrastructure of manufacturers exists to supply the extrusions, heat pipe assemblies, fans and blowers. However, this architecture has one significant shortcoming in its thermal performance capability. For example, the forced airflow thermal bus (Level 4) becomes infeasible for any cabinets that dissipate more than 5kW, as concluded by a recent study by a major telecommunication equipment OEM. Other shortcomings of this architecture include bulkiness, noise, power consumption and reliability problems associated with fan failures. Table 3 evaluates the performance capabilities of various thermal bus candidates by listing their order of magnitude heat transfer coefficients [Incopera & DeWitt 1990, Thermacore 1999, Thermacore 2000b, Thermacore 2001a].

Table 3. Heat transfer coefficients of air, liquid and two-phase flows

Type of thermal bus	Order-of-magnitude heat transfer coef.
Air – natural convection	$0.0005 W/cm^2$-°C
Air – forced convection	$0.005 W/cm^2$-°C
Liquid Fluorinert – pumped flow in large channels	$0.05 W/cm^2$-°C
Liquid water – pumped flow in large channels	$0.075 W/cm^2$-°C
Two-phase Fluorinert – pool boiling	$0.25 W/cm^2$-°C
Liquid Fluorinert – pumped flow in small channels	$0.5 W/cm^2$-°C
Liquid water – pumped flow in small channels	$0.75 W/cm^2$-°C
Two-phase water – pool boiling	$1 W/cm^2$-°C
Two-phase Fluorinert – spray	$1.5 W/cm^2$-°C
Two-phase water – spray	$5 W/cm^2$-°C
Two-phase water – evaporation off porous surface	$10 W/cm^2$-°C
Liquid water – pumped flow in porous media	$100 W/cm^2$-°C
Two-phase water – nucleate boiling in porous media	$100 W/cm^2$-°C
Two-phase water – pumped flow in porous media	$200 W/cm^2$-°C

As shown in Table 3, two-phase flows are generally more capable thermal buses than air and single-phase liquid flows. The performance capabilities can be further improved by incorporating enhancement structures such as porous media. The space industry has long adopted the two-phase flow heat transfer loop for thermal control because of its high performance and compactness, and the fact it maintains a constant temperature around the loop. Another major attractiveness of the two-phase flow heat transfer is that it can be realized passively by using properly designed capillary structures.

Table 4 shows the projected heat dissipations at various levels in a typical communications cabinet thermal architecture by 2003. The numbers have been rounded up to protect confidentiality of the information.

Table 4. Projected heat dissipations in a typical cabinet by 2003

Architecture level	Heat dissipation projection
Level 1 – Component	300W
Level 2 – Circuit board	800W
Level 3 – Chassis or shelf	1,500W
Level 4 – Thermal bus	20,000W
Level 5 – System heat exchanger	20,000W
Level 6 – Thermal link	300W
Level 7 – Room thermal control	N/A – outdoor application

The following section presents an advanced thermal architecture that utilizes thermally effective, passive components to meet the increasingly stringent thermal requirements as illustrated in Table 4 and other requirements on compactness, noise, power consumption and reliability.

3 AN ADVANCED THERMAL ARCHITECTURE

One of the key features of this advanced thermal architecture is its passiveness. Most components in this architecture are thermally driven (i.e. passive) heat transfer devices that have no moving parts and require no external power for their operations. This feature ensures the architecture to be highly reliable and cost effective. Additional advantages include quiet operation, compactness and packaging flexibility.

3.1 *Level 1: Heat spreading in chip package*

The heat flux exiting the die is in the 10 – 100W/cm^2 range. To minimize contact resistance losses, the heat flux exiting the chip package should be in the 1 – 10W/cm^2 range. Spreading at the chip level has generally been accomplished by simple conduction. The thin materials are highly conductive and produce little temperature drop directly under the die footprint, but are relatively inefficient in spreading a non-uniform heat flux to the dimension of the package. Since heat fluxes continue to increase, industry, government and academia have recently intensified the exploration for more capable spreading technologies that involve single/two-phase flow convection heat spreading.

The Department of Defense through DARPA (Defense Advanced Research Projects Agency) started the HERETIC (Heat Removal by Thermo-Integrated Circuits) program in 1999 to develop miniature/micro scale heat removal technologies that can be integrated with high power density electronics and optoelectronics. The technologies under investigation include integrated thermoelectric devices, miniature/micro heat pipes, micro-machined synthetic jets and micro-fluidic devices [DARPA 1999]. Among the participants are some of the major players in the industry, academia and government labs, including Hewlett Packard, Intel, Rockwell, Sandia National Labs, and a number of university research centers. Thermacore Technology Center is an active participant of the HERETIC program.

Sandia has been one of the leading institutions in developing micro-machined vapor chamber heat spreaders. Recent prototypes, made of Kovar and using methanol as the working fluid, demonstrated equivalent thermal conductivity on the order of 800W/m-K and maximum heat flux capability (before dry-out) up to 20W/cm^2 [Benson et al. 1998]. This represented a significant progress over prior arts by successfully demonstrating the micro fabrication process and the use of low CTE materials as the heat pipe envelope. Similar efforts are being carried out by several groups in Florida International University and the University of Maryland. However, the thermal performance of these micro-machined heat pipes remain less than desirable, considering that the peak heat fluxes in some computer chip packages are exceeding 150W/cm^2.

The University of California at Berkley, along with the Air Force Research Laboratory, is developing micro CPL's (capillary pumped loops) and LHP's (loop heat pipes). Prototype CPL's have been made in silicon using micro fabrication techniques. It was observed that the prototype CPL was able to establish the two-phase flow circulation [Kirshberg et al. 2000]. Similar efforts are being carried out by NASA, University of Cincinnati and Texas A&M University. Thermacore Technology Center is supporting Texas A&M in developing sub-micro silicon wick fabrication processes.

Under several programs sponsored by BMDO (Ballistic Missile Defense Organization) and NSF (National Science Foundation), Thermacore's Technology Center developed a miniature heat pipe heat spreader technology that uses micro grooves to facilitate a passive pulsating two-phase flow. This miniature pulsating heat pipe technology was demonstrated to be able to remove concentrated heat fluxes exceeding 250W/cm^2 at a thermal resistance below 0.15°C/W/cm^2 [Zuo et al. 2000]. A recent program was started jointly with NASA to apply this technology to silicon and other low CTE materials to allow direct integration of this technology into semiconductors.

Among the previously demonstrated concepts, two miniature heat spreader designs showed particular promise for successful integration into high power chip packages in the near future. Both designs effectively manage the complex single or two-phase flows on micro scale to achieve the exceptional performance capabilities. Figure 10 shows the incorporation of these designs into a chip package lid to improve the heat spreading effectiveness within the package. This approach minimizes the potential adverse impact on the chip packaging process, since the advanced lid can be pre-fabricated. Table 5 summarizes the two designs currently pending for patent approval.

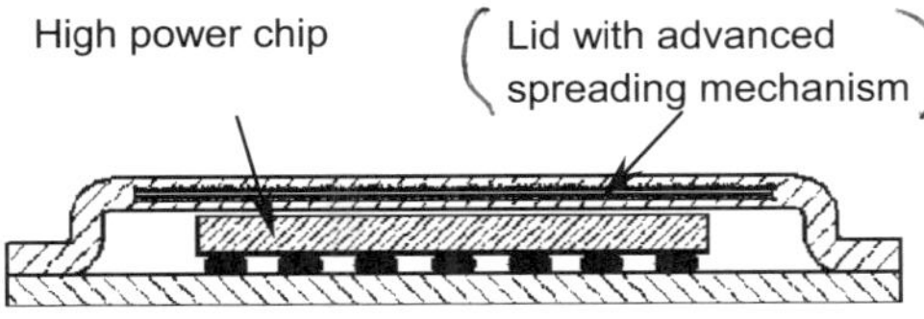

Figure 10. Achieving effective heat spreading with advanced lid designs

Table 5. Advanced heat spreading technologies at Level 1

Technology	Operating principles and main advantages
Miniature, vapor chamber heat spreader made of low CTE materials.	• Internal two-phase flow heat transfer improves heat spreading effectiveness by at least one order of magnitude. • Internal two-phase flows are thermally driven and facilitated by an optimized 3-D capillary structure. • Low CTE material envelope allows direct bonding to chip, eliminating interface thermal resistance.
MEMS actuated micro channel oscillating flow heat spreader made of low CTE materials.	• High frequency oscillating flow in micro-channels provides enhanced thermal diffusion by at least one order of magnitude. • Fluid oscillation is generated and controlled by a micro actuator that is integral to the heat spreader. • Low CTE material envelope allows direct bonding to chip, eliminating interface thermal resistance.

Figure 11 shows the junction temperature profile of the same server chip as described in Figures 4 and 8 except that a miniature vapor chamber heat spreader replaced the AlSiC spreader as the package lid. Comparisons between Figures 8 and 11 indicated a significant improvement in heat spreading effectiveness by using the two-phase flow heat transfer technology. The reduced temperature gradients on the lid surface reduce not only the junction temperatures but also the thermal stresses caused by temperature gradients.

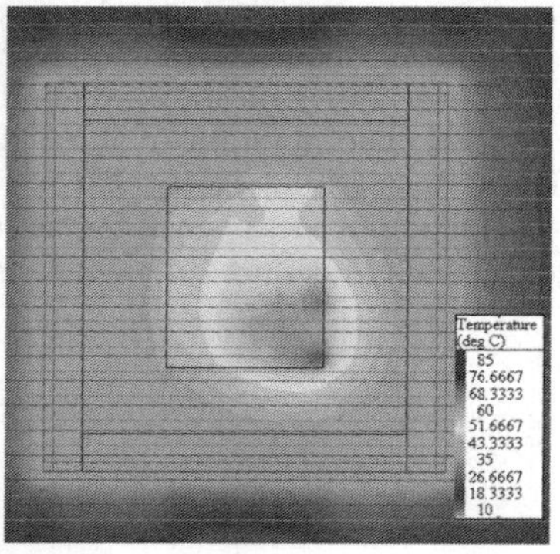

Figure 11. Junction temperature profiles with advanced heat spreading

3.2 *Level 2: Heat spreading at component-board interface*

Copper spreaders are traditional choices for heat spreading at this level. Other materials including CuW, AlSiC, SiC and AlNr have recently become increasingly popular because of their low CTE and good thermal conductivity. Aluminum plates with embedded heat pipes and vapor chamber heat spreaders (as shown in Figure 12) are higher performance alternatives. Technologies and products at this level are relatively mature. Vapor chamber heat spreaders with sintered powder wicks have been used to handle heat fluxes up to $100W/cm^2$. Several advanced wick structures have been demonstrated to dissipate heat fluxes approaching $300W/cm^2$.

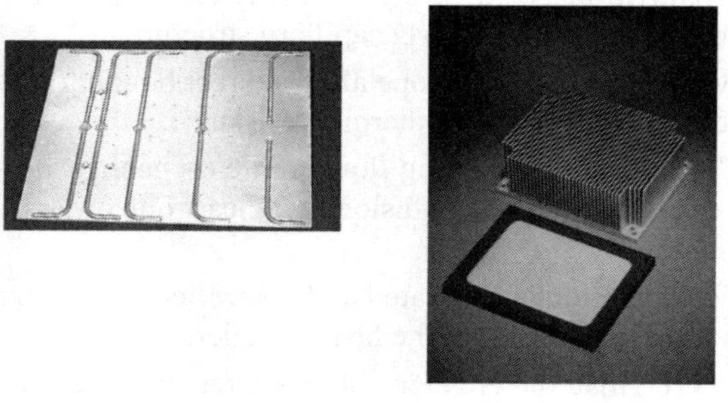

Figure 12. Embedded heat pipe cold plate and vapor chamber heat spreader

3.3 *Level 3: Cooling of rack/chassis*

Passive cooling solutions at this level come with a variety of configurations. More mature technologies include embedded heat pipe cold plates and vapor chamber cold plates. Figure 13 shows a more advanced design that integrates the evaporator of a loop thermosyphon into the card rack. A variation of this design is shown in Figure 14 where the evaporator of the loop thermosyphon is in a vapor chamber configuration and attached to one end of the circuit card. The circuit card can be either a vapor chamber heat spreader or an aluminum plate embedded with heat pipes. An important feature of both designs shown in Figures 13 and 14 is that the loop thermposyphon is both a rack for the circuit card and a thermal bus in the thermal architecture. Thermacore has initiated patent applications on both designs shown in Figures 13 and 14.

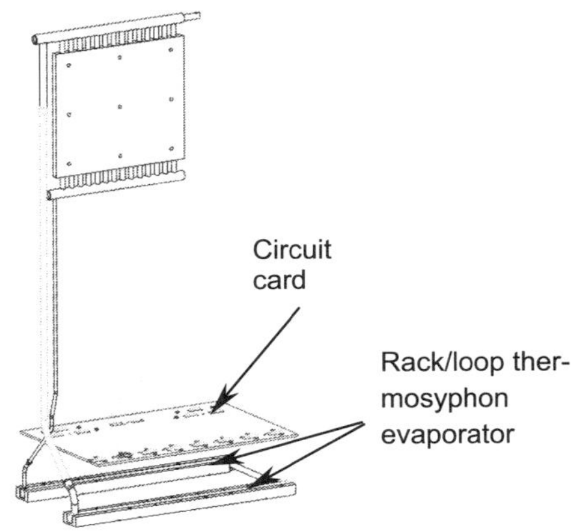

Figure 13. Loop thermosyphon evaporator as rack/chassis

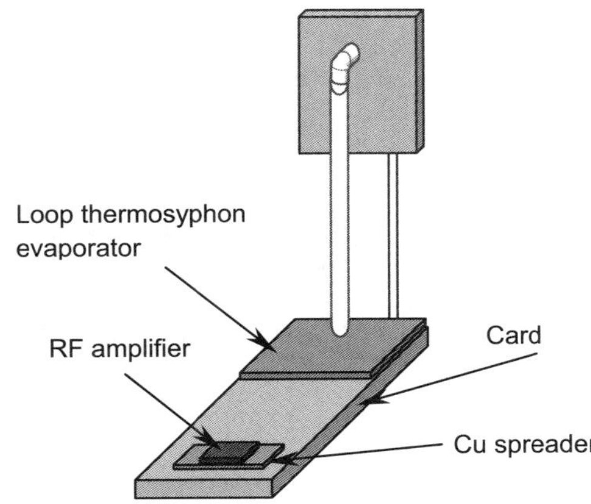

Figure 14. Loop thermosyphon evaporator on top of one end of circuit card

3.4 *Level 4: Thermal bus*

The loop thermosyphon shown in Figures 13 and 14 performs the thermal bus function by passively transporting the heat from the circuit card to a condenser at another location. Figure 15 shows twelve loop thermosyphons for cooling twelve circuit cards with a total heat dissipation exceeding 15kW. This system was designed for a telecommunication cabinet cooling application. The condenser stack is coupled with liquid cooled cold plates. The total temperature difference between the card-rack interface and the inlet cooling liquid was less than 20°C [Khrustalev 2001]. The 12-module thermal bus is an integral part of the cabinet, providing not only the transport of over 15kW of heat but also the structural support/accommodation of the 12 circuit cards.

The cooling liquid takes the heat further away from the cabinet to a "remote" heat exchanger for eventual dissipation of the heat to ambient air. This approach provides flexibility for field installation/expansion of the cabinet, since the cabinet and the heat exchanger are essentially de-coupled. It is especially beneficial for indoor, server-center type installations since the thermal bus can couple directly to the chiller circuit of the HVAC system and eliminate room air as a heat transfer medium. This eliminates the large air heat exchangers for the output of the cabinet and the input of the HVAC, and permits a more comfortable building climate.

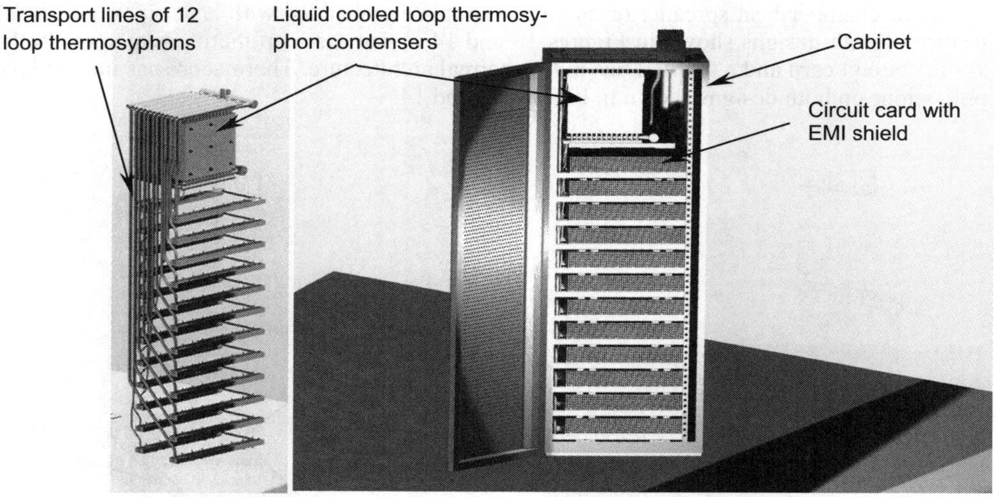

Figure 15. A loop thermosyphon thermal bus in a telecommunication cabinet

Figure 16 shows conceptual sketches of using a two-stage loop thermosyphon thermal bus for cooling of server chips. The 1st stage loop thermosyphon evaporator is in a U-shape and coupled with multiple chips in the tray. The condenser section is in a special coupling structure which allows the 1st stage loop thermosyphons to couple and de-couple (for assembly convenience) with the 2nd stage loop thermosyphons. The condensers of the 2nd stage loop thermosyphons (not shown) can be cooled by either air or liquid. Both designs shown in Figure 16 are in the process of being patented.

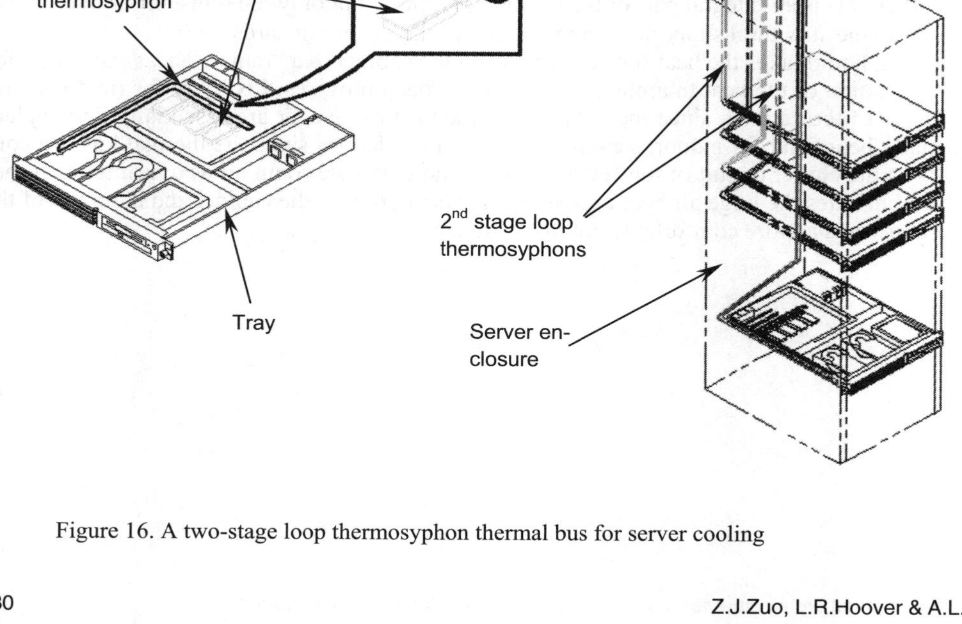

Figure 16. A two-stage loop thermosyphon thermal bus for server cooling

Another passive thermal bus technology is the loop heat pipe. Unlike the loop thermosyphon that must rely on the gravity to return the liquid, the LHP uses capillary wicks to circulate the liquid and is thus capable of operating at any orientation. This means the thermal bus can passively transport heat to a cooling circuit in the sub floor or to a sink in the floor below. The LHP is a mature and well used technology for spacecraft thermal control where the gravity effect is essentially zero. In 1997, Thermacore initiated and directed a successful space flight experiment of a loop heat pipe on board Space Shuttle Mission STS-87 [Baker et al. 1997]. The program was jointly funded by a consortium of industry, universities and government organizations. Thermacore is currently conducting a cost reduction program to make the loop heat pipe technology more suited for commercial applications.

3.5 *Level 5: System heat exchangers*

Nearly all solutions at this level require fans, blowers or pumps. The only exceptions are natural convection based systems for very low power applications, which are not in the scope of this article. For higher power applications, various air/air and liquid/air heat exchangers and even refrigeration systems have been employed to dissipate heat to ambient air – the ultimate heat sink. The heat exchangers can be inside or outside the enclosures, depending on the applications. For most telecommunication cabinets, the heat exchanger is an integral part of the cabinet infrastructure. For some server center applications, the heat exchanger is outside the server box. There are a variety of configurations of heat exchangers that are manufactured by an extensive network of suppliers. Figure 17 shows several examples of heat exchanger products currently offered by Thermacore and Modine.

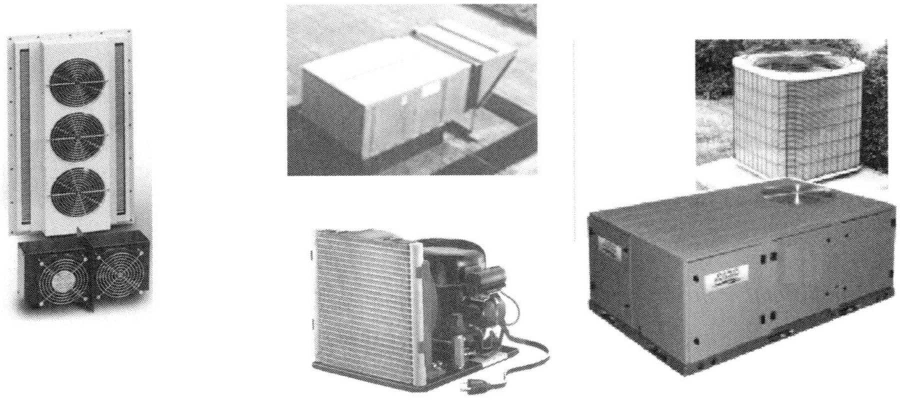

Figure 17. Examples of system heat exchangers and refrigeration systems

3.6 *Level 6: Direct thermal link between high power components and thermal bus*

Certain high power components generate orders of magnitude more heat than others on the same circuit card. Instead of treating the high power components the same way, i.e., going through Levels 2 and 3 to reach Level 4, the more effective approach is to directly link these components to the thermal bus, thus eliminating the interface thermal resistances.

An effective device/technology at Level 6 must have the following features:

(1) High heat flux capability. This is required since one end of this thermal link is attached to high power (heat flux) components. The heat flux can be as high as several hundred watts per square centimeter.

(2) Structural flexibility. The thermal link must be sufficiently flexible to link the high power components and the thermal bus without interfering with other components in between them.
(3) Passive operation. This results in not only simple structure and high reliability but also compactness, since no pumps or fans are involved.
(4) Low CTE materials. To cool some ultra-high heat flux components, it is beneficial for one end of the thermal link to be hard bonded to the component, eliminating the thermal interface between them. To achieve this, this end of the thermal link must be made of materials that have low CTE.

Figure 18 shows two examples of technologies/devices that can be used as Level 6 thermal links. The transport lines of these devices are made of flexible materials. The flexible heat pipes (on the left side in Figure 18) have been used in military avionics cooling applications. They were proven to be able to handle high heat fluxes, high-g load, shock and vibration [Gernert 1998]. Both devices can operate at any orientations. The loop heat pipe is able to work against higher gravity heads.

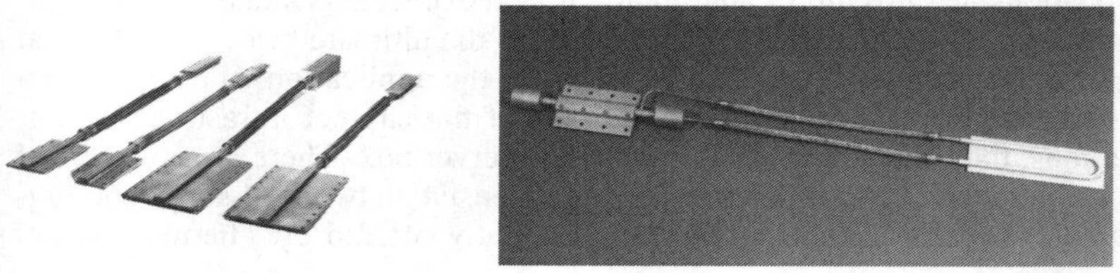

Figure 18. Flexible heat pipe and loop heat pipe

Thermacore is presently designing a high performance loop thermosyphon thermal link for a proprietary computer cooling application. Another ongoing effort is to demonstrate the flexible heat pipe and loop heat pipe technologies in low CTE materials. In particular, a joint project is to be started shortly with NASA and a computer OEM to demonstrate a miniature, flexible loop heat pipe with multiple, silicon evaporators.

3.7 *Level 7: Room thermal control*

This level is optional depending on the environment of the electronic enclosure. For outdoor applications, Level 7 is not required. For indoor applications such as data centers and server farms, Level 7 is usually represented by the air conditioning system that controls the room air at certain temperature (68°F) and humidity. Figure 19 shows a picture of a server room that is controlled by an air conditioning system. A typical data center uses 10 to 20MW of power per hour, approximately 30% of which is used to power the HVAC system [Abreu 2001]. The energy cost of running the HVAC system is approximately 20% of the total operating costs for a typical data center. Improving the Level 7 thermal control performance and efficiency will have a large positive impact on the financial bottom line as well as the environment.

Figure 19. Air conditioned server room

There are several major heat transfer processes associated with the traditional thermal control approach at Level 7. The room air is first heated by the heat dissipation at the enclosure boundary and then cooled while circulating through the HVAC system. The HVAC system transports the heat to a condenser heat exchanger for eventual dissipation to the outside ambient air. Since air is generally a poor heat transfer fluid (see Table 3 for typical airflow heat transfer coefficients), large thermal resistances incur in the heating and cooling of the room air.

Over the past several years, discussions have been centered on several concepts to eliminate the room air induced thermal resistances. One promising approach is to directly couple the enclosure heat exchanger to the HVAC system via a pumped liquid cooling loop, bypassing the room air. Figure 20 illustrates another, more passive concept of using loop thermosyphons to couple the enclosures to the HVAC system.

Figure 20. Loop thermosyphons connecting enclosure to HVAC duct

Each electronic enclosure in Figure 20 has an internal thermal bus consisting of multiple loop thermosyphons, similar to the ones shown in Figure 15. A larger size loop thermosyphon connects the condensers of the enclosure thermal bus to the HVAC system, bypassing the room air. The key to this concept is the thermal interface designs between the loop thermosyphon, the enclosure thermal bus and the HVAC system. Figure 21 shows a technician docking two loop heat pipes in a series for a space radiator application. The cylindrical docking/interfacing design as shown in Figure 21 has the advantage of being compact and easy to assemble. Other interfacing designs are being developed in Thermacore's Technology Center for various applications.

Figure 21. Docking of two loop heat pipes

4 BENEFITS AND CHALLENGES IN IMPLEMENTING THERMAL ARCHITECTURE

Figure 5 described a scenario of shrinking thermal budgets leading to a thermal crisis in the very near future when cooling to ambient air will become physically impossible. This would lead to a grim future where dedicated air conditioning systems are required to provide the needed sub-ambient temperature air for cooling server cabinets. Such AC loads would necessitate massive increases in both electrical generating capacity and fuel imports. By understanding the different levels of thermal problems in electronic enclosures, we can select technologies to address the problems at component levels and to achieve the best overall effectiveness at the system level. Figure 22 [Thermacore 2001c] projects a much brighter future by incorporating advanced thermal technologies to reverse the trends of shrinking thermal budget:

(1) Incorporating advanced heat spreading technologies in chip package (Level 1) allows the case/lid temperature to increase without overheating at the junction level, thus reversing the trend of decreasing case/lid temperature.

(2) Incorporating more capable and compact thermal bus technologies reduces the temperature rise inside the enclosure, thus reversing the trend of increasing enclosure temperature.

(3) Incorporating refrigeration systems in conjunction with the advanced thermal bus technologies further extends the available thermal budget.

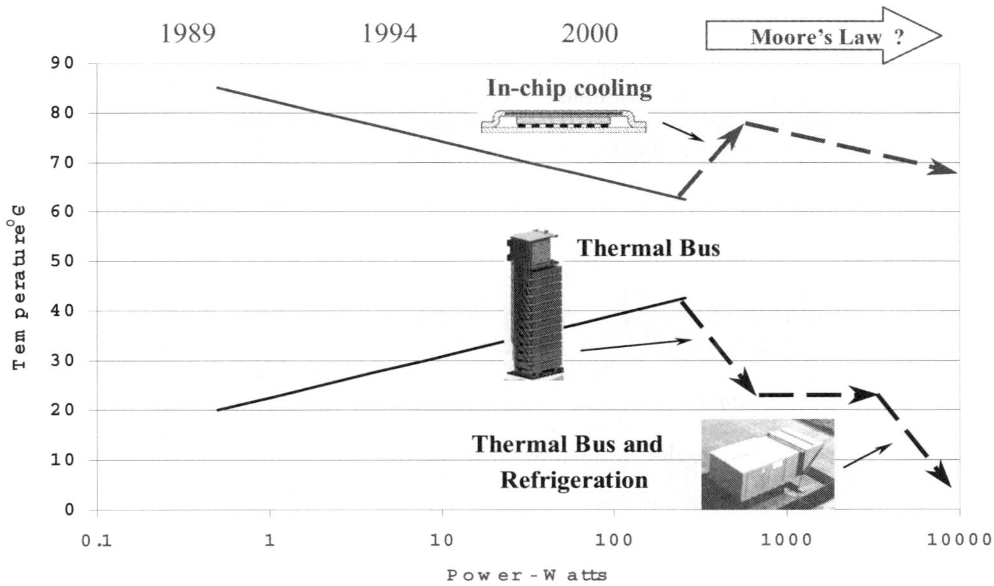

Figure 22. Bending curves with advanced thermal architecture

The incorporation of the advanced thermal architecture will delay by many years the need for dedicated refrigeration systems for electronics enclosure cooling. At such time as refrigeration systems eventually become necessary, the advanced thermal architecture will greatly reduce the required refrigeration capacity and energy requirements. By linking the loads directly to the HVAC chiller loops, the advanced thermal architecture will make the HVAC smaller and more efficient as well as improve the human comfort level within the office surrounding the electronics cabinets. Clearly, the payoffs for an advanced thermal architecture are tremendous.

A May 1999 report by analyst Mark Mills [Mills & Huber 1999] indicated that the computer equipment for Internet consumed about 295 terawatt-hours (1 tera = 10^{12}), approximately 9 percent of U.S. annual electric output. Of the 295 terawatt-hours, the Internet infrastructure such as ISP's, server farms and copper wire used for sending data accounted for 124 terawatt-hours – 3.5% of U.S. electric output. The Mills report forecasted that 30 to 50% of the electricity consumed in the U.S., within a decade, would be from the Internet. Adding to the picture is the power required to drive the thermal management devices and systems which are responsible for managing and transporting the 295 terawatt-hour power (1999 figure). If refrigeration systems must be used and assuming their average COP is 2.0, the additional power consumption by the thermal systems will be over 30 terawatt-hours per year, assuming a 23% duty cycle – typical for the Northeast. This will require the addition of 3.5GW new turbine capacity – approximately a \$1B initial investment. The cost for generating the 30 terawatt-hour electricity will be on the order of \$1.5B per year, and the NO_x emissions will be approximately 12,000 tons per year [Thermacore 2001d]. Therefore, delaying the use of refrigeration systems for one year will save costs on new turbine capacities and annual power generation. The environmental benefits will prove to be priceless.

In summary, the advanced thermal architecture technology will enable the electronics industry to continue to follow the Moore's Law without imposing huge energy and environmental penalties.

We foresee several major challenges associated with advanced thermal architecture. The technical challenges include fundamental understanding of micro-scale heat transfer and flow phenomena, materials issues such as stresses from mismatched thermal expansion, and the impact of integrating electronic packaging with thermal management. Perhaps the greatest challenges will be in the area of integration and standards. Thermal solutions must now be linked from the chip level to the HVAC system, and must therefore span components from several suppliers. Until recently, thermal issues were an afterthought, relegated to the later phases of product design. Thermal design is finally being considered from the beginning of the design process and in many cases has become the critical, enabling issue. The very near future will see thermal issues become the critical, systems-level integration issue that will determine the operating limits and lifetime of the system. Will the industry be ready?

5 REFERENCES

Abreu, E. 2001. Down on the Server Farm. *Industry Standard*, February 19, 2001.

Baker, C., Bienert, W. & Ducao, A. 1997. Loop Heat Pipe Flight Experiment Summary Report.

Benson, D., Mitchell, R., Tuck, M. & Palmer, D. 1998. Ultrahigh-Capacity Micromachined Heat Pipes. *Microscale Thermophysical Engineering* Vol.2: 21-30.

Cao, Y., Beam, J. & Donovan, B. 1996. Air-Cooling System for Metal Oxide Semiconductor Controlled Thyristors Employing Miniature Heat Pipes. *Journal of Thermophysics and Heat transfer* Vol.10: 484-489.

Corcoran, E. 2001. Too Hot to Handle. *Forbes*. April 2001.

DARPA. 1999. *HERETIC Kick-Off Meeting Proceedings*.

Fleck Research. 1999. On the Level. *Electronic News*, August 1999.

Gernert, N. 1998. Heat Pipe Reliability Documentation. *Thermacore Internal Document*.

Incopera, F. & DeWitt, D. 1990. *Fundamentals of Heat and Mass Transfer*. New York: John Wiley & Sons.

Khrustalev, D. 2001. Loop Thermosyphon Thermal Bus Test Report. *Thermacore Internal Document*.

Kirshberg, J., Yerkes, K. & Liepmann, D. 2000. Demonstration of a Micro-CPL Based on MEMS Fabrication Technologies. *AIAA-2000-3015*.

Mills, M. & Huber, P. 1999. Dig More Coal – the PCs Are Coming. *Forbes*, May 31, 1999.

Communications between Thermacore and Siemens-Westinghouse Power Corporation, April 2001.

North, M. & Rosenfeld, J. 1995. Liquid Film Evaporation Cooled Laser Diode Array. *BMDO SBIR Phase II Final Report DASG60-92-C-0093*.

Phillips, A. 2000. Thermal Bus for High Power Enclosures. *Thermacore Internal Document*.

Ponnapan, R., Leland, J. Beam, J. Fronista, G. & Weimer, J. 1995. Effective Cooling of MCT and IGBT Using Venturi Flow. *Intersociety Energy Conversion Engineering Conference Paper AP-97*.

Thermacore. 1998. Market Survey: Electronics Cooling.

Thermacore. 1999. Porous Metal Heat Exchanger Performance Data. *Thermacore Internal Document*.

Thermacore. 2000a. Communications with Hewlett Packard. August 2000.

Thermacore. 2000b. Boiling Performance Data. *Thermacore Internal Document*.

Thermacore. 2001a. Communications with Hewlett Packard. February 2001.

Thermacore. 2001b. Communications with Ericsson Radio Access AB. April 2001.

Thermacore. 2001c. International Sales Meeting. April 2001.

Thermacore. 2001d. Communications with Siemens Westinghouse Power Corporation. April 2001.

Zuo, Z., North, M. & Wert, K. 2000. High Heat Flux Heat Pipe Mechanisms for Cooling of Electronics. *ITherm 2000*.

Advanced design methodologies

Thermal Challenges in Next Generation Electronic Systems, Joshi & Garimella (eds)
 ISBN 90-77017-03-8

High-density server design methodology

D.S. De Lorenzo
Intel Corporation, DuPont, WA, USA

Keywords: Itanium™, radial blower, high-density server

ABSTRACT: Numerical simulation and experimental verification have yielded a high-density server concept dissipating a nameplate power of 700W in a 1U (1.75" or 44.45mm) rack-mount form factor. Air-cooled thermal management is achieved through a ducted radial blower, conventional heatsinks, and a partitioned airflow management strategy. It is shown that design and optimization of the air mover system requires attention not only to pressure and flow characteristics, but also to flow distribution properties. Optimizing the system for flow and thermal performance is an iterative process, most easily done before major layout or architectural properties have been fixed. Close collaboration between thermal analysts, board layout engineers, and system architects has proven the most effective method for optimizing the complete server design.

1 INTRODUCTION

Intel's Advanced Systems Lab is developing high-density hardware and software as part of an initiative to greatly increase data center compute densities. Concept systems are based on the Intel® Itanium™ processor and are designed for a 1U rack-mount form factor. The current dual-processor system concept is targeted at the front-end server market (e.g. web hosting and mail routing), at processing secure transactions, web or database searching, or acting as a "compute server" (Fig. 1, Aldridge et al. 2001). The primary architectural advantages are massive compute power, large memory address space, rich I/O connectivity, and a dense form factor. The expected deployment environment is the air-cooled, rack-mount data center.

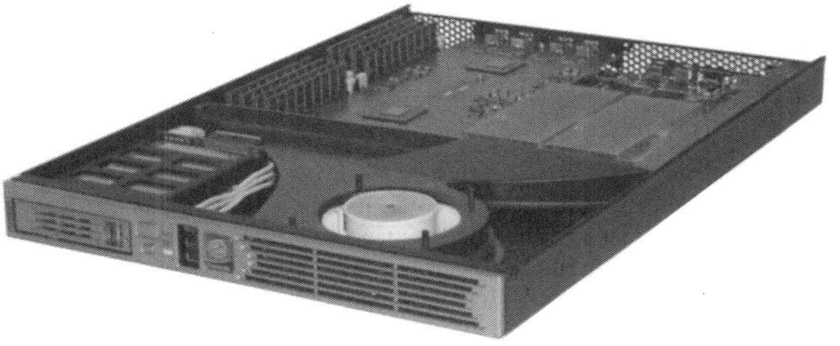

Figure 1. Intel® Itanium™ dual-processor-based 1U concept server system

However, the nameplate thermal power density, at 700W, greatly exceeds commercially available air-cooled servers – thermal optimization of this system presented significant design challenges. A careful program of numerical simulation, parametric analysis, laboratory prototypes, and experimental verification yielded a system architectural solution based on air-cooling techniques, conventional heatsinks, and the inclusion of a custom-designed radial blower. Areas to be discussed in this paper include:

- Air mover design considerations and the performance of radial blowers
- Partitioned airflow management strategy and other system-level issues
- The advantages of close collaboration between system architects and thermal analysts throughout the design process

2 SYSTEM COOLING STRATEGIES

With the extreme space premium in a 24"-depth (61cm) rack-mount system, the cooling options were severely constrained. The air mover solution would have to supply sufficient velocity and volume flow rate for heat transfer to occur, deliver adequate static pressure to drive air through the system, and direct air to the various processor, memory, and I/O components as required based on predicted power dissipation levels.

2.1 *Airflow requirements*

To predict total flow and pressure performance, an analysis of processor and system cooling requirements was made (De Lorenzo 2001). The processor heatsinks, for example, are a new design optimized to fit within the available baseboard-to-top-cover space and sink the processor thermal load as well as that of the integrated voltage regulator module (VRM). Analysis and experimental verification show that for an aluminum 30-fin heatsink, package case temperature limits were met at worst-case ambient conditions of 35°C and 5000' (1500 m) with approximately 14 cfm (1,130 LFM or 5.7 m/s) airflow at a corresponding pressure drop of 0.38 in. H_2O (95 Pa). It should be noted that this heatsink design achieves processor and VRM cooling in a cost- and mass-efficient aluminum design; very limited increases in thermal performance are realized with copper and/or vapor chamber construction.

The pressure drop noted above, 0.38 in. H_2O (95 Pa), is across the processor heatsink only; conservatively including chassis inlet and exhaust losses, a total system pressure drop of 0.6 in. H_2O (150 Pa) was predicted for initial air mover design studies. Total system airflow could then be determined based on the processor requirements and the remaining thermal load. Approximately 15 cfm (0.42 m^3/min) was required for each processor heatsink, or 30 cfm (0.85 m^3/min) total, and processor and VRM power combined to account for roughly half of the system power budget of 700W. For the remaining power dissipation, targeting a 25°F (14°C) temperature rise yielded a 70 cfm (2.0 m^3/min) total airflow requirement. In addition, airflow distribution within the chassis must direct 30 cfm (0.85 m^3/min) through the processor heatsinks, and preferably deliver a high-velocity flow plume along the memory section on the left-hand side of the baseboard.

2.2 *Air mover selection*

Current strategies in 1U system cooling for Intel® architecture or RISC servers rely either on numerous commercial off-the-shelf (COTS) axial fans, on small ducted radial impellers, or on a combination of the two. As the shortest dimension in the chassis sets the constraint on axial fan diameter, a 1U system (nominally 1.6" or 41mm floor-to-ceiling inside dimension) can only fit a 40mm axial fan. However, an analysis of expected system pressure drop characteristics indicates that 40mm fans are not a good match for this cooling scenario. For example, there is the Delta FUB0412/24VHN fan in a 40x40x28 mm size (Delta Electronics 2001); with 15.8 cfm (0.447 m^3/min) open flow at 9500 rpm and a shutoff pressure of 0.509 in. H_2O (127 Pa), this represents the current state-of-the-art in moderately priced axial fan performance. Even extending this performance envelope through a specific-speed analysis done by Torrington

Research (Aldridge et al. 2001), a solution could require as many as 12 fans throughout the system (Fig. 2). Thus it is apparent that axial fans are not a viable solution for a 1U system of this power density.

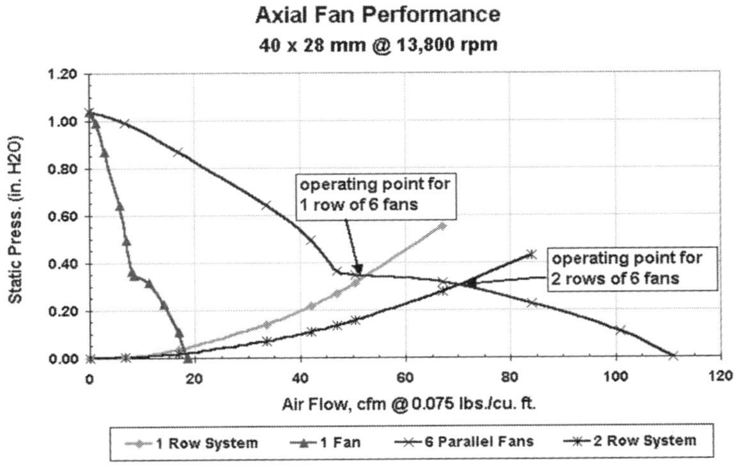

Figure 2. Fan performance vs. system operating point

In contrast, ducted impellers allow larger wheel diameters, higher shutoff pressures, lower rotational velocities, and potentially greater flow rates. Therefore, it is possible to consider a cooling strategy that combines axial fans and blowers to achieve the necessary flow, pressure, and distribution characteristics. However, there are reliability concerns with an approach that utilizes multiple fans, as the number of potential failure candidates drastically reduces MTBF (mean time before failure). The currently planned service reliability of the cooling system is three years continuous operation at an ambient air temperature of 35°C – at an 80% confidence, no more then 0.25% of cooling systems are to fail during this three year period. While this can be considered only a moderately aggressive target for a single air-mover, meeting this goal with multiple fans makes this a potentially cost-prohibitive option. Further, in a 1U chassis it was considered neither necessary nor practicable to include redundancy of the air management system.

Therefore, the selected cooling architecture is a single radial impeller drawing air from the front of the chassis (Fig. 3). A baffle housing around the impeller wheel seals the inlet from the discharge side; ducting and baffle outlet geometry distribute the air as required to the system. After passing over the processors, memory, and chipset components (in parallel paths) the air exhausts through the chassis back panel.

Figure 3. Impeller and dual-discharge baffle

3 DUCTED IMPELLER DESIGN CONSIDERATIONS

Now that the fan requirements had been specified, optimization on blade design, centerline-axis placement, duct geometry, inlet plenum height, and other factors could begin. A detailed instruction on numerical and empirical optimization of radial impellers is beyond the scope of this report; the reader is instead directed to textbooks, the research literature, or companies specializing in this area (e.g. Torrington Research [http://www.trc1.com]). The focus of this paper is on server-specific factors that affect cooling system performance, assuming in advance an impeller design that is acceptable in regards to airflow and pressure requirements.

Even with an impeller that delivers adequate performance, baffle geometry and placement within a server chassis may degrade the impeller operating characteristics to the point that they are no longer acceptable – evolution of the air mover design can require repeated iterations between simulations, prototypes, and empirical models (Fig. 4).

Figure 4. Baffle evolution – simulation, prototypes, and empirical models

3.1 *Other air-mover design considerations*

3.1.1 *Inlet plenum height*

A study of the effect of inlet plenum height on impeller performance also was conducted. The tests showed improved flow rates achievable with increasing plenum height, indicating that inlet restrictions are not trivial (Fig. 5). As a rough first approximation, the negative impact can be minimized if the product of inlet height and width is equal to or greater than the cross-sectional area of the fan inlet plane (less the cross-section of the fan motor). Another partial mitigator that was tested, space allowing, was a sloping face on the scroll baffle that smoothly transitions inlet air from rest conditions to the velocities required at the fan inlet entry plane. Keep in mind, however, that based on the limited height available in the 1U form-factor, any tradeoff between inlet plenum and fan-blade height must be carefully considered.

3.1.2 *Impeller inlet sealing*

Effective sealing between the inlet and discharge sides of the fan wheel is essential for achieving optimal flow and pressure characteristics. Any recirculation or drop in outlet pressure due to a large gap between the spinning impeller and the stationary baffle results in lost performance. Tests indicate at least a 12.5% improvement to open-flow performance just from more effective sealing at this junction – a clearance of no more than 0.050” (1.27mm) should be the design target; labyrinthine seals are another possible design solution.

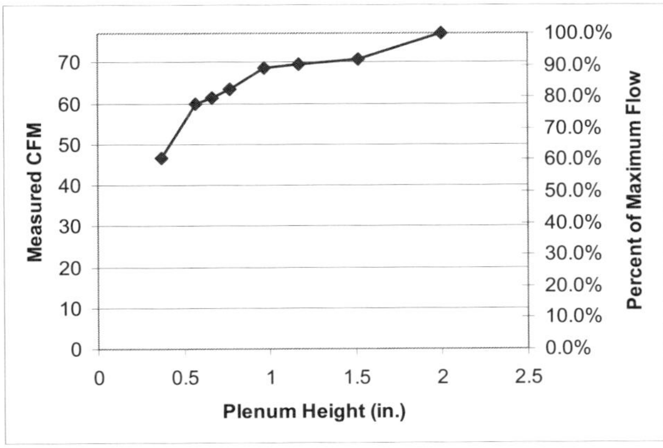

Figure 5. Flow rate reduction with decreasing inlet plenum height

3.1.3 *Intra-baffle obstructions*

While it is desirable to maintain a smooth interior profile to the impeller baffle, this is not possible in all cases. A study that populated components within the confines of the fan baffle indicates that the impact can be minimized if component heights are kept less than 20% of the baffle interior height. Assuming it to be a printed circuit board (PCB) with top-side component placement, the decrease in fan performance is more attributable to reduced inlet plenum height from PCB thickness and PCB-to-chassis standoff than from resistance effects from the components themselves.

3.1.4 *Airflow distribution methods – diverters, screens, turning vanes*

Generic and commercially available radial blowers have the discharge airstream tightly focused, commonly within a cross-section on the order of the fan height dimension. These focused flow characteristics are not suitable for the thermal management of components widely spaced in the lateral (cross-flow) dimension. In point of fact, it was the distribution properties that presented the most challenging design target for the 1U air mover solution (Fig. 6). Careful blending of the impeller baffle to the full system cross-section is the goal, minimizing sharp bends or abrupt changes of cross-section. Turning vanes, walls, or fences also are to be avoided within the scroll baffle geometry – testing has shown that they present a significant inhibitor to flow development and impeller performance. If airflow distribution balancing needs to be achieved in excess of that possible from baffle shaping alone, then the introduction of cross-flow screens or partially open grates has been shown to be effective (O'Connor & Sleicher, unpubl.).

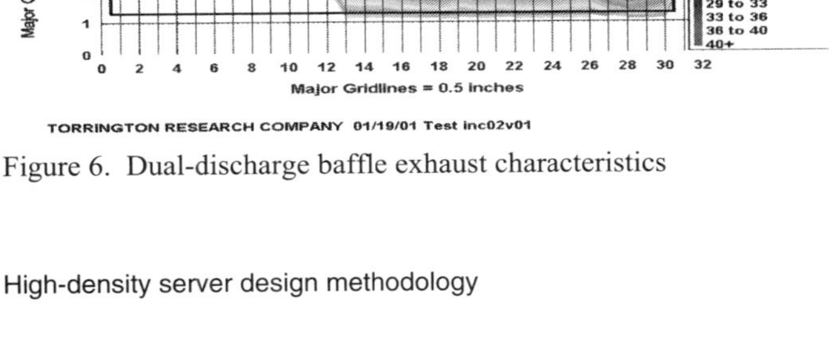

Figure 6. Dual-discharge baffle exhaust characteristics

3.1.5 *Flow path design*

A further list of general design rules for good flow path design has been compiled by Torrington Research (Aldridge et al. 2001):

- Move air through the system as directly as possible
- Keep velocities minimized for the quietest operation
- Avoid sudden expansions or contractions in transition sections
- Avoid rectangular ducts with width to height aspect ratios greater than 6 to 1
- Size filters and heat exchangers appropriately for system requirement
- Avoid obstructions in the flow path especially in high velocity streams or in close proximity to the fan
- Avoid sharp bends, use turning vanes where necessary
- Avoid leakage paths between inlet and discharge of air mover

4 SYSTEM-LEVEL DESIGN CONSIDERATIONS

Once the impeller and baffle meet design objectives for flow, pressure, and airflow distribution, successful thermal management of a 700W 1U server requires careful attention to the remainder of the system. With the packaging density and space constraints of the concept server, there is little margin for error. The complete thermal solution does not dictate non-traditional methods, but it does require significant engineering optimization of the traditional elements of its design.

4.1 *Partitioned airflow strategy*

One element of the server thermal solution that yielded benefits was a partitioned airflow management strategy. In brief, parallel airflow paths were established for the processors, memory, chipset & I/O, and power conversion regions of the chassis (Fig. 7). Given the constrained total airflow available, there wasn't an excess amount of air that could be wasted – each region receives the minimum required to effect cooling of its components. Tuning of airflow distribution was achieved through balancing of the dual-discharge baffle, introduction of flow diverters, and by placement of chassis exhaust vents. For example, there are 8 memory DIMMs along the extreme left-hand edge of the mainboard. Sufficient cool airflow is delivered to the front four elements, but analysis predicted that the rearward DIMMs would not be adequately cooled. Simulation indicated that careful placement of flow diverters would help; benefits included re-energized flow velocity, mixing with non-preheated air, and increased mass flow rate and air velocity over the rear bank of 4 DIMMs.

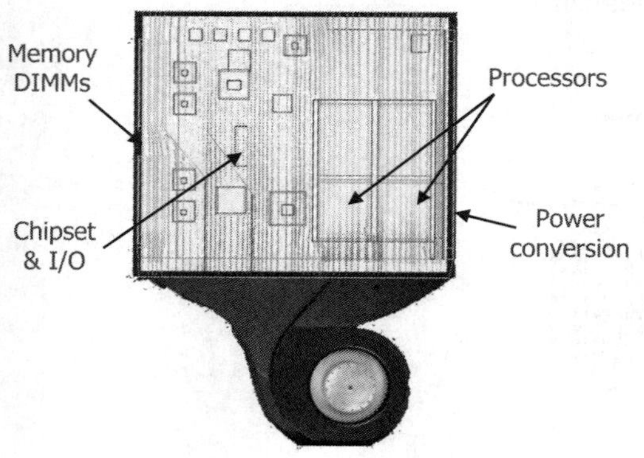

Figure 7. Flow patterns and partitioned airflow strategy

4.2 *Component shrouding and preheat*

While it may sound simplistic to discuss component shrouding, this is a design aspect that should not be ignored. Even with low-power components, low airflow velocities still can result in preheat; depending upon component height or heatsink geometry, there may be flow shrouding as well. Often parts are placed on the PCB for routing efficiency – however, with early feedback from thermal analysis some placement decisions can be modified; simulations show that even moving a component a distance less than its major dimension can still have a significant improvement on thermal performance. Some design improvements can be non-intuitive – the "bow-wake" from a tall heatsink or sheet metal obstruction can interfere with flow over an upstream component.

4.3 *PCB-to-chassis standoff*

With the extreme height limitations of a 1U server, the standoff distance between the chassis bottom tray and the main PCB becomes important. For example, with a 0.140" (3.556mm) standoff, 9% of the volume in the mainboard area is wasted. The impact is that not only do mezzanine connectors and memory DIMMs face a ""low ceiling", but also that processor and component heatsinks must be shortened accordingly. The impact on heatsink performance for a given flow velocity is nearly proportional to the reduction in height; conversely, for a given flow volume, flow resistance increases proportionately to the square of the height reduction. There may also be an airflow short circuit, where air passes under the mainboard rather than cooling top-side components. Thus, through a combination of limited back-side component placement and the selection of the shortest appropriate fasteners, PCB-to-chassis clearance should be minimized – a reduction in this dimension of 0.050" (1.27mm) gains back fully 3% of the available volume in the mainboard area. Sealing the under-board area also can help in directing airflow over top-side components.

4.4 *Mezzanine connector and cabling*

Whether an I/O mezzanine or an add-in PCI card is required in the system to achieve the desired network/storage connectivity, there is always the possibility in a 1U server that boards will be stacked 2-high. Whenever this happens, the most efficient signaling solution is a solid connector, which acts as an impenetrable barrier to airflow along the connector's length. When possible, this should be used as a tool to aid in deliberate airflow partitioning. Likewise, ribbon cables can be a nuisance from an airflow-efficiency perspective. The best bet is to route cables along the chassis sheetmetal when possible; if they must stick up, it is best done in areas of low flow velocity. Keep in mind also that drive, indicator, and power cables going to the front of the server may need to pass through the baffle area; follow the above guidelines to minimize their impact.

4.5 *Chassis outlet vent distribution*

The characteristics of chassis air discharge also should not be neglected. For EMI containment, some sort of perforated metal back panel will be specified. In addition, I/O and power connectors/cables will pierce the back plane. Within the constraints of board layout and routing, shuffling back panel openings to influence upstream flow distribution can be an effective air management tool. While it does not offer the impact of baffle-shaping or flow diverters, at a minimum the proportion of vents in the chassis back wall should match the desired flow distribution from the previously discussed partitioned airflow strategy. The use of wave-guide material, which has not only flow impedance but also EMI containment advantages over perforated panels, should be considered.

4.6 *Flow velocity gradients*

The air mover design calls for approximately 70 cfm (2.0 m^3/min) to go through a 1U system – with a 27 in^2 (174 cm^2) cross-sectional area, this equals an average flow velocity of 370 LFM (1.9 m/s) for an empty chassis; a 33% "void-fraction" pushes average airspeeds to 550 LFM (2.8 m/s).

With directed and partitioned flow from the ducted impeller, localized velocities can be significantly higher than this (e.g. 1,130 LFM [5.7 m/s] over the processor heatsinks). The resultant flow velocity gradients create airflow shearing; even though air viscosity is low, the energy loss just adds another inefficiency to the air movement process. Thus, grouping mainboard components based on airflow requirements, or specifying heatsinks to achieve the same goal, can reduce the variation of air velocity in the cross-flow direction, yield smoother flow patterns, and minimize the acoustic signature.

4.7 *Drive cooling considerations*

Even with only one hard disk drive in the server, nearly a third of the chassis frontal area is consumed. Designs that utilize dual hard disks as well as floppy/CD-ROM drives face even greater constraints on the air inlet side. Not only do inlet area and fan starvation become concerns, but also achieving adequate drive cooling is problematic; often designs are observed that neglect the cooling requirements of the installed disk drives entirely. The approach taken here was to allow inlet air to pass over the single drive – the use of a perforated, swappable carrier enabled sufficient airflow around the drive's case. Future migration to more compact drive form factors will alleviate some of the space/volumetric concerns in the valuable area at the front of the server.

5 CONCLUSIONS

In the 1U form factor, board and component densities greatly impact airflow, and consequently thermal performance. Design and optimization of the air mover system requires attention not only to pressure and flow characteristics, but also to flow distribution properties. Also, arbitrarily high flow rates still do not deliver adequate thermal performance if the flow is not directed to the correct areas – this demonstrates the advantages of a partitioned airflow strategy. Finally, total flow rates and flow distribution properties are very sensitive to board and component configurations, fan placement, and baffle geometry. Therefore, optimizing the system for flow and thermal performance is an iterative process, most easily done before major layout or architectural properties have been fixed. Close collaboration between thermal analysts, board layout engineers, and system architects has proven the most effective method for optimizing the complete server design.

ACKNOWLEDGEMENTS

The author wishes to thank Barrett Faneuf, Ven Holalkere, and Stephen Montgomery of the Intel Corporation, as well as Torrington Research and Incep Technologies, for their collaboration on the research described in this paper.

REFERENCES

Aldridge, T., Fite, B., Holalkere, V., O'Connor, J., & DiBene, T. 2001. Concept technologies for high density Intel® Itanium™ processor family servers. *Intel Developer Forum, San Jose, CA, 26 February – 1 March 2001.*

De Lorenzo, D.S. 2001. High density server thermal research. *Intel Technology Symposium, Seattle, WA, 27-28 September 2001.*

Delta Electronics 2001. Data sheet for FUB 40x40x28 MM Series Superflo Fans.

O'Connor, J.F. & Sleicher, B. Optimized inlet/diffuser for Circusink blower and case study of blower operating in a McKinley 1U server. *Torrington Research Company, Torrington, CT.*

Thermal Challenges in Next Generation Electronic Systems, Joshi & Garimella (eds)
© 2002 Millpress, Rotterdam, ISBN 90-77017-03-8

A flexible approach to cooling system design

C.M. Hagg, H. Bucsich, M. Harrer, A. Loidl, T. Raeva, A.H. Whitehead & M. Schreiber
Funktionswerkstoffe F & E GmbH, Eisenstadt, Austria

Keywords: heat exchanger, liquid coolant, removable template, copper, galvanoforming

ABSTRACT: A technique for producing seamless copper heat exchangers was investigated. Two different types of heat exchangers were produced, for component-coolant and coolant-air heat transfer. By connecting the heat exchangers in a loop with appropriate coolant transfer lines a complete heat transfer system, suitable for cooling high power electronics, was constructed. The thermal properties of the system were investigated. In addition, the versatility of the technique was explored by preparing heat exchangers with a variety of geometries.

1 INTRODUCTION

In general the cooling needs of small, high-power electronic components are met with passive heat sinks or with forced air-cooling. However, more powerful components are continually being developed which evolve increasingly more waste heat (Costlow 1998, Hachman 1998). In addition, there is a general drive to higher component densities which aggravates the problem of thermal management (Mayer 1997). In order to prevent the components from exceeding the optimum operating temperatures more effective heat dissipation techniques are required. Circulating liquid-filled cooling systems afford one solution to this problem (Frey, Stuck & Zehringer 1999, Soule 1999). A few such systems already exist, but are generally constructed by classical engineering techniques, which do not allow for a great flexibility of design (Biancomano 1999). In contrast, the approach adopted here was based on a removable templating technique, which was anticipated to enable seamless metallic structures with complex geometries to be formed relatively easily.

One of the main objections to using water-based coolant systems with electronic components is the possibility of leakage, which could cause damage to the electronics and may represent a safety hazard (at the least the electronics could then go on to overheat if not properly cooled). The formation of seamless heat exchange bodies was therefore perceived as a critical feature in reducing the probability of a leakage occurring. A further prime requisite was that the heat exchange bodies should be coated with, in excess of, a given thickness of metal over the entire surface - to ensure no "weak spots", which might be liable to rupture.

The objectives of this study were to explore the versatility of the templating technique in creating heat-exchanger bodies and to measure the thermal performance of closed-loop systems containing assemblies of the said bodies.

In order to examine the thermal performance of the heat exchangers a commercial product was tested under similar experimental conditions.

2 EXPERIMENTAL

2.1 *Heat-exchanger production*

Heat exchanger bodies to be placed in contact with the surface to be cooled were denoted as "front ends". Liquid-air heat exchanger bodies were also constructed and denoted as "back ends". Both front and back ends were constructed using the templating technique, but with differently shaped templates. The "back end" designs were kept constant, consisting of a structure with several parallel pipes connected at both ends to single inlet and outlet tubes. The parallel pipes were of smaller radii than the inlet and outlets, such that an approximately uniform cross-sectional area was obtained for any given section perpendicular to the central axis.

Various designs of "front end" were examined by using different templates. The range of templates reflected the range of electronic component packages that may be considered for cooling, and also provided a means to test the limits of the production process.

Templates were prepared from a suitable polymer by an injection moulding technique. The templates were then cleaned by immersion in alkaline and acidic baths followed by application of a thin copper layer using a commercial chemical process with a palladium catalyst. The thin copper layer was reinforced by a subsequent conventional copper-electroplating step. Electroplating allowed the average thickness of copper to be easily controlled (simply by adjusting the amount of charge passed). A thickness of 500 μm was typically employed. After electroplating, the templates were removed to leave hollow copper structures. The voids left in place of the templates provided channels for the coolant liquid.

2.2 *Physical examination of the components*

Heat exchanger bodies were individually filled with air at a pressure of 200 kPa and then isolated from the air supply by suitably arranged taps. The pressure in the heat exchanger was continuously monitored for 15 minutes by a computerized monitoring system, connected to a pressure sensor.

In addition, the metal coatings were examined by microscopy. Optical microscopy was used to inspect the surfaces of the heat exchangers, while scanning electron microscopy (SEM) and optical microscopy were used to study cross-sectioned segments of the heat exchangers. To prepare the cross-sections, the heat exchangers were cut, embedded in resin and then polished to a smooth finish.

2.3 *Measurement of thermal parameters*

Experimentally the measurement apparatus shown schematically in Figure 1 was used to determine thermodynamic parameters of the cooling systems. T_x indicates temperature measurement by a thermistor at position x. P_{drop} is the differential pressure measurement at the inlet and outlet fluid lines to the front end, CSF. The flow of coolant liquid was maintained by a pump and monitored by an in-line sensor, indicated as "flow rate" in Figure 1. Similarly airflow across the "back end" assembly, CSB, was also monitored. CSB consisted of a parallel arrangement of "back ends", inlet and outlet manifolds and a small electric fan.

Between CSF and the heat source a thin layer of thermally conducting paste was applied to ensure a good thermal contact. In a practical application such a paste would also be used. However, it should be considered that the thermal resistance of the paste might not be insignificant in comparison with the rest of the system. Therefore, use of pastes with higher thermal conductivities may make a reasonable difference in the system performance.

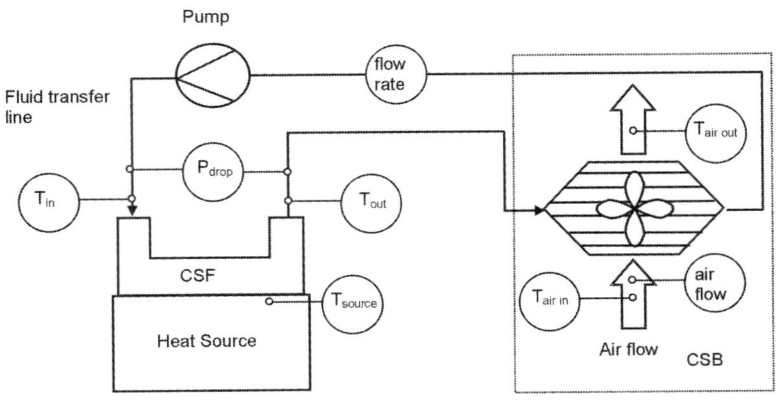

Figure 1. Schematic representation of the experimental setup used to determine thermodynamic parameters of a cooling system

The heat source was controlled via a temperature control unit (Eurotherm) which was connected to a computer. The computer also received data from each of the sensors (Fig. 1) and the pump was under software control.

In order to calculate thermal parameters the heat source temperature and coolant flow rate were set to predetermined values and the system allowed to equilibrate before data points were recorded. A series of data points were measured with different values of flow rate, from 200 - 3000 ml min^{-1}, and heat source temperature in order to define the thermal performance of a given CSF.

The fan speed was fixed and the airflow rate was found to remain constant during the measurements, at ca. 23 l s^{-1} (more than 400 times greater than the liquid flow rate).

Heat dissipation, P, was calculated according to:

$$P = (T_{out} - T_{in})\, C_{CL} v_{CL} \tag{1}$$

where C_{CL} = volumetric heat capacity of the cooling medium; T_{out} and T_{in} = the coolant temperatures at the outlet and inlet of CSF respectively; v_{CL} = the flow rate of coolant.

The heat transfer coefficient, K, of CSF was given by:

$$K = \frac{P}{\Delta T^* A} \tag{2}$$

where A = contact area between the cooler and heat source; ΔT^* = logarithmic mid-point temperature given by (Schlünder 1974):

$$\Delta T^* = \frac{T_{out} - T_{in}}{\ln\left(\dfrac{T_{HS,corr} - T_{in}}{T_{HS,corr} - T_{out}}\right)} \tag{3}$$

where $T_{HS,corr}$ = the corrected surface temperature of the heat source, which was determined by considering the temperature gradient between the thermistor position and the surface such that:

$$T_{HS,corr} = T_{HS} - \Delta T_{corr} \tag{4}$$

with

$$\Delta T_{corr} = \frac{d\,P}{A\,\lambda} \tag{5}$$

where d = distance from the measuring point to the surface; λ = Thermal conductivity of the heat source material. Such a correction was necessary, as the heat source possessed a temperature gradient due to the location of the heating element near the face opposite the heat exchanger.

The thermal resistance, R_H, of CSF was given by:

$$R_H = \frac{\Delta T^*}{P} \tag{6}$$

3 RESULTS AND DISCUSSION

Figure 2 shows three different designs of front ends, with none, one and four mounting holes, respectively. The geometric contact area, A, of the front end without mounting holes (left, Fig. 2) was 6.25 cm^2 (1 square inch).

Additional a front end with 25 cm^2 was constructed and compared with a similarly sized, commercial heat exchanger (Sockel-Kühler, LiquidPower).

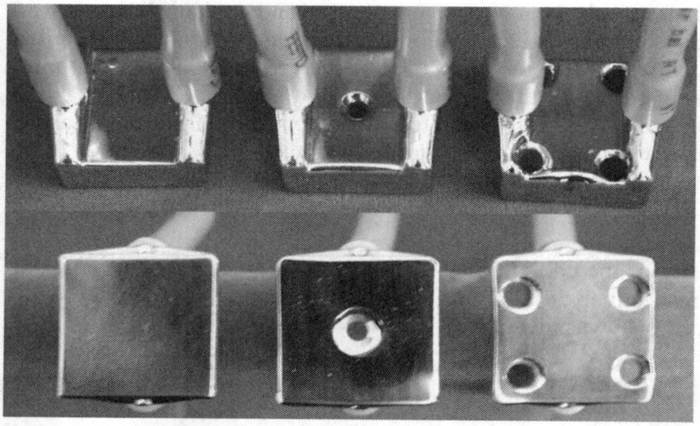

Figure 2. Front end heat exchangers produced by the templating technique

3.1 *Physical examination*

Microscopic examination also revealed that with optimized chemical coating conditions a uniform, pinhole free copper coating could be reproducibly obtained. It was found that changing template materials required slightly different treatment conditions (longer or shorter rinsing times, and so on).

SEM on many cross-sections also failed to detect any pores or inclusions in the copper. All of the samples consisted of dense copper layers, but the thickness of deposit was slightly influenced by the template geometry and location in the plating baths. To some extent these problems were overcome by additional sacrificial anodes in the plating baths. Figure 3 shows an image of a cross section with detail of the plating thickness around a mounting hole.

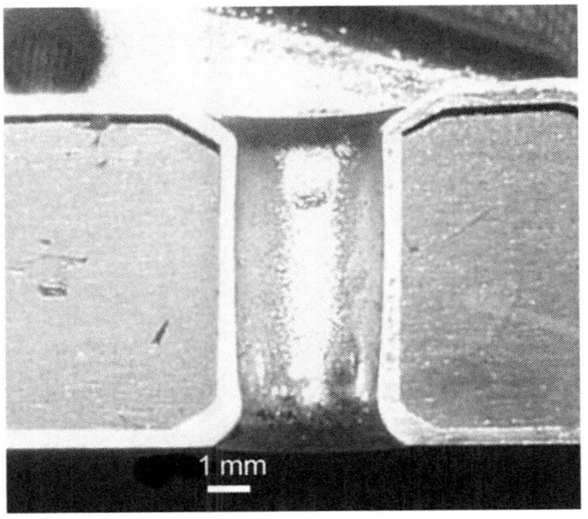

Figure 3. Image of a cross-section through a front end (template not removed), showing detail around a mounting hole in a front end

The plating thickness at various points in the hole (Fig. 3) varied from ca. 400 - 800 µm. However, the variation in plating thickness was rather non-linear with depth through the hole, being thicker at the corner and then of rather constant thickness through the hole. This suggests that holes with higher aspect ratios may successfully be plated with a sufficiently thick copper layer.

The template quality was critical in producing good-quality front ends. In particular, the template manufacturing process was optimized to provide smooth surfaces, without rims or whiskers. When produced around the optimized templates, front ends proved leak-free when filled with air at 200 kPa.

3.2 *Thermal properties*

Thermal properties of the three front ends with 6.25 cm^2 contact area, shown in Figure 2, are given in Figure 4. As can be seen the heat exchangers with mounting holes had slightly lower heat transfer coefficients than that without. However, the heat transfer coefficients were similar for the two front ends with mounting holes.

From Figure 5 it may be seen that the pressure drop across the front ends increased in a non-linear fashion with flow rate. Also, the pressure drop at any given flow rate increased with the number of mounting holes. However, the increase in pressure drop caused by the increased complexity of the front end geometry was relatively small (<2.5 kPa over the range of flow rates). An alternative representation of the data is given in Figure 6, where the thermal resistance is displayed against pressure drop.

Incorporation of mounting holes led to a higher thermal resistance and higher pressure drop at a given flow rate (from Figs 4, 5). Therefore, at a given pressure drop (above 1 kPa) there was a considerable difference in the thermal resistance (Fig. 6). However, it should be noted that the pressure drops are relatively low, especially with respect to the pressure test (up to 7 kPa in operation, compared with 200 kPa pressure testing).

What may also been seen from Figure 6 is that the three front ends had low thermal resistances even with a pressure drop of 1 kPa. For example, an electronic component with a maximum operating surface temperature of 80°C, could dissipate 350 W using the front end with no mounting holes (1 kPa pressure drop, 35°C coolant temperature), or 56 W cm^{-2}. For a front end with 4 mounting holes under similar conditions (80°C surface temperature, 1 kPa presssure drop, 35°C coolant temperature) 210 W or 33 W cm^{-2} could be dissipated.

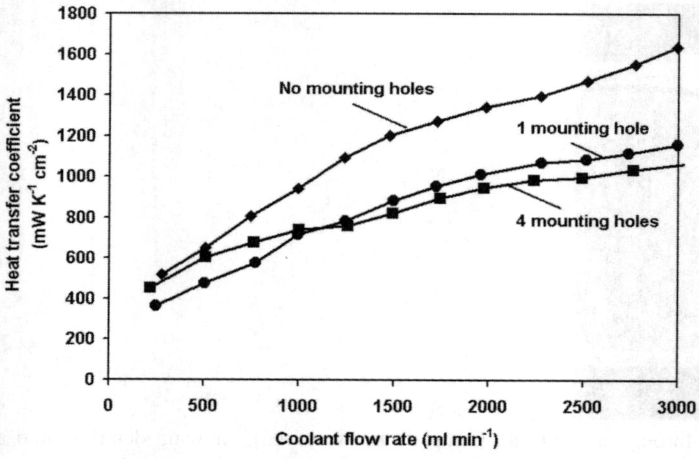

Figure 4. Graph showing the heat transfer coefficients of three front ends as a function of flow rate.

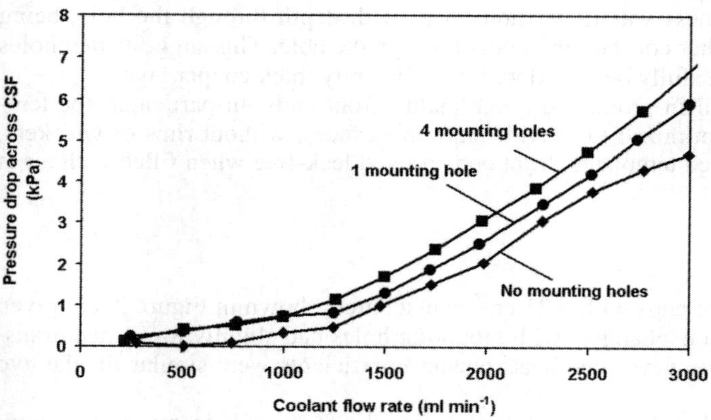

Figure 5. Pressure drop in three front ends as a function of coolant flow rate.

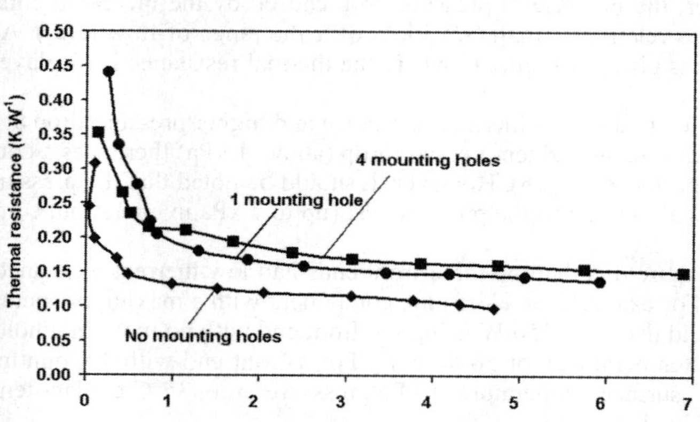

Figure 6. Thermal resistance as a function of pressure drop across three types of front end.

The performance of a 25 cm^2 front end was compared with that of a commercial product of equal contact area. The thermal resistances are shown as a function of pressure drop in Figure 7, from which it may be noted that the resistance of the commercial product at any given pressure drop was significantly higher than the front end. This probably reflects the more complex internal geometry of the commercial device, and the use of aluminum instead of copper (which has roughly twice the thermal conductivity).

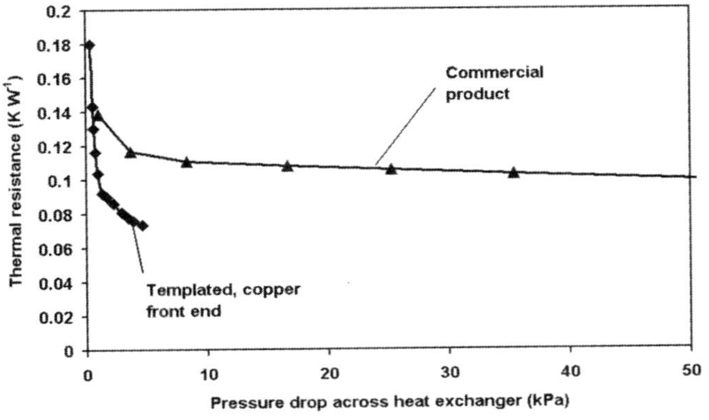

Figure 7. Thermal resistance as a function of pressure drop for two heat exchangers with 25 cm^2 contact area.

4 CONCLUSIONS

It was found that seamless heat exchangers could be made with a range of forms by a templating technique. Small diameter holes increased the probability that the copper would not be plated in sufficient thickness to prevent leakage. However, the technique may be seen as a very useful tool in producing seamless, hollow metal bodies with complex geometries.

From thermal measurements it was apparent that the inclusion of one to four mounting holes through the heat exchanger body caused only a slight decrease in heat transfer coefficient and a slight increase in pressure drop, for a given coolant flow rate. However, the thermal resistance was strongly dependent on the front end geometry as a function of pressure drop.

A 25 cm^2 contact area front end had significantly lower thermal resistance than a commercial product at a given pressure drop. This reflected the less complex internal geometry and choice of material with a higher thermal conductivity in the templated front end.

Large amounts of heat could be removed from relatively small surface areas using a system with front and back end heat exchangers and a circulating liquid coolant.

REFERENCES

Biancomano, V. 1999. Product week: liquid-cooled cold plate beats heat from IGBT modules. In *Electronic Engineering Times* 15 March: 71.

Costlow, T. 1998. Hot chips push heat sinks to way cool. In *Electronic Engineering Times* 14 September: 119.

Frey, T., Stuck, A. & Zehringer, R. 1999. *Liquid cooling device for a high-power semiconductor module*. US Patent 5978220.

Hachman, M. 1998. Thermal management: heat wave! - Today's PC chips require lots of cool. In *Electronic Buyer News* 17 August: 35.

Mayer, J. H. 1997. Thermal management: battling a longtime enemy - heat. In *Electronic Buyer News* 24 November: 75.
Schlünder, E. U. 1974. Einführung in die Lehre von der Wärmeübertragung. In *VDI-Wärmeatlas : Berechnungsblätter f. d. Wärmeübergang*. A1-A32. Düsseldorf: VDI-Verlag GmbH.
Soule, C. 1999. Heat sinks help electronic devices keep their cool. In *Machine Design* 19 August Vol. 71.

Thermal Challenges in Next Generation Electronic Systems, Joshi & Garimella (eds)
© 2002 Millpress, Rotterdam, ISBN 90-77017-03-8

Evaluation of 1-D & 2-D models for thermal insulation thickness calculation in electronic packaging

S. Moghaddam, M. Rada, A. Shooshtari, M. Ohadi
University of Maryland at College Park, College Park, MD, USA

Y. Joshi
Georgia Institute of Technology, Atlanta, GA, USA

Keywords: resistance network analysis, orthotropic material, heat spreading, numerical modeling, electronic packaging, multi-chip module (MCM)

ABSTRACT: The objective of this study is to evaluate the use of 1-D and 2-D compact heat transfer models for thermal design, optimization, and performance evaluation in electronic packaging. A model for heat spreading in orthotropic materials is developed. A resistance network analysis for calculation of heat transfer rate and junction temperatures has been carried out. A refrigeration cooled server package is used to illustrate the methodology. Heat transfer in the package components is modeled. Results of the analytical model and resistance network analysis are compared with a numerical solution. Capability of the analytical model in predicting the thermal field is discussed, and the effectiveness of using 1-D and 2-D models in thermal design and optimization of electronic packages is demonstrated.

1 NOMENCLATURE

a	Radius of circular source, [m]
A_{PCB}	Printed circuit board area, [m^2]
A_{MP}	Module projected area, [m^2]
b	Heat spreading distance, [m]
h	Convective heat transfer coefficient, [W/m^{2}C]
H	A dimensionless factor found from Kennedy chart
k	Thermal conductivity, [W/m . C]
k_z	Axial thermal conductivity, [W/m . C]
k_r	Radial thermal conductivity, [W/m . C]
l	Rectangle side length, [m]
P	Evaporator cooling capacity, [W]
Q	Chip total heat dissipation, [W]
q	Heat transfer rate, [W]
r	Radial coordinate, [m]
R	Spreading resistance, [C/W]
t	PCB thickness, [m]
T	Temperature, [C]
T_1	Circular source temperature; top insulation temperature, [C]
T_2	Side and base temperature of cylindrical model; Back insulation temperature, [C]
T_3	Board temperature at the outer edge of the module, [C]
T_∞	Ambient temperature, [C]
T_{base}	Base temperature of circular fin, [C]
T_{mean}	Circular fin mean temperature [C]
$T(r)$	Circular fin radial temperature [C]

w Cylindrical model height, [m]
z Axial coordinate, [m]
Δx_1, Δx_2 and Δx_3 Top, bottom and sides insulation thickness, [m]

Subscripts
r Radial direction
z Axial direction
o outer radius
i inner radius

Superscripts
$*$ Transformed parameters

2 INTRODUCTION

Providing a three-dimensional model for thermal and flow fields in electronic systems is time consuming, especially for early stages of design when the geometric parameters are being optimized. For quick estimate of the heat flow paths and junction temperatures in electronic packages, the concept of thermal resistance networks is commonly employed (e.g. Guenin et al. (1995)). Other authors have provided alternate modeling methodologies. Nigen (1992) has developed a thermal methodology for printed circuit boards (PCBs) in order to roughly analyse a conjugate conduction/convection problem. Furmanczyk (1998), and Franke (1999) provided similar models for thermal analysis of electronic components. Three issues of primary concern in these models are:
- Heat spreading from heat source through orthotropic media such as ceramic substrates and printed circuit boards containing layers of metallization;
- Simulation of individual heat sources on substrates;
- Modeling of board and ambient heat exchange.

A number of studies have addressed various aspects of these issues (e.g. Culham et al. (1998, 2000), Huang et al. (1993), Lemczyk et al. (1988), and Teertstra (1997)). The current study provides a model for multi-chip modules (MCMs) thermal design and optimization. In the following the details of the heat transfer model is given first, followed by a case study that demonstrates its application to a refrigeration cooled server package.

2.1 *Heat spreading in homogeneous and orthotropic materials*

Heat spreading in electronic packages refers to multi-dimensional conduction that occurs when two regions with different heat transfer cross-section areas, are in contact. A common example is the case where heat is conducted from a chip through a substrate or a PCB. Several closed form solutions have been provided to model the spreading resistance (Cooper et al. (1969), Kennedy (1959), and Lee et al. (1995)) in simplified configurations. These analytical solutions are generally restricted to homogenous, isotropic-materials. Electronic packaging structures are often anisotropic. For example, the presence of metallization layers within substrates makes the in-plane conduction properties very different from out-of plane.

Kennedy (1959) obtained the thermal resistance for a circular discrete, uniform surface heat flux source of radius a, placed at the center of a cylinder b of a homogeneous material ($K_z = K_r$=1) as:

$$R = \frac{H}{a\pi} \tag{1}$$

Where H, a function of a/b and w/b (Fig. 1), is available in closed form, and also available as the Kennedy charts (Tummala et al.). In the following, a transformation is used to extend the Kennedy solution to orthotropic materials:

$$r^* = \frac{r}{\sqrt{k_r}}, \quad z^* = \frac{z}{\sqrt{k_z}} \tag{2}$$

Where r^* and z^* are transformed coordinates. To show the applicability of this transformation, consider the heat conduction equation in an orthotropic material in cylindrical coordinates:

$$\frac{\partial}{\partial z}(k_z \frac{\partial T}{\partial z}) + \frac{1}{r}\frac{\partial}{\partial r}(k_r r \frac{\partial T}{\partial r}) = 0 \tag{3}$$

Using the above transformation, Equation 3 reduces to:

$$\frac{\partial}{\partial z^*}(\frac{\partial T}{\partial z^*}) + \frac{1}{r^*}\frac{\partial}{\partial r^*}(r^* \frac{\partial T}{\partial r^*}) = 0 \tag{4}$$

Boundary conditions are transformed accordingly. For example, for the top circular heat source in figure 1, the transformed form of boundary condition is:

$$Z^*=0,\ 0 < r^* < a^*:\ T=T_1 \tag{5}$$

Where a^* is the transformed radius of circular heat source given by:

$$a^* = \frac{a}{\sqrt{k_r}} \tag{6}$$

This transformation enables us to use the Kennedy solution in the transformed domain. The transformed thermal resistance can be determined as follows:

$$R^* = \frac{T_1 - T_2}{q^*} = \frac{H^*}{\pi a^*} \tag{7}$$

Where H^* is a function of a^*/b^* and w^*/b^*. The following expression gives the heat transfer rate through the circular source in transformed domain:

$$q^* = \int_0^{a*} \frac{\partial T}{\partial z^*} 2\pi r^* dr^* \tag{8}$$

While the actual heat transfer rate, q, is given by:

$$q = \int_o^a k_z \frac{\partial T}{\partial z} 2\pi r dr \tag{9}$$

Using Equations 2, 8, and 9, the relation between actual heat flux and transformed heat flux can be determined.

$$q = k_r \sqrt{k_z} q^* \tag{10}$$

Therefore, the relation between the actual thermal resistance, R, and the transformed thermal resistance, R^*, can be found as follows:

$$R = \frac{T_1 - T_2}{q} = \frac{1}{\sqrt{k_z k_r}} \frac{T_1 - T_2}{q^*} \tag{11}$$

Equation 11 implies the following relations between actual and transformed thermal resistances.

$$R = \frac{1}{\sqrt{k_z k_r}} R^* \tag{12}$$

$$R = \frac{H^*}{\pi \sqrt{k_z k_r a}} \tag{13}$$

2.2 *Printed circuit board and ambient heat exchange*

The extended portion of the printed circuit board (PCB) has been simulated as a circular fin. The following equations determine the equivalent internal and external radius of this circular fin that extends out of the module corners.

$$r_o = \sqrt{\frac{A_{PCB}}{\pi}} \quad , \quad r_i = \sqrt{\frac{A_{MP}}{\pi}} \tag{14}$$

Using this simplified geometry, total thermal resistance of the fin and its surrounding ambient has been determined using Bessel's solution for a circular fin with convection boundary conditions (Guenin 1995).

$$\mathrm{T(r)} - \mathrm{T}_\infty = (\mathrm{T}_{\mathrm{base}} - \mathrm{T}_\infty)\frac{\mathrm{K}_1(\mathrm{Mr_o})\mathrm{I}_0(\mathrm{Mr}) + \mathrm{I}_1(\mathrm{Mr_o})\mathrm{K}_0(\mathrm{Mr})}{\mathrm{K}_1(\mathrm{Mr_o})\mathrm{I}_0(\mathrm{Mr_i}) + \mathrm{I}_1(\mathrm{Mr_o})\mathrm{K}_0(\mathrm{Mr_i})} \tag{15}$$

Where M is defined as follows:

$$\mathrm{M} = \sqrt{\frac{2\mathrm{h}}{\mathrm{kt}}} \tag{16}$$

By integrating for heat dissipation over the surface, using Equation 15 for local temperature difference, the following expression for the thermal resistance from module to ambient yields:

$$R = \left(\frac{1}{2\pi r_i kt}\right)\frac{K_1(Mr_o)I_0(Mr_i) + I_1(Mr_o)K_0(Mr_i)}{K_1(Mr_i)I_1(Mr_o) - I_1(Mr_i)K_1(Mr_o)} \tag{17}$$

The mean temperature of the PCB is calculated using the following equation:

$$\mathrm{T_{mean}} - \mathrm{T}_\infty = \frac{2\int_{\mathrm{r_i}}^{\mathrm{r_o}}(\mathrm{T(r)} - \mathrm{T}_\infty)2\pi \mathrm{rdr}}{2\pi(\mathrm{r_o^2} - \mathrm{r_i^2})} \tag{18}$$

$$T_{mean} = T_\infty + \frac{2r_i(T_i - T_\infty)}{M(r_o^2 - r_i^2)}\left[\frac{K_1(Mr_i)I_1(Mr_o) - I_1(Mr_i)K_1(Mr_o)}{K_1(Mr_o)I_o(Mr_i) + I_1(Mr_o)K_o(Mr_i)}\right] \tag{19}$$

Flow is assumed to be laminar over the board and the natural convection heat transfer coefficient at constant wall temperature has been used in the above equations.

In order to find an appropriate magnitude of T_{mean}, a trial and error procedure has been employed. This model, along with thermal resistances for the other components that are available analytically, can be used to complete the thermal resistance network for a MCM module. An example to apply this is presented below.

3 EVALUATION OF THE DEVELOPED MODEL

In order to evaluate the developed model for MCMs' thermal analysis, a refrigeration cooled multi-chip module of an IBM CMOS Server has been considered (http://www.electronics-cooling.com/html/2000_sep_a1.html). Figure 2 shows the details of the package, in which chips have been simulated as uniform heat sources. The specifications of the components are given in Table-1.

The objective of this thermal modeling is to simplify the design procedure of a condensation free package. In order to avoid the condensation, insulation has been used on the outer surface of the evaporator/module, as well as on the board's backside. Operating ambient conditions dictate a minimum surface temperature of 21 °C (ambient dew point).

4 RESISTANCE NETWORK

A thermal resistance network, to analyze the system temperature and cooling power consumption, as a function of insulation thickness, is presented in Figure 3. Three design options were analyzed. First, a minimum insulation thickness of 3 mm was considered for the entire module as well as the backside of the board. The analysis was repeated for the second design option with 20 mm for the insulation thickness. In the third option, insulation thickness on the module was 20 mm and was 3 mm on the backside of the board. The results of the resistance network solution for temperature and cooling power consumption are given in Table 2.

5 NUMERICAL MODELING

A numerical solution is provided to evaluate the results of the analytical model. *IcePak* software (Fluent, Inc. & ICEM-CFD Engineering – Lebanon, NH) was used as the numerical solver. First, in order to verify the grid size independency of the results, airflow over the board with constant temperature was modeled numerically. The result was compared with the similarity solution for laminar flow natural convection along a vertical plate.

The circuit board with dimensions 56.0 (W) ×45.7 (H) ×0.6 (t) cm^3 was simulated as a cold vertical plate in IcePak software. The plate and ambient temperatures were set at 17.68°C and 35°C, respectively. Two solutions have been prepared in order to study the effect of the grid size on the numerical solution and the results were compared with the similarity solution. The result shows an error of 3.26 percent for the coarse mesh and 2.83 percent for the fine mesh.

In the second step the entire module and board were simulated. The evaporator was modeled as a high thermal conductivity block. A numerical grid structure with 294694 nodes was generated in the cabinet that contains the module and board assembly. Figure 4 demonstrates this configuration. The numerical solution was provided for the third design option. Using the data given in Table 3 the evaporator cooling capacity was adjusted to 934.32 W while the total heat generation rate was 925 W and the ambient temperature was 35 °C. The results of the numerical solution are presented in Figure 5.

5.1 *Numerical results*

The evaporator has an average temperature of 10.13 °C and its minimum temperature is 9.79 °C. The chip average temperature is 16.66 °C. The maximum temperature is found on the top corners of the board and is about 34 °C. Figure 5 shows the temperature contour in a cut plane of the entire module assembly. As we can see, the temperature field from the evaporator to the chips and substrate is more or less one-dimensional. The most deviation from the one-dimensional aspects occurs on the top insulation, where the temperature varies from about 24 °C on the top corner (Point

"A") to 18 °C at the center (Point "B"). The average temperature of the top insulation is about 21.34 °C. These results show that condensation will take place at the center of the top insulation. The temperature variation in the back insulation is rather small and its mean value is 26.6 °C, which is above the dew point.

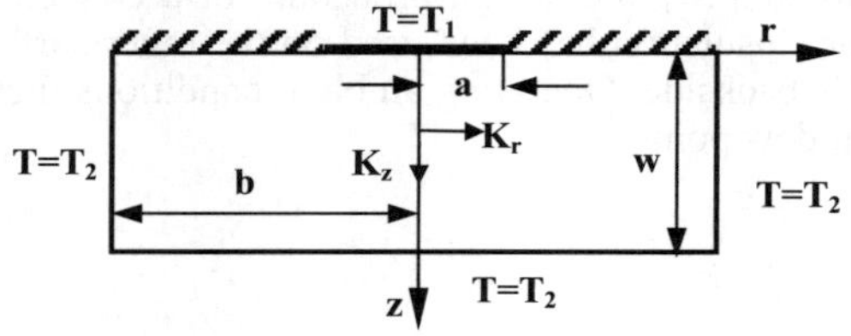

Figure 1. Schematic of the model

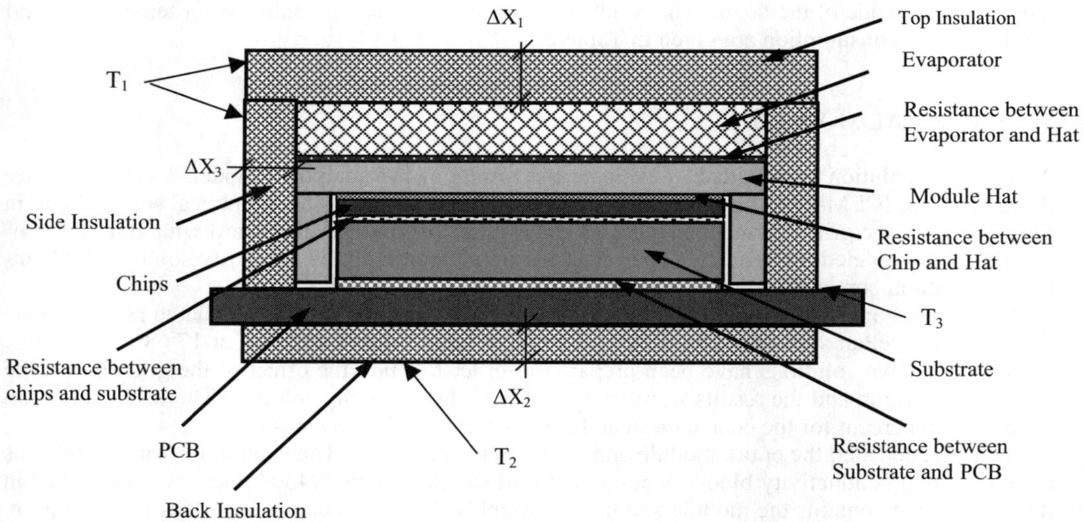

Figure 2. Multichip module (MCM) configuration

R_1: Resistance between chip and substrate
R_2: Substrate resistance
R_3: Resistance from substrate to the board
R_4: Board spreading resistance from resistance R_3 to the end of the module
R_5: Board perpendicular resistance
R_6: Insulation resistance at the back of the board
R_7: Convective resistance between the board back insulation and the ambient
R_8: Resistance from module to ambient, through the board (includes both the conductive and convective resistances)
R_9: Resistance from the chip to the hat
R_{10}: Hat thermal resistance
$R_{11,i}$: Top insulation resistance (index "i" represents the five different sides of the top insulation)
$R_{12,i}$: Average convection resistance from insulation to the ambient (index "i" represents the five different convection heat transfer coefficient on different sides of the top insulation)
R_{13}: Resistance between hat and evaporator
P: Evaporator cooling capacity
Q_{chips}: Chips' total heat dissipation

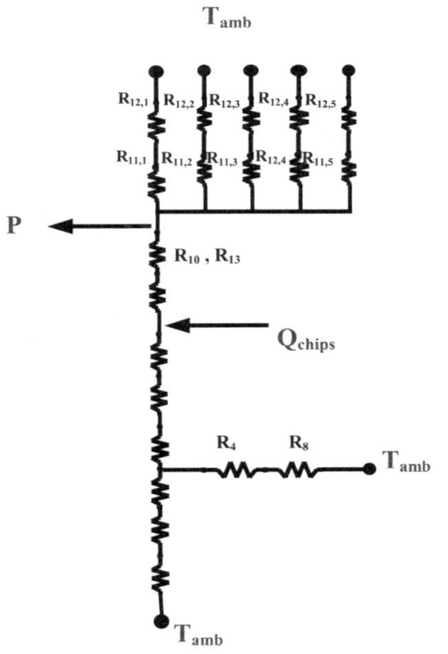

Figure 3. Resistance network for the module and board

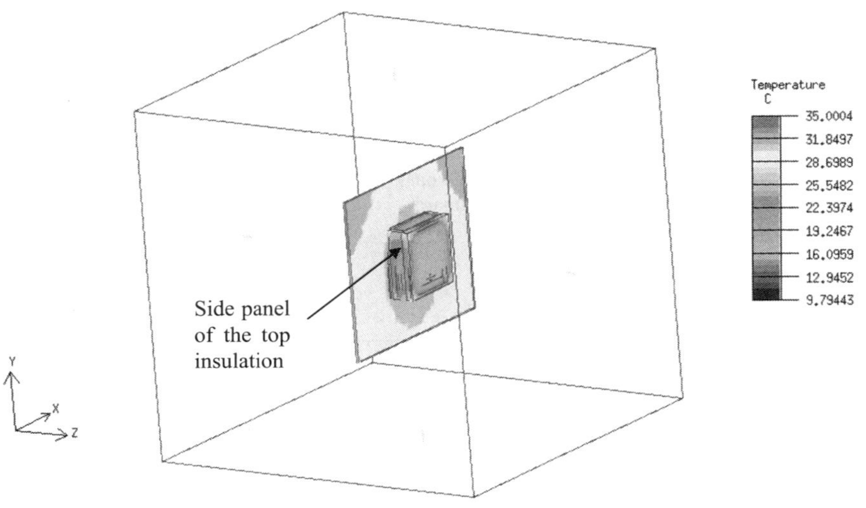

Figure 4. Numerical results for the Module/Board in natural convection

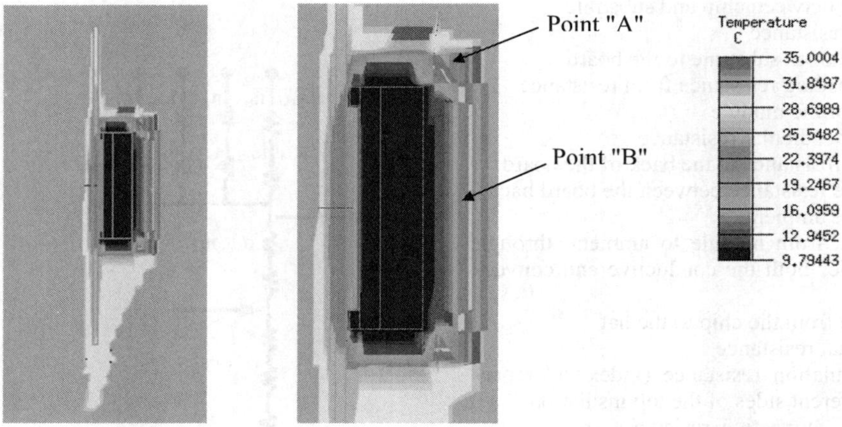

Figure 5. Temperature distribution in module/board and air

Table 1. Module components specifications

Component	Parameter	Specification
Chip	Chip size Power/chip Chip spacing (center to center) Resistance between chip and hat Resistance between chip and substrate	15 mm square x 0.38 mm thick 25×37 watts 23 mm 0.20 C/W 0.18 C/W
Substrate	Substrate size Substrate thermal conductivity Resistance between substrate and board Distance between substrate and board	127 mm square x 5 mm 0.0038 W/C-mm 0.7 C/W 2 mm
Module	Module hat thickness Module hat material	11.8 mm Pure copper
Evaporator	Evaporator size Resistance between evaporator and hat	25 mm thick and fits plan form of MCM top 0.0015 C/W
Printed Circuit Board	Board in plane thermal conductivity- Board perp. thermal conductivity Board Size	20 W/mK 0.5 W/mk 56 cm x 45.7 cm x 6 mm
Insulation	Thermal conductivity	0.1 W/mK

Table 2. Results of the resistance network analysis.

Option	$T_{(amb.)}$ °C	$T_{(evap.)}$ °C	$T_{(Chip)}$ °C	T_1 °C	T_2 °C	T_3 °C	P W
1	35	19.5	28.16	21.05	30.41	31.73	929.63
2	35	9	17.68	21.04	28.10	28.23	933.97
3	35	9	17.68	21.04	24.07	28.36	934.32

Table 3. Analytical and numerical solution results for option 3.

Model	$T_{(evap.)}$ °C	$T_{(Chip)}$ °C	T_1 °C	T_2 °C	T_3 °C
Option 3	9	17.68	21.04	24.07	28.36
Numerical	10.13	16.66	21.34	26.6	27.78*

* Average temperature around the module

6 COMPARISON AND DISCUSSION

Results of the analytical and numerical models are given in Table 3. Temperature T_1 represents the average temperature of the top insulation. A difference of 0.3 °C between the analytical model and the numerical solution is seen. However, the local temperature on the top insulation is different from the average. This difference cannot be predicted by the analytical model. In the analytical model, it is assumed that the thermal boundary layer starts at the ambient temperature on the upper corner of the top insulation vertical panel. This was not the case when the cold air from the board upper portion falls over the module. The air temperature at this point is about three degrees lower than the ambient temperature. In order to determine the local surface temperature, a numerical model or experimental results are needed.

The analytical model predicts the average bottom insulation temperature (T_2) within 2.53 °C. Also, the model predicts the board temperature at the outer edge of the module (T_3) within 0.58 °C. These differences might be quite reasonable for preliminary design evaluation purposes.

The main source of the differences is a lack of a simple model for the air side heat transfer coefficient on the board with non-uniform boundary conditions.

7 CONCLUSIONS

A model for heat spreading in orthotropic materials has been developed. In order to evaluate the developed model for MCMs' thermal analysis, the heat transfer inside a refrigeration cooled multi-chip module of an IBM CMOS Server has been considered. This study proves that the analytical models are effective tools for thermal design process of the MCMs. The developed model enables us to predict temperature distribution at entire module with an acceptable accuracy. Using a simple computer routine for solving the resistance network, one can calculate the required insulation thickness, junction temperatures, and evaporator cooling capacity for different design options. The results of the calculation for the overall geometry and sizes plus evaporator cooling capacities can be used for a more detail study of the local temperature distribution in a numerical simulation or experimental study.

Development of an appropriate model for airside heat transfer coefficient on PCB will improve the developed model in a sense that it can provide better prediction for the PCB and module contact temperature.

ACKNOWLEDGEMENTS

The parameters in Table 1 were provided by Dr. Roger Schmidt from IBM as part of a design case study conducted in the course ENME 808 W at the University of Maryland during Spring 2001. We thank Dr. P. Sathyamurthy for making the IcePack code available at the University of Maryland.

REFERENCES

Cooper, M.G., Mikic, B.B & Yovanovich, M. M. 1969. Thermal Contact Conductance. *Int. J. of Heat and Mass Transfer* 12: 279-300.

Culham, J.R., Yovanovich, M.M., Lemczyk, T.F. 2000. Thermal Characterization of Electronic Packages Using a Three Dimensional Fourier Series Solution, *Transaction of ASME, J. of Electronic Packaging,* 122 (3): 233-239.

Culham, J.R., Yovanovich, M.M. 1998. Factors Affecting The Calculation of Effective Conductivity in Printed Circuit Boards. *6th intersociety Conference on Thermal and Thermomechanical Phenomena in Electronic Systems,* xv+541: 460-467.

Furmanczyk, M., Napieralski, A., Yu, E., Przekwas, A., Turowski, M. 1998. Thermal Model Reduction for IC Packages and MCM's, *International Workshop on Thermal Investigation of ICs and Microstructures-* vii+241: 135-138.

Franke, T. 1999. Thermal Modeling of Multichip Assemblies. *Proceedings of 5th International Workshop on Thermal Investigations of ICs and Microstructures.*

Guenin, B. M., Mars, R. C., and Molnar, R. J. 1995. Analysis of a Thermally Enhanced Ball Grid Array Package. *IEEE Transaction on components, Packaging, and Manufacturing Technology*, part A 18 (4):749-756.

Huang, Y., Yovanovich, M.M. 1993. A Thermal Resistance Model for Conjugate Heat Transfer From an Isothermal Heat Source on a Vertical Plate". *Proceeding of The 1993 International Electronic Packaging Conference*: 37-57.

Kennedy, D.I. 1959. Spreading Resistance in Cylindrical Semiconductor Devices. *J. Applied Physics*, 31: 1490-1497.

Lee et al. 1995. Constriction/Spreading Resistance Model for Electronic Packaging. *ASME/JSME Thermal Engineering Conference*, 4: 199-206.

Lemczyk, T.F., Yovanovich, M.M. 1988. Thermal Constriction Resistance With Convective Boundary Conditions. *Int. J. of Heat and Mass Transfer,* 31 (9): 1873-1883.

Nigen, I.S., Amon, C.K. 1992. Concurrent Thermal Designs of PCBs, Balancing Accuracy With Time Constraints. *IEEE Transaction on components, Hybrids, and Manufacturing Technology,* 15 (5): 850-859

Teertstra, P., Culham, J.R., Yovanovich, M.M. 1997. Comprehensive Review of Natural and Mixed Convection Heat Transfer for Circuit board arrays. *J. of Electronics Manufacturing,* 7 (2): 79-92

Tummala, R.R., Rymaszewski, E.J. (eds), 1989. *Microelectronics Packaging Handbook.* New York, N.Y: Van Nostrand Reinhold.

Yovanovich, M.M. 1970. On the Temperature Distribution and Constriction Resistance in Layered Media. *J. of Composite Material*, 4: 567-570

ANALYSIS AND TESTING METHODS

Thermal Challenges in Next Generation Electronic Systems, Joshi & Garimella (eds)
 ISBN 90-77017-03-8

A study of temperature field in a GaN Heterostructure Field-Effect Transistor

M.A. Baig , M.Z. Khandkar & J.A. Khan
Dept. of Mechanical Engineering, University of South Carolina, SC, USA

M.A. Khan & G. Simin
Dept. of Electrical Engineering, University of South Carolina, SC, USA

H. Wang
Oak Ridge National Laboratory, TN, USA

Keywords: GaN, HFET, temperature

ABSTRACT: We propose in this paper a three-dimensional finite element based heat transfer model for a Gallium Nitride-based Heterostructure Field-Effect Transistor (henceforth referred to as GaN HFET). Analyses were carried out to study the distribution of temperature in the HFET under steady state conditions for two different steady-current inputs. Two different substrates for the HFET, sapphire and silicon carbide (SiC), were studied. The paper discusses the effect of using a heat sink and also that of using reasonable contact resistances on the substrate side of the HFET, on the temperature profile. In all cases, the gate region of the HFET was found to attain the highest temperature. Subsequent experiments to validate the results of the computational analysis were carried out at the Oakridge National Laboratories, Knoxville, and are also presented in this paper.

1 INTRODUCTION

Heating in electronic devices is one of the prime factors that limits their performance. With the ever-increasing level of integration and the resulting high power densities, there is a constant need to study and devise ways of dissipating the heat produced, in order to keep up the present rate of development. GaN based HFET devices offer the potential of high-power, high-speed and high-temperature performance better than that of Si and GaAs based devices. High-temperature performance of AlGaN/GaN HFETs grown on SiC substrates has been found to be much better than the performance with sapphire substrates (Gaska et al. 1997). Gaska et al. (1998) have also reported that heat dissipation strongly affects the device characteristics soon after the application of source-drain voltage. Irreversible degradation of device performance of AlGaN/GaN FETs caused by loss in channel current between 600 °C and 800 °C has been reported (Daumiller et al. 1999).

The schematic of a typical GaN HFET with its approximate dimensions is shown in Figure 1.

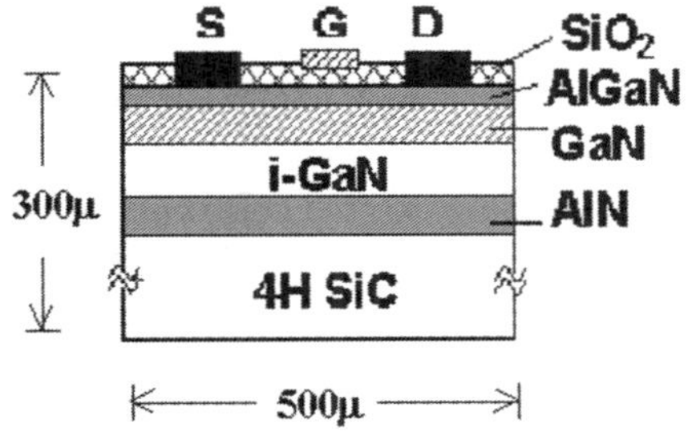

Figure 1. Schematic of a typical GaN HFET

It consists of an SiC substrate, on top of which layers of different materials are grown. SiC and GaN have different crystal structures and growing GaN directly over SiC results in cracks at the interface owing to the lattice mismatch. Therefore, an AlN layer is first grown on the SiC. The AlN layer aids in the nucleation of GaN. A thin layer of highly conducting AlGaN is grown over the GaN layer. Most of the current flow is concentrated in this layer. Current flows between the source and the drain. Applying a voltage across the gate can regulate the flow of this current. The flow of current causes the transistor to heat-up. The temperature is found to be maximum in the region around the gate. An attempt was made to simulate the effects of using different substrates, with and without a heat sink, for two different power inputs. This analysis was carried out in ABAQUS/Standard. MSC PATRAN 8.5 was used as the pre-processor and the model was post-processed in ABAQUS 5.8-16. ABAQUS is a tried and tested software that has been used extensively over the years for structural and thermal analysis (Khan et al. 1999), thus justifying it's choice in our case.

In order to simplify the numerical computations, analysis was carried out on a symmetric half of the transistor, bisected by a vertical plane through the middle of the device parallel to its depth. The face of the HFET thus exposed was made insulated. The smallest linear dimension acceptable to the program was found to be around 0.2μm. The smallest linear dimension in the actual transistor was about 0.01μm. As a result, some of the actual smaller dimensions had to be modified to accommodate this limitation. SiC was used as the substrate below which a layer of gold was attached to enhance the heat sinking. Henceforth, we will be referring to this layer of gold as the heat sink. The heat sink was maintained at a constant temperature of 25°C. The main underlying motivation was to obtain data that would provide an insight into the larger problem of managing heat at the device level itself.

2 EXPERIMENTAL ANALYSIS

Experiments were carried out at the Oakridge National Laboratories, Knoxville, Tennessee. ImageDesk II™, a Pentium PC hosted desktop image acquisition, processing, and analysis system, interfaced with Amber's Galileo imaging IR camera, was used to acquire a sequence of images of similar transistors that were powered to different levels. The Raytheon/Amber Galileo (Radiance HS) is a 12-bit IR camera (the digital output has a 12-bit dynamic range), which uses a cooled InSb 256x256 focal plane array as a detector. In snapshot mode the camera allows stop-action analysis of events as short as 2μs. Working in the 3-5μm spectral range, it monitors temperatures within a wide range at a resolution of 12 bits (Krehl et al. 1999). Favro et al. (2001) have reported using this camera for sonic IR imaging of cracks and delaminations. In order to obtain a microscopic resolution of the temperature distribution within the HFET, the thermographic camera was coupled to a 4X infrared sensitive objective, which allowed a spatial resolution of about 5μm. Figure 2 shows a snapshot of the HFET under steady-state conditions as taken by the camera. The calibration curve correlating the camera count values (proportional to the intensity of emitted IR radiation) with the actual temperatures is shown in Figure 3.

3 GOVERNING EQUATIONS

Basic energy balance equations were used in the computational analysis. Free convection was assumed to take place at the free surfaces. An explanation of the symbols used is given below:

n = each of the three coordinate axes x, y and z
α = thermal diffusivity
k = thermal conductivity, h = convection coefficient

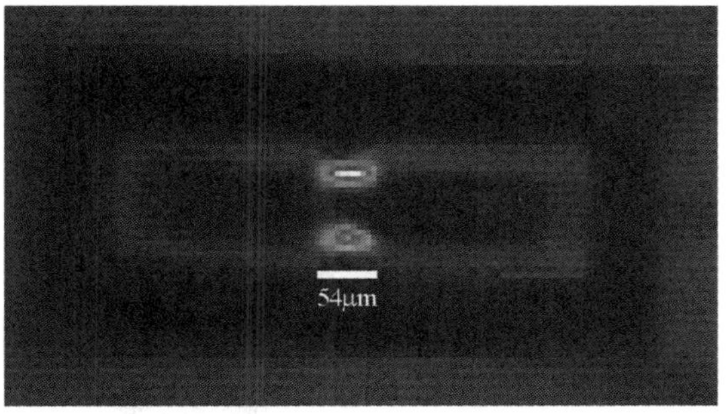

Figure 2. Thermographic image of an HFET when powered up.

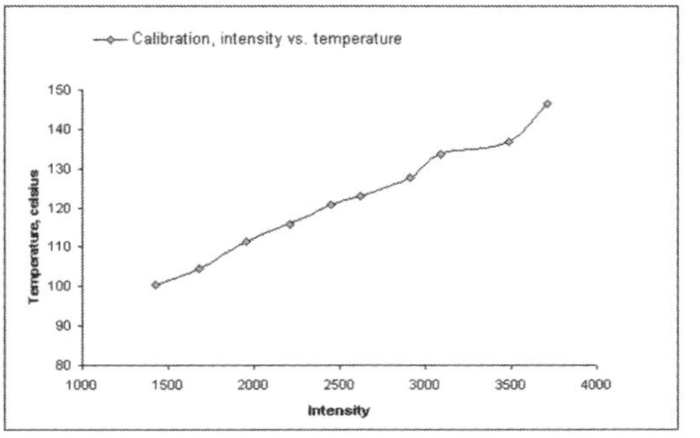

Figure 3. Calibration curve used to relate IR counts to temperature.

T_s = Surface temperature

T_∞ = Ambient temperature

$\dot{q}$ = Heat generation per unit volume of the gate material = Power input to the HFET per unit volume of the gate

3.1 *Conduction*

For each of the interior elements, the governing equation at any instant t is

$$\frac{\partial^2 T}{\partial x^2}+\frac{\partial^2 T}{\partial y^2}+\frac{\partial^2 T}{\partial z^2}=\frac{1}{\alpha}\frac{\partial T}{\partial t}$$

3.2 *Convection at the surface elements*

$$k\frac{\partial T}{\partial n}=h(T_s-T_\infty)$$

3.3 *Heat generation at the gate*

$$\frac{\partial^2 T}{\partial x^2}+\frac{\partial^2 T}{\partial y^2}+\frac{\partial^2 T}{\partial z^2}+\frac{\dot{q}}{k}=\frac{1}{\alpha}\frac{\partial T}{\partial t}$$

4 COMPUTATIONAL ANALYSIS

As mentioned earlier, the finite-element analysis was conducted on a half-section of the device. An isometric view of the model is shown in Figure 4. The geometry actually modeled was a simplified form of the schematic discussed earlier. The model consisted of three layers: the substrate (sapphire or SiC), a GaN layer and an AlGaN layer. Gold was used as the material for the source and for the sink. The objective was to obtain the temperature gradients in the x and y co-ordinate directions.

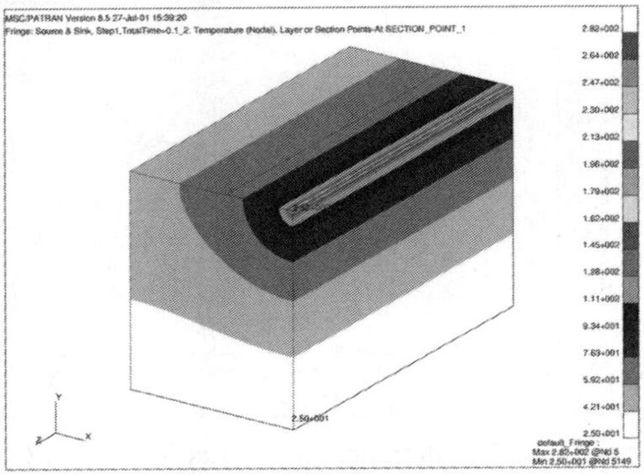

Figure 4. An isometric view of the finite-element model

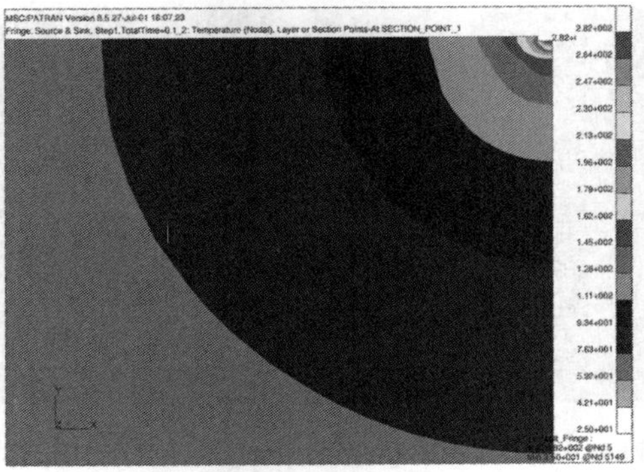

Figure 5. Magnified front view showing the peak temperature in the region of the gate.

The device was assumed to be very long in the z direction. The figure above shows a steady-state maximum temperature of 282°C, which is in the region of the gate, for a heat flux of $8x10^9$ W/m^2 and an SiC substrate. A layer of gold was used as a heat sink and was maintained at 25°C, which is the minimum temperature in the figure. Figure 5, which is a magnified front view, gives a clearer picture of the temperature maps.

Two types of cases were simulated: one with a heat sink and the other without one. The choices of these cases was driven by the motivation to simulate the temperature distribution in the transistor for an actual case using different substrates, and then predict the distribution for a case with additional heat sinking.

4.1 *Case 1: Analysis with no heat sink*

A very high value for the convective coefficient was applied to the bottom surface of the substrate. A value of 25,000 W/m^2 °C was chosen, based on the values recommended by Incropera & DeWitt (1996). As in the case with heat sink, sapphire and SiC were used as the substrates and the same conductivity values were used. The same boundary conditions were also applied.

4.2 *Case 2: Analysis with heat sink*

A thin layer of gold was added to the bottom of the substrate to act as a heat sink. The sink was maintained at a constant temperature of 25°C. The surface of the transistor parallel to the yz plane and adjoining the gate was thermally insulated. A heat flux was applied to the gate and convective boundary conditions were applied to all remaining exposed surfaces. The case with no convective boundary conditions applied was also considered. Sapphire and SiC were used as the substrates. Typical values of thermal conductivity of sapphire and SiC at room temperature are 25-50 W/m°C (Cooke 1993) and 500W/m°C (Morkoc et al. 1994), respectively. A conductivity value of 40 W/m°C was used for sapphire. In case of SiC, a value of 490 W/m°C was used for a very pure material while a general value of 300 W/m°C was used for ordinary SiC.

5 RESULTS

First, the experimental results are compared with those obtained from simulation. For an input power of 0.75W, the temperature against time plot from the experiment is shown in Figure 6. The two peaks represent the temperatures at the two gates. The average peak temperature is seen to be around 140°C. The accuracy of these results was limited by the resolution of the lenses used to obtain the thermographic data. The FEM simulation of this experiment predicted similar results (Figure7). The model did not include a heat sink. Instead, a contact resistance approximated by an h value of 12,500 W/m^2 °C was applied to the base of the SiC substrate, with conductivity of SiC taken as 490 W/m°C.

Although the peak temperature obtained from simulation corresponds closely to the actual peak temperature for the given value of the convective coefficient, a temperature gradient of only about 5° was a lot different from the gradient of around 50° to 55° plotted from the experimental data. This points to inherent defects in the simplifications done in the model. The presence of a temperature gradient seems to suggest that the assumption of a heat sink may be more reasonable than that of a contact resistance at the substrate base. Also, it is difficult to determine accurately the thermal conductivity of SiC, as it seems to be a strong function of the level of impurity. Thus, there is almost certainly an offset between the conductivity value we used and the actual value. Later in the paper we discuss how the thermal conductivity of SiC may affect the overall heat transfer process.

In order to simplify the analysis, the transistor was assumed to be exposed to the ambient, except at the base, where a contact resistance was assumed. This assumption may not correctly represent the actual scenario of the experiment where the device that was tested was surrounded by other devices. At a given time only one device was powered up.

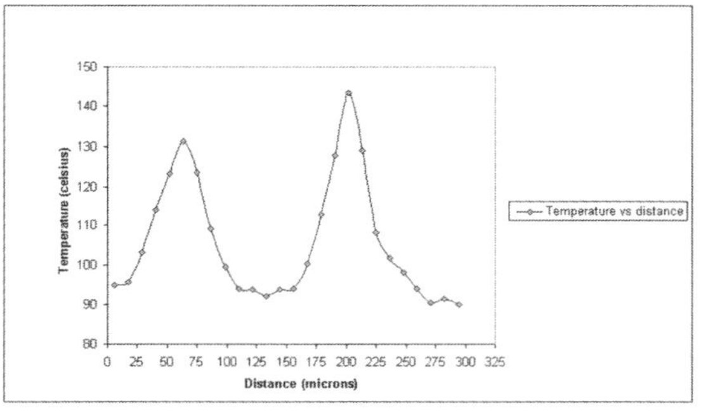

Figure 6. Temperature vs. distance plot obtained from experiment for an input power of 0.75W.

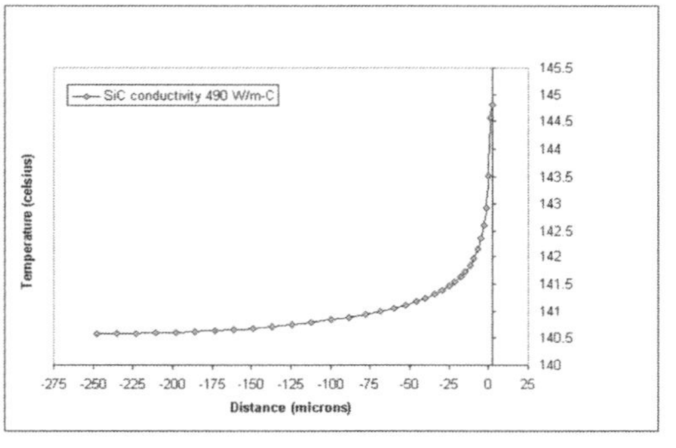

Figure 7. Temperature vs. distance plot obtained from simulation for an input power of 0. 75W and an h value of 12,500 W/m^2 °C.

6 PARAMETRIC STUDIES

Studies were carried out to see the effects on heat transfer of the following parameters:

(i) Different materials and properties for the substrate with no heat sink.
(ii) Heat sink attached to the substrate.

Sapphire and SiC were used as the substrates. In case of SiC, two different values for the conductivity, corresponding to those of impure and pure SiC, were used. Cases with no heat sink attached were also studied. Figure 8a below shows the temperature variation along the x-axis for a heat flux $4x10^9$ W/m^2 and a heat sink maintained at 25°C. As can be seen, the highest temperature is obtained in the gate region. Another noticeable aspect is the existence of a temperature gradient. For a sapphire substrate, the peak temperature was about 700°C. The peak temperature was found to be significantly lower when SiC was used as the substrate. As seen in Figure 8b, for an SiC conductivity of 300 W/m°C, it was about 190°C while for a conductivity of 490 W/m°C the highest temperature was around 150°C.

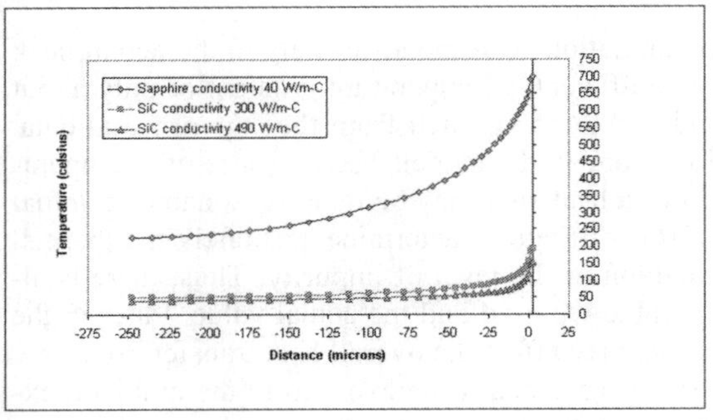

Figure 8a. Variation of temperature along the x-axis for a heat flux of $4x10^9$ W/m^2 with a heat sink maintained at 25°C.

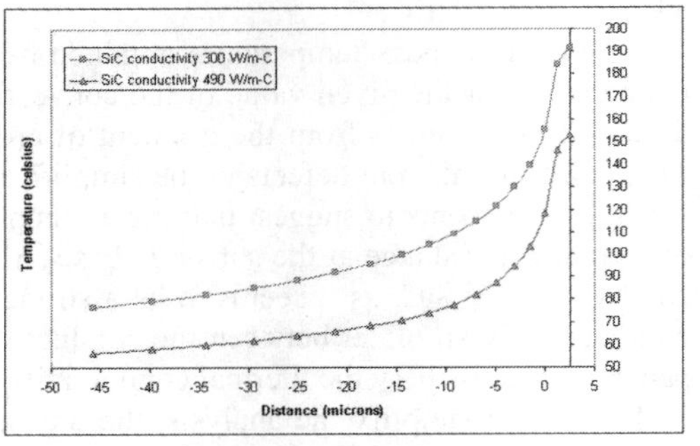

Figure 8b. Magnified view of a portion of Figure 8a showing the variation of temperature near the gate for two different thermal conductivities of SiC.

Thus, it can be inferred that the thermal conductivity of the substrate seems to have a significant bearing on the peak temperature. This has been shown to be true experimentally (Gaska et al. 1998). Figures 9a and 9b show the results obtained for the same power input, but with no heat sink attached. Instead, an h value of 25,000 W/m^2 °C was applied to the base of the substrate. The temperatures obtained were extremely high. The peak temperature approached 2300°C at the gate for Sapphire. For SiC, the temperatures were close to 1800°C for a conductivity of 300 W/m°C and 1750°C for a conductivity of 490 W/m°C.

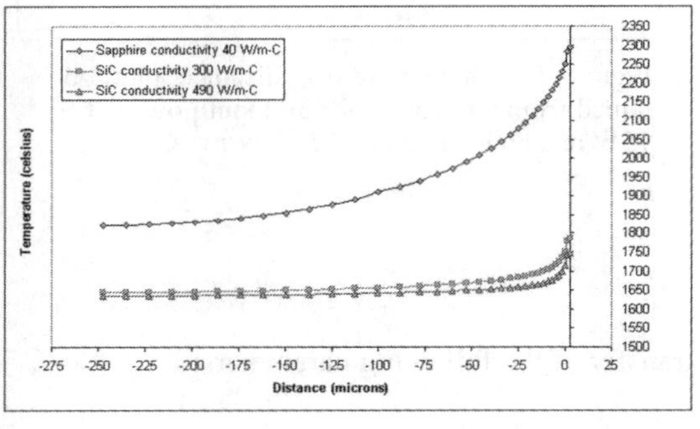

Figure 9a. Temperature graph along the x-axis for a heat flux of $4x10^9$ W/m^2, with a contact resistance applied to the substrate base instead of a heat sink.

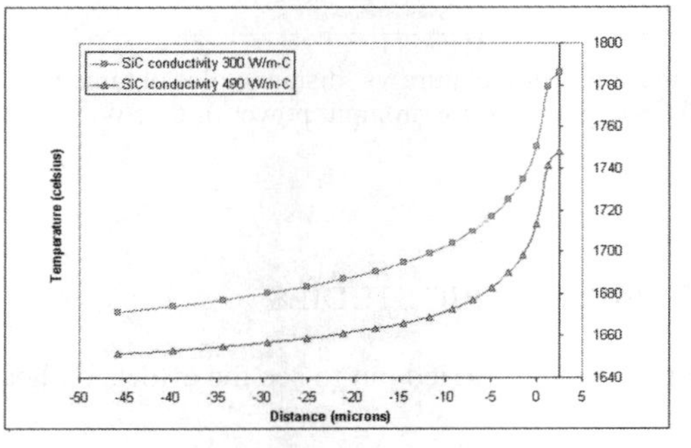

Figure 9b. A closer view of the region of the plot in Fig. 9.a. showing the temperature variation near the gate.

These temperatures are much above the Debye temperature for GaN, which is about 750°C (Manasreh 1997). The Debye temperature may be used as a criterion for the mechanical and chemical stability of semiconductor materials (Daumiller et al. 1999). Thus, beyond 750°C, GaN would become unstable. Next, an analysis with a heat sink was done for a heat flux of 8×10^9 W/m^2. The results obtained are presented in Figures 10a and 10b.

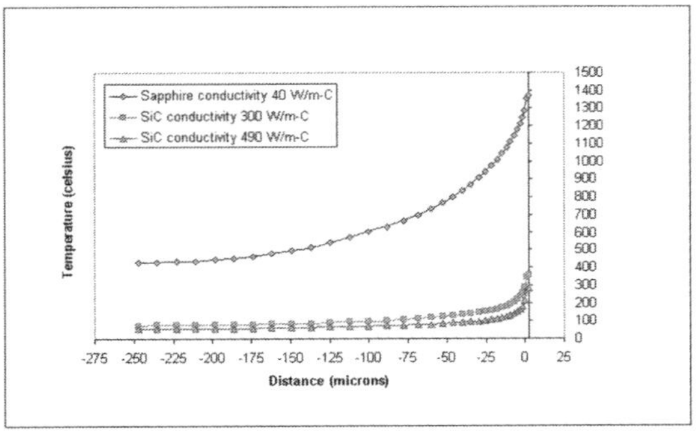

Figure 10a. Temperature gradient along the x-axis for a heat flux of 8×10^9 W/m^2, with a heat sink used.

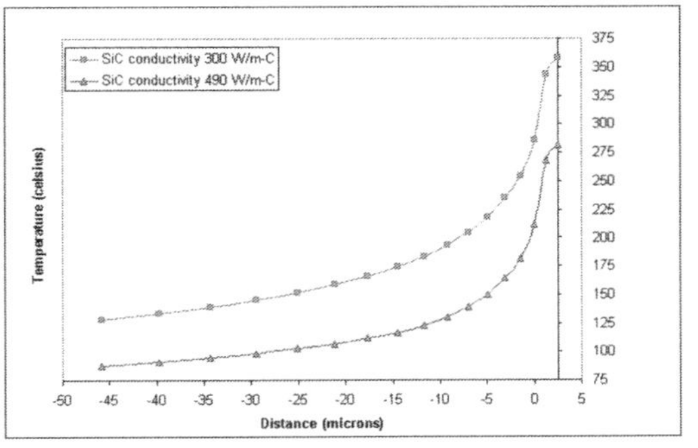

Figure 10b. Zoomed in view of Fig. 9.a. showing the temperature distribution near the gate.

7 CONCLUSION

A 3-D finite-element-based model was proposed to simulate the temperature distribution in GaN based HFETs. Data obtained from experiments performed at the Oakridge National Lab. and those obtained from the simulated analysis agreed on the peak temperature reached for the given power input. In both cases, this temperature was found to occur in the region of the gate. But the simulation did not predict the temperature gradient as found in the experimental analysis. As a result, some of the assumptions and simplifications made while developing the model would need re-evaluation.

A number of conclusions can be drawn from the parametric studies carried out. These studies confirm the fact that SiC, owing to its higher thermal conductivity, is a much better substrate for GaN-based microelectronic devices than sapphire. Or stated another way, the thermal conductivity of the substrate was found to have a significant effect on the peak temperature attained by the device. Also, the heat sinking from the transistor is predicted to be greatly enhanced by including a layer of a high thermally conducting material (gold was used in our case) right below the substrate, and maintain it at a temperature close to the ambient using some cooling mechanism.

REFERENCES

Cooke, H.F. 1993. High-Power GaAs FET Amplifiers. Norwood: Artech House.

Daumiller, I., Kirchner, C., Kamp, M., Ebeling, K.J. & Kohn, E. 1999. Evaluation of the temperature stability of AlGaN/GaN heterostructure FET's. IEEE Electron Device Letters 20 (9): 448-450.

Favro, L.D., Xiaoyan, Z.O., Sun, G. & Thomas, R.L. 2001. Sonic IR imaging of cracks and delaminations. Analytical Sciences 17 (special issue): 451-453.

Gaska, R., Chen, Q., Yang, J., Khan, M.A. & Shur, M.S. 1997. High-temperature performance of AlGaN/GaN HFET" on SiC substrates. IEEE Electron Device Letters 18 (10): 492-494.

Gaska, R., Osinsky, A., Yang, J.W. & Shur, M.S. 1998. Self-heating in high-power AlGaN-GaN HFET's. IEEE Electron Device Letters 19 (3): 89-91.

Incropera, F.P. & DeWitt, D.P. (4th) 1996. Fundamentals of heat and mass transfer. New York: John Wiley & Sons.

Khan, J.A., Xu, L. & Chao, Y.J. 1999. Nugget development during resistance spot welding using a coupled thermal-electrical-mechanical model. Journal of Science and Technology of Welding and Joining 4 (4): 201-207.

Krehl, P., Engemann, S., Rembe, C. & Hofer, E.P. 1999. High-speed visualization, a powerful diagnostic tool for microactuators – retrospect and prospect. Microsystem Technologies 5: 113-132.

Manasreh, M.O. 1997. GaN and related materials in Optoelectronic properties of semiconductors and superlattices (vol. 2). New York: Gordon & Breach.

Morkoc, H., Strite, S., Gao, G.B., Lin, M.E., Sverdlov, B. & Burns, M. 1994. Large band-gap SiC, III-V nitride, and II-VI ZnSe-based semiconductor device technologies. Journal of Applied Physics 76 (3): 1363-1398.

Thermal Challenges in Next Generation Electronic Systems, Joshi & Garimella (eds)
© 2002 Millpress, Rotterdam, ISBN 90-77017-03-8

Transient performance of multi-stack cold plates

S.L. Beh
Faculty of Engineering and Technology, Multimedia University, Melaka, Malaysia.

G.A. Quadir, K.N. Seetharamu & A.Y. Hassan
School of Mechanical Engineering, University of Science Malaysia, Nibong Tebal, Penang, Malaysia.

Keywords: cold plates, finite element method, fin theory, heat losses

ABSTRACT: This paper illustrates a simple and elegant way of analysing the cold plates, both in steady state and transient state, to know the time taken to reach the steady state. Transient analysis is also required for control purposes, where the product is very sensitive to fluctuations in temperature. The finite element method is used to solve the governing equations of single, double and triple stack cold plates. The results compare well for single and double stacks while new results are provided for triple stacks. The results of the analyses are useful in a general way as the governing equations are solved in a non-dimensionalised form. The analyses clearly show that the single unit cell performance, used in the earlier investigations, does not represent the stack performance, especially with heat losses to the ambient.

1 INTRODUCTION

As a result of miniaturisation, electronic products are shrinking in size and weight but with greater pressure for cost reduction. Heat fluxes have increased considerably and hence thermal management becomes crucial from the reliability point of view. Cold plates are used to remove heat from the assembly of printed circuit boards (Fig. 1). A cold plate consists of an array of rectangular fins, attached between two exterior plates. Fluid passes through the spaces between the fins, to help increase the rate of heat transfer from the fin. A cold plate is called a single stack cold plate if there is no splitter plate between the two exterior plates; double stack if there is one splitter plate and triple stack if there are two splitter plates.

A stack is a combination of fins connected together. Thus, the whole stack is made of repeating arrays of fins. A generalized formulation of the analysis of cold plates, using dimensionless parameters to replace the dimensional parameters used in the earlier investigations, is carried out. By non-dimensionalizing the governing equation, the analysis is not restricted to a particular set of geometry of a stack. The analysis starts by taking a single unit cell of the cold plates for the purpose of verifying results with those from Pieper & Kraus (1998). In addition to that, several unit cells assembled together vertically are analysed using the finite element method with and without heat losses at the top and bottom of the stack as it happens in reality. The transient response is observed and the steady state results of such analysis will determine if a single unit cell analysis is adequate in representing the performance of a stack under all operating conditions. The objective of this paper is to generate new and additional data, which will be helpful in the design of cold plates used for cooling of electronic systems.

2 THE GOVERNING EQUATION AND FINITE ELEMENT FORMULATION

A stack is considered as a combination of fins connected together (Fig. 2). A simple one-dimensional fin theory is then applied to the stack under investigation.

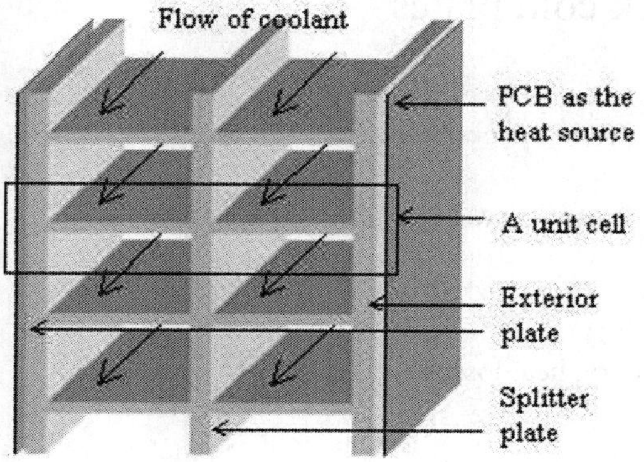

Figure 1. A double stack cold plate.

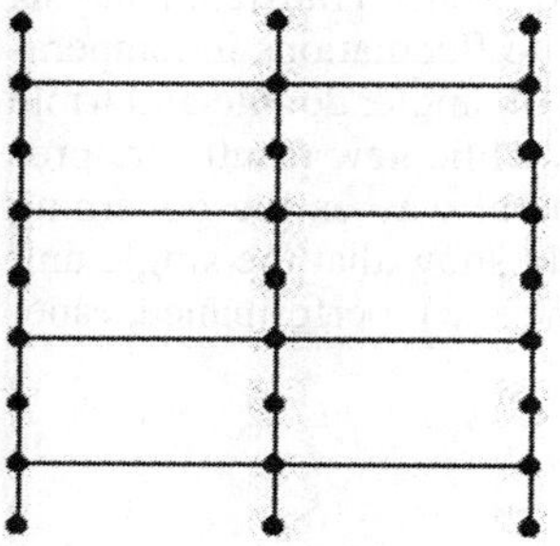

Figure 2. Finite element discretisation of a double stack cold plate.

The governing equation for a steady state, one-dimensional fin, with conduction in the solid and forced convection in the passage is given by:

$$kA\frac{\partial^2 T}{\partial x^2} - hP(T - T_\infty) = c\rho A\frac{\partial T}{\partial t} \tag{1}$$

where k = thermal conductivity of the fin material, W/m °C; A = cross sectional area perpendicular to the direction of conduction, m^2; h = convection coefficient, W/m^2 °C; P = perimeter of the surface where convection takes place, m; T_∞ = ambient temperature, °C; c = specific heat of the fin material, J/kg °C; ρ = density of the fin material, kg/m^3; T = temperature of the fin at a given location, °C; x = distance measured from the base of a fin, m; and t = time, s.

As a generalisation, the present analysis will be carried out using dimensionless parameters. This is carried out after a transformation of the field variables to their corresponding dimensionless parameters as follows:

$$Q^* = \frac{\text{Heat loading on the right exterior plate}}{\text{Heat loading on the left exterior plate}} = \frac{Q_{TR}}{Q_{TL}} \tag{2}$$

$$\theta = \frac{kA\left(T - T_\infty\right)}{Q_{TL}B} \tag{3}$$

$$X = \frac{x}{B}, \text{ where } B = \text{reference dimension, m} \tag{4}$$

$$M = \frac{hPB^2}{kA} \tag{5}$$

$$\tau = \frac{kt}{\rho c B^2} \tag{6}$$

Thus, equation (1) reduces to

$$\frac{\partial^2 \theta}{\partial X^2} - M\theta = \frac{\partial \theta}{\partial \tau} \tag{7}$$

M is the governing parameter in the analysis that takes into account the variation in h (free convection, mixed convection and forced convection including the developing flow), the variation in k (different fin materials), geometric factor (A/P ratio of the flow passage) and finally the distance between the exterior plates. Thus the introduction of the parameter M does not restrict the present analysis to a particular set of geometry or operating condition.

The variation of temperature along the fin is assumed to be linear as

$$\theta = [N]\{\theta\} \tag{8}$$

where,

$$[N] = [1 - X \quad X] \tag{9}$$

$$\{\theta\} = \begin{Bmatrix} \theta_i \\ \theta_j \end{Bmatrix} \tag{10}$$

θ_i, θ_j = dimensionless temperatures at nodes i and j respectively.

By using Galerkin's method, as explained in Segerlind (1984) and Lewis et al. (1996), the finite element formulation of equation (7) is obtained as

$$[C]\left\{\frac{d\theta}{d\tau}\right\} + [K]\{\theta\} = 0 \tag{11}$$

where,

$$[C] = \frac{1}{6}\begin{bmatrix} 2 & 1 \\ 1 & 2 \end{bmatrix} \tag{12}$$

$$[K] = \begin{bmatrix} 1 & -1 \\ -1 & 1 \end{bmatrix} + \frac{M}{6}\begin{bmatrix} 2 & 1 \\ 1 & 2 \end{bmatrix}. \tag{13}$$

Equation (11) is a transient heat conduction equation. To solve this equation using FEM, the temperature variation with time is assumed to be linear and is represented as

$$\{\theta\} = N_{it}\{\theta_{it}\} + N_{jt}\{\theta_{jt}\} \tag{14}$$

where,

$$N_{it} = 1 - \frac{\tau}{\Delta\tau} \text{ and } N_{jt} = \frac{\tau}{\Delta\tau} \tag{15}$$

$\Delta\tau$ is the time step, and

θ_{it}, θ_{jt} = temperatures at a node for two time intervals represented by *it* and *jt*

Galerkin's method is applied in time domain here, to obtain the time stepping scheme as

$$\left(\frac{[C]}{2\Delta\tau} + \frac{[K]}{6}\right)\{\theta_{jt}\} = \left(\frac{[C]}{2\Delta\tau} - \frac{[K]}{3}\right)\{\theta_{it}\} \tag{16}$$

Equation (16) is used to find the temperature distribution $\{T_{jt}\}$ at a new time step from the known temperature distribution $\{T_{it}\}$. The iteration stops when the temperature between two time steps does not change by more than 1.0E-6. The above theory is applied to single, double and triple stack cold plates.

3 RESULTS AND DISCUSSIONS

Kraus & Pieper (1995) and Pieper & Kraus (1995) performed the analyses of single and double stack cold plates using a single unit cell. Pieper & Kraus (1997) also performed a transient analysis of double stack cold plate for a specific geometry of a single unit cell of the stack. Pieper & Kraus (1998) assumed that if the cold plate is made up of more than 20 repeating segments, a single unit cell is adequate for the purpose of analysis. Moreover, they neglected the edge effect of the stack. In present work, the analyses of a single unit cell for single and double stack cold plates without heat loss and under the same operating conditions as given by Pieper & Kraus (1998) are presented first. Steady state results from the present analysis are then compared to those of Pieper & Kraus (1998) for the purpose of verification. Next, several unit cells of single, double and triple stack cold plates are assembled in the vertical direction and analysed for the same operating conditions to determine if the assumption made by Pieper & Kraus (1998) regarding the analysis based on a single cell is correct. Furthermore, the edge effect is brought into account by allowing heat losses from the top and bottom surface of the stack.

3.1 *Analysis of a single unit cell*

The heat loading for the whole stack at the left exterior plate is Q_{TL}, while the same is represented by Q_{TR} for the right exterior plate. The heat loading per unit cell on either side of the exterior plates is divided equally at the nodes of the elements considered. The heat loadings are non-dimensionalised as given in equation (2). The reference dimension in equation (4) is the total width of the stack, which remains as the reference dimension for the analyses of double and triple stack cold plate as well. NUC is used as the abbreviation for the number of unit cells. NUC = 44.4 and M = 0.55 for the case considered in Pieper & Kraus (1998). The steady state results obtained can then be compared to those of Pieper & Kraus (1998), who presented the results of excess temperatures for a single stack cold plate for different heat loading on the right exterior plate and a constant heat loading on the left exterior plate.

Figures 3 and 4 shows the steady state maximum dimensionless temperatures on the left (θ_l) and right (θ_r) exterior plates for single and double stack plate respectively. The highest value among the three nodes on each exterior plate is taken as the representative value. The results of Pieper & Kraus (1998) are also shown in figures 3 and 4. There is a good agreement of steady state results between the present analysis and the results reported by Pieper & Kraus (1998). It shows that the present analysis applied to the single and double stack cold plates is correct.

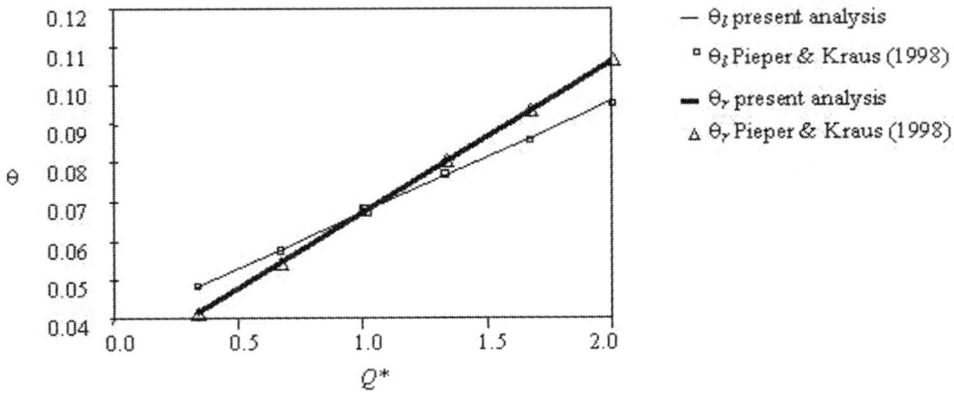

Figure 3. Steady state temperature versus heat loading for a single unit cell analysis of a single stack cold plate.

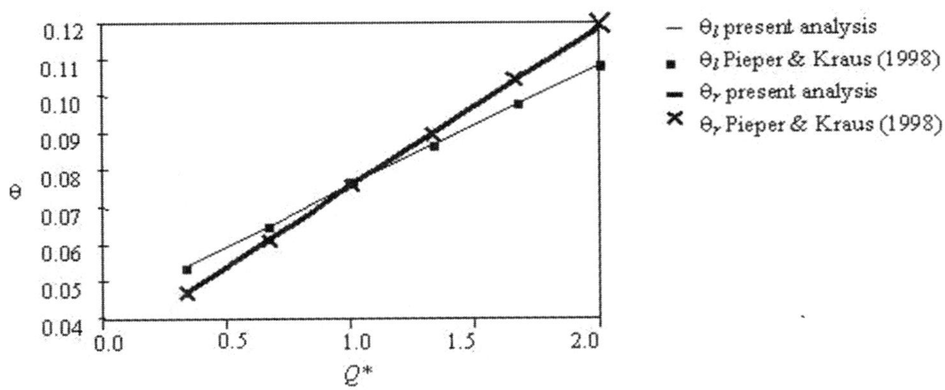

Figure 4. Steady state temperature versus heat loading for a single unit cell analysis of a double stack cold plate.

3.2 *Analysis of stack with varying NUC*

Analyses of the assembled unit cells of varying number, with and without heat losses, are carried out. The heat loss is defined as a percentage of the total heat loading at the left exterior plate Q_{TL}. Since Q_{TL} increases proportionally with NUC, the heat loss also increases accordingly. This also implies that a heat loss of 10% for NUC = 20 is twenty times larger than a heat loss of 10% for NUC = 1. This definition of heat loss also implies that for a fixed NUC, a heat loss of 10% is the same irrespective of Q^*.

Figure 5 shows the effect of the number of unit cells on the maximum temperature as a function of time for single (SS), double (DS) and triple (TS) stack cold plates for $M = 0.55$ and heat loss equal to 0.2. From Figure 3, it can be observed that as the number of unit cells decreases, the maximum temperature also decreases. This implies that the actual maximum temperature of the system cannot be obtained from the analysis of a single unit cell, if there is a heat loss from the top and bottom surface of the stack. However, it can be noted that the time taken to reach the steady state is almost the same regardless of NUC. Figure 5 also shows that for a particular value of NUC, single stack cold plate has the highest temperature.

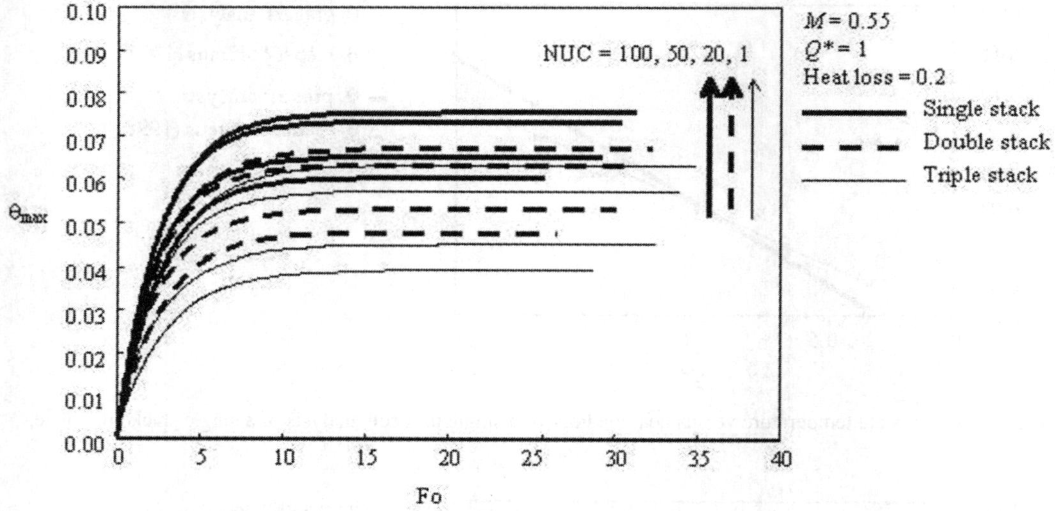

Figure 5. Effect of the number of unit cells on maximum temperature of different stacks.

Figure 6 shows the effect of different heat losses for a given stack of 50 unit cells for $M = 0.55$ and $Q^* = 1$. It is seen that as the heat loss from the stack to the ambient increases, the maximum temperature decreases. When there is no heat loss, the steady state maximum temperature is similar to the case of a single unit cell analysis for single and double stack cold plates. This further strengthens the fact that a single unit cell analysis is not a representative of the analysis of the whole stack, and only holds good for the case when there is no heat loss. It can also be noticed from Figure 6 that the temperature drop when heat loss occurs is more in triple stack cold plate compared to single stack cold plate. However, the time taken to reach the steady state is not very much affected by the incorporation of heat loss into the system.

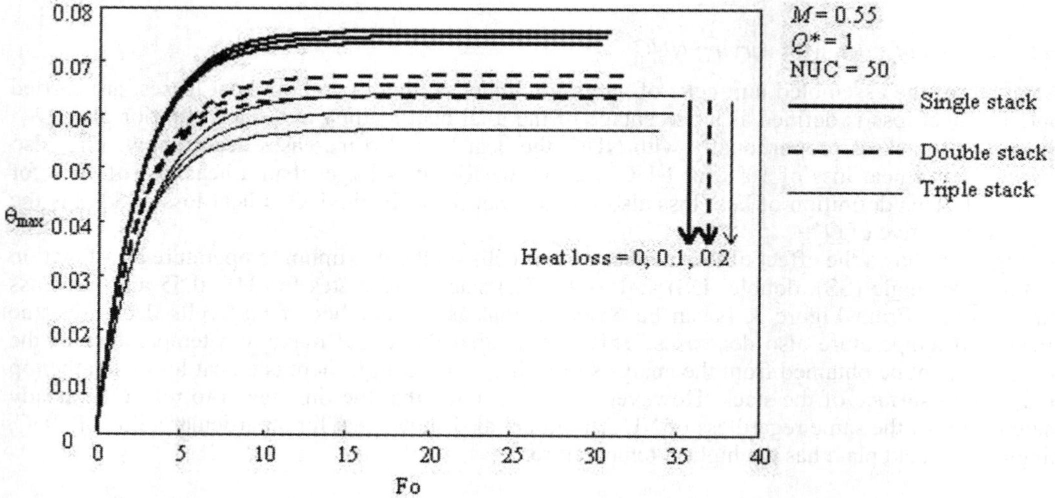

Figure 6. Effect of heat losses on maximum temperature of different stacks.

Figure 7 shows the effect of non-uniform heat loading on the left and right exterior plates of the stacks. It is obvious that when heat loading increases, the maximum temperature also increases. The single stack cold plate has the highest temperature among the single, double and triple stack for a given heat loading. It is also observed that the non-uniformity of heat loading does not change the time taken to reach the steady state.

Figure 8 shows the effect of the dimensionless parameter M on the maximum temperature. When M increases, the maximum temperature decreases, as expected. But the time taken to reach the steady state reduces considerably. This is because a large value of M implies a high value of heat transfer coefficient, which means more heat is transferred from the cold plate to the coolant more quickly. As before, single stack cold plate has the highest temperature among the cold plates at a particular value of M.

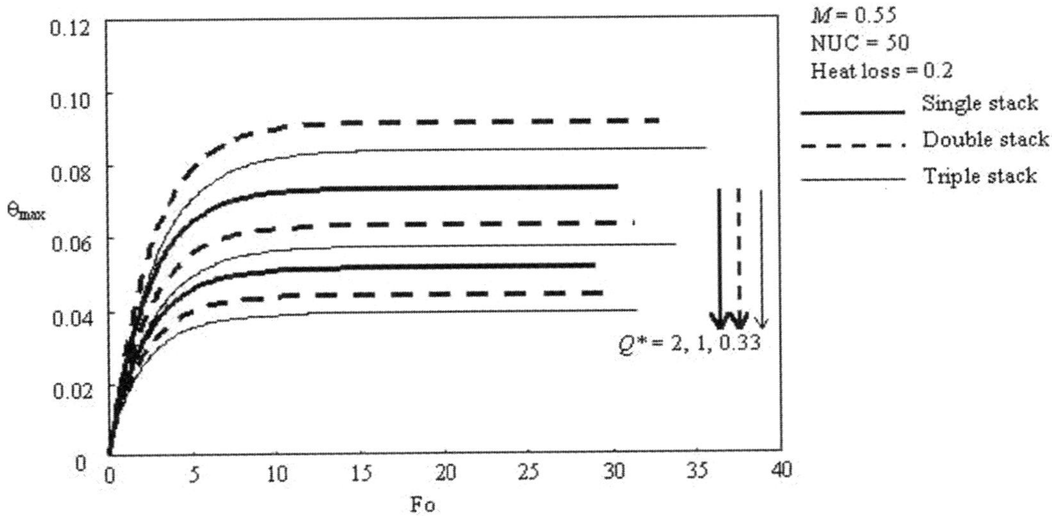

Figure 7. Effect of non-uniform distribution of heat loading on the exterior plates.

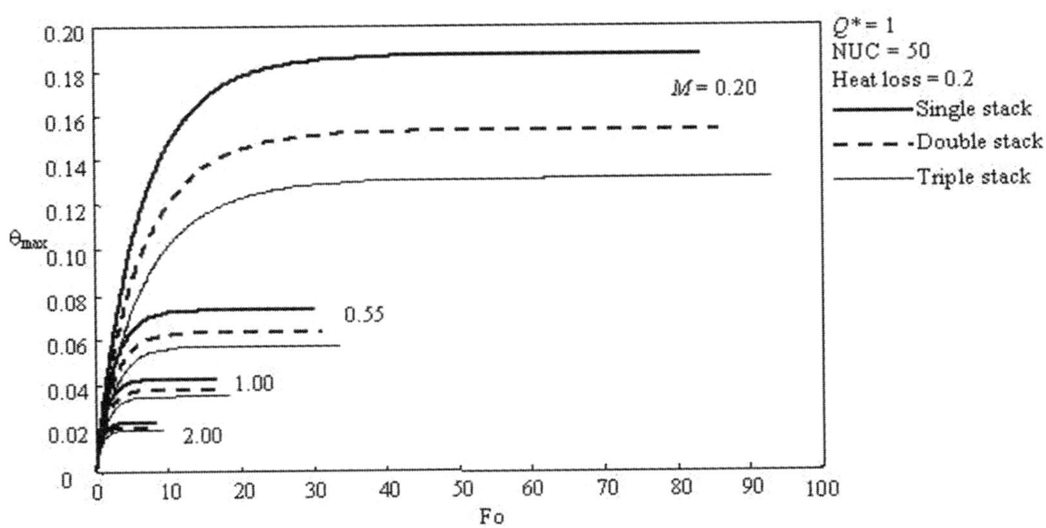

Figure 8. Effect of the dimensionless parameter M on the maximum temperature of the stacks.

4 CONCLUSIONS

The finite element method is used to analyse single, double and triple stack cold plates in transient state. A single one-dimensional fin theory is applied to the discretised elements in the above analyses. First, as verification, a single unit cell of single and double stack cold plates are analysed without heat losses from their top and bottom surfaces. Next, the analysis is carried out for an assembly of different number of unit cells (NUC) under the same boundary conditions as imposed for a single unit cell. Furthermore, the results for all the stacks for different values of M, a dimensionless parameter involving dimension, properties of the stack material and heat transfer coefficient are also presented which will be helpful in the design of cold plates used for the cooling of electronic systems. The following conclusions are made.

The steady state results obtained from the analysis of a single unit cell without heat losses from the top and bottom surfaces agree well with those available in the literature for single and double stack cold plates. Thus, the single unit cell can be taken as the representative of the stack only when there are no heat losses from the top and bottom surfaces of the stack. This is valid for a triple stack cold plate as well although no data is available in the literature to compare.

Analyses of the assembled unit cells of varying number with and without heat losses are carried out. Single stack cold plate has the highest maximum temperature among the single, double and triple stack cold plates at a particular working condition.

The time taken to reach the steady state is very much affected by M. There is a considerable reduction in time taken to reach the steady state as M increases. However, other factors like NUC, heat loss and non-uniformity of heat loading do not affect the time very much.

REFERENCES

Kraus, A.D. & Pieper, R.J. 1995. Cold plates with asymmetric heat loading, Part 2 – the double stack. *ASME Advances in Electronic Packaging*. EEP-10-2: 871-878.

Lewis, R.W., Morgan, K., Thomas, H.R. & Seetharamu, K.N. 1996. *Finite Element Method in Heat Transfer Analysis*. John Wiley.

Pieper, R.J. & Kraus, A.D. 1995. Cold plates with asymmetric heat loading, Part 1 – the single stack. *ASME Advances in Electronic Packaging*. EEP-10-2: 865-870.

Pieper, R.J. & Kraus, A.D. 1997. A transient analysis of heat transfer in double stack cold plates. *Proc 29th Southern Symposium on System Theory (SSST), Cookeville, TN, USA*. 348-352.

Pieper, R.J. & Kraus, A.D. 1998. Performance Analysis of Double Stack Cold Plates Covering All Conditions of Asymmetric Heat Loading. *ASME Journal of Electronic Packaging*. 20: 296-301.

Segerlind, L.J. 1984. *Applied finite element methods*. John Wiley & Sons.

Thermal Challenges in Next Generation Electronic Systems, Joshi & Garimella (eds)
© 2002 Millpress, Rotterdam, ISBN 90-77017-03-8

Integrated simulation toolkit for electronically controlled motor systems

Y. Liu & J. H. Lienhard V
Dept. of Mechanical Engineering, MIT, Cambridge, MA, USA

J. D. Booth & R. W. Stairs
Robert Bosch Corp, Waltham, MA, USA

Keywords: thermal management, heat transfer, thermal network modeling, system-level simulation, electronically controlled motor

ABSTRACT: The purpose of this study is to establish an analytical-numerical hybrid technique for system-wide thermal modeling of electronically controlled motors. The thermal models are fully coupled with an electronic network model, which makes the simulation closer to real working conditions. It employs the finite volume concept to handle heat conduction in solid media and adopts analytical equations to calculate velocity and temperature profiles for convection. The experimental data from various tested motors showed that good accuracy could be reached using much less time and effort compared to full numerical simulation, such as FVM or FEA. Simplicity, ease-of-use, coupling with electronic signal models, and moderate accuracy make this research attractive in solving complicated electrical-thermal systems. The model can be used not only for rapid thermal evaluation in the early design stage, but also for detailed system or physical simulation.

1 INTRODUCTION

The growing use of electronic controls demands that an increasing number of products show high reliability in harsh environments. Thermal issues now stand out at all levels of electronic product development. In order to meet thermal management challenges and accelerate the design-market-deliver cycle for the next generation of electronically controlled motors, fully coupled thermal-electrical characterization is needed. This requires integrated simulation tools.

Motors used in automobile engines face an especially challenging environment. Ambient air temperatures may reach 125°C and steep thermal and electrical transients may occur. With the growing use of electronic commutation, motors now include electronic modules whose service temperature should stay below 150°C. Motor designers thus need software capable of concurrent electrical and thermal simulation in order to minimize the cycle of thermal testing and redesign.

Many computational fluid dynamics (CFD) and finite element analysis (FEA) software packages are available to simulate heat transfer through solid media and fluid flow. However, few of them can be coupled with circuit simulation. Moreover, even with the latest computing power, these simulations are still time-consuming in terms of geometry modeling, computing, calibration, and experimental support (Igic et al. 2001, Kraus et al. 1983). The present research employs the finite volume method (FVM) concept to calculate heat conduction for motor components while using a flow network approach to model fluid convection calculations. This is very similar to the flow network modeling (FNM) described by other authors (Belady et al. 1999, Minichiello 2000). The present method conserves mass and energy at each node. Due to the limitations of using a commercial matrix solver momentum is conserved by restricting certain behavior. Flow is only driven by an external pressure source and there is no possibility for pressure recovery or buoyancy driven flows. This method represents the overall flow system as a network of components and flow paths to calculate flow rates and temperature profiles. Analytical functions for certain flow patterns, such as pipe flow and internal/external natural convection flow, are used to predict the thermal performance over the flow network (Rohsenow et al. 1998). FNM is able to analyze system-level interaction of the individual components in a rapid and accurate manner. This analytical-numerical hybrid scheme (FNM-FVM) not only simplifies the complicated flow simulation but also accounts for the motors' geometry and increases the accuracy of the simulation.

The present research aims to build a suite of tools and a methodology to assess thermal issues at all stages of motor design from concept through application. Because the systems are applied under a large range of external conditions (such as temperature, cooling air flow rate, and mechanical and electrical load), an analytical-numerical hybrid thermal management system for complex electronic-mechanical system is established. The simulation process is capable of modeling steady state and transient thermal issues within the system, including conduction from electronic and mechanical components to the convecting surfaces, natural convection, radiation, and bulk temperature variations. Electronic control signals for motor performance are handled using a separate time integration scheme. Coupled thermal effects for various electronic properties and electrically generated heat are considered through a thermal-electronic simulation interface during computation.

The software package Saber (by Avant! Corp) for mixed domain network analysis is selected to implement the system simulation model. Three-dimensional CFD simulations are performed to study the different environmental variables that dictate the heat transfer behavior. Combined with the experimental data, these simulation results are used to calibrate network models that capture all of the physics (Kang 2000). Complexity rises dramatically as more details are considered in the thermal network topology and the electronic system. In order to cope with this situation, a thermal block based on the control-element concept of the finite volume method is adopted, which considers heat conduction, heat sources, and heat capacitance. Blocks can be assembled to form components and then connected to a higher-level hierarchy. Combined with a convection network and a motor control section, the model can simulate complex thermal behaviors with the flexibility to optimize shape, materials, and performance during the design stage of motor development.

In fact, one of the model's advantages is the ability to quickly generate new system models for each design, which may otherwise take a long time to develop. Various physical arrangements of electronic board were tested based on this simulation toolkit and the results showed significant system effects on the thermal behavior. Further development includes building a user-friendly GUI (graphical user interface) that encapsulates detailed Saber programming from designers and integrates motor models into higher-level system models - such as an entire car, while still accurately capture the thermal characteristics.

2 THERMAL ELEMENT

A hierarchy of thermal component topologies has been developed. The basic component is a heat conduction block, which considers heat conduction, heat sources, and heat capacitance. The block links with other elements via its thermal connection "pins". Sets of blocks are assembled to represent any particular part. Furthermore, by combining with a fluid network and motor performance models, parts can be assembled to form a whole motor system.

2.1 *Heat conduction block*

The approach is based on the control block concept of the finite volume method. A cylindrical thermal conduction block is used during the current simulation, shown as Figure 1. However, other geometry, such as one or two dimensional block, three-dimensional rectangular cube, and spherical block can also be constructed using the same principle (Anderson et al 1984).

Heat transfer is by conduction in a solid:

$$\frac{\partial T}{\partial t} = \alpha \nabla^2 T + \frac{\dot{q}}{\rho c_p}, \quad \alpha = \frac{k}{\rho c_p} \tag{1}$$

where the thermal conductivity k is constant, $\dot{q}$ is a volume heat source, ρ is density, Cp is specific heat (constant pressure), T is temperature, and α is thermal diffusivity. For a cylindrical thermal conduction block, the discrete form of heat conduction equation is shown as below:

$$\begin{aligned}
&k\frac{(r-\Delta r/4)\Delta\phi\Delta z}{\Delta r/2}\left[T_a - T_{center}\right] + k\frac{(r+\Delta r/4)\Delta\phi\Delta z}{\Delta r/2}\left[T_b - T_{center}\right] + \\
&k\frac{\Delta r\Delta z}{r\Delta\phi/2}\left[T_c - T_{center}\right] + k\frac{\Delta r\Delta z}{r\Delta\phi/2}\left[T_d - T_{center}\right] + \\
&k\frac{r\Delta r\Delta\phi}{\Delta z/2}\left[T_e - T_{center}\right] + k\frac{r\Delta r\Delta\phi}{\Delta z/2}\left[T_f - T_{center}\right] + \dot{q}r\Delta r\Delta\phi\Delta z \\
&= \left[\frac{\rho c_p T_{center}\big|_{t+\Delta t} - \rho c_p T_{center}\big|_t}{\Delta t}\right] r\Delta r\Delta\phi\Delta z
\end{aligned} \tag{2}$$

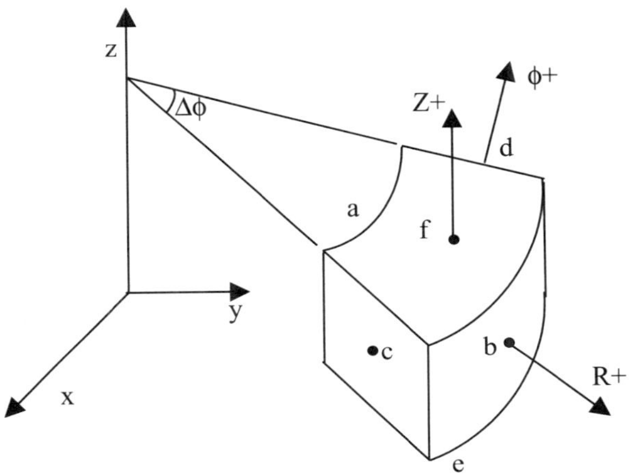

Figure 1. Cylindrical thermal conduction block with six thermal pins (a through f) on the surfaces.

2.2 *Convection heat loss*

The forced convection model is based on established results for heat transfer coefficients. For example, the exact analytical solution for laminar flow over a flat plate is expressed as:

$$\overline{Nu_L} = 0.664 \cdot \mathrm{Pr}^{1/3} \cdot \mathrm{Re}_L^{\;1/2} \tag{3}$$

where Pr is Prandtl number, $\mathrm{Re_L}$ is Reynolds number, and $\overline{\mathrm{Nu_L}}$ is average Nusselt number for laminar flow. The natural convection Nusselt number can be expressed as:

$$\overline{Nu_L} = C \cdot (Gr_L \cdot \mathrm{Pr})^m \tag{4}$$

where C and m have been experimentally determined for different geometries, $\mathrm{Gr_L}$ is the Grashof number which relates the strength of the buoyancy force to the viscous force in a natural convection system (Bar-Cohen 1998). Mixed convection formulas are used when $\mathrm{Gr/Re^2} \approx 1$. Motor geometry effects and rotor rotation are also considered when implementing detailed internal flow formulas (Booth 1999). Figure 2 shows the flow passages for a water pump with an integral electronically controlled motor. The flow passages are divided into elements, each of which has an associated velocity and temperature. Nusselt numbers are evaluated fro each element. Each element has velocity and temperature pins that are connected to the other thermal components. Many other basic thermal elements have been developed, such as contact surface elements, radiation elements, and so on.

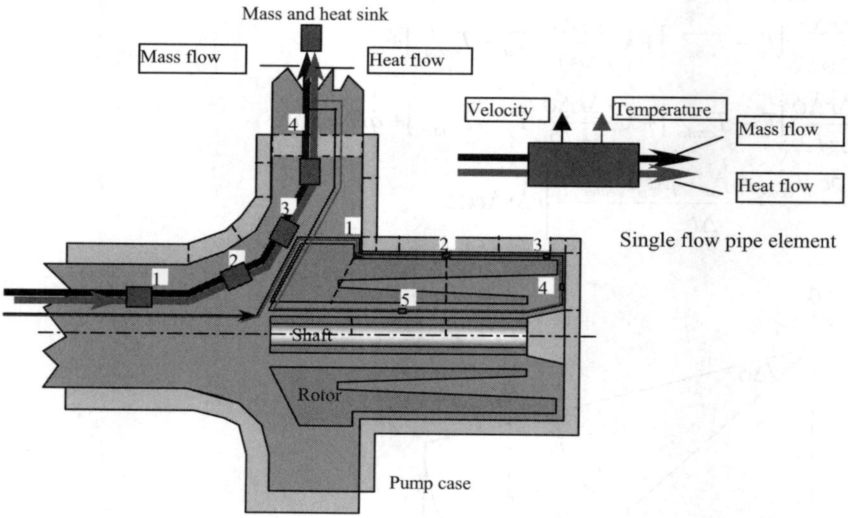

Figure 2. Flow passage elements and the flow passage network for water pump with an integral electronically controlled motor.

3 MODEL GENERATION

Cylindrical and rectangular thermal conduction blocks are used to form water pump motor parts which are combined with a fluid flow network and Saber motor performance models, as shown in Figure 3. Different grayscale coloring represents different motor parts. Some parts are simplified in the figure for clarity. The dark lines between liquid and motor parts are layers where motor parts are in contact with the coolant and where, consequently, the convection coefficient for forced convection is calculated.

A component-layer thermal "wrapper" has been developed using Saber templates to go between potentially detailed internal models of components, the printed circuit board (PCB), natural convection in internal air pockets, heat sinks, and the interface to the motor case, shown in Figure 4. Designer has flexibility to lay out electronic components in any location and the wrapper model will take care of the thermal issues.

The model uses parameterized parts, dimension, and material property information, which is essential for geometry and material optimization during various design stages. The material properties are stored in a common library that can be used for all the motor models. The information flow chart of the thermal network for a water pump motor is shown in Figure 5.

The time needed for simulation can be long for mixed-signal simulations because the motor electronic models have very small time constants (less than 0.01 second) while the thermal models have large time constants (around 60 seconds), meaning that a simulation in the range of 3600 seconds can take several hours to run. Therefore, an improved solution is adopted to speed up the simulation by sampling the electronics every few seconds to determine an "average" current, or rate of heat generation, and then applying this average to the thermal model until the next electronics sampling is performed. At that point, the component temperatures are updated and a new average current is calculated.

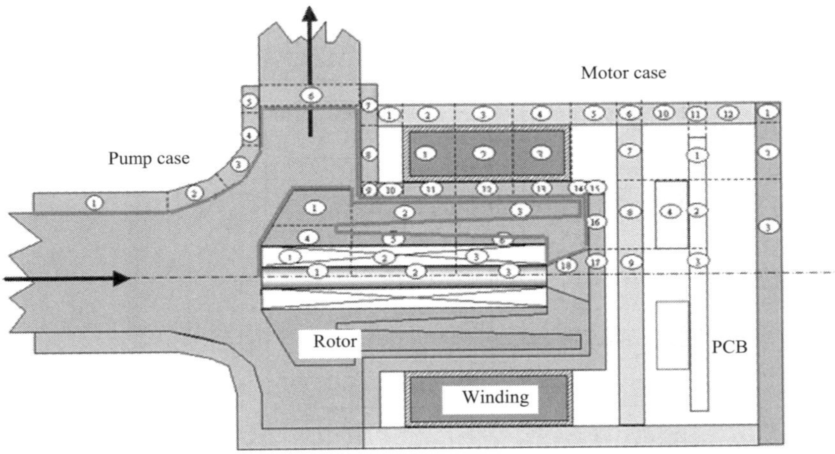

Figure 3. Thermal network model for water pump motors.

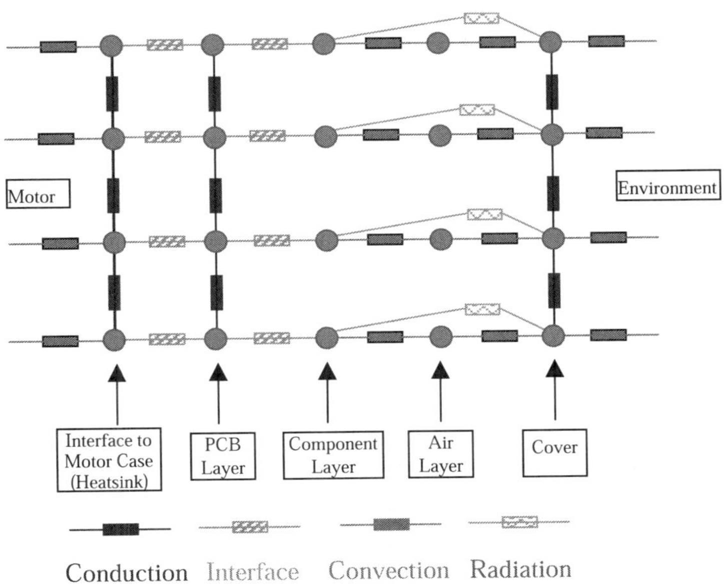

Figure 4. Thermal network strategy for electronic components.

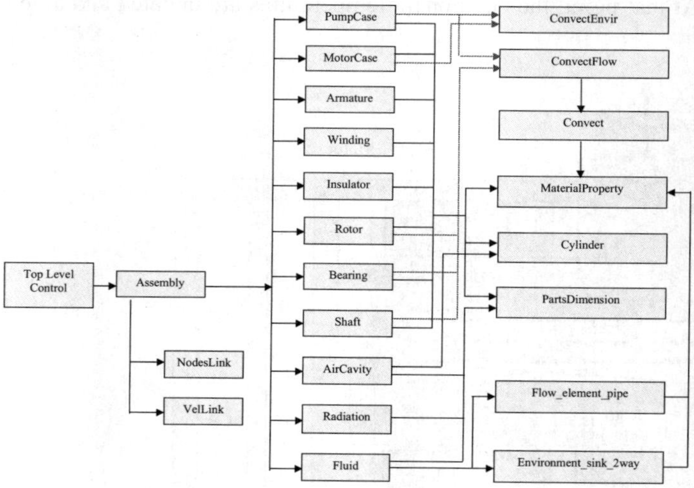

Figure 5. The information flow chart for thermal network modeling of water pump motor.

In order to enhance the usability of this thermal design toolkit and isolate the motor designers from most of the details of thermal modeling, a CAD-like GUI is under development where the geometry of the motor and a drawing of the network are presented showing temperatures, heat fluxes, and thermal resistance values.

4 EXPERIMENTAL VERIFICATION

Several electronically controlled motors have been tested. The measured and simulated temperature profiles at certain locations, as well as the winding and electronic power loss, of a water pump motor are shown in Figure 6. During the experiments, a DC current in a winding is used to represent the winding power loss. (Further testing in real working condition will be performed in the future). The ambient temperature and the coolant inlet temperature are each about 25°C. Good agreement is obtained. The corresponding steady state temperatures are listed in Table 1, where the positions, winding inner and winding outer, are located on the core and on the surface of the winding, respectively. The position, water in gap, is located between the side surface of rotor and pump case. The position, water by FETS, is located between rotor and the back of pump case (near MOSFETs).

Table 1. Comparison between test data and simulation results.

position	At 1080 seconds		At 5660 seconds	
	test temp, °C	simulated temp, °C	test temp, °C	simulated temp, °C
winding (inner)	48.00	48.76	130.00	131.88
winding (outer)	42.00	42.61	104.00	102.66
armature	39.60	39.50	88.20	87.88
water in gap	28.00	28.07	36.30	35.36
water by FETS	26.50	27.81	33.80	35.16

As expected, parameter tuning needs to be performed for the models, especially for the flow network as well as some contact surfaces, where accurate thermal resistance data is difficult to obtain. However, once the experimental correlation has been performed, the model can be easily used for motor prototype design and geometry and material optimization in similar motors.

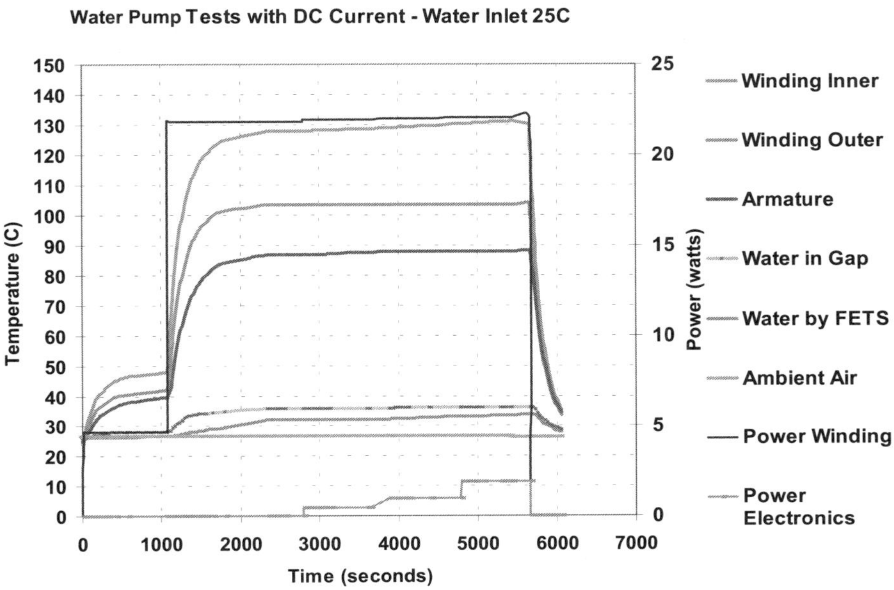

(a)

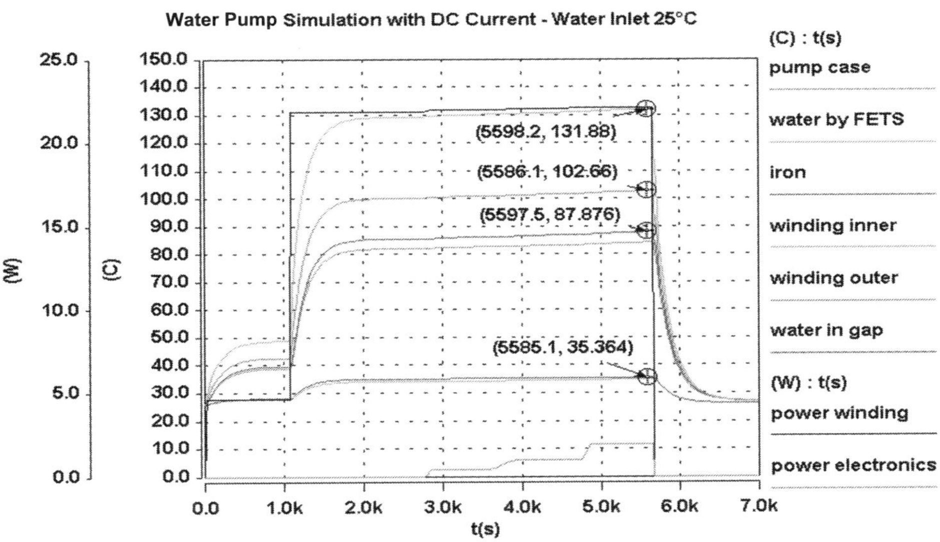

(b)

Figure 6. Motor experimental results (a) and simulation results (b) for a water pump motor.

5 CONCLUSIONS

Analytical-numerical hybrid, electronic-thermally coupled models for various electronically controlled motors have been presented in this study. A finite volume control block is created as a basic thermal component for heat conduction, and flow network modeling is used to handle heat convection. Combined with other thermal components, a hierarchy of thermal networks is established to simulate the thermal behavior at the system level. A special timing scheme has been adopted due to the large difference of time step size between electronic models and thermal models. A wrapping template for electronic components has been implemented, which enables the circuit designers to place electronic components in various locations on the motor; the thermal templates will handle the thermal issues. All the models are implemented as Saber templates using MAST scripting language (SaberDesigner 1999). A CAD-like GUI is being implemented using Tk/TCL scripting language. Accurate temperature prediction in comparison to test data has been obtained for both steady state and transient simulations.

The thermal modeling tools and methodology can be used not only for certain types of motors, but also for other electronic-thermally coupled heat transfer simulations, such as electrical power devices, automobiles, and semiconductor systems.

6 ACKNOWLEDGEMENTS

Special thanks go to Roberto Felicetti for his motor concept design, Dirk Buehler for his electronic circuit modeling, and John Murin for his experimental support. The study is funded by Robert Bosch Corporation – Waltham, Massachusetts, USA.

REFERENCES

Anderson, D.A., Tannehill, J.C. & Pletcher, R.H. 1984. Computational Fluid Mechanics and Heat Transfer. New York: Hemisphere Publishing Corp.

Bar-Cohen, A. & Kraus, A.D. 1998. Advances in Thermal Modeling of Electronic Components and Systems Vol. 4. New York: ASME Press.

Belady, C.L., Kelkar, K.M. & Patankar, S.V. 1999. Improving Productivity in Electronic Packaging with Flow Network Modeling (FNM). Electronics Cooling Vol. 5 (No. 1): 36-40.

Booth D.J. 1999. Small PMDC Motor Airflow Design Guide - Technical Report. Waltham: Robert Bosch Corp.

Igic, P.M., Mawby, P.A., Towers M.S. & Batcup, S. 2001. Dynamic Electro-Thermal Physically Based Compact Models of the Power Devices for Device and Circuit Simulations. Semiconductor Thermal Measurement and Management Symposium. San Jose, CA, USA, March 20-22, 2001: 35-42.

Kang, S.S. 2000. Application of Flow Network Modeling and CFD to Computer System Design. 2000 Intersociety Conference on Thermal and Thermomechanical Phenomena in Electronic Systems Vol. 1: 82-89.

Kraus, A.D. & Bar-Cohen, A. 1983. Thermal Analysis and Control of Electronic Equipment. New York: Hemisphere Publishing Corp.

Minichiello, A. 2000. Flow Network Modeling: A Case Study in Expedient System Prototyping. 2000 Intersociety conference on thermal and thermomechanical phenomena in Electronic Systems Vol. 1: 70-77.

Rohsenow, W.M., Hartnett, J.P. & Cho, Y.I. 1998. Handbook of Heat Transfer, 3rd Edition, New York: McGraw-Hill.

SaberDesigner User Manual. 1999. Avant! Corporation, Fremont, CA, USA.

Thermal Challenges in Next Generation Electronic Systems, Joshi & Garimella (eds)
© 2002 Millpress, Rotterdam, ISBN 90-77017-03-8

Finite elements or finite volumes: Which to choose for airborne electronic packaging thermal analysis

Frank Wang
Eldec Co., Lynnwood, WA, USA

Tel. (425) 743-8478; fwang@eldec.com

Keywords: CFD, Computational Fluid Dynamics, Finite Element Analysis, FEA, airborne, electronics packaging, thermal analysis

ABSTRACT: Thermal analysis provides information on the thermal environment parts and materials will experience in a particular application. This information is used in the electrical and mechanical design of the product and forms a critical input to defining design margins. Thermal analysis is usually performed using commercially available finite element, finite difference, and/or finite volume computational fluid dynamics analysis tools. There are three types of analysis tools on the market to assist mechanical engineer to perform thermal analysis of electronic packaging. Equation calculator, Finite Element Analysis, and Finite Volume Computational Fluid Dynamics. To choose the proper analysis tools will save a lot of time and money in electronic packaging thermal analysis. This paper will discuss the pros and cons of the above items in detail and give examples against testing data.

1 INTRODUCTION

The idea of analysis is to utilize computer calculation power to save time and money over iterative testing of real products. Building one design evaluation unit to calibrate the CAD and analysis model can lessen or eliminate the need for building several costly test units in the traditional way. Engineers should learn to evaluate design options on computer models as much as possible instead of on real products. A mockup unit is sometimes necessary to calibrate the CAD and analysis model. If similar product models have been calibrated before, some stages of mockup can be eliminated and more time and money savings will be achieved.

The importance of thermal analysis in airborne electronics packaging design is driven by the following factors.

Reliability: Automation in Flight-control, Navigation, and Communication puts more safety stake in Avionics.

Cost: Shrinking Non-Recursive Engineering budgets; Widely usage of COTS components; Push components to their spec limits; Customers demand better maintainability and lower life cycle cost.

Time: Shorter Development Schedule; Eliminating Re-design and Re-testing; Qualification by Analysis or Similarity.

On the market, there are many commercial analysis tools packages to assist engineers to perform thermal analysis on electronic packaging design. Each analysis tools has its strength and weakness. Most of these analysis tools packages can be grouped in three category, equation calculator, Finite Element Analysis (FEA), and Computational Fluid Dynamics Finite Volume Analysis (CFD).

Equation calculator is defined as non-discrete calculation method. It uses traditional experimental and theoretical formulas calculates thermal information from few components to the whole system.

Equation calculator treats the target system as a whole. Finite element and Finite volume analysis segregate the target system into small finite discrete segments. Thermal calculations are performed at the segment level. The basic difference between FEA and CFD is that CFD will calculate mass transfer in addition to the heat transfer.

Equation calculator is, easy to use, good for pre-design calculation, abstract, fast, working with partial thermal information, and technically understandable. It also has limited thermal stress capabilities. But the equation calculator is also, not as accurate, weak dealing with complicated problems, hard to check errors, weak in presentation graphics, hard for non-technical people to understand the results, weak calculating specific component temperatures, weak dealing with complicated geometry, hard to obtain local temperature distribution.

Computational Fluid Dynamics is using finite volume equations to solve the flow variables as well as heat transfer variables. There are two basic equations used in differential forms. They are continuity equation and temperature equation.

Computational Fluid Dynamics is, calculating flow pattern and related thermal data, best for calculating complicated convection problem, best for calculating complicated view factor radiation heat transfer problem, relative easier to use than FEA, less heat transfer knowledge demanding, capable generating motion pictures, and easy to detect errors. But it also has the following weaknesses: Model has to be simplified or may have to be rebuild; Slow calculation; Rough temperature data; Not as accurate for conduction cooled electronics; Requires detailed thermal information; Cannot calculate thermal stress.

Finite Element Analysis is first developed for structure analysis. It uses an assembly of discrete elements to approximate the behavior of an irregular, continuous structure under general loading and constraints. The thermal behavior of the structure is obtained by calculating the collective behavior of the discrete elements. FEA does not using continuity equation, therefore cannot calculate mass transfer and its related heat transfer.

Finite Element Analysis has the following advantages: Model Imported Directly From CAD; Fast Calculation; More detailed temperature results; Good for Conduction problem; Good for Simple Radiation problem; Good for Simple Convection problem; Good for calculating thermal stress; But it also has the following weaknesses: Needs more thermal knowledge to operate; Weak at calculating flow pattern and fluid data; Weak at calculating complicated convection; Weak at generating motion fluid pictures; Requires detailed thermal information.

2 METHOD

There are many different levels of analysis that can be done to predict the reliability of the PCB. Choosing analysis method depends on the level of analysis, and stages of the design. They are based on the information available and the precision requirement. The more detailed results required, the more information required.

In planing, proposal or preliminary design stage, thermal data is usually not available, assumptions have to be made; component heat dissipation or locations are not defined; precision results are not expected. In that level of analysis, a quick compute using equation calculator is the best way to get answers or directions.

Equation calculator normally requires fewer resources than other detailed analysis package. User usually types in a few data, answers will appear magically. It can be done by computer software, calculator, or pencil and paper. Many people use spreadsheet software.

User experience is very important. Answers could vary greatly from person to person, software to software, equation to equation. However, it is quick.

In a better-informed proposal stage or design verification stage, component heat dissipations and locations are defined. Component sizes, packaging and mounting structure are defined. Customer's thermal environment is defined. Chassis is designed. Chassis CAD 3D model can be obtained to determine the chassis temperature rise. A more detailed FEA or CFD analysis should be used in this level of analysis. More detailed design analysis is performed using detailed FEA or/and CFD software.

Thermal problem with complicated airflow in consideration prefers CFD analysis. There is either forced or buoyancy driven airflow to provide convection heat transfer. The CFD model is constructed with many simplifications. If the models are too complicated, CFD analysis tools package will slow substantially and may not converge quickly. The current CFD software Flotherm requires rebuilding many of the geometry. The calculation could take hours and even days of computer time. The result generated is worth the effort. May be because of the software is newer, it is relatively easy to use, minimum thermal knowledge is required to run the software. Result is graphical. Airflow motion picture generation is impressive to many none technical people.

FEA software package models can be directly imported from CAD software, which possesses great details. A crude convection heat transfer FEA analysis can be performed by calculating fluid property and temperature on the side, and using that information to feed the calculation of the convection. Current FEA software package can read 3D CAD model directly from Mechanical Desktop, Inventor, PADS, Pro-E, CATIA, etc. Therefore, model building can be performed much faster. The more detailed model is build, the more detailed results will be produced. Computing time could vary from a few minutes to few hours depends on the number of elements involved.

Conduction heat transfer analysis is performed in conduction only situation. There is minimum airflow to provide convection heat transfer. Unless there are some critique component in question, there is no need to perform convection heat transfer analysis. FEA is preferred, because of the fast model building technique.

Combined FEA and CFD approach is also used in many situations. The CFD software package is used to calculate the inside ambient temperature and component surface temperature. FEA analysis uses results like ambient temperature and airflow data from CFD software package. FEA is used to calculate detailed thermal information, like junction temperature etc.

Thermal stress analysis applies to high temperature practice. If there are two different kinds of material connected rigidly, like the rectifier part of the Transformer Rectifier Unit, material will be stressed greatly by temperature cycles. Forced air convection is always expected unless the unit is liquid cooled. Thermal and vibration will be calculated together. Therefore, vibration required information is also needed. FEA software is the only software capable of doing this analysis.

Radiation is secondary in most cases. CFD package will calculate the radiation automatically by turn on one of the switches. Surface emissivity information is needed. If enough surrounding structures are constructed in CFD model, view factor will be calculated by the software. On the other hand, FEA with assumed or hand calculated view factor will be faster to calculate the radiation heat transfer. Especially, if the surrounding structure is unknown and the model is designed in CAD.

2.1 *Non-technical concerns*

There are many concerns of choosing an analysis tool other than technical preference, like financial, personnel, usability, etc. If a company can afford three types of tools, that is great. However, are there enough people trained to use them?

FEA can be used for thermal analysis as well as structure analysis. CFD could only perform thermal analysis, but the result may worth the money. Cost of CFD is comparable to FEA.

Every engineer should perform equation calculator analysis. It is cheap, quick, understandable. But engineers need more experience and thermal science knowledge. All qualified thermal engineers should be able to perform this type of analysis to a certain level. Self-made equation calculator using spreadsheet works just as fine as the commercial one.

2.2 *Analysis steps*

First step is always to use equation calculator type of the analysis tools to perform a rough calculation. The results serve as sanity check and as a base to perform Finite Element Analysis. If the re-

2.2 *Analysis steps*

First step is always to use equation calculator type of the analysis tools to perform a rough calculation. The results serve as sanity check and as a base to perform Finite Element Analysis. If the result were not satisfactory, the change would be easy and quick, since no CAD model had been created at this stage.

Then use detailed analytical modeling to estimate feasibility of various concepts and potential thermal problems at the various boundary conditions. The results point out the potential hot spots inside the unit. If the results are not satisfactory, the change could involve changing the CAD design model. Compare with reproducing the hardware, the change is still easier, faster and cheaper.

Next, it is necessary to validate a computationally efficient (compact) model of the unit with experimental data. The unit temperature affects the reliability of the components. The unit model includes all the critical heat sources inside the unit to simulate heat sources from components. The model boundary conditions were equivalent to validation test conditions.

Then, the next step is to evaluate the different design variations. The general preference is to eliminate the need for pin heat sink. Only if these concepts did not prove out would it be necessary to evaluate the pin fin concept. CFD software package is used for all evaluations in conjunction with analytical calculations. CFD software package modeling narrows down possible design alternatives. A final design option is chosen based on the result of CFD software package modeling.

2.3 *Analysis error study*

It is normal to have three different kinds of results by using three different calculation methods. The largest discrepancy is the air velocity and temperature calculation.

Air velocity determines the convection heat transfer coefficient, local air temperature, frictions, volume flow rate, etc. It effects the final result significantly.

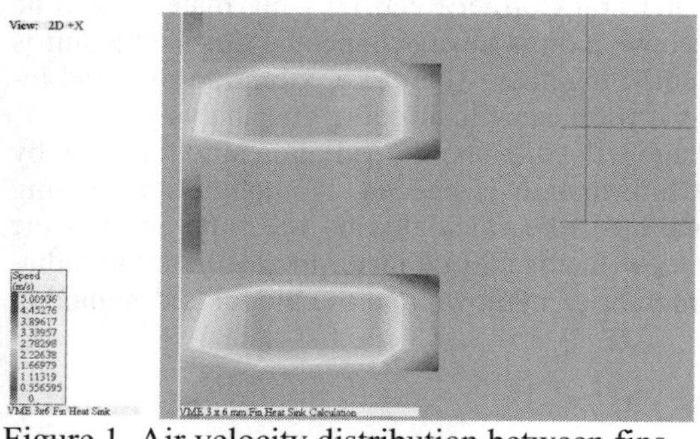

Figure 1. Air velocity distribution between fins

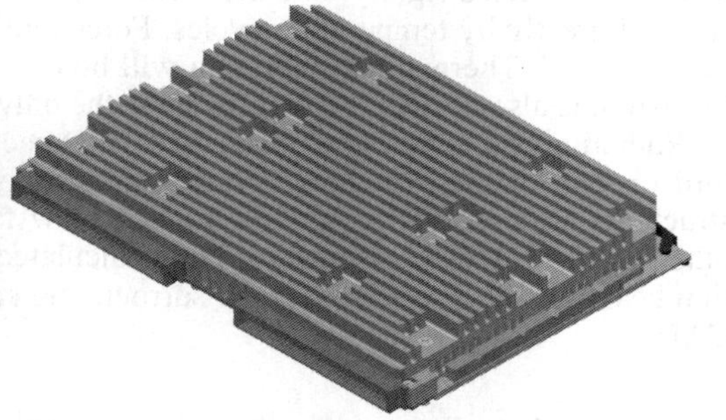

Figure 2. Avionics Power Supply Module

CFD package treats air velocity as variables, calculates the boundary layer's air velocity distribution. Although in FEA package, a sophisticated function can be used to simulate the variable air velocity. That excise is too complicated to be used by engineer. FEA and Equation calculator is usually done with assumption that air velocity is constant.

FEA will calculate detailed temperature distribution. Depends on the mesh size, fin temperature efficiency will take into account in convection heat transfer calculation.

Analysis errors are inevitable. Therefore, design should leave enough margins for errors. Some of the error can be corrected by taking measurement to certain parameters. The analysis model should be calibrated with realistic measurement. Correction should be recorded for future analysis. If enough historic correction data exist, the analysis error margin can be controlled within 10% of the testing data. Total temperature error between analysis results and test data consists of the following parts.

$E = Ee + Etest + Econv + Erad + Ek$

E = Total temperature error

Ee = Electrical Error

Etest = Testing Error

Econv = Convection Error

Erad = Radiation Error

Ek = Thermal Conductance Error

Electrical Error is error caused by incorrect current and voltage prediction. This part of error is 40% of the total error. Voltage drop of the components is not usually agree with the catalog value which is used in analysis. Ee can be corrected by measuring the true current and voltage drop of the component.

Testing error accounts for 10% of the total error. It is caused mainly by improperly mounted thermal couple, location of the thermal couple and the testing equipment. Careful handling of the testing equipment and thermal couple will reduce the error. Etest is hard to eliminate.

Convection Error is error caused by incorrect air velocity and convection coefficient calculation and/or prediction. It counts 30% of the total error. Surface roughness, dimension tolerance, pressure, temperature fluctuation, etc, all effect air velocity and convection coefficient. Convection coefficients may vary +/- 25%. Taking measurement of the air velocity will correct a large portion of the error. No matter what you do, Econv will always be there.

Radiation error is an error caused by incorrect radiation emissitivity. Surface condition, surrounding structure temperature, surrounding structure surface condition, true view factor, etc all contribute to the errors in radiation calculation. Heavily oxidized aluminum has 300% more emissitivity than a polished one.

Thermal conduction error is error caused by incorrect conductance. Aluminum's thermal conductivity can vary from 137 W/mK to 204 W/mK. Dimension variation also changes the total conductance of the parts. Obtain proper alloy thermal conductance, and the geometric dimension is very important to reduce that error.

The most cost-effective way in reducing errors is to find the correct electrical data and correct air flow information.

2.4 *Error corrections*

If realistic data like component heat dissipation, ambient temperature, ambient airflow, and ambient radiation, etc, are different from the modeling data, the CFD software package model and FEA software package models should be corrected based on the measured input data. Then the calculated data was compared with the testing results. Mismatched data is followed with explanations or calibration of the computer model.

After confidence in the computer model is achieved, design variances are tested on the computer. The optimal solution was found and the product was modified based on final data.

If similar products have gone through the above process and computer models have been proved, design can be evaluated by analysis only. The first mockup could be the qualification unit. Many weeks or months of development testing stages may be eliminated.

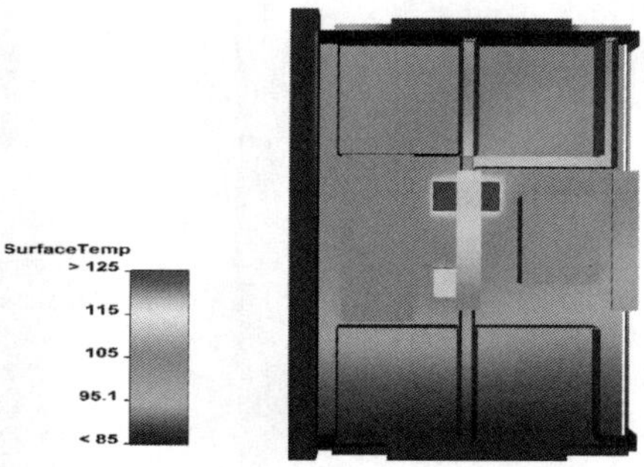

Figure 3. CFD Model

Figure 4. FEA Model

3 EXAMPLE

Figure 2, 3 and 4 show an example product, Avionics Power Supply Module (PSM), design used this method and verified through extensive testing. CFD and FEA software simulation models were corrected by Design Verification Test data of the existing similar product.

The Power Supply Module is a Line Replaceable Module that conditions and filters input 28 VDC or 115 V 400 Hz AC power sources and provides regulated low-voltage DC outputs to the Avionics.

The first design is based on equation calculator analysis tool developed on spreadsheet. It calculates the air velocity, basic heat exchange, exit air temperature, etc. to verify that the design requirement is reasonable.

Detailed analysis model is first build to simulate 280 W PSM. Analysis results compared with experimental data. Correction factor or margin is developed by analyzing errors. Then model is modified with minimum changes to simulate 160 W PSM. Analysis results are corrected by correction factors or correction margins.

	PSM
Pitch	25.00000
PSU Thickness	24.00000
Screw Head	1.19380
Total Thickness	25.19380
Gap	1.00000
Width	160.00000
Length	214.00000
Fin Gap Width	3.00000
Fin Height	6.00000
Fin Gap area mm2	18.00000
Number of Fins	26.00000
Total Fin Gap Area mm2	486.00000
Total Gap Area mm2	160.00000
Cross Section M^2	0.00065
mass flow (kg/s)	0.00240
Density kg/m^3	1.17740
Vol. Flow Rate m^3/s	0.00204
CFM	4.31910
Speed m/s	3.15540

Figure 5. Equation Calculator

cv = hA(Tskin-Tamb)						**Convection**
hcv	L	W	Area	Tskin	Tamb	Qcv
W/m^2C	mm	mm	m^2	C	C	W
11.50	233	160	0.05592	108	75	**21.22**
*A(Tskin^4-Tamb^4)						
hr=Qr/A(Tskin-Tamb)						**Radiation**
hr	SB	Emissivity	Area	Tskin	Tamb	Qrad
W/m^2C	W/(m^2K^4)		m^2	K	K	W
2.24	5.76E-08	0.2	0.05592	381.16	348.16	**4.13**
= hcd*A*(Tpcb-Trail)						**Conduction**
hcd	Rail Length	Contact Width	Area	Tpcb	Trail	Qcd
W/m^2C	mm	mm	m^2	C	C	W
1,784.5	160	5.715	0.0013716	108	85	**56.30**

Total Q (W)
81.64

Figure 6. Equation Calculator

3.1 *Design verification test*

After computer analysis was finished, decisions would be made to modify the design or proceed with Design Verification Testing (DVT).

The DVT contains selective testing to demonstrate that the design of Avionics Power Supply Module will meet the requirements defined in design specification. The DVT serves the following purposes:

To demonstrate direct compliance to applicable specification requirements, or

To demonstrate robustness of design by showing that important design-specific parameters meet expectations, or

To lower integration risk in cases such as EMC, Thermal and Vibration, where Power Supply Module performance is specified at the system level, but no system or chassis is available for direct performance verification.

To find out if pin heat sinks are needed for thermal performance.

To verify or to calibrate the CFD software package results.

3.2 *Test set ups*

Unit was mounted on the fixture as shown. Unit is fastened to the chassis by wedge-lock. Back of the chassis will be enclosed after the picture was taken.

Figure 7. & 8. PSM Thermal Design Verification Testing Thermal Test Fixtures

Fixture was designed to simulate the Avionics chassis with dimension supplied by Avionics Manufacturer. Ventilating hole diameter, location and pattern are provided by Avionics Manufacturer. Chassis dimensions are to simulate the Avionics chassis as much as possible.

Front of the chassis is connected to a flow regulating device through the flexible flow duct shown. Flow regulator will extract 2.5 g/s air from the fixture during the thermal test.

Figure 9. Avionics Power Supply Module Thermal Design Verification Testing

Mounting Fixture was positioned inside Thermal Chamber and covered by perforated box to provide Mil-spec required airflow rate around the chassis. A non-perforated plate placed on top of the perforated box to prevent direct airflow from the thermal chamber blower above.

3.3 *Results analysis*

	280 W VME PSU Test Data (1)	280 W VME PSU Test Data (2)	280 W VME PSU Test Data (3)	280 W VME PSU Test Data (4)	280 W VME PSU Test Data (5)	2D FEA Estimation	3D FEA Estimation	CFD Estimation
Center Rail	101 C	96 C	96 C	99 C	104 C	120 C	97 C	110 C
Rail_ Sink Junction	102 C	96 C	96 C	98 C	101 C	100 C	97 C	105 C
Ambient inside Unit	103 C	100 C	100 C	106 C	107 C	100 C	95 C	100 C
Chassis @ 5V module	98 C	94 C	94 C	94 C	99 C	98 C	94 C	100 C
Rail near 5V Chassis	91 C	83 C	84 C	82 C	85 C	85 C	85 C	85 C
Q22 Tab	128 C	141 C	145 C	142 C	125 C	130 C	120 C	125 C
CR1 Tab	131 C	146 C	150 C	145 C	124 C	130 C	120 C	125 C
Q23 Tab	129 C	136 C	145 C	123 C	122 C	130 C	120 C	115 C
Sink @CR1	121 C	130 C	132 C	126 C	119 C	130 C	120 C	115 C
Chassis @ 12V	101 C	92 C	92 C	95 C	98 C	100 C	94 C	105 C
Chassis @ center Rail	100 C	94 C	94 C	97 C	103 C	120 C	97 C	105 C
L6 Winding	114 C	117 C	118 C	136 C	125 C	120 C	120 C	110 C
L1 Winding	108 C	110 C	112 C	106 C	121 C	110 C	110 C	110 C
L6 Core	106 C	104 C	104 C	106 C	105 C	110 C	110 C	115 C
L1 Core	104 C	103 C	103 C	115 C	106 C	110 C	110 C	115 C

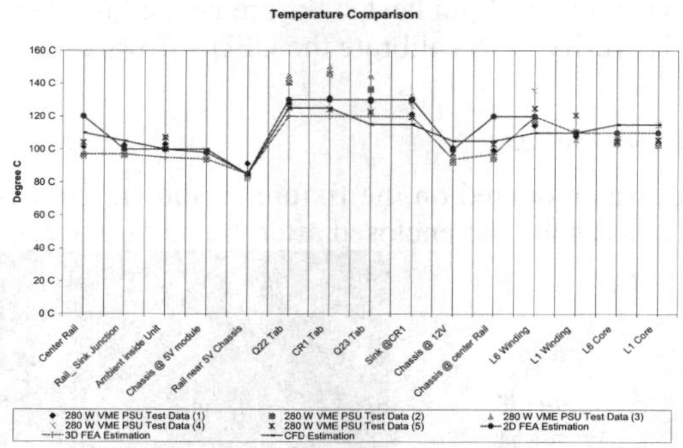

Figure 10. Results analysis

Figure 10. show the comparison between analysis results to five sets of testing data. A few TO-220 packages' analysis temperature is apparently lower than the test data. The error is caused by the unknown contact resistance between the TO-220 tab and the metal heat spreader, as well as the metal heat sink. Instead of trying to model the contact resistance, error margin correction is used to correct the present analysis results and any other future development results.

4 CONCLUSION

In contrast to many traditional approaches in electronics industry, thermal analysis saves time and money in electronic product design and testing. Obtaining correct electrical data is the key to have closer analytical results. Calibrating the analysis model is the key to obtaining results that are more realistic. Greater savings will be achieved when several similar analysis models are calibrated and proved by testing. An intelligent proof by similarity can be used to reduce costly tests and mockups.

5 ACKNOWLEDGMENTS

Andrew Xing, David Cooper, and David Wiley of Eldec Corporation have contributed a great deal to this article.

REFERENCES

Assouad, Y. , Caplot, M., Gautier, T. 1997, "Advanced Cooling Techniques Comparison for Airborne Electronic Circuits", Semitherm, Austin, TX

F. Wang, 2001 "Avionics Thermal Analysis & Testing Strategy", Proc. of ASME/InterPACK, Kauai, Hawaii.

Holman, J.P., 1997, "Heat Transfer", McGraw-Hill, 8th Edition, USA.

Plunkett, J, 2000, "An Airspeed Measurement Technique for Accurate Correlation to CFD Simulation Results", International Conference on High-Density interconnect and Systems Packaging.

Steinberg, D., 1991, "Cooling Techniques for Electronic Equipment", John Wiley & Son, Second Edition, NY, NY.

MISCELLANEOUS

Keyword index

Thermal Challenges in Next Generation Electronic Systems, Joshi & Garimella (eds)
© 2002 Millpress, Rotterdam, ISBN 90-77017-03-8

Author index